UNDERSTANDING
PHYSICS

PART 3

Karen Cummings
Rensselaer Polytechnic Institute
Southern Connecticut State University

Priscilla W. Laws
Dickinson College

Edward F. Redish
University of Maryland

Patrick J. Cooney
Millersville University

GUEST AUTHOR

Edwin F. Taylor
Massachusetts Institute of Technology

ADDITIONAL MEMBERS OF ACTIVITY BASED PHYSICS GROUP

David R. Sokoloff
University of Oregon

Ronald K. Thornton
Tufts University

Understanding Physics is based on *Fundamentals of Physics*
by David Halliday, Robert Resnick, and Jearl Walker.

WILEY

John Wiley & Sons, Inc.

This book is dedicated to Arnold Arons, whose pioneering work in physics education and reviews of early chapters have had a profound influence on our work.

SENIOR ACQUISITIONS EDITOR	Stuart Johnson
SENIOR DEVELOPMENT EDITOR	Ellen Ford
MARKETING MANAGER	Bob Smith
SENIOR PRODUCTION EDITOR	Elizabeth Swain
SENIOR DESIGNER	Kevin Murphy
INTERIOR DESIGN	Circa 86, Inc.
COVER DESIGN	David Levy
COVER PHOTO	© Antonio M. Rosario/The Image Bank/Getty Images
ILLUSTRATION EDITOR	Anna Melhorn
PHOTO EDITOR	Hilary Newman

This book was set in 10/12 Times Ten Roman by Progressive and printed and bound by Von Hoffmann Press. The cover was printed by Von Hoffmann Press.

This book is printed on acid free paper. ∞

Library of Congress Cataloging in Publication Data:

Understanding physics / Karen Cummings . . . [et al.]; with additional members of the
  Activity Based Physics Group.
    p. cm.
  Includes index.
  ISBN 0-471-46437-6 (pt. 3 : pbk. : acid-free paper)
  1. Physics.   I. Cummings, Karen.   II. Activity Based Physics Group.

QC23.2.U54 2004
530—dc21   2003053481

L.C. Call no.   Dewey Classification No.   L.C. Card No.

Printed in the United States of America

10 9 8 7 6 5 4 3 2

Preface

Welcome to *Understanding Physics*. This book is built on the foundations of the 6th Edition of Halliday, Resnick, and Walker's *Fundamentals of Physics* which we often refer to as HRW 6th. The HRW 6th text and its ancestors, first written by David Halliday and Robert Resnick, have been best-selling introductory physics texts for the past 40 years. It sets the standard against which many other texts are judged. You are probably thinking, "Why mess with success?" Let us try to explain.

Why a Revised Text?

A physics major recently remarked that after struggling through the first half of his junior level mechanics course, he felt that the course was now going much better. What had changed? Did he have a better background in the material they were covering now? "No," he responded. "I started reading the book before every class. That helps me a lot. I wish I had done it in Physics One and Two." Clearly, this student learned something very important. It is something most physics instructors wish they could teach all of their students as soon as possible. Namely, no matter how smart your students are, no matter how well your introductory courses are designed and taught, your students will master more physics if they learn how to read an "understandable" textbook carefully.

We know from surveys that the vast majority of introductory physics students do not read their textbooks carefully. We think there are two major reasons why: (1) many students complain that physics textbooks are impossible to understand and too abstract, and (2) students are extremely busy juggling their academic work, jobs, personal obligations, social lives and interests. So they develop strategies for passing physics without spending time on careful reading. We address both of these reasons by making our revision to the sixth edition of *Fundamentals of Physics* easier for students to understand and by providing the instructor with more **Reading Exercises** (formerly known as Checkpoints) and additional strategies for encouraging students to read the text carefully. Fortunately, we are attempting to improve a fine textbook whose active author, Jearl Walker, has worked diligently to make each new edition more engaging and understandable.

In the next few sections we provide a summary of how we are building upon HRW 6th and shaping it into this new textbook.

A Narrative That Supports Student Learning

One of our primary goals is to help students make sense of the physics they are learning. We cannot achieve this goal if students see physics as a set of disconnected mathematical equations that each apply only to a small number of specific situations. We stress conceptual and qualitative understanding and continually make connections between mathematical equations and conceptual ideas. We also try to build on ideas that students can be expected to already understand, based on the resources they bring from everyday experiences.

In *Understanding Physics* we have tried to tell a story that flows from one chapter to the next. Each chapter begins with an introductory section that discusses why new topics introduced in the chapter are important, explains how the chapter builds on previous chapters, and prepares students for those that follow. We place explicit emphasis on basic concepts that recur throughout the book. We use extensive forward and backward referencing to reinforce connections between topics. For example, in the introduction of Chapter 16 on Oscillations we state: "Although your study of simple harmonic motion will enhance your understanding of mechanical systems it is also vital to understanding the topics in electricity and magnetism encountered in Chapters 30-37. Finally, a knowledge of SHM provides a basis for understanding the wave nature of light and how atoms and nuclei absorb and emit energy."

Emphasis on Observation and Experimentation

Observations and concrete everyday experiences are the starting points for development of mathematical expressions. Experiment-based theory building is a major feature of the book. We build ideas on experience that students either already have or can easily gain through careful observation.

Whenever possible, the physical concepts and theories developed in *Understanding Physics* grow out of simple observations or experimental data that can be obtained in typical introductory physics laboratories. We want our readers to develop the habit of asking themselves: What do our observations, experiences and data imply about the natural laws of physics? How do we know a given statement is true? Why do we believe we have developed correct models for the world?

Toward this end, the text often starts a chapter by describing everyday observations with which students are familiar. This makes *Understanding Physics* a text that is both relevant to students' everyday lives and draws on existing student knowledge. We try to follow Arnold Arons' principle "idea first, name after." That is, we make every attempt to begin a discussion by using everyday language to describe common experiences. Only then do we introduce formal physics terminology to represent the concepts being discussed. For example, everyday pushes, pulls, and their impact on the motion of an object are discussed before introducing the term "force" or Newton's Second Law. We discuss how a balloon shrivels when placed in a cold environment and how a pail of water cools to room temperature before introducing the ideal gas law or the concept of thermal energy transfer.

The "idea first, name after" philosophy helps build patterns of association between concepts students are trying to learn and knowledge they already have. It also helps students reinterpret their experiences in a way that is consistent with physical laws.

Examples and illustrations in *Understanding Physics* often present data from modern computer-based laboratory tools. These tools include computer-assisted data acquisition systems and digital video analysis software. We introduce students to these tools at the end of Chapter 1. Examples of these techniques are shown in Figs. P-1 and P-2 (on the left) and Fig. P-3 on the next page. Since many instructors use these computer tools in the laboratory or in lecture demonstrations, these tools are part of the introductory physics experience for more and more of our students. The use of real data has a number of advantages. It connects the text to the students' experience in other parts of the course and it connects the text directly to real world experience. Regardless of whether data acquisition and analysis tools are used in the student's own laboratory, our use of realistic rather that idealized data helps students develop an appreciation of the role that data evaluation and analysis plays in supporting theory.

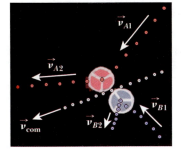

FIGURE P-1 ■ A video analysis shows that the center of mass of a two-puck system moves at a constant velocity.

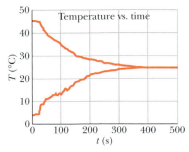

FIGURE P-2 ■ Electronic temperature sensors reveal that if equal amounts of hot and cold water mix the final temperature is the average of the initial temperatures.

FIGURE P-3 ■ A video analysis of human motion reveals that in free fall the center of mass of an extended body moves in a parabolic path under the influence of the Earth's gravitational force.

Using Physics Education Research

In re-writing the text we have taken advantage of two valuable findings of physics education research. One is the identification of concepts that are especially difficult for many students to learn. The other is the identification of active learning strategies to help students develop a more robust understanding of physics.

Addressing Learning Difficulties

Extensive scholarly research exists on the difficulties students have in learning physics.[1] We have made a concerted effort to address these difficulties. In *Understanding Physics,* issues that are known to confuse students are discussed with care. This is true even for topics like the nature of force and its effect on velocity and velocity changes that may seem trivial to professional physicists. We write about subtle, often counter-intuitive topics with carefully chosen language and examples designed to draw out and remediate common alternative student conceptions. For example, we know that students have trouble understanding passive forces such as normal and friction forces.[2] How can a rigid table exert a force on a book that rests on it? In Section 6-4 we present an idealized model of a solid that is analogous to an inner spring mattress with the repulsion forces between atoms acting as the springs. In addition, we invite our readers to push on a table with a finger and experience the fact that as they push harder on the table the table pushes harder on them in the opposite direction.

FIGURE P-4 ■ Compressing an innerspring mattress with a force. The mattress exerts an oppositely directed force, with the same magnitude, back on the finger.

Incorporating Active Learning Opportunities

We designed *Understanding Physics* to be more interactive and to foster thoughtful reading. We have retained a number of the excellent Checkpoint questions found at the end of HRW 6th chapter sections. We now call these questions **Reading Exercises.** We have created many new Reading Exercises that require students to reflect on the material in important chapter sections. For example, just after reading Section 6-2 that introduces the two-dimensional free-body diagram, students encounter Reading Exercise 6-1. This multiple-choice exercise requires students to identify the free-body diagram for a helicopter that experiences three non-collinear forces. The distractors were based on common problems students have with the construction of free-body diagrams. When used in "Just-In-Time Teaching" assignments or for in-class group discussion, this type of reading exercise can help students learn a vital problem solving skill as they read.

---

[1] L. C. McDermott and E. F. Redish, "Resource Letter PER-1: Physics Education Research," *Am. J. Phys.* **67**, 755-767 (1999)

[2] John J. Clement, "Expert novice similarities and instruction using analogies," *Int. J. Sci. Ed. 20*, 1271-1286 (1998)

We also created a set of **Touchstone Examples.** These are carefully chosen sample problems that illustrate key problem solving skills and help students learn how to use physical reasoning and concepts as an essential part of problem solving. We selected some of these touchstone examples from the outstanding collection of sample problems in HRW 6th and we created some new ones. In order to retain the flow of the narrative portions of each chapter, we have reduced the overall number of sample problems to those necessary to exemplify the application of fundamental principles. Also, we chose touchstone examples that require students to combine conceptual reasoning with mathematical problem-solving skills. Few, if any, of our touchstone examples are solvable using simple "plug-and-chug" or algorithmic pattern matching techniques.

Alternative problems have been added to the extensive, classroom tested end-of-chapter problem sets selected from HRW 6th. The design of these new problems are based on the authors' knowledge of research on student learning difficulties. Many of these new problems require careful qualitative reasoning. They explicitly connect conceptual understanding to quantitative problem solving. In addition, estimation problems, video analysis problems, and "real life" or "context rich" problems have been included.

The organization and style of *Understanding Physics* has been modified so that it can be easily used with other research-based curricular materials that make up what we call *The Physics Suite*. The *Suite* and its contents are explained at length at the end of this preface.

Reorganizing for Coherence and Clarity

For the most part we have retained the organization scheme inherited from HRW 6th. Instructors are familiar with the general organization of topics in a typical course sequence in calculus-based introductory physics texts. In fact, ordering of topics and their division into chapters is the same for 27 of the 38 chapters. The order of some topics has been modified to be more pedagogically coherent. Most of the reorganization was done in Chapters 3 through 10 where we adopted a sequence known as *New Mechanics*. In addition, we decided to move HRW 6th Chapter 25 on capacitors so it becomes the last chapter on DC circuits. Capacitors are now introduced in Chapter 28 in *Understanding Physics*.

The New Mechanics Sequence

HRW 6th and most other introductory textbooks use a familiar sequence in the treatment of classical mechanics. It starts with the development of the kinematic equations to describe constantly accelerated motion. Then two-dimensional vectors and the kinematics of projectile motion are treated. This is followed by the treatment of dynamics in which Newton's Laws are presented and used to help students understand both one- and two-dimensional motions. Finally energy, momentum conservation, and rotational motion are treated.

About 12 years ago when Priscilla Laws, Ron Thornton, and David Sokoloff were collaborating on the development of research-based curricular materials, they became concerned about the difficulties students had working with two-dimensional vectors and understanding projectile motion before studying dynamics.

At the same time Arnold Arons was advocating the introduction of the concept of momentum before energy.[3] Arons argued that (1) the momentum concept is simpler than the energy concept, in both historical and modern contexts and (2) the study

---

[3] Private Communication between Arnold Arons and Priscilla Laws by means of a document entitled "Preliminary Notes and Suggestions," August 19, 1990; and Arnold Arons, *Development of Concepts of Physics* (Addison-Wesley, Reading MA, 1965)

of momentum conservation entails development of the concept of center-of-mass which is needed for a proper development of energy concepts. Additionally, the impulse-momentum relationship is clearly an alternative statement of Newton's Second Law. Hence, its placement immediately after the coverage of Newton's laws is most natural.

In order to address these concerns about the traditional mechanics sequence, a small group of physics education researchers and curriculum developers convened in 1992 to discuss the introduction of a new order for mechanics.[4] One result of the conference was that Laws, Sokoloff, and Thornton have successfully incorporated a new sequence of topics in the mechanics portions of various curricular materials that are part of the Physics Suite discussed below.[5] These materials include *Workshop Physics*, the *RealTime Physics Laboratory Module in Mechanics*, and the *Interactive Lecture Demonstrations*. This sequence is incorporated in this book and has required a significant reorganization and revisions of HRW 6th Chapters 2 through 10.

The New Mechanics sequence incorporated into Chapters 2 through 10 of understanding physics includes:

- Chapter 2: One-dimensional kinematics using constant horizontal accelerations and vertical free fall as applications.

- Chapter 3: The study of one-dimensional dynamics begins with the application of Newton's laws of motion to systems with one or more forces acting along a single line. Readers consider observations that lead to the postulation of "gravity" as a constant invisible force acting vertically downward.

- Chapter 4: Two-dimensional vectors, vector displacements, unit vectors and the decomposition of vectors into components are treated.

- Chapter 5: The study of kinematics and dynamics is extended to two-dimensional motions with forces along only a single line. Examples include projectile motion and circular motion.

- Chapter 6: The study of kinematics and dynamics is extended to two-dimensional motions with two-dimensional forces.

- Chapters 7 & 8: Topics in these chapters deal with impulse and momentum change, momentum conservation, particle systems, center of mass, and the motion of the center-of-mass of an isolated system.

- Chapters 9 & 10: These chapters introduce kinetic energy, work, potential energy, and energy conservation.

Just-in-Time Mathematics

In general, we introduce mathematical topics in a "just-in-time" fashion. For example, we treat one-dimensional vector concepts in Chapter 2 along with the development of one-dimensional velocity and acceleration concepts. We hold the introduction of two- and three-dimensional vectors, vector addition and decomposition until Chapter 4, immediately before students are introduced to two-dimensional motion and forces in Chapters 5 and 6. We do not present vector products until they are needed. We wait to introduce the dot product until Chapter 9 when the concept of physical work is presented. Similarly, the cross product is first presented in Chapter 11 in association with the treatment of torque.

---

[4] The New Mechanics Conference was held August 6-7, 1992 at Tufts University. It was attended by Pat Cooney, Dewey Dykstra, David Hammer, David Hestenes, Priscilla Laws, Suzanne Lea, Lillian McDermott, Robert Morse, Hans Pfister, Edward F. Redish, David Sokoloff, and Ronald Thornton.

[5] Laws, P. W. "A New Order for Mechanics" pp. 125-136, *Proceedings of the Conference on the Introductory Physics Course*, Rensselaer Polytechnic Institute, Troy New York, May 20-23, Jack Wilson, Ed. 1993 (John Wiley & Sons, New York 1997)

Notation Changes

Mathematical notation is often confusing, and ambiguity in the meaning of a mathematical symbol can prevent a student from understanding an important relationship. It is also difficult to solve problems when the symbols used to represent different quantities are not distinctive. Some key features of the new notation include:

- We adhere to recent notation guidelines set by the U.S. National Institute of Standard and Technology Special Publication 811 (SP 811).

- We try to balance our desire to use familiar notation and our desire to avoid using the same symbol for different variables. For example, p is often used to denote momentum, pressure, and power. We have chosen to use lower case p for momentum and capital P for pressure since both variables appear in the kinetic theory derivation. But we stick with the convention of using capital P for power since it does not commonly appear side by side with pressure in equations.

- We denote vectors with an arrow instead of bolding so handwritten equations can be made to look like the printed equations.

- We label each vector component with a subscript that explicitly relates it to its coordinate axis. This eliminates the common ambiguity about whether a quantity represents a magnitude which is a scalar or a vector component which is not a scalar.

- We often use subscripts to spell out the names of objects that are associated with mathematical variables even though instructors and students will tend to use abbreviations. We also stress the fact that one object is exerting a force on another with an arrow in the subscript. For example, the force exerted by a rope on a block would be denoted as $\vec{F}_{\text{rope}\rightarrow\text{block}}$.

Our notation scheme is summarized in more detail in Appendix A4.

Encouraging Text Reading

We have described a number of changes that we feel will improve this textbook and its readability. But even the best textbook in the world is of no help to students who do not read it. So it is important that instructors make an effort to encourage busy students to develop effective reading habits. In our view the single most effective way to get students to read this textbook is to assign appropriate reading, reading exercises, and other reading questions after every class. Some effective ways to follow up on reading question assignments include:

1. Employ a method called "Just-In-Time-Teaching" (or JiTT) in which students submit their answers to questions about reading before class using just plain email or one of the many available computer based homework systems (Web Assign or E-Grade for example). You can often read enough answers before class to identify the difficult questions that need more discussion in class;

2. Ask students to bring the assigned questions to class and use the answers as a basis for small group discussions during the class period;

3. Assign multiple choice questions related to each section or chapter that can be graded automatically with a computer-based homework system; and

4. Require students to submit chapter summaries. Because this is a very effective assignment, we intentionally avoided doing chapter summaries for students.

Obviously, all of these approaches are more effective when students are given some credit for doing them. Thus you should arrange to grade all, or a random sample, of the submissions as incentives for students to read the text and think about the answers to Reading Exercises on a regular basis.

The Physics Suite

In 1997 and 1998, Wiley's physics editor, Stuart Johnson, and an informally constituted group of curriculum developers and educational reformers known as the *Activity Based Physics Group* began discussing the feasibility of integrating a broad array of curricular materials that are physics education research-based. This led to the assembly of an *Activity Based Physics Suite* that includes this textbook. The *Physics Suite* also includes materials that can be combined in different ways to meet the needs of instructors working in vastly different learning environments. The *Interactive Lecture Demonstration Series*[6] is designed primarily for use in lecture sessions. Other *Suite* materials can be used in laboratory settings including the *Workshop Physics Activity Guide*,[7] the *Real Time Physics Laboratory* modules,[8] and *Physics by Inquiry*.[9] Additional elements in the collection are suitable for use in recitation sessions such as the University of Washington *Tutorials in Introductory Physics* (available from Prentice Hall)[10] and a set of *Quantitative Tutorials*[11] developed at the University of Maryland. The *Activity Based Physics Suite* is rounded out with a collection of thinking problems developed at the University of Maryland. In addition to this **Understanding Physics** text, the Physics Suite elements include:

1. **Teaching Physics with the Physics Suite** by Edward F. Redish (University of Maryland). This book is not only the "Instructors Manual" for *Understanding Physics*, but it is also a book for anyone who is interested in learning about recent developments in physics education. It is a handbook with a variety of tools for improving both teaching and learning of physics—from new kinds of homework and exam problems, to surveys for figuring out what has happened in your class, to tools for taking and analyzing data using computers and video. The book comes with a Resource CD containing 14 conceptual and 3 attitude surveys, and more than 250 thinking problems covering all areas of introductory physics, resource materials from commercial vendors on the use of computerized data acquisition and video, and a variety of other useful reference materials. (Instructors can obtain a complimentary copy of the book and Resource CD, from John Wiley & Sons.)

2. **RealTime Physics** by David Sokoloff (University of Oregon), Priscilla Laws (Dickinson College), and Ronald Thornton (Tufts University). *RealTime Physics* is a set of laboratory materials that uses computer-assisted data acquisition to help students build concepts, learn representation translation, and develop an understanding of the empirical base of physics knowledge. There are three modules in the collection: Module 1: Mechanics (12 labs), Module 2: Heat and Thermodynamics (6 labs), and Module 3: Electric Circuits (8 labs). (Available both in print and in electronic form on *The Physics Suite CD*.)

[6]David R. Sokoloff and Ronald K. Thornton, "Using Interactive Lecture Demonstrations to Create an Active Learning Environment." *The Physics Teacher*, **35**, 340-347, September 1997.

[7]Priscilla W. Laws, *Workshop Physics Activity Guide*, Modules 1-4 w/ Appendices (John Wiley & Sons, New York, 1997).

[8]David R. Sokoloff, *RealTime Physics*, Modules 1-2, (John Wiley & Sons, New York, 1999).

[9]Lillian C. McDermott and the Physics Education Group at the University of Washington, *Physics by Inquiry* (John Wiley & Sons, New York, 1996).

[10]Lillian C. McDermott, Peter S. Shaffer, and the Physics Education Group at the University of Washington, *Tutorials in Introductory Physics*, First Edition (Prentice-Hall, Upper Saddle River, NJ, 2002).

[11]Richard N. Steinberg, Michael C. Wittmann, and Edward F. Redish, "Mathematical Tutorials in Introductory Physics," in, *The Changing Role Of Physics Departments In Modern Universities*, Edward F. Redish and John S. Rigden, editors, AIP Conference Proceedings **399**, (AIP, Woodbury NY, 1997), 1075-1092.

3. **Interactive Lecture Demonstrations** by David Sokoloff (University of Oregon) and Ronald Thornton (Tufts University). ILDs are worksheet-based guided demonstrations designed to focus on fundamental principles and address specific naïve conceptions. The demonstrations use computer-assisted data acquisition tools to collect and display high quality data in real time. Each ILD sequence is designed for delivery in a single lecture period. The demonstrations help students build concepts through a series of instructor led steps involving prediction, discussions with peers, viewing the demonstration and reflecting on its outcome. The ILD collection includes sequences in mechanics, thermodynamics, electricity, optics and more. (Available both in print and in electronic form on *The Physics Suite CD.*)

4. **Workshop Physics** by Priscilla Laws (Dickinson College). *Workshop Physics* consists of a four part activity guide designed for use in calculus-based introductory physics courses. Workshop Physics courses are designed to replace traditional lecture and laboratory sessions. Students use computer tools for data acquisition, visualization, analysis and modeling. The tools include computer-assisted data acquisition software and hardware, digital video capture and analysis software, and spreadsheet software for analytic mathematical modeling. Modules include classical mechanics (2 modules), thermodynamics & nuclear physics, and electricity & magnetism. (Available both in print and in electronic form on *The Physics Suite CD.*)

5. **Tutorials in Introductory Physics** by Lillian C. McDermott, Peter S. Shaffer and the Physics Education Group at the University of Washington. These tutorials consist of a set of worksheets designed to supplement instruction by lectures and textbook in standard introductory physics courses. Each tutorial is designed for use in a one-hour class session in a space where students can work in small groups using simple inexpensive apparatus. The emphasis in the tutorials is on helping students deepen their understanding of critical concepts and develop scientific reasoning skills. There are tutorials on mechanics, electricity and magnetism, waves, optics, and other selected topics. (Available in print from Prentice Hall, Upper Saddle River, New Jersey.)

6. **Physics by Inquiry** by Lillian C. McDermott and the Physics Education Group at the University of Washington. This self-contained curriculum consists of a set of laboratory-based modules that emphasize the development of fundamental concepts and scientific reasoning skills. Beginning with their observations, students construct a coherent conceptual framework through guided inquiry. Only simple inexpensive apparatus and supplies are required. Developed primarily for the preparation of precollege teachers, the modules have also proven effective in courses for liberal arts students and for underprepared students. The amount of material is sufficient for two years of academic study. (Available in print.)

7. **The Activity Based Physics Tutorials** by Edward F. Redish and the University of Maryland Physics Education Research Group. These tutorials, like those developed at the University of Washington, consist of a set of worksheets developed to supplement lectures and textbook work in standard introductory physics courses. But these tutorials integrate the computer software and hardware tools used in other Suite elements including computer data acquisition, digital video analysis, simulations, and spreadsheet analysis. Although these tutorials include a range of classical physics topics, they also include additional topics in modern physics. (Available only in electronic form on *The Physics Suite CD.*)

8. **The Understanding Physics Video CD for Students** by Priscilla Laws, et. al.: This CD contains a collection of the video clips that are introduced in *Understanding Physics* narrative and alternative problems. The CD includes a number of Quick-Time movie segments of physical phenomena along with the QuickTime player

software. Students can view video clips as they read the text. If they have video analysis software available, they can reproduce data presented in text graphs or complete video analyses based on assignments designed by instructors.

9. **WPTools** by Priscilla Laws and Patrick Cooney: These tools consist of a set of macros that can be loaded with Microsoft Excel software that allow students to graph data transferred from computer data acquisition software and video analysis software more easily. Students can also use the *WPTools* to analyze numerical data and develop analytic mathematical models.

10. **The Physics Suite CD.** This CD contains a variety of the Suite Elements in electronic format (Microsoft Word files). The electronic format allows instructors to modify and reprint materials to better fit into their individual course syllabi. The CD contains much useful material including complete electronic versions of the following: *RealTime Physics*, *Interactive Lecture Demonstrations*, *Workshop Physics*, *Activity Based Physics Tutorials.*

A Final Word to the Instructor

Over the past decade we have learned how valuable it is for us as teachers to focus on what most students actually need to do to learn physics, and how valuable it can be for students to work with research-based materials that promote active learning. We hope you and your students find this book and some of the other *Physics Suite* materials helpful in your quest to make physics both more exciting and understandable to your students.

Supplements for Use with Understanding Physics

Instructor Supplements

1. **Instructor's Solution Manual** prepared by Anand Batra (Howard University). This manual provides worked-out solutions for most of the end-of-chapter problems.

2. **Test Bank** by J. Richard Christman (U. S. Coast Guard Academy). This manual includes more than 2500 multiple-choice questions adapted from HRW 6th. These items are also available in the *Computerized Test Bank* (see below).

3. **Instructor's Resource CD.** This CD contains:
   - The entire *Instructor's Solutions Manual* in both Microsoft Word© (IBM and Macintosh) and PDF files.
   - A *Computerized Test Bank*, for use with both PCs and Macintosh computers with full editing features to help you customize tests.
   - All text illustrations, suitable for classroom projection, printing, and web posting.

4. **Online Homework and Quizzing:** *Understanding Physics* supports WebAssign and eGrade, two programs that give instructors the ability to deliver and grade homework and quizzes over the Internet.

Student Supplements

1. **Student Study Guide** by J. Richard Christman (U. S. Coast Guard Academy). This student study guide provides chapter overviews, hints for solving selected end-of-chapter problems, and self-quizzes.

2. **Student Solutions Manual** by J. Richard Christman (U. S. Coast Guard Academy). This manual provides students with complete worked-out solutions for approximately 450 of the odd-numbered end-of-chapter problems.

Acknowledgements

Many individuals helped us create this book. The authors are grateful to the individuals who attended the weekend retreats at Airlie Center in 1997 and 1998 and to our editor, Stuart Johnson and to John Wiley & Sons for sponsoring the sessions. It was in these retreats that the ideas for *Understanding Physics* crystallized. We are grateful to Jearl Walker, David Halliday and Bob Resnick for graciously allowing us to attempt to make their already fine textbook better.

The authors owe special thanks to Sara Settlemyer who served as an informal project manager for the past few years. Her contributions included physics advice (based on her having completed Workshop Physics courses at Dickinson College), her use of Microsoft Word, Adobe Illustrator, Adobe Photoshop and Quark XPress to create the manuscript and visuals for this edition, and skillful attempts to keep our team on task—a job that has been rather like herding cats.

Karen Cummings: I would like to say "Thanks!" to: Bill Lanford (for endless advice, use of the kitchen table and convincing me that I really could keep the same address for more than a few years in a row), Ralph Kartel Jr. and Avery Murphy (for giving me an answer when people asked why I was working on a textbook), Susan and Lynda Cummings (for the comfort, love and support that only sisters can provide), Jeff Marx, Tim French and the poker crew (for their friendship and laughter), my colleagues at Southern Connecticut and Rensselaer, especially Leo Schowalter, Jim Napolitano and Jack Wilson (for the positive influence you have had on my professional life) and my students at Southern Connecticut and Rensselaer, Ron Thornton, Priscilla Laws, David Sokoloff, Pat Cooney, Joe Redish, Ken and Pat Heller and Lillian C. McDermott (for helping me learn how to teach).

Priscilla Laws: First of all I would like thank my husband and colleague Ken Laws for his quirky physical insights, for the Chapter 11 Kneecap puzzler, for the influence of his physics of dance work on this book, and for waiting for me countless times while I tried to finish "just one more thing" on this book. Thanks to my daughter Virginia Jackson and grandson Adam for all the fun times that keep me sane. My son Kevin Laws deserves special mention for sharing his creativity with us—best exemplified by his murder mystery problem, *A(dam)nable Man*, reprinted here as problem 5-68. I would like to thank Juliet Brosing of Pacific University who adapted many of the Workshop Physics problems developed at Dickinson for incorporation into the alternative problem collection in this book. Finally, I am grateful to my Dickinson College colleagues Robert Boyle, Kerry Browne, David Jackson, and Hans Pfister for advice they have given me on a number of topics.

Joe Redish: I would like to thank Ted Jacobsen for discussions of our chapter on relativity and Dan Lathrop for advice on the sources of the Earth's magnetic field, as well as many other of my colleagues at the University of Maryland for discussions on the teaching of introductory physics over many years.

Pat Cooney: I especially thank my wife Margaret for her patient support and constant encouragement and I am grateful to my colleagues at Millersville University: John Dooley, Bill Price, Mike Nolan, Joe Grosh, Tariq Gilani, Conrad Miziumski, Zenaida Uy, Ned Dixon, and Shawn Reinfried for many illuminating conversations.

We also appreciate the absolutely essential role many reviewers and classroom testers played. We took our reviewers very seriously. Several reviewers and testers deserve special mention. First and foremost is Arnold Arons who managed to review 29 of the 38 chapters either from the original HRW 6th material or from our early drafts before he passed away in February 2001. Vern Lindberg from the Rochester

Institute of Technology deserves special mention for his extensive and very insightful reviews of most of our first 18 chapters. Ed Adelson from Ohio State did a particularly good job reviewing most of our electricity chapters. Classroom tester Maxine Willis from Gettysburg Area High School deserves special recognition for compiling valuable comments that her advanced placement physics students made while class testing Chapters 1-12 of the preliminary version. Many other reviewers and class testers gave us useful comments in selected chapters.

Class Testers

Gary Adams
Rensselaer Polytechnic Institute

Marty Baumberger
Chestnut Hill Academy

Gary Bedrosian
Rensselaer Polytechnic Institute

Joseph Bellina,
Saint Mary's College

Juliet W. Brosing
Pacific University

Shao-Hsuan Chiu
Frostburg State

Chad Davies
Gordon College

Hang Deng-Luzader
Frostburg State

John Dooley
Millersville University

Diane Dutkevitch
Yavapai College

Timothy Hayes
Rensselaer Polytechnic Institute

Brant Hinrichs
Drury College

Kurt Hoffman
Whitman College

James Holliday
John Brown University

Michael Huster
Simpson College

Dennis Kuhl
Marietta College

John Lindberg
Seattle Pacific University

Vern Lindberg
Rochester Institute of Technology

Stephen Luzader
Frostburg State

Dawn Meredith
University of New Hampshire

Larry Robinson
Austin College

Michael Roth
University of Northern Iowa

John Schroeder
Rensselaer Polytechnic Institute

Cindy Schwarz
Vassar College

William Smith
Boise State University

Dan Sperber
Rensselaer Polytechnic Institute

Roger Stockbauer
Louisiana State University

Paul Stoler
Rensselaer Polytechnic Institute

Daniel F. Styer
Oberlin College

Rebecca Surman
Union College

Robert Teese
Muskingum College

Maxine Willis
Gettysburg Area High School

Gail Wyant
Cecil Community College

Anne Young
Rochester Institute of Technology

David Ziegler
Sedro-Woolley High School

Reviewers

Edward Adelson
Ohio State University

Arnold Arons
University of Washington

Arun Bansil
Northeastern University

Chadan Djalali
University of South Carolina

William Dawicke
Milwaukee School of Engineering

Robert Good
California State University-Hayware

Harold Hart
Western Illinois University

Harold Hastings
Hofstra University

Laurent Hodges
Iowa State University

Robert Hilborn
Amherst College

Theodore Jacobson
University of Maryland

Leonard Kahn
University of Rhode Island

Stephen Kanim
New Mexico State University

Hamed Kastro
Georgetown University

Debora Katz
U. S. Naval Academy

Todd Lief
Cloud Community College

Vern Lindberg
Rochester Institute of Technology

Mike Loverude
California State University-Fullerton

Robert Luke
Boise State University

Robert Marchini
Memphis State University

Tamar More
Portland State University

Gregor Novak
U. S. Air Force Academy

Jacques Richard
Chicago State University

Cindy Schwarz
Vassar College

Roger Sipson
Moorhead State University

George Spagna
Randolf-Macon College

Gay Stewart
University of Arkansas-Fayetteville

Sudha Swaminathan
Boise State University

We would like to thank our proof readers Georgia Mederer and Ernestine Franco, our copyeditor Helen Walden, and our illustrator Julie Horan.

Last but not least we would like to acknowledge the efforts of the Wiley staff; Senior Acquisitions Editor, Stuart Johnson, Ellen Ford (Senior Development Editor), Justin Bow (Program Assistant), Geraldine Osnato (Project Editor), Elizabeth Swain (Senior Production Editor), Hilary Newman (Senior Photo Editor), Anna Melhorn (Illustration Editor), Kevin Murphy (Senior Designer), and Bob Smith (Marketing Manager). Their dedication and attention to endless details was essential to the production of this book.

Brief Contents

Contents

CHAPTER 29 Magnetic Fields 829

CHAPTER 30 Magnetic Fields Due to Currents 861

CHAPTER 31 Induction and Maxwell's Equations 888

CHAPTER 32 Inductors and Magnetic Materials 922

CHAPTER 33 Electromagnetic Oscillations and Alternating Current 954

Appendices

22 | Electric Charge

Nothing happens if you place a plastic comb near tiny scraps of paper, but immediately after you comb your hair or stroke the comb with fur, it will attract the paper scraps. In fact, the attractive force exerted on the paper by the small comb is so strong that it overcomes the opposing gravitational pull of the entire Earth. This phenomenon, commonly called "static cling," occurs between many different objects and is especially easy to observe during cold dry weather.

What causes these pieces of paper to stick to the comb and to one another?

The answer is in this chapter.

22-1 The Importance of Electricity

If you walk across a carpet when it's cold and dry outside, you can produce a spark by bringing your finger close to a metal doorknob. Television advertisements alert us to the problem of "static cling" in clothing. On a grander scale, lightning is familiar to everyone. These phenomena represent a tiny glimpse into the vast number of electric interactions that occur every day.

The phenomenon of electricity plays a major role in modern life. Less than two hundred years ago, fire was almost the only source of heat, the only source of light when the sun or moon was not up, and the only way to cook food. Without electric water pumps, most people did not even have indoor plumbing. It's hard to imagine life without electric lights (not even flashlights), stoves, refrigerators, air conditioners, computers, telephones, radios, televisions, CD players, and a host of other electrical devices. We make extensive use of electricity, but *what is it*? In this chapter we consider this very important question.

So far in our study of the physical world we have learned how the forces acting on objects affect motion. We have also learned about the gravitational force, an action-at-a-distance force, that objects can exert on each other without touching. In this chapter, we will investigate another action-at-a-distance force—the *electrostatic* interaction force. Studying the electrostatic force will provide a foundation for our understanding of the phenomenon of electricity.

We begin our study by looking at the nature of electrical interaction forces between some everyday objects. We then develop the concepts of charging and electric charge as tools for explaining our observations of electrostatic forces on a macroscopic level. However, to obtain a more coherent understanding of electrostatic phenomena, we must turn to the findings of atomic theory.

An understanding of electrostatic interactions will give you insight into the fundamental relationship between electricity and magnetism. In Chapter 30, which is about magnetic fields due to currents, you will discover that although magnetic forces are generated by the interaction between moving charges, they have distinctly different properties than do electrostatic forces. Later, in Chapter 32, you will see how electricity and magnetism are fundamentally related to each other.

READING EXERCISE 22-1: List all the electrical devices that you use in a typical week. ■

22-2 The Discovery of Electric Interactions

FIGURE 22-1 ■ Fossilized resin, known as amber, is popular both for its beauty and for its ability to preserve ancient vegetation and insects, like the bees, wasps, ants, flies and mosquitos seen here. Amber also has electrical properties.

Amber, which is resin that oozed from trees long ago and hardened, has been admired both for its beauty and its ability to preserve early life forms mired in it (Fig. 22-1). Amber has electrical properties of interest to scientists as well. The early Greeks knew that if one rubbed a yellow-brown piece of amber with fur, it would attract bits of straw. The strength of the attraction decreased as the distance between the amber and the straw was increased. The strength of the attraction was also known to fade over time, especially in damp weather.

By the 1600s, this strange force due to amber that was sometimes present and sometimes not, prompted more careful studies. It was subsequently discovered that other materials such as glass can also attract small bits of matter after being rubbed with silk. As was the case for amber, this attractive force diminished with time, especially on humid days, and was not present if the glass had not been rubbed. Additionally, the strength of the attraction decreased as the distance between the glass and small bits of matter increased, just as was the case with the force associated with rubbed amber.

This interaction phenomenon, created by rubbing certain materials with cloth, was named *electrification*. The term is derived from the Greek word for amber, which is *electron*. Any object (not just glass or amber) is defined as becoming *electrified* if:

1. There is an interaction force between this object and another that is present after the objects have been in very close contact, usually through rubbing;

2. The magnitude of this interaction force diminishes with time and is affected by humid weather; and

3. The magnitude of the force decreases with increasing distance between the objects.

Although the similarities between electrified glass and electrified amber were interesting, it was not until 1733 that a French scientist, Charles DuFay, published articles presenting evidence that:

Two amber rods stroked with fur always repel one another.

Two glass rods stroked with silk always repel one another.

A stroked amber rod *attracts* a stroked glass rod (Fig. 22-2).

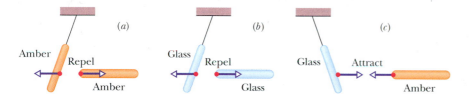

FIGURE 22-2 ■ (*a*) Two amber rods electrified in the same way repel each other. (*b*) Two glass rods electrified in the same way also repel each other. (*c*) An electrified glass rod and an electrified amber rod attract each other.

Provided the weather is not too humid, you may be able to repeat DuFay's observations yourself by replacing the amber and glass rods (which are difficult to find outside of a physics laboratory) with Styrofoam cups and plastic sandwich bags as shown in Fig. 22-3. Place your hand inside a plastic bag and use a rubbing motion to assure that the entire surface of the Styrofoam cup comes in contact with the entire surface of the plastic bag. Then rub another Styrofoam cup with a second sandwich bag in the same manner. If you put one of the cups on its side on a smooth, level, nonmetallic surface and bring the other cup near it, the first cup should roll away as shown in Fig. 22-3a. Note that after the two cups have been electrified in a *like manner* they *repel* one another just like DuFay's rods. Now hold the two plastic bags together at the top end. Both plastic bags have also been electrified in a like manner and they repel one another as well as shown in Fig. 22-3b. However, an electrified sandwich bag and an electrified Styrofoam cup will be *attracted* to each other just as electrified amber attracts electrified glass. Think about these observations carefully and you must conclude that there are two classes of materials that behave differently when electrified.

Not all types of materials can be electrified. Nevertheless, additional observations with electrified materials lead to the following general statements:

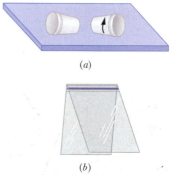

FIGURE 22-3 ■ (*a*) Two Styrofoam cups electrified in the same way repel each other. (*b*) The two sandwich bags used to electrify the cups also repel each other.

OBSERVATION 1. Two identical objects electrified by the same process always repel one another.

OBSERVATION 2. Two different electrified objects will always interact, but they may either repel or attract one another.

OBSERVATION 3. Any two objects that have not been electrified will neither repel nor attract one another. (They interact only by means of an imperceptibly small gravitational force.)

Suppose you have electrified two Styrofoam cups so they repel each other. What happens when you give one of the cups extra stroking? The magnitude of the interaction forces between the cups increases. This means that if we think of the first cup (that did not receive extra stroking) as a "standard object," we can determine the degree of electrification of any other object by measuring the magnitude of the electric force exerted on it by the standard object.

One logical way of interpreting our observations regarding electrification is to assume that a substance is added or removed from an object during the stroking process. Extensive experiments done at the end of the 18th century by Benjamin Franklin and others indicate that this is correct. They also found that there are actually two types of the substance involved. Today we call these substances **electrical charge** and say that there are two types of (electrical) charge. When an object contains more of one type of charge than the other, the object is electrified or **electrically charged.** Furthermore, any process of electrification (not just rubbing) is called **charging.** Thus, in the example above, the cup that was stroked for the longer time gained the greater quantity of excess charge. (Quantity of charge is often called **amount** of charge). An object with a greater amount or quantity of charge is observed to experience more force in the presence of a standard electrified object than one with a smaller amount of excess charge.

READING EXERCISE 22-2: The creation of electrified objects can also be done with strips of Scotch™ Magic Tape using a peeling action rather than stroking. In order to charge the tape, cut 2 strips about 10 cm long. (a) If you were to stick the tapes side by side on a table and peel them both off, what do you predict would happen if you then brought the tapes close together? Explain the reasoning for your prediction. (b) Perform the experiment and describe what happens. Is this consistent with your prediction? If not, explain what you think is going on. ■

22-3 The Concept of Charge

Various observations, including our observations using Styrofoam cups and plastic bags, indicate that interaction forces between charged objects can be explained in terms of two (and only two) different kinds of charged matter. The type associated with glass rubbed with silk is one and the type associated with amber rubbed with fur is the other. We cannot prove directly that there are no other types of charge. However, the fact that no one has found a charged object that attracts both charged glass and charged amber leads us to believe that there is no third type of charge.

Today, the terms we associate with these two types of charged matter are *positive* and *negative*. Benjamin Franklin is responsible for assigning these names. He introduced the following definitions:

OBSERVATION 1. An object that is repelled by a glass rod stroked with silk is **positively charged.**

OBSERVATION 2. An object that is repelled by amber (or plastic) stroked with fur is **negatively charged.**

OBSERVATION 3. Any two nonmagnetic objects that do not interact with each other except by gravitational forces are electrically neutral.

The names given to the two varieties of charge are arbitrary. Benjamin Franklin could just as easily have used other words, such as light and dark, to distinguish between the two types of charges. However, we observe that equal amounts of the positive and negative charges combine to produce nonelectrified (i.e., **electrically neutral**) matter. That is, the two types of charge combine algebraically—like positive and negative numbers. So, positive and negative are convenient and appropriate names.

Applying this new terminology to our previous observations leads us to say that if two objects are each repelled by a piece of glass that has been rubbed with silk, then both objects must be positively charged. Furthermore, we hypothesize that these objects repel *because* they contain the same kind of charge or **like charges.** On the other hand, if we find that two objects made of different materials attract after being stroked, we hypothesize that one object has a positive charge while the other has a negative charge. We conclude that objects with **unlike** or **opposite charges** attract.

READING EXERCISE 22-3: Suppose you stroked a smooth wooden rod with a linen cloth and announced that you had created a new type of charge you decided to call *woodolin charge.* (a) If a skeptic asked you to prove that woodolin was really a new type of charge, how would you do it? Specifically what would have to happen if you were to bring two wooden rods together that had both been rubbed with linen? If you were to bring a charged wooden rod near a charged glass rod? Near a charged amber (or plastic) rod? (b) Why do you think most observers agree that there are only two types of known charge? ■

22-4 Using Atomic Theory to Explain Charging

How can we account for the fact that when certain objects are rubbed together they acquire opposite types of charge? One way to make sense of this observed fact is to use a contemporary understanding of the atomic structure of matter. The atomic model that we discuss here has been developed over the past century. We will use it as an explanatory tool without presenting evidence for it.

The Atomic Model

According to modern atomic structure theory, atoms consist of positively charged *protons,* negatively charged *electrons,* and electrically neutral *neutrons.* Electrons and protons have the same amount (although with opposite sign) of charge. We often represent this amount of charge, called the **elementary charge,** with an e (Table 22-1). Hence an electron has a charge of $-1e$ and a proton has a charge of $+1e$.

Protons and neutrons are packed tightly together in a central *nucleus.* They are much more massive than electrons, which lie outside the nucleus as depicted in Fig. 22-4. Most of the atoms that are contained in matter have equal numbers of electrons and protons, so whenever a charged object is at some distance away from the atom, the atom appears to be electrically neutral.

According to contemporary atomic theory, electric charge is an intrinsic characteristic of electrons and protons. You often encounter casual phrases—such as "the

TABLE 22-1

Charges of the Three Fundamental Atomic Particles

Particle	Symbol	Charge
Electron	e or e⁻	$-e$
Proton	p	$+e$
Neutron	n	0

Note: The symbols for electron and for electronic charge are the same. This can be confusing.

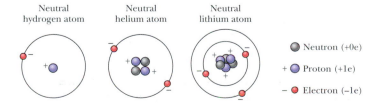

Neutral hydrogen atom   Neutral helium atom   Neutral lithium atom

Neutron (+0e)
Proton (+1e)
Electron (−1e)

FIGURE 22-4 ■ The structure of the atoms representing the three lightest chemical elements, H, He, and Li. The number of protons that define the element along with the typical number of neutrons in each element's nucleus are shown. The darker circles represent protons, the lighter circles neutrons, and the white circles electrons. The diagram is simplified, as physicists do not actually believe that electrons orbit nuclei in nice neat circles and that the nuclei are much smaller relative to the size of their atoms.

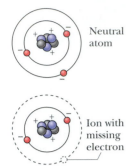

FIGURE 22-5 ■ The upper diagram shows a neutral lithium atom with its full complement of electrons. The bottom atom has lost an outer electron and is now an ion with a net charge of +1e because it has three protons and only two electrons.

charge on a sphere," "the amount of charge transferred," and "the charge carried by the electron." Such phrases can be misleading because they suggest that charge is a substance. You should, however, keep in mind that experiments show electrons and protons are the basic substances. Charge and mass are simply two of their fundamental properties.

The mass of an electron is about 2000 times smaller than that of a neutron or proton. Electrons are attracted to the nucleus because electrons and the protons within the nucleus have opposite charges. However, the electrons that are farthest away from the nucleus are only weakly attracted to the protons within the nucleus and so they don't always remain associated with individual atoms. In many types of materials the electrons are free to wander within the material if they experience forces. If the atom loses an electron it is no longer neutral, but has a net positive charge because there are now more protons than electrons, as seen in Fig. 22-5. Charged atoms are called **ions.** We call mobile electrons **conduction electrons.** If an electric or other force is applied to the atom, only the conduction electrons, with their negative charges, move appreciably. The much more massive positive ions stay fixed in place.

Charge Is Quantized

In Benjamin Franklin's day, electric charge was thought to be a continuous fluid that could "contain" any arbitrary amount of charge. Today we know that fluids, such as air and water, are not continuous but are made up of atoms and molecules. Matter is discrete. In 1909 an American physicist, Robert Millikan, used opposing electric and gravitational forces to balance drops of oil between two electrified metal plates. His famous oil drop experiment and others that followed showed that the "electrical fluid" is not continuous either but is made up of multiples of the elementary charge.

Any positive or negative charge q that has ever been detected as a free particle can be written as

$$q = ne, \quad n = \pm 1, \pm 2, \pm 3, \ldots, \tag{22-1}$$

in which e, the elementary charge, has the value

$$e = 1.60 \times 10^{-19} \text{ C}. \tag{22-2}$$

The SI unit of charge is the **coulomb** (C), named for Charles Augustin Coulomb, who studied electric forces in the late 1700s. When a physical quantity such as charge can have only discrete values rather than any arbitrary value, we say that the quantity is **quantized.** It is possible, for example, to find a free particle that has no charge at all or a charge of $+10e$ or $-6e$, but not a free particle with a charge of, say, $3.57e$. Modern studies of the structure of neutrons and protons have produced strong evidence that neutrons and protons are made up of tightly bound particles with charges $+2/3e$ and $-1/3e$ that we call quarks, but quarks do not seem to be able to exist as free particles. Hence, extensive experimentation confirms that:

> Charge is quantized. In free particles charge has never been measured to have an amount other than an integer multiple of 1.60×10^{-19} C.

As we noted in the introduction, if you drag your feet as you walk across a carpet, you can produce a spark caused by moving electric charge by bringing your finger close to a metal doorknob. This is a demonstration of a small sample of the vast amount of electric charge that is stored in electrically neutral objects. However, that charge is usually hidden because the object contains equal amounts of positive and negative charge. This was hinted at above when we stated that electrically neutral atoms contain equal numbers of protons and electrons. With such an equality—or *balance*—of the amounts charge there is no *net* charge. If the two types of charge are not in balance, we say an object is *charged* to indicate that it has a charge imbalance, or nonzero net charge.

> Macroscopic objects that are electrically neutral are *not* devoid of charge. Instead, they contain equal numbers of positive protons and negative electrons. This results in a cancellation of their electrical effects.

Charging Is Transferring Electrons

Glass and silk or Styrofoam and plastic become oppositely charged when they are brought into contact and we can use our modern understanding of the atom to explain why. Suppose we observe that Styrofoam becomes positively charged and plastic becomes negatively charged. It is logical to assume that outer electrons associated with atoms in the Styrofoam are attracted to the atoms in the plastic and move over to the plastic. The Styrofoam is now missing electrons so there is a net positive charge on the Styrofoam. The plastic now has excess electrons and has a net negative charge.

In general, experiment shows that an object becomes charged when a very tiny fraction of the mobile electrons with their negative charge are transferred from one object to another. This is why we must rub, stroke, or otherwise make significant contact between two objects for the objects to become charged. Thus, when a Styrofoam cup is stroked with a plastic bag, a very tiny fraction of the electrons near the surface of the Styrofoam cup are transferred to the plastic bag.

Why doesn't the plastic bag get heavier when electrons are transferred to it? Using modern atomic theory, we understand that even if we transfer a lot of electrons to the plastic bag (a typical number might be between 10^9 and 10^{12}) the increase in mass would be less than 10^{-10} kg—not measurable. In ordinary matter, positive charge is much less mobile than negative charge. For this reason, an object becomes positively charged through the *removal of negatively charged electrons* rather than through the addition of positively charged protons.

Charge Is Conserved

Careful measurements reveal that whenever there is excess charge on one of the objects after contact, there are excess charges on the other object too. These charges are equal in amount but opposite in sign. This demonstrates that when electrons are transferred from one object to another, no electrons are destroyed or created in the process. The amount of charge contained in the two objects is constant or conserved.

This hypothesis of **conservation of charge** was first proposed by Benjamin Franklin based on his experiments. It is observed to hold both for large-scale charged bodies and for atoms, nuclei, and elementary particles. No exceptions have ever been found. So, we add electric charge to our list of quantities—including energy, linear momentum, and angular momentum—that are conserved quantities. In summary, extensive experimentation confirms:

> The total amount of electric charge in the universe is conserved. Although particles that carry charge can be transferred from one object to another, the charge associated with particles cannot be created or destroyed.

Force and Quantity of Charge

Because charge is conserved, we can transfer charge from one object to another without changing the total amount of charge in the system. This allows us to perform experiments that indicate how the interaction force between charged objects depends on the amount of charge on each object. These experiments lead to surprisingly simple results when the charged objects are symmetric, made of metal, and are particle-like (so that their dimensions are small compared to the distances between their centers). For example, consider the experiment shown in Fig. 22-6. Two identical uncharged metal spheres are both electrified with a charged plastic rod. We then touch the two spheres together. Since the spheres are identical and the excess electrons repel each other, we expect electrons to travel between the spheres until both spheres have the same number of excess electrons. Next we measure the force exerted by one sphere on the other and record it (Fig. 22-6a).

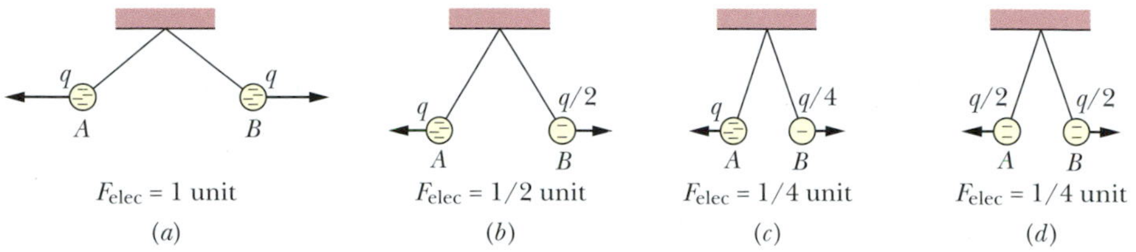

FIGURE 22-6 ▪ Depiction of an idealized experiment to measure the forces between small metal spheres that hold different fractions of charge. *Note:* In order to make force measurements for particle-like objects, the distance between the centers of the two balls of identical shape must be more than twice the diameter of a ball.

Then we leave sphere A alone and move sphere B a long distance from sphere A to place it in contact with a third sphere C that is uncharged. The excess electrons on sphere B will now be shared equally between spheres B and C so the number of excess electrons on sphere B will now be half of what it was before. If we return B to its original location and measure the magnitude $F_{elec} = |\vec{F}_{elec}|$ of the force between spheres A and B, we find that it is one-half of the force magnitude we first measured (Fig. 22-6b). If we repeat this process so we reduce the amount of charge on sphere B to one-fourth of what it was originally, then the magnitude of the interaction force between the spheres is also reduced to one-fourth of what it was originally (Fig. 22-6c). In a similar experiment we can reduce the charge on both spheres A and B to half their original values and then the force measures one-fourth the original force between them (Fig. 22-6d). These observations, which are summarized in Table 22-2, indicate that the magnitude of the interaction force is proportional to the *product* of the amounts of charge on the two spheres. This relationship is given by

$$F_{elec} \propto |q_A||q_B|, \tag{22-3}$$

where the absolute value signs denote charge amounts independent of sign.

TABLE 22-2

q_A	q_B	$q_A \cdot q_B$	F_{elec}(arbitrary units)
q	q	q^2	$1\,F$
q	$q/2$	$q^2/2$	$1/2\,F$
q	$q/4$	$q^2/4$	$1/4\,F$
$q/2$	$q/2$	$q^2/4$	$1/4\,F$

> The amount of the charge on a particle-like object can be quantified through measurement of the magnitude of the interaction force between it and a standard charged object that is also particle-like.

The Electroscope

The fact that like charges repel has been used in the development of the **electroscope,** a sensitive charge-measuring device, as seen in Fig. 22-7. A net charge can be transferred to an electroscope by stroking the metal ball with a charged rod. If the rod is negatively charged, some of its excess electrons will be transferred to the ball and then they will spread throughout the metal rod and the foil attached to the ball. If a flexible metal leaf is attached to the central conducting bar, the flexible conductor will be repelled from the central charges and rise. As more electrons are transferred to the electroscope, the metal leaf will rise higher. Alternatively if the rod is positively charged it will *attract* electrons from the electroscope, leaving a net positive charge on it. Once again the foil will rise.

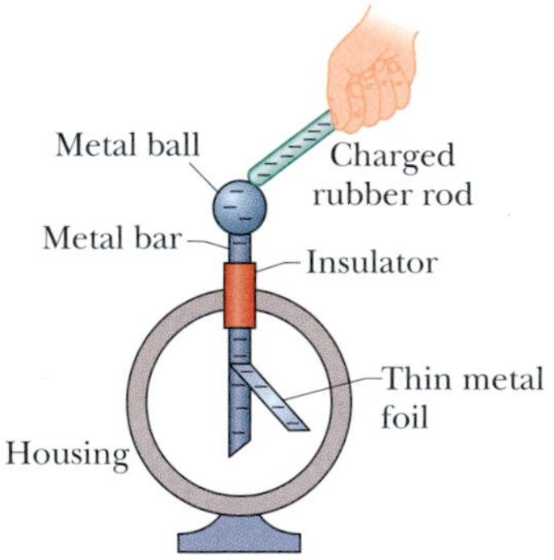

FIGURE 22-7 ▪ The electroscope can be used to measure charge. The rise of a metal foil is caused by the repulsion due to an excess of like charged particles distributed on the parts of a metal conducting system. The foil rises in proportion to the net charge contained on the conductor.

READING EXERCISE 22-4: Assuming that solid objects are made up of atoms rather than being continuous, can you think of a plausible way to explain why it is so difficult to pull solids apart or push them together? ■

READING EXERCISE 22-5: Consider the measurements depicted in Fig. 22-6. Suppose you have measured the repulsion force between two identical metal-coated spheres that each have a total negative charge q due to excess electrons. Next, you would like to measure the force on the metal-coated spheres that each have one-fourth of the excess electrons they originally had. Describe how you could use similar uncharged spheres to reduce the excess electrons on each of the original spheres to $q/4$. ■

22-5 Induction

Let's consider some additional observations involving electrical interactions. Typically, bits of straw or paper that have not been rubbed do not attract or repel one another. They are electrically neutral. Thus, it is surprising to find that a plastic comb made negatively charged by rubbing (like the comb shown in the photograph at the beginning of this chapter) can attract bits of electrically neutral paper. It is equally surprising to find that a positively charged glass rod will attract bits of paper.

> **INDUCTION** is the process that causes the attraction we observe between a charged object and an uncharged one.

How can we explain induction? We turn to atomic theory for an explanation. The idea that electrically neutral materials are not devoid of charge, but rather are composed of atoms that have the same number of positive protons and negative electrons, is the first step in developing a viable explanation for induction. The second important idea, mentioned in the last section, is that electrons are more mobile than protons.

Let's begin by considering how induction occurs when a charged object is placed near an uncharged *metal* object. Then in the next section, we will consider induction in a class of nonmetals known as insulators.

Induction in Metals

What happens when we dangle a very small metal rod from a string near a charged object as shown in Fig. 22-8? According to our atomic model, mobile negative electrons in the metal rod are repelled from the negatively charged object. When the mobile electrons in the neutral metal object are repelled they move away and unpaired protons are left behind. The unlike charges at the surfaces of the two objects will now attract as shown in Fig. 22-8a. An attraction between a neutral object and a positively charged object occurs as shown in Fig. 22-8b. The process of separation of positive protons and negative electrons in the neutral objects is known as **polarization.**

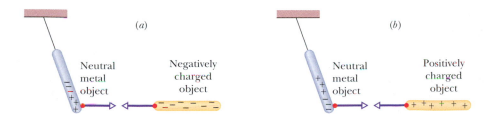

(a)

Neutral metal object — Negatively charged object

(b)

Neutral metal object — Positively charged object

FIGURE 22-8 ■ A tiny neutral metal rod is suspended on a nonconducting thread. The part of the neutral rod that is closest to a charged object will be attracted by either: (a) a negatively charged object (such as amber) or (b) a positively charged object (such as glass). The extent of the charge separation in the metal rod is exaggerated.

READING EXERCISE 22-6: (a) If we state that two bits of paper are electrically neutral, what observation can you make to verify this is the case? Explain. (b) Can induction be used to determine whether the charged object is positive or negative? Why or why not? ■

READING EXERCISE 22-7: (a) In the figure that follows, objects made of different materials are arranged in six pairs. *A* is plastic stroked with fur, *B* is glass stroked with silk, and *C* is an electrically neutral metal object. State whether the interaction between each pair will be attractive, repulsive, or nonexistent. Explain your reasoning.

(1)	(2)	(3)	(4)	(1)	(1)
A	*A*	*B*	*B*	*C*	*C*
B	*A*	*B*	*C*	*C*	*A*

(b) According to the sign convention Benjamin Franklin decided to adopt, which of the three objects carry an excess positive charge? An excess negative charge? No excess charge? Explain.

(c) Suppose you were told that *A*, *B*, and *C* are made of new materials but you were not told what those materials are. Would it be possible to determine whether they are negatively charged, positively charged, or electrically neutral by observing how the pairs shown in the diagram above interact? Why or why not? ■

22-6 Conductors and Insulators

Whenever a charged object is near an electrically neutral object, induction and polarization occur. However, the attractive forces are stronger for neutral metal objects than for nonmetal objects. Why? Let's summarize the outcomes of some important observations you can make yourself using metal rods, Styrofoam cups, and plastic bags.

OBSERVATION 1. (a) The electrification created on nonmetal objects, like plastic, does not spread out. Instead charge seems to remain in regions where the object is rubbed, and (b) touching charged nonmetal objects removes the electrification only at locations where an object is touched.

OBSERVATION 2. (a) Metal objects can be charged when mounted on nonmetal objects such as glass or plastic but they cannot be charged while being held in someone's hand, and (b) metal objects that are touched anywhere by a person will immediately lose *all* of their charge.

Even though paper is a nonmetal, the ideas of induction and polarization can be used to explain how the charged comb pictured in the puzzler at the beginning of the chapter can be used to pick up a string of paper bits. We assume the first piece of paper is attracted to the charged comb by induction and becomes polarized in the process. Then the excess negative charge at the bottom end of the first paper bit attracts the second paper bit by induction. Since the second paper bit is now polarized, the process continues. (Although experiments show that magnetic forces behave differently than do electrostatic forces, this process of electric polarization looks like a similar process in which a magnet can induce magnetic polarization in a steel paper clip, which then attracts and polarizes a second clip, and so on.)

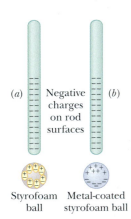

(*a*) Negative charges on rod surfaces (*b*)

Styrofoam ball   Metal-coated styrofoam ball

FIGURE 22-9 ■ According to atomic theory: (*a*) polarization induced in an insulator involves very very tiny atomic-scale charge separations as shown in exaggerated form for a Styrofoam ball, whereas (*b*) polarization in a conductor can involve a much larger scale migration of electrons as shown in a metal-coated ball.

Atomic Theory and the Behavior of Conductors and Insulators

Once again we can turn to atomic theory to develop a plausible explanation for these new observations. In materials such as metals, tap water, water droplets in air, and the human body, experiments indicate that some of the negative electrons move easily. We call such materials **conductors.** We observe that charge can flow onto or off conductors quite quickly. In other materials, such as glass, chemically pure water, and plastic, the electrons can reposition themselves within an atom but cannot migrate between atoms. We call these materials **nonconductors** or **insulators.** Electrons do not travel from atom to atom very easily in insulators. An exception is that some of the electrons at the surface of an insulator can have a greater affinity for another type of surface. For example, electrons can travel from the surface of a glass rod to the surface of a piece of silk cloth brought into contact with it, leaving the glass rod positively charged.

One implication of this difference in the mobility of charge in insulators and conductors is that the polarization process discussed in Section 22-5 is not as strong in insulators. Neutral conductors and neutral insulators both undergo charge separation (become polarized) when they are brought close to a charged object—but, the electrons in insulators are tightly bound to atoms, and the charge separation (polarization) is only a small fraction of an atomic radius. In contrast, some of the electrons in conductors can move through the material fairly freely and become separated from the atoms to which they were originally associated. This difference is shown in the comparison of the two images in Fig. 22-9. This atomic model provides a plausible explanation for the observed fact that induction is much stronger between a charged object and a conductor than between the same charged object and an insulator.

The difference in mobility of charge carriers in conductors and insulators explains why you cannot charge a metal rod by rubbing if you are holding it. Both you and the rod are conductors. Although the rubbing will cause a charge imbalance on the rod, the excess charge will immediately move from the rod through you to the floor (which is connected to Earth's surface), and the rod will quickly be neutralized. Setting up a pathway for electrons between an object and Earth's surface is called **grounding** the object, and always results in electrically neutralizing the object. If instead of holding the metal rod in your hand, you hold it by an insulating handle, you eliminate the conducting path to Earth, and rubbing can then charge the rod, as long as you do not touch it directly with your hand.

These ideas give us a very functional way to determine whether a material is a conductor or an insulator. If you have two interacting charged objects and you touch one of them, do the objects stop interacting? If so, charge must have been transferred to or from the object. Hence, it must be a conductor. If the transfer of charge does not occur, the object must be an insulator.

Charging by Induction

The rubbing methods we use to charge insulators such as glass, rubber, and amber do not work well with conductors. Fortunately, you can take advantage of the polarization model (which was used to explain induction) and a process known as **charging by induction** to accomplish this. In the example of charging by induction, an electrically neutral metal plate is brought near a negatively charged insulator shown in Fig. 22-10b. The charges in the metal plate are polarized by induction. Since the top of the metal plate now has an excess of electrons, touching it will cause these electrons to flow onto your body. If you now stop touching the rod, the return pathway for electrons is removed. So the metal plate is no longer electrically neutral. If you now move the metal plate away from the charged insulator, the polarization effect disappears and the metal plate remains positively charged due to a deficiency of electrons. Of course, it is also

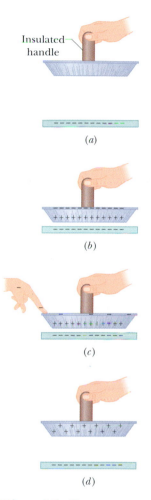

FIGURE 22-10 ■ An electrophorus is an apparatus that can be used to charge a conductor by induction. It consists of an insulated plate such as a slab of Styrofoam and a conducting plate such as an aluminum pie plate. (*a*) The insulating slab is charged negatively after being stroked with fur. (*b*) When a conducting plate is brought near the charged insulator the conductor is polarized so that its free electrons move away from the insulator. (*c*) When the top of the conductor is touched these free electrons move to the hand. (*d*) This leaves a net positive charge on the conductor.

FIGURE 22-11 ■ This is not a parlor stunt but a serious experiment carried out in 1774 to prove that the human body is a conductor of electricity. The etching shows a person suspended by nonconducting ropes while being charged by a charged rod (which probably touched flesh instead of the trousers). When the person brought his face, left hand, or the conducting ball and rod in his right hand near bits of paper on the plates, charge was induced on the paper, which flew through the intermediate air to him.

possible to develop a similar procedure that will leave an excess of electrons on the metal plate. (Figure 22-11 shows another experiment set up for inducing charge.)

READING EXERCISE 22-8: (a) Make the observation described in Reading Exercise 22-2. Is Scotch™ Magic Tape best described as an insulator or a conductor? Explain your reasoning. (b) Is a balloon an insulator or a conductor? Explain your reasoning. ■

READING EXERCISE 22-9: (a) Can you charge an insulator by induction? Explain your reasoning. (b) Describe the steps you would take to give an object excess negative charge using the process of charging by induction utilizing the electrophorus apparatus shown in Fig. 22-10. ■

22-7 Coulomb's Law

So far all our explanations of electrical phenomena have been qualitative. Can a mathematical law be formulated to quantitatively describe the interaction forces between electric charges?

The observations we depicted in Fig. 22-6 led us to the conclusion that the interaction forces are proportional to the product of the charges on the objects. So the magnitude of the force on either particle is given by $F \propto |q_A q_B|$. But, what observations have been made that would lead to a mathematical relationship that also describes how interaction forces are related to the distance between charged particle-like objects?

Benjamin Franklin observed that a small cork hanging from a silk thread is attracted by induction to the outside of a charged metal can. However, if the cork is dangled inside the can, there are no apparent forces on it. Recall that in Chapter 14 on gravitation we presented a shell theorem Newton derived from the assumption that gravitational forces fall off as the inverse square of the distance between masses. Newton's shell theorem implies that a shell of mass exerts no net gravitational force on a point mass contained within it. Joseph Priestly reasoned that since an analogous shell theorem seems to hold for electric interactions, then the inverse square law ought to hold for electric forces too.

In 1785, Priestley's hypothesis regarding the dependence of electric forces on the inverse square of distance was verified by the experiments of Charles Augustin Coulomb using a sensitive torsion balance to measure the forces between charged spheres.

Coulomb assumed the forces between charged spheres would be the same as if the charge of each object was concentrated at its center. He also found that the forces between the objects lie along a line between their centers. Coulomb used a method like the one we described in Section 22-4 to reduce the charges on the metal spheres in his torsion apparatus by known fractions. Thus, he was able to verify that the interaction forces were proportional to the product of the charges on the interacting objects.

As a result of his careful experiments, Coulomb found that the magnitude of the electric (or Coulomb) force, F^{elec}, between two stationary particle-like charged objects is

$$F^{elec} = k \frac{|q_A||q_B|}{r^2} \quad \text{(Coulomb's law)}, \quad (22\text{-}4)$$

where k is a positive constant of proportionality, r is the distance between the centers of the two objects, q_A is the charge on one of the objects, and q_B is the charge on the other object.

Using modern tools available in many introductory physics laboratories, we can verify the inverse square ($1/r^2$) relationship in Coulomb's law. The experiment is pictured in Fig. 22-12. Two Ping-Pong balls that are covered with conducting paint are stroked with a fur-charged rubber rod so they are negatively charged. One of the balls is hung as a pendulum from a long, nonconducting string. The other ball, which serves as a prod, is attached to a nonconducting rod.

As the prod is moved very slowly toward the hanging ball, the hanging ball is repelled and rises. The hanging ball is displaced further and further from its equilibrium as the prod is brought closer to it. This demonstrates qualitatively that the force exerted by the prod on the hanging ball is greater when the distance, r, between the centers of the two charged balls is smaller. It also indicates that the electrostatic force acts along the line connecting the two charges. We know this because the hanging ball is not pushed off to the side.

Video technology allows us to take this experiment a step further. The motion of the prod inching forward can be captured with a video camera and digitized. Then computer software can be used to perform a frame-by-frame analysis of the angular displacement of the hanging ball and of the distance between the balls. Figure 22-13 is an example of this digital analysis. When the ball is stationary, the net force consists of the vector sum of the gravitational force acting vertically downward, the tension force exerted by the string, and the electric force acting in the horizontal direction. Thus, the magnitude of the electric force on the ball can be calculated from the mass of the ball and its angle of rise, θ, with respect to the vertical.

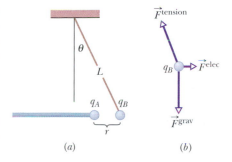

(a) (b)

FIGURE 22-12 ■ A charged metal-coated Ping-Pong ball is repelled from a charged prod. At equilibrium, the vector sum of the gravitational force, the tension in the string, and the Coulomb force on the hanging ball is zero.

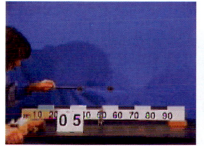

FIGURE 22-13 ■ Three of twenty-five digitized video frames depicting the forces between two charged balls. The string holding up the hanging ball is too thin to see and its point of attachment is well above the top of the video frames.

A plot of the data is shown in Fig. 22-14. If we try to fit the data, we find that the force between electrical charges falls off with distance as $1/r^2$ just as the gravitational force does. This verifies Coulomb's result summarized in Eq. 22-4.

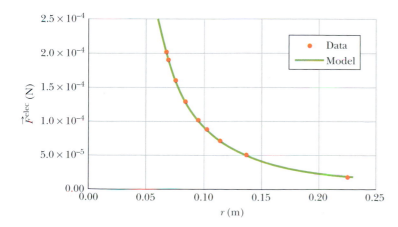

FIGURE 22-14 ■ A graph of the magnitude of the Coulomb force vs. the distance between two charged Ping-Pong balls each having a mass of 2.40 g. The green line represents an excellent inverse square fit to the red data points. The fit is given by $F^{elec} = (7.9 \times 10^{-4}\ \text{N} \cdot \text{m}^2)/r^2$. Using a Coulomb constant of $8.99 \times 10^9\ \text{N} \cdot \text{m}^2/\text{C}^2$ (as shown in Eqs. 22-6 and 22-7), it can be shown that each ball carries about 1×10^{-8} coulombs of excess charge. VideoPoint® software was used to obtain the data from video frames (Dson015.mov).

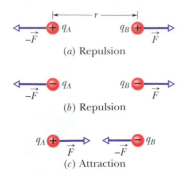

(a) Repulsion

(b) Repulsion

(c) Attraction

FIGURE 22-15 ■ Two charged particles, separated by distance r, repel each other if their charges are (a) both positive or (b) both negative. (c) They attract each other if their charges are of opposite signs. In each of the three situations, the force acting on one particle is equal in magnitude to the force acting on the other particle but has the opposite direction.

We use the absolute value signs on the force and charges in Eq. 22-4 to remind ourselves that the sign of the force (a vector) indicates the *direction* of the force and not simply whether we are multiplying like or unlike charges. Hence, we should calculate the magnitude of the force between two charged objects using the amounts of the charges. We determine the direction of the force using the attraction and repulsion rules we discussed earlier, remembering that the force always acts along the line connecting the two charges. If the particles *repel* each other, the force on each particle is directed *away from* the other particle (as in Figs. 22-15a and b). If the particles *attract* each other, the force on each particle is directed *toward* the other particle (as in Fig. 22-15c).

As an example of the need for absolute values in the equation

$$F_{A \to B}^{elec} = k \frac{|q_A||q_B|}{r^2},$$

consider the two unlike charges in Fig. 22-15c. Inspection of the expression for force above indicates that each particle exerts a force of the same magnitude on the other particle,

$$F_{A \to B} = F_{B \to A}. \tag{22-5}$$

But, electrostatic interactions satisfy Newton's Third Law. The force on the positive charge (in Fig. 22-15c) due to the negative charge points to the right. The force on the negative charge due to the positive charge points to the left. If we use explicit positive and negative signs on the charges and don't make use of absolute values, the product $q_A q_B$ (or $q_B q_A$) is always negative and so the force would be negative, regardless of whether we are calculating the force on the positive charge or the force on the negative charge. This cannot be correct, since these two forces point in opposite directions and the sign denotes direction. The force cannot be negative for both charges. In this and every other situation, we avoid this pitfall if we use the absolute values of the charges in our calculations and then determine the sign associated with the force by *thinking* about our coordinate system and the issues of attraction and repulsion.

Curiously, the form of Eq. 22-4 is the same as Newton's law of gravitation presented in Chapter 14 that relates the gravitational force between two particles with masses m_A and m_B to the distance, r, between their centers,

$$F_{A \to B}^{grav} = G \frac{m_A m_B}{r^2}, \tag{Eq. 14-1}$$

in which G is the gravitational constant.

Coulomb's law has survived every experimental test; no exceptions to it have ever been found. It holds even within the atom, correctly describing the force between the positively charged nucleus and each of the negatively charged electrons. This is true even though classical Newtonian mechanics fails in that realm and is replaced there by quantum physics. This simple law also correctly accounts for the forces that bind atoms together to form molecules and for the forces that bind atoms and molecules together to form solids and liquids.

The Coulomb constant k is often replaced by a factor $1/4\pi\varepsilon_0$ where $\varepsilon_0 = 4\pi k$. As you will see later this more complicated expression for the electrostatic constant simplifies many related equations that we have not yet introduced. Substituting the $1/4\pi\varepsilon_0$ term for k gives an alternate form of Coulomb's law as

$$F_{A \to B}^{elec} = \frac{1}{4\pi\varepsilon_0} \frac{|q_A||q_B|}{r^2} \quad \text{(Coulomb's law)}, \tag{22-6}$$

where k has the value

$$k = \frac{1}{4\pi\varepsilon_0} = 8.99 \times 10^9 \text{ N}\cdot\text{m}^2/\text{C}^2 \qquad \text{(Coulomb constant).} \qquad (22\text{-}7)$$

The quantity ε_0, known as the **electric constant** (sometimes called the *permittivity constant* or simply *epsilon sub zero*), often appears separately in equations and is given by

$$\varepsilon_0 = 8.85 \times 10^{-12} \text{ C}^2/\text{N}\cdot\text{m}^2 \qquad \text{(electric constant).} \qquad (22\text{-}8)$$

READING EXERCISE 22-10: Use the information provided at the end of Section 22-4 and in Fig. 22-6 to explain why the following statements cannot be true: (a) The force between two charged particle-like objects is independent of the charge on the objects. (b) The magnitude of the force between two charged particle-like objects is proportional to $1/|q_A||q_B|$. (c) The force between two charged objects is proportional to $|q_A| + |q_B|$. ■

22-8 Solving Problems Using Coulomb's Law

Coulomb's law can be used to find the forces on particle-like objects having excess charge on them. When solving quantitative (numerical) problems using Coulomb's law, there are several issues to keep in mind. For example, we must be sure to express the charges in coulombs and the distance between the charges in meters. These are the SI units for distance and charge and are required if we are to use the standard value for the Coulomb constant shown in Eq. 22-7. As we discussed in Section 22-7, we must calculate the magnitude of the force using the (positive) absolute value of the charges. We should then make a sketch of the situation, showing the direction of the force and adopting a coordinate system. If the force acts in the negative direction, then we associate a negative sign with the magnitude of the force.

What happens if there are more than two charges interacting as shown in Fig. 22-16?

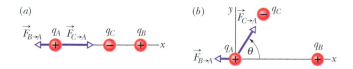

FIGURE 22-16 ■ (a) A set of particle-like objects with excess charge on them that lie along a line. The force vectors depict the forces of charges B and C on charge A. (b) A similar diagram showing the force vectors on charge A when the other charges do not lie along the same line.

Superposition of Forces

As is the case with all other forces (including the gravitational force), the electrostatic force obeys the principle of superposition. For example, if we have 4 charged particles, they interact *independently in pairs,* and the force on any one of the charges, let us say particle A, is given by the vector sum

$$\vec{F}_A^{\text{net}} = \vec{F}_{B\to A} + \vec{F}_{C\to A} + \vec{F}_{D\to A}, \qquad (22\text{-}9)$$

in which, for example, $\vec{F}_{D\to A}$ is the force acting on particle A due to the presence of particle D.

Often, as is the case for the charges in Fig. 22-16b, the various forces acting on a particle do not all act along the same line. We know how to combine forces such as these, but let's review the process. First we must calculate the magnitudes of the individual forces (in this case, using Coulomb's law), adopt a coordinate system, and determine the directions of the forces. We then calculate the orthogonal (perpendicular) components of each force and determine the direction of these components. For example, this might mean determining the x- and y-components of each of the forces as well as determining whether those components are in the positive or negative direction. We then add or subtract all of the components of forces that act along the same line. We add or subtract depending on whether the components are in the same or opposite directions. This gives us the components of the net (resultant) force. We then use trigonometry to get the magnitude and direction of the resultant force. These steps are presented in brief below.

Steps to Solving Quantitative Problems Using Coulomb's Law

Although the steps that follow are illustrated using the configuration of charges in Fig. 22-17a, they should work for any combination of charged particles.

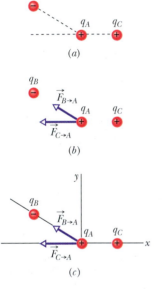

1. Sketch the array of charged objects and draw arrows to represent the anticipated directions of forces on the charged object of interest due to the surrounding charges. An example is shown in Fig. 22-17b where q_A is the charge of interest.

2. Use Coulomb's law to calculate the *magnitudes* of the individual forces on the charged particle of interest.

3. Determine the directions of the forces and create a free-body diagram like that shown in Fig. 22-17b.

4. Choose a coordinate system and sketch it on your diagram as shown in Fig. 22-17c.

5. Calculate the orthogonal (perpendicular) vector components of each force along the coordinate directions using the expressions $F_x = F\cos\theta$ and $F_y = F\sin\theta$, (or equivalent expressions) where F is the magnitude of the force on particle A.

6. Determine the sign of these components based on their directions.

7. Combine all the force components that act along the same line.

8. Combine the resultant components to get the magnitude of the resultant force using the expression

$$[F^{net}]^2 = [F_x^{net}]^2 + [F_y^{net}]^2. \tag{22-10}$$

9. Determine the angle at which the force acts (relative to the positive x axis) using

$$\tan\theta = \frac{F_y^{net}}{F_x^{net}}, \tag{22-11}$$

or alternatively express the force in vector notation as $\vec{F}^{net} = F_x^{net}\hat{i} + F_y^{net}\hat{j}$.

FIGURE 22-17 ■ Diagrams used to illustrate steps in problem solving using Coulomb's law.

READING EXERCISE 22-11: The figure shows two protons (symbol p) and one electron (symbol e) on an axis. What are the directions of (a) the electrostatic force on the central proton due to the electron, (b) the electrostatic force on the central proton due to the other proton, and (c) the net electrostatic force on the central proton? Are there any points along the line connecting the three charges where the central proton can be moved so that the net force on it is zero? Explain your reasoning and how your answers relate to superposition for forces. ■

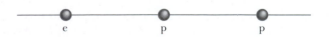

TOUCHSTONE EXAMPLE 22-1: Force on a Charge

(a) Figure 22-18a shows two positively charged particles fixed in place on an x axis. The charges are $q_A = 1.60 \times 10^{-19}$ C and $q_B = 3.20 \times 10^{-19}$ C, and the particle separation is $R = 0.0200$ m. What are the magnitude and direction of the electrostatic force $\vec{F}_{B \to A}$ on particle A from particle B?

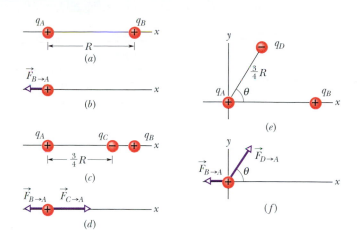

FIGURE 22-18 ■ (a) Two charged particles of charges q_A and q_B are fixed in place on an x axis, with separation R. (b) The free-body diagram for particle A, showing the electrostatic force on it from particle B. (c) Particle C is now fixed in place on the x axis between particles A and B. (d) The free-body diagram for particle A due to particles B and C. (e) Particle D is fixed in place on a line at angle θ to the x axis with just particles A and B present. (f) The new free-body diagram for particle A due to particles B and D.

SOLUTION ■ The **Key Idea** here is that, because both particles are positively charged, particle A is repelled by particle B, with a force magnitude given by Eq. 22-4. Thus, the direction of force $\vec{F}_{B \to A}$ on particle A is *away from* particle B, in the negative direction of the x axis, as indicated in the free-body diagram of Fig. 22-18b. Using Eq. 22-4 with separation R substituted for r, we can write the magnitude $F_{B \to A}$ of this force as

$$F_{B \to A} = k \frac{|q_A||q_B|}{R^2}$$

$$= (8.99 \times 10^9 \text{ N} \cdot \text{m}^2/\text{C}^2) \times \frac{(1.60 \times 10^{-19} \text{ C})(3.20 \times 10^{-19} \text{ C})}{(0.0200 \text{ m})^2}$$

$$= 1.15 \times 10^{-24} \text{ N}.$$

Thus, force $\vec{F}_{B \to A}$ has the following magnitude and direction (relative to the positive direction of the x axis):

$$1.15 \times 10^{-24} \text{ N} \quad \text{and} \quad 180°. \qquad \text{(Answer)}$$

We can also write $\vec{F}_{B \to A}$ in unit-vector notation as

$$\vec{F}_{B \to A} = -(1.15 \times 10^{-24} \text{ N})\hat{i}. \qquad \text{(Answer)}$$

(b) Figure 22-18c is identical to Fig. 22-18a except that particle C now lies on the x axis between particles A and B. Particle C has charge $q_C = -3.20 \times 10^{-19}$ C and is at a distance $\frac{3}{4}R$ from particle A. What is the net electrostatic force $\vec{F}_A^{net}$ on particle A due to particles B and C?

SOLUTION ■ One **Key Idea** here is that the presence of particle C does not alter the electrostatic force on particle A from particle B. Thus, force $\vec{F}_{B \to A}$ still acts on particle A. Similarly, the force $\vec{F}_{C \to A}$ that acts on particle A due to particle C is not affected by the presence of particle B. Because particles A and C have charge of opposite sign, particle A is attracted to particle C. Thus, force $\vec{F}_{C \to A}$ is directed *toward* particle C, as indicated in the free-body diagram of Fig. 22-18d.

To find the magnitude of $\vec{F}_{C \to A}$, we can rewrite Eq. 22-4 as

$$F_{C \to A} = k \frac{|q_A||q_C|}{(\frac{3}{4}R)^2}$$

$$= (8.99 \times 10^9 \text{ N} \cdot \text{m}^2/\text{C}^2) \times \frac{(1.60 \times 10^{-19} \text{ C})(3.20 \times 10^{-19} \text{ C})}{(\frac{3}{4})(0.0200 \text{ m})^2}$$

$$= 2.05 \times 10^{-24} \text{ N}.$$

We can also write $\vec{F}_{C \to A}$ in unit-vector notation:

$$\vec{F}_{C \to A} = +(2.05 \times 10^{-24} \text{ N})\hat{i}.$$

A second **Key Idea** here is that the net force $\vec{F}_A^{net}$ on particle A is the vector sum of $\vec{F}_{B \to A}$ and $\vec{F}_{C \to A}$; that is, from Eq. 22-9, we can write the net force $\vec{F}_A^{net}$ on particle A in unit-vector notation as

$$\vec{F}_A^{net} = \vec{F}_{B \to A} + \vec{F}_{C \to A}$$

$$= -(1.15 \times 10^{-24} \text{ N})\hat{i} + (2.05 \times 10^{-24} \text{ N})\hat{i}$$

$$= (9.00 \times 10^{-25} \text{ N})\hat{i}. \qquad \text{(Answer)}$$

Thus, $\vec{F}_A^{net}$ has the following magnitude and direction (relative to the positive direction of the x axis):

$$9.00 \times 10^{-25} \text{ N} \quad \text{and} \quad 0°. \qquad \text{(Answer)}$$

(c) Figure 22-18e is identical to Fig. 22-18a except that particle D is now positioned as shown. Particle D has charge $q_D = -3.20 \times 10^{-19}$ C, is at a distance $\frac{3}{4}R$ from particle A, and lies on a line that makes an angle $\theta = 60°$ with the x axis. What is the net electrostatic force $\vec{F}_A^{net}$ on particle A due to particles B and D?

SOLUTION ■ The **Key Idea** here is that the net force $\vec{F}_A^{net}$ is the vector sum of $\vec{F}_{B \to A}$ and a new force $\vec{F}_{D \to A}$ acting on particle A due to particle D. Because particles A and D have charges of opposite sign, particle A is attracted to particle D. Thus, force $\vec{F}_{D \to A}$ on particle A is directed *toward* particle D, at angle $\theta = 60°$, as indicated in the free-body diagram of Fig. 22-18f.

To find the magnitude of $\vec{F}_{D \to A}$, we can rewrite Eq. 22-4 as

$$|\vec{F}_{D \to A}| = k \frac{|q_A||q_D|}{(\frac{3}{4}R)^2}$$

$$= (8.99 \times 10^9 \text{ N} \cdot \text{m}^2/\text{C}^2) \times \frac{(1.60 \times 10^{-19} \text{ C})(3.2 \times 10^{-19} \text{ C})}{(\frac{3}{4})^2(0.0200 \text{ m})^2}$$

$$= 2.05 \times 10^{-24} \text{ N}.$$

Then from Eq. 22-9, we can write the net force $\vec{F}_A^{\text{net}}$ on particle A as

$$\vec{F}_A^{\text{net}} = \vec{F}_{B \to A} + \vec{F}_{D \to A}.$$

To evaluate the right side of this equation, we need another **Key Idea**: Because the forces $\vec{F}_{B \to A}$ and $\vec{F}_{D \to A}$ are not directed along the same axis, we *cannot* sum simply by combining their magnitudes. Instead, we must add them as vectors, using one of the following methods.

Method 1. *Summing directly on a vector-capable calculator.* For $\vec{F}_{B \to A}$, we enter the magnitude 1.15×10^{-24} and the angle $180°$. For $\vec{F}_{D \to A}$, we enter the magnitude 2.05×10^{-24} and the angle $60°$. Then we add the vectors.

Method 2. *Summing in unit-vector notation.* First we rewrite $\vec{F}_{D \to A}$ as

$$\vec{F}_{D \to A} = (F_{D \to A}\cos\theta)\hat{i} + (F_{D \to A}\sin\theta)\hat{j}.$$

Substituting 2.05×10^{-24} N for $F_{D \to A}$ and $60°$ for θ, this becomes

$$\vec{F}_{D \to A} = (1.025 \times 10^{-24} \text{ N})\hat{i} + (1.775 \times 10^{-24} \text{ N})\hat{j}.$$

Then we sum:

$$\vec{F}_A^{\text{net}} = \vec{F}_{B \to A} + \vec{F}_{D \to A}$$

$$= -(1.15 \times 10^{-24} \text{ N})\hat{i} + (1.025 \times 10^{-24} \text{ N})\hat{i} + (1.775 \times 10^{-24})\hat{j}$$

$$\approx (-1.25 \times 10^{-25} \text{ N})\hat{i} + (1.78 \times 10^{-24} \text{ N})\hat{j}. \qquad \text{(Answer)}$$

Method 3. *Summing components axis by axis.* The sum of the x-components gives us

$$\vec{F}_{Ax}^{\text{net}} = F_{B \to Ax} + F_{D \to Ax} = F_{B \to A} + F_{D \to A}\cos 60°$$

$$= -1.15 \times 10^{-24} \text{ N} + (2.05 \times 10^{-24} \text{ N})(\cos 60°)$$

$$= -1.25 \times 10^{-25} \text{ N}.$$

The sum of the y-components gives us

$$F_{Ay}^{\text{net}} = F_{B \to Ay} + F_{D \to Ay} = 0 + F_{D \to A}\sin 60°$$

$$= (2.05 \times 10^{-24} \text{ N})(\sin 60°)$$

$$= 1.78 \times 10^{-24} \text{ N}.$$

The net force $\vec{F}_A^{\text{net}}$ has the magnitude

$$F_A^{\text{net}} = \sqrt{(F_{Ax}^{\text{net}})^2 + (F_{Ay}^{\text{net}})^2} = 1.78 \times 10^{-24} \text{ N}.$$

To find the direction of $\vec{F}_A^{\text{net}}$, we take

$$\theta = \tan^{-1} \frac{F_{Ay}^{\text{net}}}{F_{Ax}^{\text{net}}} = -86.0°.$$

However, this is an unreasonable result because $\vec{F}_A^{\text{net}}$ must have a direction between the directions of $\vec{F}_{B \to A}$ and $\vec{F}_{D \to A}$. To correct θ, we add $180°$, obtaining

$$-86.0° + 180° = 94.0°. \qquad \text{(Answer)}$$

TOUCHSTONE EXAMPLE 22-2: Equilibrium Point

Figure 22-19a shows two particles fixed in place: a particle of charge $q_A = +8q$ at the origin and a particle of charge $q_B = -2q$ at $x = L$. At what point (other than infinitely far away) can a proton of charge q_p be placed so that it is in *equilibrium* (meaning that the net force on it is zero)? Is that equilibrium *stable* or *unstable*?

FIGURE 22-19 ■ (a) Two particles of charges q_A and q_B are fixed in place on an x axis, with separation L. (b)–(d) Three possible locations P, S, and R for a proton. At each location, $\vec{F}_{A \to p}$ represents the force on the proton from particle A and $\vec{F}_{B \to p}$ represents the force on the proton from particle B.

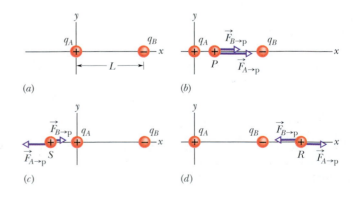

SOLUTION ■ The **Key Idea** here is that, if $\vec{F}_{A \to p}$ is the force on the proton due to charge q_A and $\vec{F}_{B \to p}$ is the force on the proton due to charge q_B, then the point we seek is where $\vec{F}_{A \to p} + \vec{F}_{B \to p} = 0$. This condition requires that

$$\vec{F}_{A \to p} = -\vec{F}_{B \to p}. \tag{22-12}$$

This tells us that at the point we seek, the forces acting on the proton due to the other two particles must be of equal magnitudes,

$$|\vec{F}_{A \to p}| = |\vec{F}_{B \to p}|, \tag{22-13}$$

and that the forces must have opposite directions.

A proton has a positive charge. Thus, the proton and the particle of charge q_A are of the same sign, and force $\vec{F}_{A \to p}$ on the proton must point away from q_A. Also, the proton and the particle of charge q_B are of opposite signs, so force $\vec{F}_{B \to p}$ on the proton must point toward q_B. "Away from q_A" and "toward q_B" can be in opposite directions only if the proton is located on the x axis.

If the proton is on the x axis at any point between q_A and q_B, such as P in Fig. 22-19b, then $\vec{F}_{A \to p}$ and $\vec{F}_{B \to p}$ are in the same direction and not in opposite directions as required. If the proton is at any point on the x axis to the left of q_A, such as point S in Fig. 22-9c, then $\vec{F}_{A \to p}$ and $\vec{F}_{B \to p}$ are in opposite directions. However, Eq. 22-4 tells us that $\vec{F}_{A \to p}$ and $\vec{F}_{B \to p}$ cannot have equal magnitudes there: $|\vec{F}_{A \to p}|$ must be greater than $|\vec{F}_{B \to p}|$, because $|\vec{F}_{A \to p}|$ is produced by a closer charge (with lesser r) of greater magnitude ($8q$ versus $2q$).

Finally, if the proton is at any point on the x axis to the right of q_B, such as point R in Fig. 22-19d, then $\vec{F}_{A \to p}$ and $\vec{F}_{B \to p}$ are again in opposite directions. However, because now the charge of greater amount (q_A) is *farther* away from the proton than the charge of lesser amount, there is a point at which $|\vec{F}_{A \to p}|$ is equal to $|\vec{F}_{B \to p}|$. Let x be the coordinate of this point, and let q_p be the charge of the proton. Then with the aid of Eq. 22-4, we can rewrite Eq. 22-13 as

$$k\frac{8|q||q_p|}{x^2} = k\frac{2|q||q_p|}{(x-L)^2}. \tag{22-14}$$

(Note that only the charge amounts appear in Eq. 22-14.) Rearranging Eq. 22-14 gives us

$$\left(\frac{x-L}{x}\right)^2 = \frac{1}{4}.$$

After taking the square roots of both sides, we have

$$\frac{x-L}{x} = \frac{1}{2},$$

which gives us

$$x = 2L. \qquad \text{(Answer)}$$

The equilibrium at $x = 2L$ is unstable; that is, if the proton is displaced leftward from point R, then $|\vec{F}_{A \to p}|$ and $|\vec{F}_{B \to p}|$ both increase but $|\vec{F}_{B \to p}|$ increases more (because q_B is closer than q_A), and a net force will drive the proton farther leftward. If the proton is displaced rightward, both $|\vec{F}_{A \to p}|$ and $|\vec{F}_{B \to p}|$ decrease but $|\vec{F}_{B \to p}|$ decreases more, and a net force will then drive the proton farther rightward. In a stable equilibrium, each time the proton was displaced slightly, it would return to the equilibrium position.

22-9 Comparing Electrical and Gravitational Forces

Consider two point-like objects, A and B, separated by a distance r. What are the magnitudes of the electrical and gravitational forces between them?

$$F_{A \to B}^{\text{elec}} = k\frac{|q_A||q_B|}{r^2} \quad \text{(Coulomb's Law),} \tag{Eq. 22-4}$$

has the same form as that of Newton's equation for the gravitational force between two particles with masses m_A and m_B that are separated by a distance r:

$$F_{A \to B}^{\text{grav}} = G\frac{m_A m_B}{r^2} \quad \text{(Newton's Law of gravitation).} \tag{14-2}$$

Both of these equations have the distance between the two interacting objects squared and in the denominator of the fraction. That is, they are both "inverse square laws." Both also involve a property of the interacting particles — the mass in one case and the charge in the other. Both the gravitational force and the electrostatic force are conservative forces — the work done by these forces around a closed path is zero. Both forces act along the line connecting the two objects — such forces are called "central" forces.

However, as similar as these forces are, they are not the same force. They are not even different aspects of one force. How do we know this? Electrostatic forces are intrinsically much stronger than gravitational forces. For example, the gravitational attraction between a plastic comb and a small piece of paper is not large enough to overcome the opposing gravitational attraction of Earth on the paper. However, if you rub the comb with fur, the resulting electrostatic force *is* large enough to overcome the gravitational attraction of Earth. Furthermore, the electrostatic force differs from the gravitational forces because the gravitational force is always attractive but the electrostatic force may be *either* attractive or repulsive, depending on the signs of the two charges. This difference arises because although there is only one kind of mass, there are two kinds of charge. That is why absolute value signs are needed in

$$F_{A \to B}^{elec} = k \frac{|q_A||q_B|}{r^2}$$

but not in

$$F_{A \to B}^{grav} = G \frac{m_A m_B}{r^2}.$$

Before concluding our discussion of the electrostatic or Coulomb force, let's compare it to another somewhat similar force—the force associated with magnets.

In addition to amber, the early Greeks knew of another special material that had the ability to attract other objects. They recorded the observation that some naturally occurring "lodestones," known today as the mineral magnetite, would attract iron. Lodestones were the first known magnets. Could the phenomena of amber (electricity) and lodestones (magnetism) be related?

Observation of the interactions between two magnets and two electrified objects shows that the phenomena of electricity and magnetism are *not* the same. Two magnets will either attract or repel one another, depending on their orientation. Two pieces of rubbed amber (or glass) always repel one another, regardless of their orientation.

Hence, the study of electricity and magnetism developed separately for centuries—until 1820, in fact, when Hans Christian Oersted found a connection between them: an electric current in a wire can deflect a magnetic compass needle. The new science of *electromagnetism* (the combination of electrical and magnetic phenomena) was developed further by Michael Faraday, a truly gifted experimenter with a talent for physical intuition and visualization. In fact, Faraday's laboratory notebooks do not contain a single equation. In the mid-19th century, James Clerk Maxwell put Faraday's ideas into mathematical form, introduced many new ideas of his own, and put electromagnetism on a sound theoretical basis.

Table 32-1 shows the basic laws of electromagnetism, now called Maxwell's equations. We plan to work our way through them in the chapters between here and there, but you might want to glance at them now, to see where we are heading.

TOUCHSTONE EXAMPLE 22-3: Nuclear Repulsion

The nucleus in an iron atom has a radius of about 4.0×10^{-15} m and contains 26 protons.

(a) What is the magnitude of the repulsive electrostatic force between two of the protons that are separated by 4.0×10^{-15} m?

SOLUTION ■ The **Key Idea** here is that the protons can be treated as charged particles, so the magnitude of the electrostatic force on one from the other is given by Coulomb's law. Table 22-1 tells us that their charge is $+e$. Thus, Eq. 22-4 gives us

$$F^{elec} = \frac{ke^2}{r^2}$$

$$= \frac{(8.99 \times 10^9 \text{ N} \cdot \text{m}^2/\text{C}^2)(1.60 \times 10^{-19} \text{ C})^2}{(4.0 \times 10^{-15} \text{ m})^2}$$

$$= 14 \text{ N}. \quad \text{(Answer)}$$

This is a small force to be acting on a macroscopic object like a cantaloupe but an enormous force to be acting on a proton. Such forces should blow apart the nucleus of any element but hydrogen (which has only one proton in its nucleus). However, they don't, not even in nuclei with a great many protons. Therefore, there must be some enormous attractive force to counter this enormous repulsive electrostatic force.

(b) What is the magnitude of the gravitational force between those same two protons?

SOLUTION ■ The **Key Idea** here is like that in part (a): Because the protons are particles, the magnitude of the gravitational force on one from the other is given by Newton's equation for the gravitational force (Eq. 14-2). With $m_p (= 1.67 \times 10^{-27}$ kg) representing the mass of a proton, Eq. 14-2 gives us

$$F = G \frac{m_p^2}{r^2}$$

$$= \frac{(6.67 \times 10^{-11} \text{N} \cdot \text{m}^2/\text{kg}^2)(1.67 \times 10^{-27} \text{ kg})}{(4.0 \times 10^{-15} \text{ m})^2}$$

$$= 1.2 \times 10^{-35} \text{ N}. \qquad \text{(Answer)}$$

This result tells us that the (attractive) gravitational force is far too weak to counter the repulsive electrostatic forces between protons in a nucleus. Instead, the protons are bound together by an enormous force aptly called the *strong nuclear force*—a force that acts between protons (and neutrons) when they are close together, as in a nucleus.

Although the gravitational force is many times weaker than the electrostatic force, it is more important in large-scale situations because it is always attractive. This means that it can collect many small bodies into huge bodies with huge masses, such as planets and stars, that then exert large gravitational forces. The electrostatic force, on the other hand, is repulsive for charges of the same sign, so it is unable to collect either positive charge or negative charge into large concentrations that would then exert large electrostatic forces.

22-10 Many Everyday Forces Are Electrostatic

In Chapter 6 we presented an idealized model of a solid as an array of atoms held together by forces that act like tiny springs that resist both stretching and compression forces. We then used this spring model to help explain the nature of most of the everyday forces encountered in the study of motion including normal forces, friction forces, and tension forces. We made the claim that all of these forces are basically electrical.

Let's look once again at our spring model of solids in light of our new understanding of the nature of the electrostatic forces between protons and electrons in atoms. Since protons and electrons have opposite charges they attract each other. This is what holds individual atoms together and causes a tension force to arise in a string as it resists stretching. Under compression, the outer electrons in the atoms of one object repel the outer electrons in the other object. This is the origin of the normal force. Although we imagine this repulsion starting at the surface, it is happening in other layers of atoms as well. As the electrons from one layer of atoms are being moved closer to those in the next layer, the repulsion forces increase sharply as the electrons are forced closer together.

Thus, we think of a solid as having a delicately balanced equilibrium in which the electron glue holds the atoms together at just the right spacing. Although more detailed analysis of these phenomena requires quantum mechanics, all of the everyday forces we encounter appear to be either gravitational, electrical, or magnetic. We explore the relationship between electrostatic and magnetic interactions further in Chapter 32.

Problems

SEC. 22-4 ■ USING ATOMIC THEORY TO EXPLAIN CHARGING

1. A Large Charge What is the total charge in coulombs of 75.0 kg of electrons?

2. How Many? How many megacoulombs of positive (or negative) charge are in 1.00 mol of neutral molecular-hydrogen gas (H_2)?

3. How Many Electrons How many electrons would have to be removed from a coin to leave it with a charge of $+1.0 \times 10^{-7}$ C?

4. Glass of Water Calculate the number of coulombs of positive charge in 250 cm^3 of (neutral) water (about a glassful).

5. Cosmic Ray Protons Earth's atmosphere is constantly bombarded by *cosmic ray protons* that originate somewhere in space. If the protons all passed through the atmosphere, each square meter of Earth's surface would intercept protons at the average rate of 1500 protons per second. What would be the corresponding rate of charge flow intercepted by the total surface area of the planet?

6. Fibrillation A charge flow of 0.300 *C*/s through your chest can send your heart into fibrillation, disrupting the flow of blood (and thus oxygen) to your brain. If that current persists for 2.00 min, how many conduction electrons pass through your chest?

7. Beta Decay In *beta decay* a massive fundamental particle changes to another massive particle, and either an electron of charge $-e$ or a positron of charge $+e$ (positive particle with the same amount of charge and mass as an electron) is emitted. (a) If a proton undergoes beta decay to become a neutron, which particle is emitted? (b) If a neutron undergoes beta decay to become a proton, which particle is emitted?

8. Identify X Identify X in the following nuclear reactions (in the first, n represents a neutron): (a) ^{1}H + ^{9}Be → X + n; (b) ^{12}C + ^{1}H → X; (c) ^{15}N + ^{1}H → ^{4}He + X. Appendix F will help.

SEC. 22-8 ■ SOLVING PROBLEMS USING COULOMB'S LAW

9. What Distance At what distance between point charge $q_A = 26.0\ \mu$C and point charge $q_B = -47.0\ \mu$C will the electrostatic force between them have a magnitude of 5.70 N?

10. Force on Each A point charge of $+3.00 \times 10^{-6}$ C is 12.0 cm from a second point charge of -1.50×10^{-6}C. Calculate the magnitude of the force on each charge.

11. Two Equally Charged Two equally charged particles, held 3.2×10^{-3} m apart, are released from rest. The initial acceleration of the first particle is observed to be 7.0 m/s^2 and that of the second to be 9.0 m/s^2. If the mass of the first particle is 6.3×10^{-7} kg, what are (a) the mass of the second particle and (b) the amount of charge on each particle?

12. Isolated Conducting Spheres Identical isolated conducting spheres *A* and *B* have the same excess charges and are separated by a distance that is large compared with their diameters (Fig. 22-20*a*). The electrostatic force acting on sphere *B* due to sphere *A* is $\vec{F}_{A \to B}$. Suppose now that a third identical sphere *C*, having an insulating handle and initially neutral, is touched first to sphere *A* (Fig. 22-20*b*), then to sphere *B* (Fig. 22-20*c*), and finally removed (Fig. 22-20*d*). In terms of the force magnitude $F_{A \to B}$, what is the magnitude of the electrostatic force $\vec{F}'_{A \to B}$ that now acts on sphere *B*?

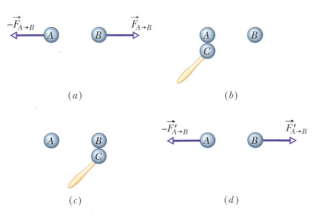

FIGURE 22-20 ■ Problem 12.

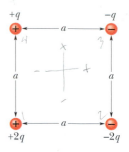

13. The Square In Fig. 22-21, what are the (a) horizontal and (b) vertical components of the net electrostatic force on the charged particle in the lower left corner of the square if $q = 1.0 \times 10^{-7}$ C and $a = 5.0$ cm?

FIGURE 22-21 ■ Problem 13.

14. Where Along the Line Point charges q_A and q_B lie on the *x* axis at points $x = -d$ and $x = +d$, respectively. (a) How must q_A and q_B be related for the net electrostatic force on point charge $+Q$, placed at $x = +d/2$, to be zero? (b) Repeat (a) but with point charge $+Q$ now placed at $x = +3d/2$.

15. Two Identical Spheres Two identical conducting spheres, fixed in place, attract each other with an electrostatic force of 0.108 N when separated by 50.0 cm, center to center. The spheres are then connected by a thin conducting wire. When the wire is removed, the spheres repel each other with an electrostatic force of 0.0360 N. What were the initial charges on the spheres?

16. Three Charges In Fig. 22-22, three charged particles lie on a straight line and are separated by distances *d*. Charges q_A and q_B are held fixed. Charge q_C is free to move but happens to be in equilibrium (no net electrostatic force acts on it). Find q_A in terms of q_B.

FIGURE 22-22 ■ Problem 16.

17. Two Free Particles Two free particles (that is, free to move) with charges $+q$ and $+4q$ are a distance L apart. A third charge is placed so that the entire system is in equilibrium. (a) Find the location, amount and sign of the third charge. (b) Show that the equilibrium is unstable.

18. Two Fixed Particles Two fixed particles, of charges $q_A = +1.0\ \mu$C and $q_B = -3.0\ \mu$C, are 10 cm apart. How far from each should a third charge be located so that no net electrostatic force acts on it?

19. A Certain Charge Q A certain charge Q is divided into two parts q and $Q - q$, which are then separated by a certain distance. What must q be in terms of Q to maximize the electrostatic repulsion between the two charges?

20. Charges and Coordinates The charges and coordinates of two charged particles held fixed in the *xy* plane are $q_A = +3.0\ \mu$C, $x_A = 3.5$ cm, $y_A = 0.50$ cm, and $q_B = -4.0\ \mu$C, $x_B = -2.0$ cm, $y_B = 1.5$ cm. (a) Find the magnitude and direction of the electrostatic force on q_B. (b) Where could you locate a third charge $q_C = +4.0\ \mu$C such that the net electrostatic force on q_B is zero?

21. Identical Ions The magnitude of the electrostatic force between two identical ions that are separated by a distance of 5.0×10^{-10} m is 3.7×10^{-9} N. (a) What is the charge of each ion? (b) How many electrons are "missing" from each ion (thus giving the ion its charge imbalance)?

22. Salt Crystal What is the magnitude of the electrostatic force between a singly charged sodium ion (Na$^+$, of charge $+e$) and an adjacent singly charged chlorine ion (Cl$^-$, of charge $-e$) in a salt crystal if their separation is 2.82×10^{-10} m?

23. Cesium Chloride In the basic CsCl (cesium chloride) crystal structure, Cs$^+$ ions form the corners of a cube and a Cl$^-$ ion is at the cube's center (Fig. 22-23). The edge length of the cube is 0.40 nm. The Cs$^+$ ions are each deficient by one electron (and thus each has a charge of $+e$), and the Cl$^-$ ion has one excess electron (and thus has a charge of $-e$). (a) What is the magnitude of the net

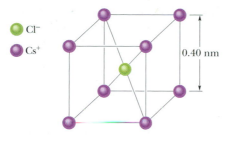

FIGURE 22-23 ▪ Problem 23.

electrostatic force exerted on the Cl$^-$ ion by the eight Cs$^+$ ions at the corners of the cube? (b) If one of the Cs$^+$ ions is missing, the crystal is said to have a *defect;* what is the magnitude of the net electrostatic force exerted on the Cl$^-$ ion by the seven remaining Cs$^+$ ions?

24. Water Drops Two tiny, spherical water drops, with identical charges of -1.00×10^{-16} C, have a center-to-center separation of 1.00 cm. (a) What is the magnitude of the electrostatic force acting between them? (b) How many excess electrons are on each drop, giving it its charge imbalance?

25. Beads Figure 22-24 shows four tiny charged beads that can be slid or fixed in place on wires that stretch along x and y axes. A central bead at the crossing point of the wires (the origin) has a charge of $+e$. The other beads each have a charge of $-e$. Initially beads A, B, and C are at distance $d = 10.0$ cm from the central bead, and bead D is at a distance of $d/2$. (a) How far from the central

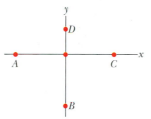

FIGURE 22-24 ▪ Problem 25.

bead must you position bead A so that the direction of the net electrostatic force $\vec{F}^{\,net}$ on the central bead rotates counterclockwise by 30°? (b) With bead A still in its new position, where must you slide bead C so that the direction of $\vec{F}^{\,net}$ rotates back by 30°?

26. Two Copper Coins We know that the negative charge on the electron and the positive charge on the proton are equal in amount. Suppose, however, that these amounts differ from each other by 0.00010%. With what force would two copper coins, placed 1.0 m apart, repel each other? Assume that each coin contains 3×10^{22} copper atoms. (*Hint:* A neutral copper atom contains 29 protons and 29 electrons.) What do you conclude?

27. Particles A and B Figure 22-25a shows charged particles A and B that are fixed in place on an x axis. Particle A has an amount of charge of $|q_A| = 8.00e$. Particle C, with a charge of $q_C = +8.00e$, is

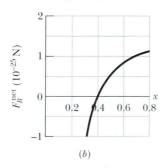

(a)

(b)

FIGURE 22-25 ▪ Problem 27.

initially on the x axis near particle B. Then particle C is gradually moved in the positive direction of the x axis. As a result, the magnitude of the net electrostatic force $\vec{F}_B^{\,net}$ on particle B due to particles A and C changes. Figure 22-25b gives the x-component of that net force as a function of the position x of particle C. The plot has an asymptote of $\vec{F}_B^{\,net} = 1.5 \times 10^{-25}$ N as $x \to \infty$. As a multiple of e, what is the charge q_B of particle B?

28. Above the Floor In Fig. 22-26, a particle of charge $+4e$ is above a floor by distance $d_1 = 2.0$ mm and a particle of charge $+6e$ is on the floor at horizontal distance $d_2 = 6.0$ mm from the first particle. What is the x-component of the electrostatic force on the second particle due to the first particle?

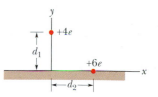

FIGURE 22-26 ▪ Problem 28.

29. Fixed on the x Axis In Fig. 22-27a, particle A (with charge q_A) and particle B (with charge q_B) are fixed in place on an x axis, 8.00 cm apart. Particle C with a charge $q_C = +5e$ is to be placed on the line between particles A and B, so that they produce a net electrostatic force $\vec{F}_C^{\,net}$ on it. Figure 22-27b gives the x-component of that force versus the coordinate x at which particle C is placed. What are (a) the sign of charge q_A and (b) the ratio q_B/q_A?

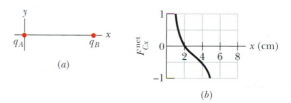

(a)

(b)

FIGURE 22-27 ▪ Problem 29.

30. Four Charged Particles Figure 22-28 shows four charged particles that are fixed along an axis, separated by distance $d = 2.00$ cm. The charges are indicated. Find the magnitude and direction of the net

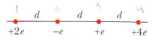

FIGURE 22-28 ▪ Problem 30.

electrostatic force on (a) the particle with charge $+2e$ and (b) the particle with charge $-e$, due to the other particles.

31. Split in Two A charge of 6.0 μC is to be split into two parts that are then separated by 3.0 mm. What is the maximum possible magnitude of the electrostatic force between those two parts?

32. Two on the Axis Figure 22-29 shows two particles, each of charge $+2e$, that are fixed on a y axis, each at a distance $d = 17$ cm from the x axis. A third particle, of charge $+4e$, is moved slowly along the x axis, from $x = 0$ to $x = +5.0$ m. At what values of x will the magnitude of the electrostatic force on the third particle from the other two particles be (a) minimum and (b) maximum? What are (c) the minimum magnitude and (d) the maximum magnitude?

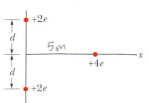

FIGURE 22-29 ▪ Problem 32.

33. How Far In Fig. 22-30, how far from the charged particle on the right and in what direction is there a point where a third charged particle will be in balance?

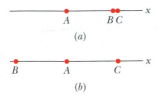

FIGURE 22-30 ■ Problem 33.

34. Three Positive Charges In Fig 22-31*a*, three positively charged particles are fixed on an *x* axis. Particles *B* and *C* are so close to each other that they can be considered to be at the same distance from particle *A*. The net force on particle *A* due to particles *B* and *C* is 2.014×10^{-23} N in the negative direction of the *x* axis. In Fig. 22-31*b*, particle *B* has been moved to the opposite side of *A* but is still at the same distance from it. The net force on *A* is now 2.877×10^{-24} N in the negative direction of the *x* axis. What is the ratio of the charge of particle *C* to that of particle *B*?

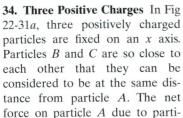

FIGURE 22-31 ■ Problem 34.

35. Fixed at the Origin A particle of charge *Q* is fixed at the origin of an *xy* coordinate system. At $t = 0$ a particle ($m = 0.800$ g, $q = 4.00$ μC) is located on the *x* axis at $x = 20.0$ cm, moving with a speed of 50.0 m/s in the positive *y* direction. For what value of *Q* will the moving particle execute circular motion? (Assume that the gravitational force on the particle may be neglected.)

36. Seven Charges Figure 22-32 shows an arrangement of seven positively charged particles that are separated from the central particle by distances of either *d* (= 1.0 cm) or 2*d*, as drawn. The charges are indicated. What are the magnitude and direction of the net electrostatic force on the central particle due to the other six particles?

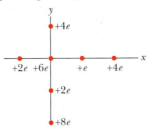

FIGURE 22-32 ■ Problem 36.

37. What is q In Fig. 22-33, what is *q* in terms of *Q* if the net electrostatic force on the charged particle at the upper left corner of the square array is to be zero?

38. Charges Figure 22-34*a* shows an arrangement of three charged particles separated by distance *d*. Particles *A* and *C* are fixed on the *x* axis, but particle *B* can be moved along a circle centered on particle *A*. During the

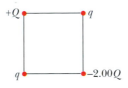

FIGURE 22-33 ■ Problem 37.

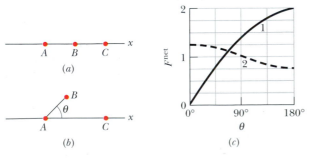

FIGURE 22-34 ■ Problem 38.

movement, a radial line between *A* and *B* makes an angle *θ* relative to the positive direction of the *x* axis (Fig. 22-34*b*). The curves in Fig. 22-34*c* give, for two situations, the magnitude F^{net} of the net electrostatic force on particle *A* due to the other particles. That net force magnitude is given as a function of angle *θ* and as a multiple of a basic force magnitude *F*. For example on curve 1, at $\theta = 180°$, we see that $F^{net} = 2F$. (a) For the situation corresponding to curve 1, what is the ratio of the charge of particle *C* to that of particle *B* (including sign)? (b) For the situation corresponding to curve 2, what is that ratio?

39. Two Electrons–Two Ions Figure 22-35 shows two electrons (charge $-e$) on an *x* axis and two negative ions of identical charges $-q$ and at identical angles *θ*. The central electron is free to move; the other particles are fixed in place at horizontal distances *R* and are intended to hold the free electron in place. (a) Plot the required amount of *q* versus angle *θ* if this is to happen. (b) From the plot, determine which values of *θ* will be needed for physically possible values of $q \leq 5e$.

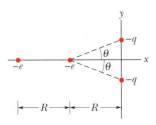

FIGURE 22-35 ■ Problem 39.

40. Diamond Figure 22-36 shows an arrangement of four charged particles, with angle $\theta = 30°$ and distance $d = 2.00$ cm. The two negatively charged particles on the *y* axis are electrons that are fixed in place. The particle at the right has a charge $q_B = +5e$. (a) Find distance *D* such that the net force on q_A, the particle at the left, due to the three other particles, is zero. (b) If the two electrons were moved closer to the *x* axis, would the required value of *D* be greater than, less than, or the same as in part (a)?

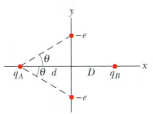

FIGURE 22-36 ■ Problem 40.

41. Each Positive Two particles, each of positive charge *q*, are fixed in place on an *x* axis, one at $x = 0$ and the other at $x = +d$. A particle of positive charge *Q* is to be placed along that axis at locations given by $x = \alpha d$. (a) Write expressions, in terms of *α*, that give the net electrostatic force $\vec{F}^{elec}$ acting on the third particle when it is in the three regions $x < 0$, $0 < x < d$, and $d < x$. The expressions should give a positive result when $\vec{F}^{elec}$ acts in the positive direction of the *x* axis and a negative result when $\vec{F}^{elec}$ acts in the negative direction. (b) Graph the magnitude of $\vec{F}^{elec}$ versus *α* for the range $-2 < \alpha < 3$.

42. Particles A and B In Fig. 22-37, particles *A* and *B* are fixed in place on an *x* axis, at a separation of $L = 8.00$ cm. Their charges are $q_A = +e$ and $q_B = -27e$. Particle *C* with charge $q_C = +4e$ is to be placed on the line between particles *A* and *B*, so that they produce a net electrostatic force $\vec{F}_C^{net}$ on it. (a) At what coordinate should particle *C* be placed to minimize the magnitude of that force? (b) What is that minimum magnitude?

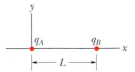

FIGURE 22-37 ■ Problem 42.

SEC. 22-9 ■ COMPARING ELECTRICAL AND GRAVITATIONAL FORCES

43. Earth and Moon (a) What equal positive charges would have to be placed on Earth and on the Moon to neutralize their gravitational attraction? Do you need to know the lunar distance to solve this problem? Why or why not? (b) How many kilograms of hydrogen would be needed to provide the positive charge calculated in (a)?

44. A Particle with Charge Q A particle with charge Q is fixed at each of two opposite corners of a square, and a particle with charge q is placed at each of the other two corners. (a) If the net electrostatic force on each particle with charge Q is zero, what is Q in terms of q? (b) Is there any value of q that makes the net electrostatic force on each of the four particles zero? Explain.

45. Hang from Thread In Fig. 22-38, two tiny conducting balls of identical mass m and identical charge q hang from nonconducting threads of length L. Assume that θ is so small that $\tan \theta$ can be replaced by its approximate equal, $\sin \theta$. (a) Show that, for equilibrium,

$$x = \left(\frac{2kq^2L}{mg} \right)^{1/3},$$

where x is the separation between the balls. (b) If $L = 120$ cm, $m = 10$ g, and $x = 5.0$ cm, what is q?

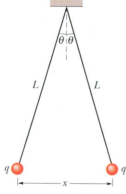

FIGURE 22-38 ■ Problem 45.

46. What Happens? Explain what happens to the balls of Problem 45b if one of them is discharged (loses its charge q to, say, the ground), and find the new equilibrium separation x, using the given values of L and m and the computed value of q.

47. Pivot Figure 22-39 shows a long, nonconducting, massless rod of length L, pivoted at its center and balanced with a block of weight W at a distance x from the left end. At the left and right ends of the rod are attached small conducting spheres with positive charges q and $2q$, respectively. At distance h directly beneath each of these spheres is a fixed sphere with positive charge Q. (a) Find the distance x when the rod is horizontal and balanced. (b) What value should h have so that the rod exerts no vertical force on the bearing when the rod is horizontal and balanced?

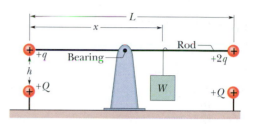

FIGURE 22-39 ■ Problem 47.

48. An Electron in a Vacuum An electron is in a vacuum near the surface of Earth. Where should a second electron be placed so that the electrostatic force it exerts on the first electron balances the gravitational force on the first electron due to Earth?

Additional Problems

49. Opposites Attract It is said that unlike charges attract. You can observe that after the sticky side of a piece of scotch tape is pulled quickly off the smooth side of another piece of tape the tapes attract each other. Perhaps each tape has a like charge and the rule has been stated backwards. Why do you believe the charges on the two tapes are different? *Note:* It is not acceptable to answer "because unlike charges attract and I observed the attraction."

50. Hanging Ball of Foil (a) Explain how a metal conductor such as a hanging ball of aluminum foil can be attracted to a charged insulator *even though the ball of foil has no net charge so that it is electrically neutral.* (b) Can two metal balls with no net charge attract each other? Explain. (c) Can the process of induction cause a neutral conductor to be *repelled* from a charged insulator? Explain.

51. Four Balls Consider four lightweight metal-coated balls suspended on nonconducting threads as shown in Fig. 22-40. Suppose ball A is stroked with a plastic rod that has been rubbed with fur. When you observe interactions between pairs of balls one at a time you find that:

    1. B, C, and D are each attracted to A.

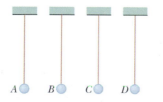

FIGURE 22-40 ■ Problem 51.

    2. B and C seem to have no effect on each other.

    3. B and C are both attracted to D.

Use the concept of electric induction to figure out what type of net charge is on each of the balls: Negative charge? Positive charge? No charge at all? Explain your reasoning. [Based on question 5.9, Arons, *Homework and Test Questions for Introductory Physics Teaching* (Wiley, New York, 1994).]

52. Plastic Rubbed with Fur Suppose you rub a plastic rod with fur that gives it a negative charge. You then bring it close to an uncharged metal coated Styrofoam ball that is suspended from a string. (a) When the rod gets close to the ball, the ball starts moving toward it. Use the concept of induction to explain what happens to the atomic electrons and protons in the ball. Include a sketch of the ball and the rod that shows the excess negative charges on the rod. Also show how the charges are distributed on the ball just before it touches the rod. (b) After the ball touches the rod, it moves away from the rod quickly. Explain why.

53. Small Charged Sphere A small, charged sphere of mass 5.0 g is released 32 cm away from a fixed point charge of $+5.0 \times 10^{-9}$ C. Immediately after release, the sphere is observed to accelerate toward the charge at 2.5 m/s². What is the charge on the sphere? *Hint:* The force of gravity can be ignored in your calculation.

54. Lightning Bolt In a lightning bolt electrons travel from a thundercloud to the ground. If there are 1.0×10^{20} electrons in a lightning bolt, how many coulombs of charge are dumped onto the ground?

55. Estimating Charge Two hard rubber spheres of mass ~10 g are rubbed vigorously with fur on a dry day. They are then suspended from a rod with two insulating strings. They are observed to hang at equilibrium as shown in Fig. 22-41, which is drawn approximately to scale. Estimate the amount of charge that is found on each sphere.

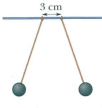

FIGURE 22-41 ■
Problem 55.

56. Various Arrangements Various arrangements of two fixed charges are shown in Fig. 22-42 along with a point labeled P. The amount of each charge is the same but the charges are positive or negative as indicated. All the distances between charges and between point P and the charge nearest to it are the same. Rank these arrangements in order of the strength of the force (that is, its magnitude) on a tiny positive test charge located at point P in each case. Go from greatest to least and indicate when force magnitudes are the same using an equal to sign. For example, if the force magnitudes at (d) and (c) were the same as each other and were the greatest and (b) was less than (d) and (c) with (a) being the least, your answer would be (d) = (b) > (c) > (a). Include a diagram and sketch the individual force vectors in each case. [Based on Ranking Task 128 O'Kuma, et. al., *Ranking Task Exercises in Physics* (Prentice Hall, Upper Saddle River NJ, 2000).]

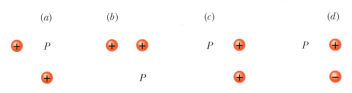

FIGURE 22-42 ■ Problem 56.

23 | Electric Fields

Coal-burning power plants account for 56% of the electricity generated in the United States. But coal is dirty, and smokestack emissions containing sulfur dioxide, nitrogen oxides, and fine particles are a health hazard. A demonstration model of a new Advanced Hybrid Particulate Collector (AHPC) was recently installed at South Dakota's Big Stone plant. This new collector virtually eliminates particulate emissions. Although there are filters in the collector, 90% of the smoke particles are removed using another method.

What technology is used to remove most of the pollutants in the AHPC?

The answer is in this chapter.

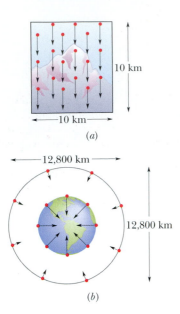

FIGURE 23-1 ■ (*a*) Force field maps for gravitational forces near the Earth are relatively simple: This map shows the uniform nature of the gravitational force on a "test mass" placed at different locations within a vertical and horizontal distance of 10 km. Mt. Everest is in the background. The tails of the force vectors have been placed at various possible test mass locations. (*b*) When distances become much larger than 10 km, the map is not quite so simple. The direction "downward" changes at different points. Also, the gravitational force on a test mass can decrease significantly at altitudes comparable to the Earth's radius.

23-1 Implications of Strong Electric Forces

In Section 3-9 we discuss the experimental fact that the gravitational force the Earth exerts on a mass has essentially the same magnitude and acts in a downward direction for all locations near the surface of the Earth as shown in Fig. 23-1*a*. By comparison the gravitational force exerted on the mass by other objects is negligible. In contrast, we learned in Section 22-2 that the electric forces between small objects are so strong that even Styrofoam cups rubbed with plastic can exert observable forces on each other. Also, it is not easy to calculate the net electric force exerted on a charged object by a complex array of charges in its vicinity. We would have to use the techniques introduced in Section 22-8 that combine Coulomb's law and the principle of superposition to obtain a vector sum of forces. To make matters worse, as our charged object (which we'll call a *test charge*) is moved, we often find the electric forces on it vary in direction and magnitude from point to point. The same can be said for the variation of gravitational forces on a space vehicle when large distances are involved as shown in Fig. 23-1*b*.

How can we describe the effect of the Earth's gravity or the electric force when a small test object is placed at various locations? What is the net effect of a collection of charges on a small test charge located in their vicinity? Answers lie in the concept of a force field.

We begin this chapter by creating a map of the force on a point-like test mass (due to gravitational interactions) or test charge (due to electrostatic interactions) at various locations in space. These maps introduce the concept of a *force field*. We then refine the field concept to define *electric fields* and *gravitational fields*, which are properties of a local space. Knowledge of a gravitational or electric field is useful because it allows us to determine the net force on a small object regardless of its mass or charge.

23-2 Introduction to the Concept of a Field

The temperature at every point in a room can be measured. If the room contains both a good heater and a window open to cold winter air, the temperature at each point in the room might be different. We call the resulting distribution of temperatures around the room a *temperature field*. In much the same way, you can imagine a *pressure field* in the atmosphere. Temperature and pressure are scalar quantities because they have no direction associated with them, so both temperature fields and pressure fields are *scalar fields*. In general,

> A **field** is defined as a representation of any physically measurable quantity that can vary in space.

Vector Fields

We can also have fields associated with vector quantities like force. In contrast to scalar fields like temperature, forces have direction, so it's not good enough to attach just a number to each point in space—the quantity must be identified with a vector. For example, let's ask, "What force would an object feel if it were placed at various locations in space?" and then represent the result pictorially. Consider the simple example of the gravitational force shown in Fig. 23-1*b*. We can calculate (or measure) the gravitational force exerted by the Earth on another object of known mass at several locations. The magnitude of each force vector is directly proportional to both the mass of the Earth and the mass of the object of interest. That is, the gravitational force of the Earth on different objects is different. Objects are pulled more strongly when they have more mass. A similar representation of the electrostatic forces on a test charge at various locations due to a source charge is shown in Fig. 23-2.

Note that there are two separate aspects to creating a map of the fields we discuss above. The first aspect involves the sources of the quantity being measured. For example, the heater and window can be thought of as stimulating thermal energy transfer that can affect the temperature at different locations in the room. In the gravitational case, the Earth and other large astronomical bodies are the sources of gravitational forces. The second aspect involved in creating a field map is a single measurement device that can be moved to different locations. In the case of the pressure field, it is a pressure gauge. In the case of gravitational force, our measurement tool is the motion of the test mass being acted on. Without some kind of sensing tool, we could not "know" the force fields.

In mapping force fields, the fixed objects are called **source objects** because they are the objects that exert forces on another object of interest. We use this other object that the forces are exerted *on* as our measurement tool. We refer to this other object as the **test object.**

FIGURE 23-2 ■ Vectors representing the direction and relative magnitude of the force that would be exerted *by* a fixed positive source charge shown on a second positive test charge if the second charge was placed at the various locations of the vector tails. The magnitude of the force is given by the length of the vector.

Test Objects Should Not Be Large Enough to Move Source Objects

Since all forces occur in pairs, a test object exerts forces on the source objects, trying to change their locations, and hence to change the nature of the field. This issue makes the determination of the field values very complex unless the amount of a source object is much greater than the amount of a test object or the source object is somehow fixed in space. The fact that the field concept is only useful in cases where the test object does not move the source objects is an example of a universal issue involved in making measurements. Namely, the measurement device or tool should not change the value of the quantity being measured. For example, one should not use a large bathtub thermometer to measure the temperature of water in a tiny cup. The presence of the thermometer itself would affect the temperature of the water. Here are several examples of circumstances in which a field can be mapped:

1. In the Earth's gravitational interactions, the location of a much smaller test mass has a negligible effect on the Earth's location and hence on its force field (Fig. 23-3).

2. A conducting sphere with billions of excess electrons that can act as source charges that can exert a net force on a test charge consisting of a single proton. Each excess electron stays put because the net force from all the other charges in its vicinity is much stronger than that from the single proton (unless the proton gets very close to a particular electron).

3. A few electrons on the surface of an insulator such as a piece of Styrofoam can act as source charges that exert a net force on a single electron. The source charges are trapped on the insulator and do not reconfigure themselves.

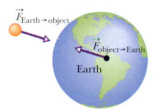

FIGURE 23-3 ■ An Earth–object system in which the interaction forces between a much less massive object and the Earth have, as always, the same magnitude. As the object falls, the movement of the Earth is negligible, so the gravitational force field surrounding the Earth does not change.

> The field concept is useful when the test object is too small to change the location of the source objects or if the source objects are fixed.

Mapping a Vector Field

How can we map the force field for a test object at various locations in space? The procedure that follows can be used to create a field map for any kind of force. The force on the test object could be the result of its interaction with a single source object or a collection of source objects at several locations. The procedure is as follows:

1. Choose a grid or array of points in the vicinity of the sources. The grid should be fine enough to give you a good idea of how the force field looks. However, if the

grid is too fine the procedure will be tedious and the arrows will clutter up the page.

2. Determine the magnitude and direction of the force on the test object at each point on the grid. (Note that we place our test object at one location at a time. If we put down lots of test objects at the same time, they would exert forces on each other and the source object and mess things up.) We must ensure that our test object is not so large that it disturbs the locations of the source objects. We might determine the force at the various points in space experimentally by making direct measurements. Alternatively, if we know the field sources, we can use theoretical relationships such as the law of gravitation or Coulomb's law to make theoretical calculations of the forces on our test object at the different grid locations.

3. Place an arrow representing the force vector at each of the grid locations. Each arrow should point in the direction of the force with a length that is proportional to the force magnitude at that location. It is conventional to locate the tail of the arrow (rather than its tip) at the point for which we have calculated (or measured) the force on the test charge.

Note: We are free to choose a convenient length for the first arrow we draw on our map. Once the length of the first arrow is chosen, a second arrow can be drawn at another point in space. However, the ratio of the lengths of the two arrows must be the same as the ratio of the magnitudes of the two forces. For example, according to Coulomb's law, if the distance between our test charge and the center of the source charge is doubled, the new force on it has only one-fourth the magnitude. This is shown in Fig. 23-4, where the arrows at twice the distance have one-fourth the length.

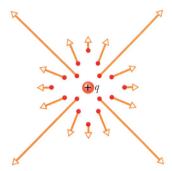

FIGURE 23-4 ■ Here the positive test charge is twice what it was in Fig. 23-2. Coulomb's law tells us that if the test charge is doubled, the electrostatic force it experiences at each location will be doubled.

This type of *vector field plot* is valuable when used to map forces because it immediately tells us important information about characteristics of the forces. For example, Figs. 23-1*b* and 23-4 allow us to infer that the gravitational force exerted by the Earth is everywhere attractive and the electrostatic (or Coulomb) force exerted by one positively charged object on another is everywhere repulsive. These two figures also show us that both of these forces act along the line connecting the centers of the two objects (they are "central forces"). They immediately remind us that these forces are large close to the source (long arrows) and small farther away (short arrows).

READING EXERCISE 23-1: In Fig. 23-4, suppose we chose an arrow that had a length of 36 mm to represent the electrostatic force at a distance of 2 cm from the source charge. What would the length of the arrows representing the magnitude of the force on the same test charge be if it was (a) 4 cm away from the center of the source charge? (b) 6 cm away from the center of the source charge? ■

READING EXERCISE 23-2: In measuring a field that has different values at each point in space, the "test" or measurement device must be small relative to the region of space over which the measurements are to be made. Why? ■

23-3 Gravitational and Electric Fields

Although the lengths of the arrows shown in Figs. 23-1 and 23-4 represent the magnitudes of the gravitational and electrostatic forces experienced by a test object at various locations, we note that the force magnitudes (and hence lengths of the arrows) are different for different test objects.

For example, Figure 23-4 shows the force field of a test object with twice as much charge as the one depicted in Fig. 23-2. Note that every arrow shown in the figure doubles in length. That is, each arrow is scaled by the same factor based on how much larger or smaller the test object's charge is. The force vectors scale as they do because for a given source charge, the electrostatic force is directly proportional to the amount of the test charge. The same is true for the gravitational force due to a given source mass. These changes in the force fields with test object strength are difficult, because we need an infinite number of field plots to represent all different test objects.

Actually, we have already dealt with the problem of test masses for gravitational forces. Recall from Section 14-2 that the magnitude of the gravitational force exerted by a spherical source mass m_s such as the Earth on a test mass m_t (perhaps a ball) is given by

$$F_{s \to t}^{\text{grav}} = |\vec{F}_{s \to t}^{\text{grav}}| = \left(\frac{Gm_s}{r^2} \right) m_t, \tag{23-1}$$

where G is the gravitational constant and r is the distance between the objects. To facilitate calculation of the force *exerted on* various objects by the same source object (typically the Earth) we took advantage of this proportionality to define the *local gravitational strength* g_s as a scalar given by

$$g_s = |\vec{g}_s| \equiv \left(\frac{F_{s \to t}^{\text{grav}}}{m_t} \right) = \left(\frac{Gm_s}{r^2} \right) \qquad \text{(spherical source).} \tag{23-2}$$

Using the field concept that we have now developed, we can call the vector $\vec{g}_s$ the **local gravitational field vector.** Combining Eqs. 23-1 and 23-2, we see that the gravitational force exerted by a source mass m_s on a test mass m_t can be determined using the simple expression

$$\vec{F}_{s \to t}^{\text{grav}} = m_t \vec{g}_s. \tag{23-3}$$

In other words, the gravitational force on an object at a certain point in space is the product of its mass and the gravitational field vector at that point. The gravitational field vector is especially convenient because it is solely a property of space. It is completely determined by locations and the masses of the source objects. It is independent of the mass of the test object we might choose to investigate in any given instance.

Similarly, we can take advantage of the direct proportionality involved in electrostatic forces. That is, there is a similar proportionality between the amount of a test charge and the forces exerted on it by a source or a set of sources. Our approach is to define a new field, the *electric field*, which allows us to focus on the influence of the source of the force. To do this, we define the **electric field vector** due to one or more source charges as

$$\vec{E}_s \equiv \frac{\vec{F}_{s \to t}^{\text{elec}}}{q_t} \qquad \text{(definition of electric field),} \tag{23-4}$$

where $\vec{F}_{s \to t}^{\text{elec}}$ represents the net electrostatic force a test charge q_t experiences from a set of source charges.

The force in Eq. 23-4 is the vector sum of the electrostatic forces on the test charge q_t from all the source charges. Since the force from each source charge is proportional to q_t, dividing the force by q_t cancels it out in each term, leaving the electric field vector $\vec{E}_s$ independent of q_t. Thus, the electric field vector shares the convenient characteristics of the gravitational field vector. It is solely a property of space and source and is independent of the test charge one might choose to probe it. *The electric field depends only on the source charges, not on the test charge.*

TABLE 23-1
Some Electric Field Magnitudes

Field Location or Situation	Value (N/C)
At the surface of a uranium nucleus	$\sim 3 \times 10^{21}$
Within a hydrogen atom, at a radius of 5.29×10^{-11} m	$\sim 5 \times 10^{11}$
Nerve cell membrane	$\sim 1 \times 10^{7}$
Electric breakdown in air (sparking)	$\sim 3 \times 10^{6}$
Near the charged drum of a photocopier	$\sim 1 \times 10^{5}$
Near a charged plastic comb	$\sim 1 \times 10^{3}$
In the lower atmosphere	$\sim 1 \times 10^{2}$
Inside the copper wire of household circuits	$\sim 1 \times 10^{-2}$

FIGURE 23-5 ■ The electric field associated with a negatively charged source object points inward at all locations outside of the source. Since a legitimate test charge does not noticeably influence the electric field created by the source charges, we think of the electric field as existing whether or not a small test charge is present to experience a force.

According to this definition, the **electric field** is the ratio of the electrostatic force on a test charge to the amount of that test charge. In other words, in SI units the electric field is the force per unit of test charge. This gives an SI unit for electric field as newtons per coulomb (N/C). In comparison, recall that the gravitational field was a measure of force per unit of *mass* and the SI unit for the gravitational field g was the N/kg.

The number given for the electric field can also be thought of as the force exerted on a 1 C test charge. To give you some idea of how much force is exerted on a one coulomb charge in various circumstances, Table 23-1 shows the electric fields that occur in a few physical situations. Remember, though, that one coulomb is a very large amount of charge and not an appropriate test charge. For example, an object with 10 000 more protons than electrons would have only a charge on the order of 10^{-15} C.

Although we use a test charge to determine the electric field associated with a charged object, remember that the electric field's existence is independent of the test charge just as the temperature in a room's existence is independent of whether or not there is a thermometer present to detect it. The test charge is simply the measurement device. The field at point P in Fig. 23-5 exists both before and after the test charge shown in the figure is put there. (We must always assume the test charge does not alter the electric field we are defining.)

The remainder of this chapter is primarily devoted to exploring how to use Coulomb's law and the principle of superposition to find the electric fields associated with a single point charge and with relatively simple arrangements of charged objects. We will also explore the concept of *electric field lines* as an alternative to electric field vectors to represent electric fields visually.

TOUCHSTONE EXAMPLE 23-1: Predicting Forces on Charges

Consider a set of hidden source charges that cannot move. Suppose you are trying to explore the nature of electrical forces in the vicinity of the source charges using a positive test charge given by $q_t = 14$ nC. You discover that at a point A shown in Fig. 23-6, the test charge experiences a force given by $\vec{F}_{s \to tA}^{\,elec} = +(2.8 \text{ N})\hat{i}$ and at point B it experiences a force given by $\vec{F}_{s \to tB}^{\,elec} = -(4.2 \text{ N})\hat{j}$.

(a) What will the forces at points A and B be on a different point charge given by $q_1 = -15$ nC? Or on another given by $q_2 = +25$ nC? *Note*: nC stands for nanocoulomb, which is 10^{-9} C.

SOLUTION ■ The **Key Idea** here is that you can use the information about the electrostatic forces experienced by the test

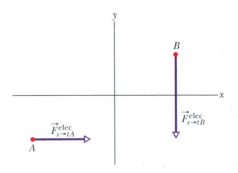

FIGURE 23-6 ■ Two points A and B are in the vicinity of a set of hidden fixed charges. A test charge experiences different electrostatic forces at each location.

charge to map the electric field in the region of our source charges. Once that is done you can then use your knowledge of the electric field to determine the forces on other charges (as long as the source charges are fixed or are large relative to any additional charges you are bringing into their vicinity). Using the definition of electric field given by Eq. 23-4, we see that the electric field at points A and B are

$$\vec{E}_A \equiv \frac{\vec{F}_{s \to t}^{elec}}{q_t} = \frac{+(2.8 \text{ N})\hat{i}}{14 \times 10^{-9} \text{C}} = +(2.0 \times 10^8 \text{ N/C})\hat{i}$$

$$\vec{E}_B \equiv \frac{\vec{F}_{s \to t}^{elec}}{q_t} = \frac{-(4.2 \text{ N})\hat{i}}{14 \times 10^{-9} \text{C}} = -(3.0 \times 10^8 \text{ N/C})\hat{j}.$$

(23-5)

Now that we know the values of the electric field at points A and B we can find the forces on other charges by rearranging Eq. 23-5 to solve for electrostatic force and using the values of the new charges q_1 and q_2 in our calculations.

For q_1:

$$\vec{F}_{s \to q_1}^{elec} = q_1 \vec{E}_A = (-15 \times 10^{-9} \text{ C})[+(2.0 \times 10^8 \text{ N/C})\hat{i}] = (-3.0 \text{ N})\hat{i}$$

$$\vec{F}_{s \to q_1}^{elec} = q_1 \vec{E}_B = (-15 \times 10^{-9} \text{ C})[-(3.0 \times 10^8 \text{ N/C})\hat{j}] = (+4.5 \text{ N})\hat{j}$$

(Answer)

For q_2:

$$\vec{F}_{s \to q_2}^{elec} = q_2 \vec{E}_A = (+25 \times 10^{-9} \text{ C})[+(2.0 \times 10^8 \text{ N/C})\hat{i}] = (+5.0 \text{ N})\hat{i}$$

$$\vec{F}_{s \to q_2}^{elec} = q_2 \vec{E}_B = (+25 \times 10^{-9} \text{ C})[-(3.0 \times 10^8 \text{ N/C})\hat{j}] = (-7.5 \text{ N})\hat{j}$$

(Answer)

(b) Explain how the direction and magnitude of the forces on small charges in an electric field are related to the direction and magnitude of the electric field vector at point A and at point B. *Hint:* Sketching the force vectors will help you visualize the situation.

SOLUTION ■ The **Key Idea** here is that charge is a scalar so when you multiply it by the electric field vector to determine a

force, the force vector must either be in exactly the same direction as the electric field vector (for a positive charge) or opposite to it (negative charge). For example, the equations just above show that the force on q_1, which is negative, is in the *opposite* direction from the electric field at both points A and B. On the other hand, the equations just above show that the force on q_2, which is positive, is in the *same* direction as the electric field at points A and B. This is illustrated in Fig. 23-7.

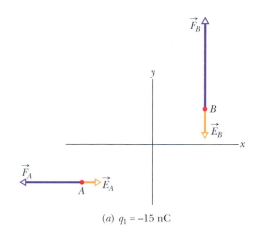

(a) $q_1 = -15$ nC

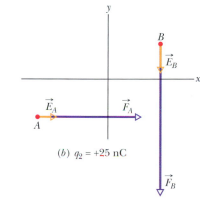

(b) $q_2 = +25$ nC

FIGURE 23-7 ■ A diagram showing the force vectors on two charges at points A and B due to an electric field acting on them. (a) $q_1 = -15$ nC and (b) $q_2 = +25$ nC.

23-4 The Electric Field Due to a Point Charge

The simplest of all possible charge distributions is a charge that can be approximated by a point with zero size. Protons, electrons, nuclei, and ions can all be considered to be point charges. Understanding the interactions of these objects is vital to our understanding of the physical world. In fact, even large charged objects can be viewed as point charges when considered from afar.

In this section, we determine the magnitude of the electric field due to a point-like charge q_s. We do this with the understanding that if we know the mathematical expression for the electric field of the charge, we know the mathematical expression for the force exerted per unit charge on any other charge that we might bring into the region surrounding it.

Here we discuss how to calculate the theoretical value of the electric field created by a positive charge q_s. We can also use another positive charge q_t to probe the region around q_s, testing the magnitude and direction of the force exerted on q_t by q_s. The

FIGURE 23-8 ■ When a source charge and a point-like test charge are both positive, the force between them is repulsive and both the vector representing the force on the test charge q_t and the electric field vector at its location point radially outward in the same direction. Since the units of $\vec{E}$ and $\vec{F}$ are different, the relative lengths of the vectors are arbitrary.

force vector and electric field vector for two positive point-like charges separated by a distance r are shown in Fig. 23-8. We develop a mathematical expression for the electric field associated with a positive charge below. As we do so, we consider how the situation would be different for negative charges.

We know that q_t experiences a repulsive force caused by q_s because they are like charges. The magnitude of the force $F_{s \to t}^{elec} = |\vec{F}_{s \to t}^{elec}|$ on our test charge q_t can be found using Coulomb's law:

$$F_{s \to t}^{elec} = k\frac{|q_s||q_t|}{r^2}. \tag{23-6}$$

The absolute value signs on q_s and q_t serve as a reminder that the magnitude of the force is independent of the type (sign) of charge we have chosen to use in this development. Using our concept of the electric field vector from Section 23-3 above, we can express the magnitude of the force on the test charge q_t due to the electric field created by the source charge q_s as

$$F_{s \to t}^{elec} = |q_t|E_s, \tag{23-7}$$

where E_s is the magnitude of the electric field due to the source charge.

By combining Eq. 23-6 with Eq. 23-7, we can express the magnitude of the electric field at a distance r from a point charge of magnitude q_s as

$$E_s = k\frac{|q_s|}{r^2} \quad \text{(magnitude of the electric field due to a point charge).} \tag{23-8}$$

In agreement with our definition of the electric field as the force *per unit* charge, this expression is independent of the amount (and sign) of the test charge q_t we use as the probe. This expression is valid everywhere around the point charge q_s.

The magnitude of the electric field due to a positive or negative point charge is given by the expression above. However, the electric field is a vector. Hence, we must still determine the direction of the electric field associated with our positive point charge q_s. Recall the definition of electric field magnitude in Eq. 23-4,

$$\vec{E}_s \equiv \frac{\vec{F}_{s \to t}^{elec}}{q_t}. \tag{23-9}$$

For a positive test charge q_t, this means that the direction of the field vector $\vec{E}_s$ is the same as the electrostatic force vector $\vec{F}_{s \to t}^{elec}$. This force points radially away from q_s as shown in Fig. 23-8. Since the direction of the field is the same as the direction of the force for the positive charge q_t, we know the electric field created by the positive point charge q_s must also point radially away from the charge q_s as shown in Fig. 23-8.

Would the direction of the field change if the charge q_s producing the field was negative rather than positive? The answer is yes. Consider a positive test charge q_t. The vector relationship between force and field (Eq. 23-4) tells us that since q_t is positive, the direction (sign) of the field is still the same as the direction of the force on q_t. However, now q_s is negative and q_t is positive. These unlike charges will attract one another. Hence, as shown in Fig. 23-9, the direction of the force on q_t due to q_s points radially *toward* q_s. Since the directions of the force and the field are the same, so does the electric field. According to the force-field relationship, if we used a negative test charge, the electric field would not change but the force on a negative test charge would act in the opposite direction.

FIGURE 23-9 ■ When the charge on a source charge is negative while a point-like test charge is positive, the forces between them are attractive and both the vector representing the force on the test charge and the electric field vector at its location point radially inward in the same direction.

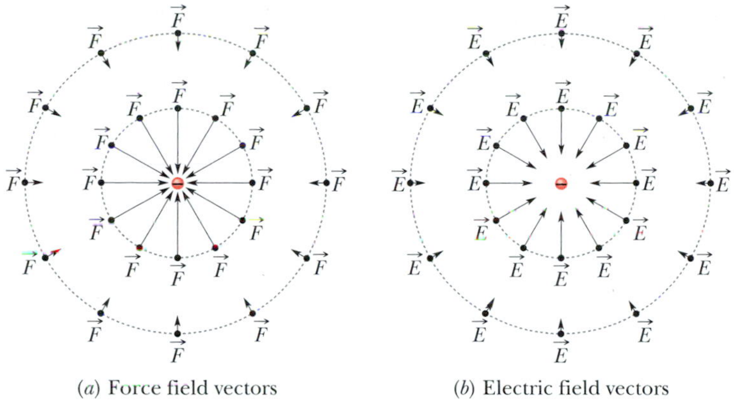

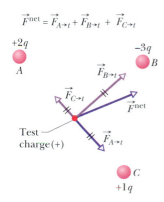

(*a*) Force field vectors   (*b*) Electric field vectors

FIGURE 23-10 ■ There are two methods commonly used to depict the pattern of forces a test charge might experience at different locations in space. For the special case of forces associated with a single negative source charge: (*a*) shows a vector force field on a positive test charge, and (*b*) shows a vector electric field. The vector electric field is solely a property of space, independent of the test charge used to generate (*a*).

FORCE AND ELECTRIC FIELD DIRECTIONS: The direction of the force on a positive test charge is always the same as the electric field at the location of the positive test charge. So, at any point in space, the direction of the electric field produced by a positive point charge is radially away from the charge. The direction of the electric field produced by a negative point charge is radially toward the charge. "Radially" means along the line connecting the charge and the point of evaluation.

The Electric Field Vector Representation

The electric field vector representation is extremely useful when we want to determine the force on a charge placed at a given location (Fig. 23-10). We merely have to multiply the electric field vector at the location by the value of the charge. *This method of using the electric field to find force on a charge is valid for any charge, q, that is not large enough to disturb the electric field, regardless of the source of the field.*

READING EXERCISE 23-3: Rewrite the discussion of how we determine the direction of the field using a negative test charge rather than a positive test charge. Does it make any difference which type of charge we decide to use in determining the direction of the field? ■

23-5 The Electric Field Due to Multiple Charges

In the real world, problems are seldom as simple as one charged object exerting a force on another. It is more common for several charges to be present and the force exerted on the test charge to be the net result of the forces due to each of the source charges. As we mention in Section 22-8, experiments involving both gravitational and electrical forces have confirmed that the net force exerted on a test object by a collection of source objects is the vector sum of the forces exerted by each individual source object.

If we place a positive test charge q_t near n point charges $q_A, q_B, \ldots, q_n$, as shown for only three charges in Fig. 23-11, the forces exerted by the individual charges superimpose so that the net force $\vec{F}_t^{\text{net}}$ from the n point charges acting on the test charge is

$$\vec{F}_t^{\text{net}} = \vec{F}_{A \to t} + \vec{F}_{B \to t} + \cdots + \vec{F}_{n \to t}. \qquad (23\text{-}10)$$

For each of the terms in the expression above, we can replace the individual forces with the equivalent expressions based on the definition of the electric field. For example,

$$\vec{F}^{\text{net}} = \vec{F}_{A \to t} + \vec{F}_{B \to t} + \vec{F}_{C \to t}$$

FIGURE 23-11 ■ Three point charges exert forces on a small positive test charge at a point in space. These force vectors superimpose to yield a net force.

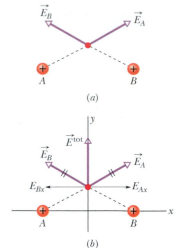

(a)

(b)

FIGURE 23-12 ■ When charges are arranged symmetrically it is often the case that electric field components cancel each other. In this case the x-components cancel everywhere along a line bisecting the line connecting the charge centers.

$$\vec{F}_{A \to t} = q_t \vec{E}_A, \qquad (23\text{-}11)$$

where $\vec{E}_A$ is the electric field associated with charge q_A. If we make such replacements for each force term in the expression above we have:

$$\vec{F}_t^{\text{net}} = q_t \vec{E}^{\text{net}} = q_t \vec{E}_A + q_t \vec{E}_B + \cdots + q_t \vec{E}_n. \qquad (23\text{-}12)$$

Here $\vec{E}^{\text{net}}$ is the resultant electric field associated with the entire group of charges. If we divide both sides of the expression

$$q_t \vec{E}^{\text{net}} = q_t \vec{E}_A + q_t \vec{E}_B + \cdots + q_t \vec{E}_n \qquad (23\text{-}13)$$

by q_t, the result is an expression for the net electric field associated with a group of charges. Namely,

$$\vec{E}^{\text{net}} = \vec{E}_A + \vec{E}_B + \cdots + \vec{E}_n. \qquad (23\text{-}14)$$

This expression shows us that the principle of superposition applies to electric fields as well as to electrostatic forces. When doing calculations, however, it is important to remember that we are adding vectors here. Hence, the addition is more complex than simply adding numbers together.

If the array of point charges is symmetric, sometimes the addition of vectors at certain points in the vicinity of the charges is simplified. In Figs. 23-12 and 23-13 we show examples of symmetric situations for which the net electric field only has one component everywhere on a line bisecting the line that connects the two charges.

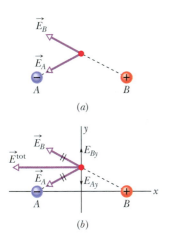

(a)

(b)

FIGURE 23-13 ■ A symmetrical arrangement for which y-components of the electric field cancel everywhere along a line, bisecting the line connecting the charge centers.

READING EXERCISE 23-4: The figure here shows a proton p and an electron e on an x axis. Draw vectors indicating the direction of the electric field due to the electron and describe the direction in words at (a) point S and (b) point R. Draw vectors indicating the direction of the electric field due to both charges and describe the direction in words at (c) point R and (d) point S.

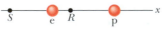

TOUCHSTONE EXAMPLE 23-2: Three Charges

Figure 23-14 shows three particles with charges $q_A = +2Q$, $q_B = -2Q$, and $q_C = -4Q$, each a distance d from the origin. We assume Q is positive. What net electric field $\vec{E}^{\text{net}}$ is produced at the origin?

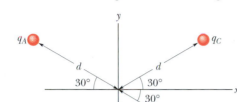

FIGURE 23-14 ■
Three particles with charges q_A, q_B, and q_C are at the same distance d from the origin.

SOLUTION ■ We need to find the electric field vectors $\vec{E}_A$, $\vec{E}_B$, and $\vec{E}_C$ that act at the origin. The **Key Idea** is that we can pick a more convenient coordinate system to describe these electric field vectors. An x'–y' coordinate system that is rotated by $30°$ in a clockwise direction has q_A and q_B lying along its x' axis, as shown in Fig. 23-15a.

Another **Key Idea** is that charges q_A, q_B, and q_C produce electric field vectors $\vec{E}_A$, $\vec{E}_B$, and $\vec{E}_C$, respectively, at the origin, and the net electric field is the vector sum $\vec{E}^{\text{net}} = \vec{E}_A + \vec{E}_B + \vec{E}_C$. To find this sum, we first must find the magnitudes and orientations of the three field vectors. To find the magnitude of $\vec{E}_A$, which is due to q_A, we use Eq. 23-8, substituting d for r and $2Q$ for $|q|$ and obtaining

$$E_A = k\frac{2Q}{d^2}.$$

Similarly, we find the magnitudes of the fields $\vec{E}_B$ and $\vec{E}_C$ to be

$$E_B = k\frac{2Q}{d^2} \quad \text{and} \quad E_C = k\frac{4Q}{d^2}.$$

We next must find the orientations of the three electric field vectors at the origin. Because q_A is a positive charge, the field

vector it produces points directly *away* from it, and because q_B and q_C are both negative, the field vectors they produce point directly *toward* each of them. Thus, the three electric fields produced at the origin by the three charged particles are oriented as in Fig. 23-15b. (*Caution:* Note that we have placed the tails of the vectors at the point where the fields are to be evaluated; doing so decreases the chance of misinterpretation.)

We can now add the fields vectorially as outlined for forces in Touchstone Example 22-1c. However, here we can use symmetry to simplify the procedure. From Fig. 23-15b, we see that $\vec{E}_A$ and $\vec{E}_B$ have the same direction. Hence, their vector sum points along the positive x' axis and has the magnitude

$$\vec{E}_A + \vec{E}_B = k\frac{2Q}{d^2}\hat{i}' + k\frac{2Q}{d^2}\hat{i}' = k\frac{4Q}{d^2}\hat{i}'$$

or

$$E_{Ax'} + E_{Bx'} = k\frac{4Q}{d^2},$$

where $\hat{i}'$ and $\hat{j}'$ are unit vectors in the x'–y' coordinate system. This sum happens to equal the magnitude of $\vec{E}_C$.

We must now combine two vectors, $\vec{E}_C$ and the vector sum $\vec{E}_A + \vec{E}_B$, that have the same magnitude. We do this by resolving $\vec{E}_C$ into its x' and y' components.

$$E_{Cx'} = E_C \cos 60° = \frac{1}{2}E_C$$

$$E_{Cy'} = E_C \sin 60° = \frac{\sqrt{3}}{2}E_C.$$

Then we find $E_{x'}^{\text{net}}$ and $E_{y'}^{\text{net}}$ components.

$$E_{x'}^{\text{net}} = E_{Ax'} + E_{Bx'} + E_{Cx'}$$

$$= k\frac{4Q}{d^2} + \frac{1}{2}k\frac{4Q}{d^2}$$

$$= \frac{3}{2}\left(k\frac{4Q}{d^2}\right).$$

$$E_{y'}^{\text{net}} = \frac{\sqrt{3}}{2}\left(k\frac{4Q}{d^2}\right).$$

Using vector notation we get

$$\vec{E}^{\text{net}} = \left(k\frac{4Q}{d^2}\right)\left(\frac{3}{2}\hat{i}' + \frac{\sqrt{3}}{2}\hat{j}'\right). \qquad \text{(Answer)}$$

The magnitude of $\vec{E}^{\text{net}}$ is given by

$$E^{\text{net}} = |\vec{E}^{\text{net}}| = \sqrt{(E_{x'}^{\text{net}})^2 + (E_{y'}^{\text{net}})^2}$$

$$= k\frac{6.93Q}{d^2}.$$

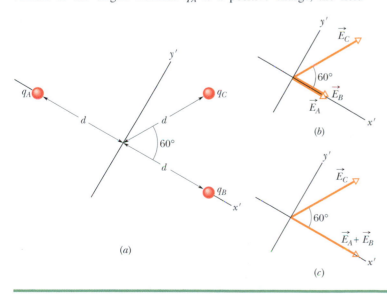

FIGURE 23-15 ■ (a) The same three charges in the new x' − y' coordinate system. (b) The electric field vectors $\vec{E}_A$, $\vec{E}_B$, and $\vec{E}_C$ at the origin due to the three particles. (c) The electric field vector $\vec{E}_C$ and the vector sum $\vec{E}_A + \vec{E}_B$ at the origin.

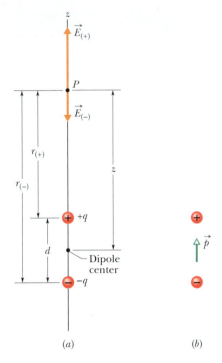

FIGURE 23-16 ■ (a) An electric dipole. The electric field vectors $\vec{E}_{(+)}$ and $\vec{E}_{(-)}$ at point P on the dipole axis result from the dipole's two charges. P is at distances $r_{(+)}$ and $r_{(-)}$ from the individual charges that make up the dipole. (b) By definition, the dipole moment $\vec{p}$ is a vector that points from the negative to the positive charge of the dipole.

23-6 The Electric Field Due to an Electric Dipole

Figure 23-16a shows two charged particles of amount $|q|$ but of opposite sign, separated by a distance d. We call this configuration an **electric dipole.** Separation of positive and negative charge in an electrically neutral object occurs quite naturally. For example, recall the discussion of polarization in Chapter 22. As a result, true electric dipoles and approximations of electric dipoles are reasonably common. Hence, we take some time to develop an expression for the electric field due to a dipole. We start with the idea of superposition of electric fields that we discuss in Section 23-5.

Let us find the electric field due to the dipole of Fig. 23-16a at a point P, a distance z from the midpoint of the dipole and on the axis through the particles, which is called the dipole axis. From symmetry, the electric field $\vec{E}$ at point P—and also the fields $\vec{E}_{(+)}$ and $\vec{E}_{(-)}$ due to the separate charges that make up the dipole—must lie along the dipole axis, which we have taken to be a z axis. Applying the superposition principle for electric fields, we find that the magnitude $E = |\vec{E}|$ of the electric field at P is

$$
\begin{aligned}
E &= |\vec{E}_{(+)}| - |\vec{E}_{(-)}| \\
&= k\frac{|q|}{r_{(+)}^2} - k\frac{|q|}{r_{(-)}^2} \\
&= \frac{k|q|}{(z - \tfrac{1}{2}d)^2} - \frac{k|q|}{(z + \tfrac{1}{2}d)^2},
\end{aligned}
\tag{23-15}
$$

where k is the Coulomb constant. Using algebra, we can rewrite this equation as

$$
E = \frac{k|q|}{z^2}\left[\left(1 - \frac{d}{2z}\right)^{-2} - \left(1 + \frac{d}{2z}\right)^{-2}\right].
\tag{23-16}
$$

We are usually interested in the electrical effect of a dipole only at distances that are large compared with the dimensions of the dipole—that is, at distances such that $z \gg d$. At such large distances, we have $d/z \ll 1$ in the expression above. We can then expand the two quantities in the brackets in that equation by the binomial theorem (Appendix E), obtaining for those quantities

$$
\left[\left(1 + \frac{2d}{2z(1!)} + \cdots\right) - \left(1 - \frac{2d}{2z(1!)} + \cdots\right)\right].
$$

Thus,

$$
E = \frac{k|q|}{z^2}\left[\left(1 + \frac{d}{z} + \cdots\right) - \left(1 - \frac{d}{z} + \cdots\right)\right].
\tag{23-17}
$$

The unwritten terms in these two expansions involve d/z raised to progressively higher powers. Since $d/z \ll 1$, the contributions of those terms are progressively less, and to approximate the electric field magnitude, E, at large distances, we can neglect them. Then, in our approximation, we can rewrite this expression as

$$
E \approx \frac{k|q|}{z^2}\frac{2d}{z} = 2k\frac{|q|d}{z^3} \qquad \text{(for } d/z \ll 1\text{)}.
\tag{23-18}
$$

The product $|q|d$, which involves the two intrinsic properties of charge q and separation d of the dipole, is the magnitude $|\vec{p}|$ of a vector quantity known as the **electric dipole moment** $\vec{p}$ of the dipole. (The unit of $\vec{p}$ is the coulomb-meter and should not be confused with either momentum or pressure.) Thus, we can rewrite Eq. 23-18 as

$$E = 2k\frac{|\vec{p}|}{z^3} \qquad \text{(electric field magnitude for a dipole along axis with } d/z <\!\!< 1). \qquad (23\text{-}19)$$

The direction of $\vec{p}$ is taken to be from the negative to the positive end of the dipole, as indicated in Fig. 23-16b. We can use $\vec{p}$ to specify the orientation of a dipole.

The expression for the electric field due to a dipole shows that if we measure the electric field of a dipole only at distant points, we can never find both $|q|$ and d separately, only their product. The field at distant points would be unchanged if, for example, $|q|$ were doubled and d simultaneously halved. Thus, the dipole moment is a basic property of a dipole.

Although Eq. 23-19 holds only for distant points along the dipole axis, it turns out that $\vec{E}$ for a dipole varies as $1/r^3$ for all distant points, regardless of whether they lie on the dipole axis. Here r is the distance between the point in question and the dipole center.

Inspection of the electric field vectors in Fig. 23-16 shows that the direction of $\vec{E}$ for distant points on the dipole axis is always the direction of the dipole moment vector $\vec{p}$. This is true whether point P in Fig. 23-16a is on the upper or the lower part of the dipole axis.

Inspection of Eq. 23-19 shows that if you double the distance of a point from a dipole, the electric field at the point drops by a factor of 8. If you double the distance from a single point charge, however (see Eq. 23-8), the electric field drops only by a factor of 4. Thus the electric field of a dipole decreases more rapidly with distance than does the electric field of a single charge. The physical reason for this rapid decrease in electric field for a dipole is that from distant points a dipole looks like two equal but opposite charges that almost—but not quite—coincide. Thus, their electric fields at distant points almost—but not quite—cancel each other.

23-7 The Electric Field Due to a Ring of Charge

So far we have considered the electric field produced by one or, at most, a few point charges. We now consider charge distributions consisting of a great many closely spaced point charges (perhaps billions) spread along a line, a curve, over a surface, or within a volume. Such distributions can be treated as if they were **continuous** rather than discrete. Since these distributions can include an enormous number of point charges, we find the electric fields that they produce using integral calculus rather than by considering the point charges one by one. In this section we discuss the electric field caused by a ring of charge. In the next chapter, we shall find the field inside a uniformly charged sphere.

When we deal with continuous charge distributions, it is most convenient to express the charge on an object as a *charge density* rather than as a total charge. For a line of charge, for example, we would report the linear charge density (or charge per unit length) λ, whose SI unit is the coulomb per meter. If we have a total amount of charge q distributed uniformly along a line or curve then the **linear charge density** is defined as

$$\lambda \equiv q/L$$

where L is the total length of the path taken by the line or curve. For example, the curve subscribed by the uniformly charged circular ring of radius R shown in Fig. 23-17 has a total length of $L = 2\pi R$.

TABLE 23-2			
Some Measures of Charge Density			
Name	**Symbol**	**Definition***	**SI Unit**
Charge	q	q	C
Linear charge density	λ	q/L	C/m
Surface charge density	σ	q/A	C/m^2
Volume charge density	ρ	q/V	C/m^3

*These definitions assume a uniform charge density. Otherwise the charge densities depend on location so that $\lambda(x) = dq/dx$, $\sigma(x, y) = dq/dA(x, y)$, $\lambda(x, y, z) = dq/dV(x, y, z)$.

Table 23-2 summarizes information about the types of charge densities we use in this text.

We may imagine the ring in Fig. 23-17 to be made of plastic or some other insulator, so the charges can be regarded as fixed in place. What is the electric field $\vec{E}$ at point P, a distance z from the plane of the ring along its central axis?

To answer this question, we cannot just use the expression for the electric field set up by a point charge, because the ring is obviously not a point charge. However, we can mentally divide the ring into differential elements of charge. If these charge elements are small they act like point charges, and then we can use Eq. 23-8 to find the electric field magnitude contributed by a single element. This gives us

$$E_q = k\frac{|q|}{r^2},$$

where the value of the coulomb constant k is given in Eq. 22-7. Next, we can add the electric fields set up at location P due to all the differential elements. The vector sum of all those fields gives us the net electric field set up at P by the entire ring.

Let ds be the (arc) length of any differential element of the ring. Since λ is the charge per unit length, the amount of charge in the element is given by

$$|dq| = |\lambda|\,ds. \qquad (23\text{-}20)$$

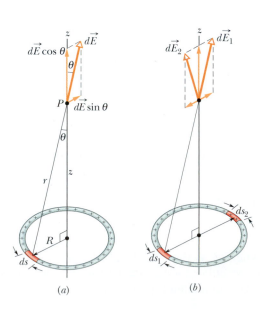

FIGURE 23-17 ■ (*a*) A ring of uniform positive charge. A differential element of charge occupies a length ds (greatly exaggerated for clarity). This element sets up an electric field $d\vec{E}$ at point P. The magnitude of the component of $d\vec{E}$ along the central axis of the ring is $|d\vec{E}|\cos\theta$. (*b*) Each ring element ds_1 has an opposite element ds_2. As a result of this symmetry the on-axis $d\vec{E}$ components (along z) add while those perpendicular to the axis cancel (in the x-y plane).

This differential charge sets up a differential electric field $d\vec{E}$ at point P, which is a distance r from the element. Treating the element as a point charge and using the equation above for dq, we can express the magnitude of $d\vec{E}$ as

$$|d\vec{E}| = k\frac{|dq|}{r^2} = k\frac{|\lambda|\,ds}{r^2}. \tag{23-21}$$

From Fig. 23-17a, we see we can use the Pythagorean theorem to rewrite the equation above as

$$|d\vec{E}| = k\frac{|\lambda|\,ds}{z^2 + R^2}. \tag{23-22}$$

Figure 23-17a also shows us that the vector $d\vec{E}$ is at an angle θ to the central axis (which we have taken to be a z axis) and has components perpendicular to and parallel to that axis.

Every element of charge in the ring sets up a differential field $d\vec{E}$ at P, with magnitude given by the expression above. All the $d\vec{E}$ vectors have identical components parallel to the central axis. All these $d\vec{E}$ vectors also have components perpendicular to the central axis. However, these perpendicular components are identical in magnitude but point in different directions. In fact, for any perpendicular component that points in a given direction, there is another one that points in the opposite direction as shown in Fig. 23-17b. The sum of this pair of components, like the sum of all other pairs of oppositely-directed components, is zero. Thus, the perpendicular components cancel and we need not consider them further. This leaves only the parallel components. They all have the same direction, so the net electric field at P is just their algebraic sum.

The parallel component of $d\vec{E}$ shown in Fig. 23-17a has magnitude $|d\vec{E}|(\cos\theta)$. The figure also shows us that

$$\cos\theta = \frac{z}{r} = \frac{z}{(z^2 + R^2)^{1/2}}. \tag{23-23}$$

Then combining our expressions for $d\vec{E}$ and $\cos\theta$ gives us the magnitude of the parallel component of $d\vec{E}$,

$$|d\vec{E}_\parallel| = |d\vec{E}|\cos\theta = \frac{k|z\lambda|\,ds}{(z^2 + R^2)^{3/2}}. \tag{23-24}$$

To add the parallel components, $|d\vec{E}|\cos\theta$, produced by all the elements, we integrate this expression around the circumference of the ring, from $s = 0$ to $s = 2\pi R$. Since the only quantity that varies during the integration is s, the other quantities can be moved outside the integral sign. The integration then gives us an electric field magnitude of

$$E = |\vec{E}| = \int |d\vec{E}|\cos\theta = \frac{k|z\lambda|}{(z^2 + R^2)^{3/2}}\int_0^{2\pi R} ds$$
$$= \frac{k|z\lambda|(2\pi R)}{(z^2 + R^2)^{3/2}}. \tag{23-25}$$

Since λ is the charge per unit length of the ring, the term $\lambda(2\pi R)$ is q, the total charge on the ring. We can then rewrite this expression as

$$E = \frac{k|qz|}{(z^2 + R^2)^{3/2}} \qquad \text{(electric field magnitude of a charged ring).} \tag{23-26}$$

If the charge on the ring is negative, rather than positive as we have assumed, the magnitude of the field at P is still given by this expression. However, the electric field vector then points toward the ring instead of away from it.

Let us evaluate this equation for the electric field for a point on the central axis so far away that $z \gg R$. For such a point, the expression $z^2 + R^2$ can be approximated as z^2, and Eq. 23-26 becomes

$$E = \frac{k|q|}{z^2} \qquad \text{(on central axis for } z \gg R \text{ at large distance).} \qquad (23\text{-}27)$$

This is a reasonable result, because from a large distance, the ring simply "looks" like a point charge. So, if we replace z with r then Eq. 23-27 becomes the expression for the electric field due to a point charge.

Let us next check Eq. 23-26 for a point at the center of the ring—that is, for $z = 0$. At that point, this expression tells us that $\vec{E} = 0$. This is a reasonable result, because if we were to place a test charge at the center of the ring, there would be no net electrostatic force acting on it. The force due to any element of the ring would be canceled by the force due to the element on the opposite side of the ring. If the force at the center of the ring is zero, the electric field there also has to be zero.

TOUCHSTONE EXAMPLE 23-3: Charged Arc

Figure 23-18a shows a plastic rod having a uniformly distributed charge $-Q$. We assume Q is positive, so $-Q$ is negative. The rod has been bent in a 120° circular arc of radius r. We place coordinate axes such that the axis of symmetry of the rod lies along the x axis and the origin is at the center of curvature P of the rod. In terms of Q and r, what is the electric field $\vec{E}$ due to the rod at point P?

SOLUTION ■ The **Key Idea** here is that, because the rod has a continuous charge distribution, we must find an expression for the electric fields due to differential elements of the rod and then sum those fields via integration. Consider a differential element having arc length ds and located at an angle θ above the x axis (Fig. 23-18b). If we let λ represent the linear charge density of the rod, our element ds has a differential charge of magnitude

$$dq = \lambda \, ds. \qquad (23\text{-}28)$$

Our element produces a differential electric field $d\vec{E}$ at point P, which is a distance r from the element. Treating the element as a point charge, we can rewrite Eq. 23-21 to express the magnitude of $d\vec{E}$ as

$$|d\vec{E}| = k\frac{|dq|}{r^2} = k\frac{|\lambda|\,ds}{r^2}. \qquad (23\text{-}29)$$

The direction of $d\vec{E}$ is toward ds, because charge dq is negative.

Our element has a symmetrically located (mirror image) element ds' in the bottom half of the rod. The electric field $d\vec{E}'$ set up at P by ds' also has the magnitude given by Eq. 23-29, but the field vector points toward ds' as shown in Fig. 23-18b. If we resolve the electric field vectors due to ds and ds' into x- and y-components as shown in Fig. 23-18b, we see that their y-components cancel

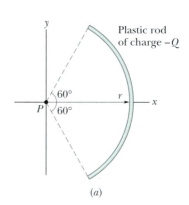

(a)

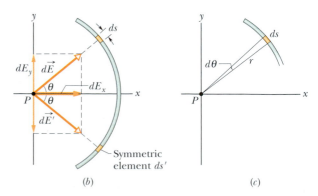

(b)   (c)

FIGURE 23-18 ■ (a) A plastic rod of charge $-Q$ in a circular section of radius r and central angle 120°; point P is the center of curvature of the rod. (b) A differential element in the top half of the rod, at an angle θ to the x axis and of arc length ds, sets up a differential electric field $d\vec{E}$ at P. An element ds', symmetric to ds about the x axis, sets up a field $d\vec{E}'$ at P with the same magnitude. (c) Arc length ds makes an angle $d\theta$ about point P.

(because they have equal magnitudes and are in opposite directions). We also see that their x-components have equal magnitudes and are in the same direction.

Thus, to find the electric field set up by the rod, we need sum (via integration) only the x-components of the differential electric fields set up by all the differential elements of the rod. From Fig. 23-18b and Eq. 23-29, we can write the component dE_x set up by ds as

$$|dE_x| = |d\vec{E}|\cos\theta = k\frac{|\lambda|}{r^2}\cos\theta\,ds. \qquad (23\text{-}30)$$

Equation 23-30 has two variables, θ and s. Before we can integrate it, we must eliminate one variable. We do so by replacing ds, using the relation

$$ds = r\,d\theta,$$

in which $d\theta$ is the angle at P that includes arc length ds (Fig. 23-18c). *Note*: We choose to replace ds here rather than $d\theta$ because we know the angle into which the arc is bent. With this replacement, we can integrate Eq. 23-30 over the angle made by the rod at P, from $\theta = -60°$ to $\theta = 60°$. That will give us E, the magnitude of the electric field at P due to the rod:

$$E = \int |dE_x| = \int_{-60°}^{60°} \frac{k|\lambda|}{r^2}\cos\theta\,r\,d\theta$$

$$= \frac{k|\lambda|}{r}\int_{-60°}^{60°}\cos\theta\,d\theta = \frac{k|\lambda|}{r}\left[\sin\theta\right]_{-60°}^{60°}$$

$$= \frac{k|\lambda|}{r}[\sin 60° - \sin(-60°)]$$

$$= \frac{1.73k|\lambda|}{r}. \qquad (23\text{-}31)$$

(If we had reversed the limits on the integration, we would have gotten the same result but with a minus sign. Since the integration gives only the magnitude of $\vec{E}$, we would then have discarded the minus sign.)

To evaluate the amount of charge per unit length, $|\lambda|$, we note that the rod has an angle of 120° and so is one-third of a full circle. Its arc length is then $2\pi r/3$, and its linear charge density must be

$$|\lambda| = \frac{\text{charge}}{\text{length}} = \frac{Q}{2\pi r/3} = \frac{0.477Q}{r}.$$

Substituting this into Eq. 23-31 and simplifying give us an electric field magnitude of

$$E = |\vec{E}| = \frac{(1.73)(0.477)kQ}{r^2}$$

$$= \frac{0.83kQ}{r^2}. \qquad \text{(Answer)}$$

The direction of $\vec{E}$ is toward the rod, along the axis of symmetry of the charge distribution. We can write $\vec{E}$ in unit-vector notation as

$$\vec{E} = \frac{0.83kQ}{r^2}\,\hat{\imath}.$$

23-8 Motion of Point Charges in an Electric Field

So far we have concentrated on finding electric field values by doing theoretical calculations. However, the electric field concept is especially valuable when we have little or no knowledge about source charges. In such a case, we can use measured values of the forces on our test charge to create an experimentally determined map of the field. In either case, once the electric field is known, we can determine forces on any pointlike object with a known quantity of excess charge at any location in the field (provided that our test charge doesn't disturb the source charges). Assuming we know the mass of our charge, knowing the force means we can calculate the magnitude and direction of the particle's acceleration. Knowing the acceleration allows us to accurately predict its subsequent motion.

In the preceding sections we worked at the first of two tasks: given a charge distribution, find the electric field it produces in the surrounding space. Here we begin the second task: to determine what happens to a charged particle when it is in an electric field. When a charged particle is placed in an electric field, an electrostatic force acts on the particle. This force, a vector quantity, is given by

$$\vec{F}_t^{\text{elec}} = q_t\vec{E}_s \qquad (23\text{-}32)$$

in which q_t is the charge of the test particle (including its sign) and $\vec{E}_s$ is the electric field that source charges have produced at the location of the particle. (The field is not the field set up by the test particle itself. A charged particle is not affected by its own electric field.) Equation 23-32 tells us:

> The electrostatic force $\vec{F}^{\text{elec}}$ acting on a charged test particle located in an external electric field $\vec{E}_s$ has the direction of $\vec{E}_s$ if the charge q_t of the test particle is positive and is opposite the direction of $\vec{E}_s$ if q_t is negative.

Knowing the force on the particle allows us to directly calculate the acceleration, but determining the exact motion of the object is more complicated. The electric field determines the force that a charged particle feels. That, in turn, determines its acceleration, *not its velocity*. So at first a charged particle starting from rest follows the direction of the field. This is because without an initial velocity, the direction of the force and acceleration are in the direction of the velocity. However, if the field changes direction, the path of the particle quickly deviates from the direction of the field. If the charged particle is given an initial velocity that is not aligned with the field, it may never follow the direction of the field. A good analogy to this situation is the path of a projectile in Earth's gravitational field. The gravitational field is uniformly directed downward. Yet this is the direction of the force (and acceleration), not necessarily the direction of the velocity at any given moment. The path of a launched projectile may never follow the direction of the gravitational field.

Electrostatic Precipitation

A few years ago, an Advanced Hybrid Particulate Collector (AHPC) was added to the Big Stone coal-fired power plant shown in the chapter's opening photograph. This hybrid collector eliminates essentially all the particulates in the smoke by combining the best features of filtration systems with a new type of **electrostatic precipitation** system. When the smoke from the coal boiler enters the device, more than 90% of the tiny smoke particles become electrically charged and then are attracted to one of the collection plates. The other 10% of particles flow through holes in the collection plates and are trapped by tubular filter bags that are especially efficient at removing extremely small particles (Fig. 23-19a).

The use of strong electric fields is a key factor in the effectiveness of electrostatic precipitation. Discharge electrodes (Fig. 23-19b) consisting of metal wires are negatively charged. The electric fields surrounding the electrodes are so intense that electrons are discharged. When the electrons fly away from the electrodes and encounter smoke particles, negative ions are formed. These ions are repelled from the electrodes, attracted to neutral metal collector plates (Fig. 23-19c), and captured. In short, these electrons and ions act like point charges in the presence of the electrodes' electric field.

The configuration of components in an AHPC system is shown in Fig. 23-20. A collector plate with holes in it is installed between wire discharge electrodes and filters to protect the filters. On the other side of the electrodes, another collector plate is installed to yield an arrangement in which each row of filters has collector plates on both sides of it. With this arrangement, the collector plates with holes function both as the primary collection surface and as a protective shield for the filters.

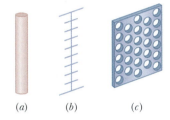

(a)   (b)   (c)

FIGURE 23-19 ■ An end-on side view of AHPC components. (*a*) Tubular filter. (*b*) Discharge electrode. (*c*) Collector plate with holes.

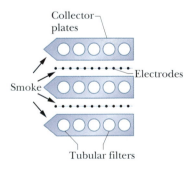

Collector plates

Electrodes

Smoke

Tubular filters

FIGURE 23-20 ■ Simplified AHPC top view showing arrangement of components with collector plates surrounding the filters.

READING EXERCISE 23-5: (a) In the figure, what is the direction of the electrostatic force on the electron due to the uniform electric field shown? (b) In which direction does the electron accelerate if it is moving parallel to the *y* axis before it encounters the electric field? What path does it follow? (c) If, instead, the electron is initially moving rightward, does its speed increase, decrease, or remain constant? What path will it follow in this case?

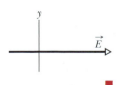

TOUCHSTONE EXAMPLE 23-4: Deflecting an Ink Drop

Figure 23-21 shows the deflecting plates of an ink-jet printer, with superimposed coordinate axes. An ink drop with a mass m of 1.3×10^{-10} kg and an amount of negative charge $|Q| = 1.5 \times 10^{-13}$ C enters the region between the plates, initially moving along the x axis with speed $v_x = 18$ m/s. The length L of the plates is 1.6 cm. The plates are charged and thus produce an electric field at all points between them. Assume that field $\vec{E}$ is downward directed, uniform, and has a magnitude of 1.4×10^6 N/C. What is the vertical deflection of the drop at the right edge of the plates? (The gravitational force on the drop is small relative to the electrostatic force acting on the drop and can be neglected.)

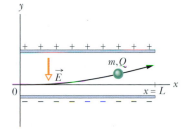

FIGURE 23-21 ■ An ink drop of mass m and an amount of charge $|Q|$ is deflected in the electric field of an ink-jet printer.

SOLUTION ■ The drop is negatively charged and the electric field is directed *downward*. The **Key Idea** here is that, from Eq. 23-4, a constant electrostatic force of magnitude $|Q|E$ acts *upward* on the charged drop. Thus, as the drop travels parallel to the x axis

at constant speed v_x, it accelerates upward with some constant acceleration of magnitude a_y where a_y is positive. Applying Newton's Second Law ($F_y^{net} = ma_y$) for components along the y axis, we find that the y-component of the acceleration is directly proportional to the y-component of the electrostatic force so that

$$a_y = \frac{F_y^{elec}}{m} = + \frac{|Q|E}{m}. \qquad (23\text{-}33)$$

Let Δt represent the time required for the drop to pass through the region between the plates. In Chapter 4, we found that during the time Δt the vertical and horizontal displacements of the drop are

$$\Delta y = \tfrac{1}{2} a_y \Delta t^2 \quad \text{and} \quad \Delta x = L = v_x \Delta t, \qquad (23\text{-}34)$$

respectively. Eliminating Δt between these two equations and substituting Eq. 23-33 for a_y, we find

$$\Delta y = \frac{|Q|EL^2}{2mv_x^2}$$

$$= \frac{(1.5 \times 10^{-13}\,\text{C})(1.4 \times 10^6\,\text{N/C})(1.6 \times 10^{-2}\,\text{m})^2}{(2)(1.3 \times 10^{-10}\,\text{kg})(18\,\text{m/s})^2}$$

$$= 6.4 \times 10^{-4}\,\text{m}$$

$$= 0.64\,\text{mm}. \qquad \text{(Answer)}$$

23-9 A Dipole in an Electric Field

Many electrical effects in matter can be understood by considering matter to be made up of many little electric dipoles. When an electric field is applied to that matter, the dipoles change their orientation in a consistent way. Although each dipole is small, since they all do the same thing, they can produce a substantial electrical effect. In this section, we consider the torque that can be exerted on a dipole that is placed in a uniform electric field.

We have defined the electric dipole moment $\vec{p}$ of an electric dipole to be a vector pointing from the negative to the positive end of the dipole. It turns out that behavior of a dipole in a uniform external electric field $\vec{E}_s$ can be described completely in terms of the two vectors $\vec{E}_s$ and $\vec{p}$, with no need of any details about the dipole's structure.

A molecule of water (H_2O), as shown in Fig. 23-22, is an electric dipole. There the black dots represent the oxygen nucleus (having eight protons) and the two hydrogen nuclei (having one proton each). The colored enclosed areas represent the region in which electrons can be located around the nuclei.

In a water molecule, the two hydrogen atoms and the oxygen atom do not lie on a straight line but form an angle of about 105°. As a result, the molecule has a definite "oxygen side" and "hydrogen side." Moreover, the 10 electrons of the molecule tend to remain closer to the oxygen nucleus than to the hydrogen nuclei. This makes the oxygen side of the molecule slightly more negative than the hydrogen side and creates an electric dipole moment $\vec{p}$ that points along the symmetry axis of the molecule as shown. If the water molecule is placed in an external electric field, it behaves like the idealized electric dipole shown in Fig. 23-16.

To examine this behavior, we now consider what happens to an idealized electric dipole placed in a *uniform* **external electric field** $\vec{E}_s$. This is shown in Fig. 23-23a.

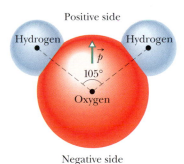

FIGURE 23-22 ■ This H_2O molecule has 3 nuclei (shown as dots). The electrons orbiting the nuclei spend more time near the oxygen nucleus, so the molecule behaves like a dipole. Its moment $\vec{p}$ points from the (negative) oxygen side to the (positive) hydrogen side.

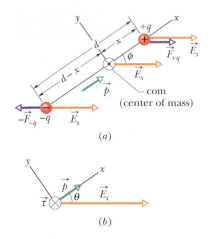

FIGURE 23-23 ■ (a) An electric dipole in a uniform electric field $\vec{E}$. Two equal but opposite charges are separated by a distance d. The line between them represents their rigid connection. (b) Field $\vec{E}$ causes a torque $\vec{\tau}$ on the dipole. The direction of $\vec{\tau}$ is into the plane of the page, as represented by the symbol ⊗.

Assume that the dipole is a rigid structure that consists of two charges of opposite sign, each having an amount $|q|$, separated by a distance d. Assume that the dipole moment $\vec{p}$ makes an angle ϕ with field $\vec{E}_s$. Recall from our discussion in Section 23-6 above that the electric field far from the dipole depends only on $\vec{p}$ (the product of charge q and separation d). The detailed structure of the water molecule is not important as long as the interaction between it and the electric field isn't strong enough to change the shape of either the molecule or the electric field.

Electrostatic forces act on the charged ends of the dipole. Because the electric field is uniform, those forces act in opposite directions (as shown in Fig. 23-23) and with the same magnitude $|\vec{F}_t^{\text{elec}}| = |q_t\vec{E}_s|$. Thus, because the field is uniform, the net force on the dipole from the field is zero and the center of mass of the dipole does not move. However, the forces on the charged ends do produce a net torque $\vec{\tau}$ on the dipole about its center of mass. Since the charges on a dipole do not necessarily have the same mass, we assume the center of mass lies on the line connecting the charged ends, at some distance x from one end and thus a distance $d - x$ from the other end. From Eq. 11-29 ($|\vec{\tau}| = |\vec{r}||\vec{F}|\sin\phi$), we can express the net torque magnitude $\tau = |\vec{\tau}|$ as

$$\tau = F^{\text{elec}}|x|\sin\phi + F^{\text{elec}}|(d - x)|\sin\phi = F^{\text{elec}}d\sin\phi. \tag{23-35}$$

We can also write the magnitude of the torque in terms of the magnitudes of the electric field $E = |\vec{E}|$ and the dipole moment $p = |\vec{p}| = |q|d$. To do so, we substitute $|q|E$ for the magnitude of the electrostatic force, F^{elec}, and $p/|q|$ for the dipole spacing, d, to find an expression for the magnitude of the torque. This magnitude is given by

$$\tau = pE\sin\phi. \tag{23-36}$$

We know the direction of the vector $\vec{\tau}$ is given by the right-hand rule. So, we see the result for both the magnitude and direction can be written in terms of the cross product as

$$\vec{\tau} = \vec{p} \times \vec{E} \quad \text{(torque on a dipole).} \tag{23-37}$$

Vectors $\vec{p}$ and $\vec{E}$ are shown in Fig. 23-23b. The torque acting on a dipole tends to rotate $\vec{p}$ (hence the dipole) into the direction of $\vec{E}$, thereby reducing ϕ. In Fig. 23-23, such rotation is clockwise. As we discuss in Chapter 11, we can represent a torque component τ that gives rise to such a rotation as

$$\vec{\tau} = -\tau\hat{k} = -(pE\sin\phi)\hat{k}, \tag{23-38}$$

where $\hat{k}$ is a unit vector pointing along the z axis (not the Coulomb constant).

READING EXERCISE 23-6: The figure shows four orientations of an electric dipole in an external electric field. Rank the orientations according to the magnitude of the torque on the dipole.

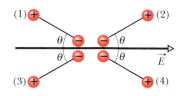

23-10 Electric Field Lines

So far in this chapter we have represented electric fields using vector arrows and creating force vector field plots. There is another common method for creating a visual representation of information about electric fields in a region of space. It involves drawing *electric field lines*.

In the electric field line representation, we use continuous lines to convey information about the direction of the field at different points. Since the magnitude and direction of the electric field usually changes smoothly, this turns out to be rather convenient. Michael Faraday, who introduced the idea of electric fields in the 19th century, thought of the space around a charged body as filled with *lines of force*. Although we attach no reality to these lines, they provide a nice way to visualize patterns of changing force. The field line representation is used in Chapter 24 where we introduce Gauss' law. The field line representation is also used in Chapters 29–33 to describe magnetic fields.

Field lines are a good way to visualize the directions of a vector field in a region of space. To draw an **electric field line,** we start at any point and look at the direction of the field at that point. We then draw a short line in that direction. We determine the field direction at the new location and draw another short line in that direction. We continue this process until we reach a charge or get to infinity. Compare Figs. 23-24 and 23-25 for an example. Note that field *lines* shown in Fig. 23-25 and Fig. 23-26 differ from the short straight field *vectors* shown in Fig. 23-24 because they always start or end on the source charge(s). The direction of a straight field line or the direction of the tangent to a curved field line gives the direction of the electric field vector $\vec{E}_s$ at that point. Because the field lines point in the direction of the field, field lines must originate on positive charges and terminate on negative charges.

It is important to note we could draw field lines through every point in space. However, this would not be very helpful since our paper would be totally filled with field lines and we couldn't distinguish one from another. Instead, we choose to draw a few field lines, with the number of lines leaving each positive charge (or ending on each negative charge) being proportional to the amount of each charge. If we choose to have 16 lines originating on a $+4\,\mu C$ charge, then we should have 8 lines ending on a $-2\,\mu C$ charge. This scaling of the number of field lines with amount of the charge turns out to be quite convenient since then the field lines are forced to be closely packed together where the field is strong and far apart where it is weak. We can see this in Figs. 23-25 and 23-26. In other words, the average density of field lines (the number of lines crossing through a small area perpendicular to their direction) is proportional to the strength of the field. We then have the following rules:

> At any point along an electric field line, the direction of the corresponding electric field vector is always tangent to the line at that point. **Electric field lines** extend away from positive charge (where they originate) and toward negative charge (where they terminate). The density of field lines is proportional to the strength of the field.

Electric Field Near a Nonconducting Sheet

Figure 23-27a shows part of an infinitely large, nonconducting *sheet* (or plane) with a uniform distribution of positive charge on its right side. The electric field lines shown in Fig. 23-27b are uniformly spaced and always perpendicular to the sheet. Why? If the sheet can be treated as if it is infinitely large, then it looks the same in any direction. The pulls or pushes sideways from any bit of the sheet to one side of the test charge are cancelled by those from a symmetric bit of the sheet on the opposite side of the charge. As a result, the electric force vector at that point must point directly toward the sheet (if the sheet is negative) or directly away from the sheet (if the sheet is positive).

Since the field lines are perpendicular to the sheet and have to start and end on charges, they don't diverge or get closer as you move farther from the sheet. This suggests the surprising result that the field should not get weaker or stronger as you get

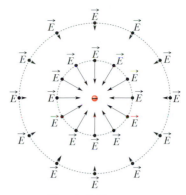

Electric field vectors

FIGURE 23-24 ■ There are two methods commonly used to depict an electric field. This figure shows a vector electric field map (or plot). See Fig. 23-25 for a comparison method.

Electric field lines

FIGURE 23-25 ■ The second common representation of an electric field is the field line representation shown here. See Fig. 23-24 for a comparison method.

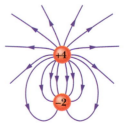

FIGURE 23-26 ■ Electric field lines for a $-2q$ and $+4q$ charge configuration.

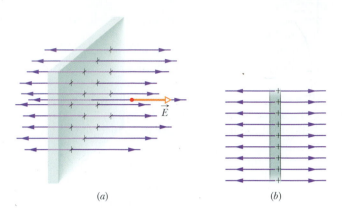

(a)   (b)

FIGURE 23-27 ■ Depictions of the electric field lines due to a very large, nonconducting sheet with uniformly distributed positive charge on one side. The vector fields shown in both figures are uniform and perpendicular to the charged sheet. (*a*) The electric field vector $\vec{E}$ is shown at the location of a test charge, and (*b*) a side view of *a* showing electric field lines pointing away from the positive charges in the space near the sheet.

farther from the sheet. This is even more strongly suggested by a dimensional analysis argument. An E-field has dimensions that look like those of kq/r^2. For an infinite sheet we can't talk about the total charge (it's infinite) but only the charge density sigma, denoted as σ. Sigma already has units of kq/r^2 so it is not possible to put in an extra distance. Doing the integral (which can be generalized from our integral for the ring) is messy but confirms this result. Because the charge is uniformly distributed along the sheet and all the field vectors have the same magnitude, the sheet creates a **uniform electric field.**

Of course, no real nonconducting sheet (such as a flat expanse of plastic) is infinitely large, but if we consider a region near the middle of a real sheet and not near its edges, the field lines through that region are arranged as in Figs. 23-27*a* and *b*.

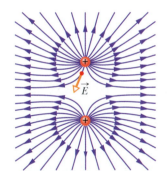

FIGURE 23-28 ■ Field lines for two equal positive point charges. The charges repel each other. (The lines terminate on distant negative charges.) To "see" the actual three-dimensional pattern of field lines, mentally rotate the pattern shown here about an axis passing through both charges in the plane of the page. The three-dimensional pattern and the electric field it represents are said to have rotational symmetry about that axis. The electric field vector at one point is shown; note that it is tangent to the field line through that point.

Field Lines for Two Positive Charges

Figure 23-28 shows the field lines for two equal positive charges. Although we do not often use field lines quantitatively, they are very useful to visualize what is going on. It takes practice to learn to draw electric field lines even for a simple array of point sources. The steps include: (1) creating an electric field map, (2) deciding how detailed the field line representation should be and assigning a certain number of lines per unit charge, (3) placing the assigned number of lines at each point source. The lines should be equally spaced at the source with initial directions that depend on the sign of the source charge. The lines pass radially out from each positive source (and should be marked with an outward arrow). Or the lines pass radially into each negative source (and should be marked with an inward arrow as shown in Figs. 23-25 and 23-26). (4) Each line through the vector field map should be drawn so it is always tangent to the electric field vectors. (5) If the net charge in the region on the map is zero, each line begins on a positive charge and ends on a negative charge. If the net charge is negative, some of the lines come from infinity (off the field line diagram). If the net charge is positive, some of the lines veer off to infinity as shown in Fig. 23-26.

READING EXERCISE 23-7: Explain why the definition that says electric field lines point in the direction of the electric field means electric field lines must originate on positive charges and terminate on negative charges. ■

READING EXERCISE 23-8: Examine Fig. 23-12 showing that the net electric field along a line that bisects the two charges is always perpendicular to the line connecting the charges. How does the construction of this diagram help explain the fact that the net electric field due to a uniformly charged sheet (shown in Fig. 23-27) is always perpendicular to the sheet of charge? ■

Problems

In the following problems, all electric fields referenced are those produced by the source charge(s). That is, $\vec{E} = \vec{E}_s$.

SEC. 23-4 ■ THE ELECTRIC FIELD DUE TO A POINT CHARGE

1. Point Charge What is the amount of charge on a small particle whose electric field 50 cm away has the amount of 2.0 N/C?

2. What Amount? What is the amount of a point charge that would create an electric field of 1.00 N/C at points 1.00 m away?

3. Plutonium-239 A plutonium-239 nucleus of radius 6.64 femptometers has an atomic number $Z = 94$. Assuming that the positive charge is distributed uniformly within the nucleus, what are the magnitude and direction of the electric field at the surface of the nucleus? Assume the influence of the positive charge at the nuclear surface is the same as that of a point charge.

SEC. 23-5 ■ THE ELECTRIC FIELD DUE TO MULTIPLE CHARGES

4. Two Particles Two particles with equal charge amounts 2.0×10^{-7} C but opposite signs are held 15 cm apart. What are the magnitude and direction of $\vec{E}$ at the point midway between the charges?

5. Two Point Charges Two point charges $q_A = 2.1 \times 10^{-8}$ C and $q_B = -4.0q_A$ are fixed in place 50 cm apart. Find the point along the straight line passing through the two charges at which the electric field is zero.

6. Two Fixed Charges In Fig 23-29, two fixed point charges $q_A = +1.0 \times 10^{-6}$ C and $q_B = +3.0 \times 10^{-6}$ C are separated by a distance $d = 10$ cm. Plot their net electric field $\vec{E}(x)$ as a function of x for both positive and negative values of x, taking $\vec{E}$ to be positive when $\vec{E}$ points to the right and negative when $\vec{E}$ points to the left.

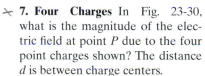

FIGURE 23-29 ■ Problems 6 and 8.

7. Four Charges In Fig. 23-30, what is the magnitude of the electric field at point P due to the four point charges shown? The distance d is between charge centers.

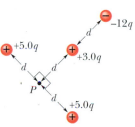

FIGURE 23-30 ■ Problem 7.

8. Separation of d In Fig. 23-29, two fixed point charges $q_A = -5q$ and $q_B = +2q$ are separated by distance d. Locate the point (or points) at which the net electric field due to the two charges is zero.

9. Square What are the magnitude and direction of the electric field at the center of the square of Fig. 23-31 if $q = 1.0 \times 10^{-8}$ C and the distance between charge centers $a = 5.0$ cm?

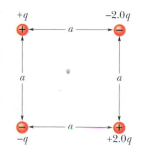

FIGURE 23-31 ■ Problem 9.

10. Three Charges Calculate the direction and magnitude of the electric field at point P in Fig. 23-32, due to the three point charges. The distance between charge centers is a.

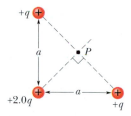

FIGURE 23-32 ■ Problem 10.

11. Equilateral Triangle Two particles, each with an amount of $|q|$ equal to charge of 12 nC, are placed at two of the vertices of an equilateral triangle. The length of each side of the triangle is 2.0 m. What is the magnitude of the electric field at the third vertex of the triangle if (a) both of the charges are positive and (b) one of the charges is positive and the other is negative?

12. Plastic Ring Figure 23-33 shows a plastic ring of radius $R = 50.0$ cm. Two small charged beads are on the ring: Bead 1 of charge $+2.00\ \mu$C is fixed in place at the left side; bead 2 of charge $+6.00\ \mu$C can be moved along the ring. The two beads produce a net electric field of magnitude E at the center of the ring. At what angle θ should bead 2 be positioned such that $E = 2.00 \times 10^5$ N/C?

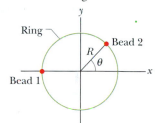

FIGURE 23-33 ■ Problem 12.

13. Three Particles Two Three particles, each with positive charge q, form an equilateral triangle, with each side of length d. What is the magnitude of the electric field produced by the particles at the midpoint of any side?

14. Separation L Figure 23-34a shows two charged particles fixed in place on an x axis with separation L. The ratio q_A/q_B of their charge amounts is 4.00. Figure 23-34b shows the x-component E_x^{net} of their

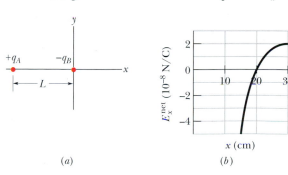

FIGURE 23-34 ■ Problem 14.

net electric field along the x axis just to the right of particle B. (a) At what value of $x > 0$ is E_x^{net} maximum? (b) If particle B has charge $-q_B = -3e$, what is the value of that maximum?

15. Two Charges Two Figure 23-35 shows two charged particles on an x axis: $-q = -3.20 \times 10^{-19}$ C at $x = -3.00$ m and $q = 3.20 \times$

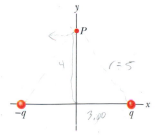

FIGURE 23-35 ■ Problem 15.

10^{-19} C at $x = +3.00$ m. What are the magnitude and direction of the net electric field they produce at point P at $y = 4.00$ m?

16. Eight Charges In Fig. 23-36, eight charged particles form a square array; charge $q = e$ and distance $d = 2.0$ cm. What are the magnitude and direction of the net electric field at the center?

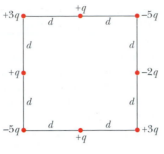

FIGURE 23-36 ■ Problem 16.

SEC. 23-6 ■ THE ELECTRIC FIELD DUE TO AN ELECTRIC DIPOLE

17. Calculate the Moment Calculate the electric dipole moment of an electron and a proton 4.30 nm apart.

18. Field at P In Fig. 23-16, let both charges be positive. Assuming $z \gg d$, show that the magnitude of the vector $\vec{E}$ at point P in that figure is then given by

$$|\vec{E}| = k\frac{2|q|}{z^2}.$$

19. Electric Quadrupole Figure 23-37 shows an electric quadrupole. It consists of two dipoles with dipole moments that are equal in magnitude but opposite in direction. Show that the magnitude of the vector $\vec{E}$ on the axis of the quadrupole for a point P a distance z from its center (assume $z \gg d$) is given by

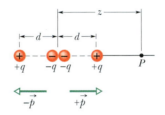

FIGURE 23-37 ■ Problem 19.

$$|\vec{E}| = \frac{3|Q|}{4\pi\varepsilon_0 z^4},$$

in which $Q(= 2qd^2)$ is known as the *quadrupole moment* of the charge distribution.

20. Electric Dipole Find the magnitude and direction of the electric field at point P due to the electric dipole in Fig 23-38. The distance between charge center is d and P is located at a distance $r \gg d$ along the perpendicular bisector of the line joining the charges. Express your answer in terms of the magnitude and direction of the electric dipole moment $\vec{p}$.

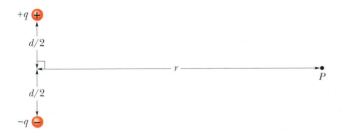

FIGURE 23-38 ■ Problem 20.

SEC. 23-7 ■ THE ELECTRIC FIELD DUE TO A RING OF CHARGE

21. Electron Constrained An electron is constrained to the central axis of the ring of charge of radius R discussed in Section 23-7. Show that the electrostatic force on the electron can cause it to oscillate through the center of the ring with an angular frequency

$$\omega = \sqrt{\frac{eq}{4\pi\varepsilon_0 mR^3}},$$

where q is the ring's charge and m is the electron's mass.

22. Two Rings Figure 23-39 shows two parallel nonconducting rings arranged with their central axes along a common line. Ring A has uniform charge q_A and radius R; ring B has uniform charge q_B and the same radius R. The rings are separated by a distance $3R$. The net electric field at point P on the common line, at distance R from ring A, is zero. What is the ratio q_A/q_B?

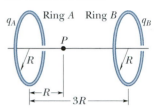

FIGURE 23-39 ■ Problem 22.

23. Thin Glass Rod A thin glass rod is bent into a semicircle of radius r. A charge $+q$ is uniformly distributed along the upper half, and a charge $-q$ is uniformly distributed along the lower half, as shown in Fig. 23-40a. Find the magnitude and direction of the electric field $\vec{E}$ at P, the center of the semicircle.

24. Two Curved Plastic Rods In Fig. 23-40b, two curved plastic rods, one of charge $+q$ and the other of charge $-q$, form a circle of radius R in an xy plane. The x axis passes through their connecting points, and the charge is distributed uniformaly on both rods. What are the magnitude and direction of the electric field $\vec{E}$ produced at P, the center of the circle?

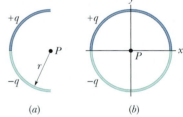

FIGURE 23-40 ■ Problems 23 and 24.

25. Nonconducting Rod In Fig. 23-41, a nonconducting rod of length L has charge $-q$ uniformly distributed along its length. (a) What is the linear charge density of the rod? (b) What is the electric

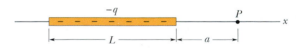

FIGURE 23-41 ■ Problem 25.

field at point P, a distance a from the end of the rod? (c) If P were very far from the rod compared to L, the rod would look like a point charge. Show that your answer to (b) reduces to the electric field of a point charge for $a \gg L$.

26. What Distance? At what distance along the central axis of a ring of radius R and uniform charge is the magnitude of the electric field due to the ring's charge maximum?

27. Semi-Infinite Rod In Fig. 23-42, a "semi-infinite" nonconducting rod (that is, infinite in one direction only) has uniform linear charge density λ. Show that the electric field at point P makes an angle of 45° with the rod and that this result is independent of the distance R. (*Hint:* Separately find

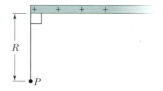

FIGURE 23-42 ■ Problem 27.

the parallel and perpendicular (to the rod) components of the electric field at P, and then compare those components.)

28. Length L Rod A thin nonconducting rod of finite length L has a charge q spread uniformly along it. Show that

$$|\vec{E}| = \frac{|q|}{2\pi\varepsilon_0 y} \frac{1}{(L^2 + 4y^2)^{1/2}}$$

gives the magnitude $|\vec{E}|$ of the electric field at point P on the perpendicular bisector of the rod (Fig. 23-43).

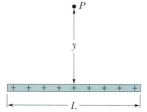

FIGURE 23-43 ■ Problem 28.

29. Density, Density, Density. (a) A charge of $-300e$ is uniformly distributed along a circular arc of radius 4.00 cm, which subtends an angle of 40°. What is the linear charge density along the arc? (b) A charge of $-300e$ is uniformly distributed over one face of a circular disk of radius 2.00 cm. What is the surface charge density over that face? (c) A charge of $-300e$ is uniformly distributed over the surface of a sphere of radius 2.00 cm. What is the surface charge density over that surface? (d) A charge of $-300e$ is uniformly spread through the volume of a sphere of radius 2.00 cm. What is the volume charge density in that sphere?

30. Nonconducting Rod Two A thin nonconducting rod with a uniform distribution of positive charge Q is bent into a circle of radius R (Fig. 23-44). The central axis through the ring is a z axis, with the origin at the center of the ring. What is the magnitude of the electric field due to the rod at (a) $z = 0$ and (b) $z = \pm \infty$? (c) In terms of R, at what values of z is that magnitude maximum? (d) If radius $R = 2.00$ cm and charge $Q = 4.00$ μC, what is the maximum magnitude?

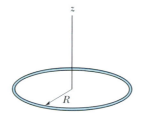

FIGURE 23-44 ■ Problem 30.

31. Circular Rod A circular rod has a radius of curvature R and a uniformly distributed charge Q and it subtends an angle θ (in radians). What is the magnitude of the electric field it produces at the center of curvature?

32. Two Concentric Rings Figure 23-45 shows two concentric rings, of radii R and $R' = 3.00R$, that lie on the same plane. Point P lies on the central z axis, at distance $D = 2.00R$ from the center of the rings. The smaller ring has uniformly distrib-

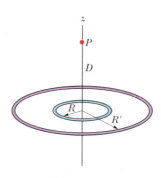

FIGURE 23-45 ■ Problem 32.

uted charge $+Q$. What must be the uniformly distributed charge on the larger ring if the net electric field at point P due to the two rings is to be zero?

33. Charge $+Q$ In Fig. 23-46a, a particle of charge $+Q$ produces an electric field with a magnitude E_{part} at point P, at distance R from it. In Fig. 23-46b, that same amount of charge is spread uniformly along a circular arc that has radius R and subtends an angle θ. The charge on the arc produces an electric field with a magnitude E_{arc} at its center of curvature P. For what value of θ does $E_{arc} = 0.500$ E_{part}? (*Hint:* You can use a graphical solution.)

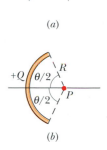

FIGURE 23-46 ■ Problem 33.

34. Half Circle Figure 23-47a shows a nonconducting rod with a uniformly distributed charge $+Q$. The rod forms a half circle with radius R and produces an electric field of magnitude E_{arc} at its center of curvature P. If the arc is collapsed to a point at distance R from P (Fig. 23-47b), by what factor is the magnitude of the electric field at P multiplied?

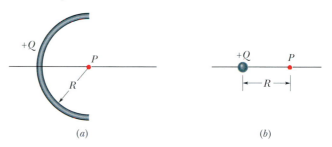

FIGURE 23-47 ■ Problem 34.

SEC. 23-8 ■ MOTION OF POINT CHARGES IN AN ELECTRIC FIELD

35. Electron Released from Rest An electron is released from rest in a uniform electric field of magnitude 2.00×10^4 N/C. Calculate the acceleration of the electron. (Ignore gravitation).

36. Accelerated Electron An electron is accelerated eastward at 1.80×10^9 m/s² by an electric field. Determine the magnitude and direction of the electric field.

37. Force Due to Dipole Calculate the magnitude of the force, due to an electric dipole of dipole moment 3.6×10^{-29} C · m, on an electron 25 nm from the center of the dipole, along the dipole axis. Assume that this distance is large relative to the dipole's charge separation.

38. Alpha Particle An alpha particle (the nucleus of a helium atom) has a mass of 6.64×10^{-27} kg and a charge of $+2e$. What are the magnitude and direction of the electric field that will balance the gravitational force on it?

39. Charged Cloud A charged cloud system produces an electric field in the air near Earth's surface. A particle of charge -2.0×10^{-9} C is acted on by a downward electrostatic force of 3.0×10^{-6} N when placed in this field. (a) What is the magnitude of the electric field? (b) What are the magnitude and direction of the electrostatic force exerted on a proton placed in this field? (c) What is the gravitational force on the proton? (d) What is the ratio of the magnitude of the electrostatic force to the magnitude of the gravitational force in this case?

40. Humid Air Humid air breaks down (its molecules become ionized) in an electric field of 3.0×10^6 N/C. In that field, what is the magnitude of the electrostatic force on (a) an electron and (b) an ion with a single electron missing?

41. High-Speed Protons Beams of high-speed protons can be produced in "guns" using electric fields to accelerate the protons. (a) What acceleration would a proton experience if the gun's electric field were 2.00×10^4 N/C? (b) What speed would the proton attain if the field accelerated the proton through a distance of 1.00 cm?

42. Floating a Sulfur Sphere An electric field E with an average magnitude of about 150 N/C points downward in the atmosphere near Earth's surface. We wish to "float" a sulfur sphere weighing 4.4 N in this field by charging the sphere. (a) What charge (both sign and magnitude) must be used? (b) Why is the experiment impractical?

43. Two Oppositely Charged Plates A uniform electric field exists in a region between two oppositely charged plates. An electron is released from rest at the surface of the negatively charged plate and strikes the surface of the opposite plate, 2.0 cm away, in a time 1.5×10^{-8} s. (a) What is the speed of the electron as it strikes the second plate? (b) What is the magnitude of the electric field $\vec{E}$?

44. Field Retards Motion An electron with a speed of 5.00×10^8 cm/s enters an electric field of magnitude 1.00×10^3 N/C, traveling along the field lines in the direction that retards its motion. (a) How far will the electron travel in the field before stopping momentarily and (b) how much time will have elapsed? (c) If the region with the electric field is only 8.00 mm long (too short for the electron to stop within it), what fraction of the electron's initial kinetic energy will be lost in that region?

45. Two Copper Plates Two large parallel copper plates are 5.0 cm apart and have a uniform electric field between them as depicted in Fig. 23-48. An electron is released from the negative plate at the same time that a proton is released from the positive plate. Neglect the force of the particles on each other and find their distance from the positive plate when they pass each other. (Does it surprise you that you need not know the electric field to solve this problem?)

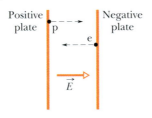

FIGURE 23-48 ■ Problem 45.

46. Velocity Components At some instant the velocity components of an electron moving between two charged parallel plates are $v_x = 1.5 \times 10^5$ m/s and $v_y = 3.0 \times 10^3$ m/s. Suppose that the electric field between the plates is given by $\vec{E} = (120 \text{ N/C})\hat{j}$. (a) What is the acceleration of the electron? (b) What will be the velocity of the electron after its x coordinate has changed by 2.0 cm?

47. Uniform Upward Field In Fig. 23-49, a uniform, upward-directed electric field $\vec{E}$ of magnitude 2.00×10^3 N/C has been set up between two horizontal plates by charging the lower plate positively and the upper plate negatively. The plates have length

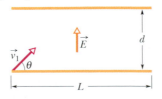

FIGURE 23-49 ■ Problem 47.

$L = 10.0$ cm and separation $d = 2.00$ cm. An electron is then shot between the plates from the left edge of the lower plate. The initial velocity $\vec{v}_1$ of the electron makes an angle $\theta = 45.0°$ with the lower plate and has a magnitude of 6.00×10^6 m/s. (a) Will the electron strike one of the plates? (b) If so, which plate and how far horizontally from the left edge will the electron strike?

48. Charge in an E Field A 10.0 g block with a charge of $+8.00 \times 10^{-5}$ C is placed in electric field $\vec{E} = (3.00 \times 10^3 \text{ N/C})\hat{i} - 600 \text{ N/C})\hat{j}$. (a) What are the magnitude and direction of the force on the block? (b) If the block is released from rest at the origin at $t = 0.00$ s, what will be its coordinates at $t = 3.00$ s?

49. Entering a Field An electron enters a region of uniform electric field with an initial velocity of 40 km/s in the same direction as the electric field, which has magnitude $E = 50$ N/C. (a) What is the speed of the electron 1.5 ns after entering this region? (b) How far does the electron travel during the 1.5 ns interval?

50. An Electron Is Shot In Fig. 23-50, an electron is shot at an initial speed of $v_1 = 2.00 \times 10^6$ m/s, at angle $\theta_1 = 40°$ from an x axis. It moves in a region with uniform electric field $\vec{E} = (5.00 \text{ N/C})\hat{j}$. A screen for detecting electrons is positioned parallel to the y axis, at distance $x = 3.00$ m. In unit-vector notation, what is the velocity of the electron when it hits the screen?

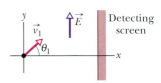

FIGURE 23-50 ■ Problem 50.

51. TV Tube Figure 23-51 shows the deflection-plate system of a conventional TV tube. The length of the plates is 3.0 cm and the electric field between the two plates is 10^6 N/C (vertically up). If the electron enters the plates with a horizontal velocity of 3.9×10^7 m/s, what is the vertical deflection Δy at the end of the plates?

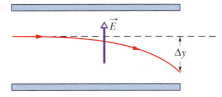

FIGURE 23-51 ■ Problem 51.

SEC. 23-9 ■ A DIPOLE IN AN ELECTRIC FIELD

52. Dipole in a Field An electric dipole, consisting of charges of magnitude 1.50 nC separated by 6.20 μm, is in an electric field of strength 1100 N/C. (a) What is the magnitude of the electric dipole moment? (b) What is the difference between the potential energies corresponding to dipole orientations parallel to and antiparallel to the field?

53. Torque on a Dipole An electric dipole consists of charges $+2e$ and $-2e$ separated by 0.78 nm. It is in an electric field of strength 3.4×10^6 N/C. Calculate the magnitude of the torque on the dipole when the dipole moment is (a) parallel to, (b) perpendicular to, and (c) antiparallel to the electric field.

54. Work Required Find the work required to turn an electric dipole end for end in a uniform electric field $\vec{E}$, in terms of the magnitude $|\vec{p}|$ of the dipole moment, the magnitude $|\vec{E}|$ of the field, and the initial angle θ_1 between $\vec{p}$ and $\vec{E}$.

55. Frequency of Oscillation Find the frequency of oscillation of an electric dipole, of dipole moment $\vec{p}$ and rotational inertia I, for small amplitudes of oscillation about its equilibrium position in a uniform electric field of magnitude $|\vec{E}|$.

56. A Certain Dipole A certain electric dipole is placed in a uniform electric field $\vec{E}$ of magnitude 40 N/C. Figure 23-52 gives the magnitude τ of the torque on the dipole versus the angle θ between $\vec{E}$ and the dipole moment $\vec{p}$. What is the magnitude of $\vec{p}$?

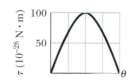

FIGURE 23-52 ■
Problem 56.

57. How Much Energy How much energy is needed to flip an electric dipole from being lined up with a uniform external electric field to being lined up opposite the field? The dipole consists of an electron and a proton at a separation of 2.00 nm, and it is in a uniform field of magnitude 3.00×10^6 N/C.

58. See Graph A certain electric dipole is placed in a uniform electric field $\vec{E}$ of magnitude 20 N/C. Figure 23-53 gives the potential energy U of the dipole versus the angle θ between $\vec{E}$ and the dipole moment $\vec{p}$. What is the magnitude of $\vec{p}$?

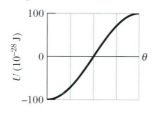

FIGURE 23-53 ■ Problem 58.

SEC. 23-10 ■ ELECTRIC FIELD LINES

59. Twice the Separation In Fig. 23-54, the electric field lines on the left have twice the separation of those on the right. (a) If the magnitude of the field at A is 40 N/C, what force acts on a proton at A? (b) What is the magnitude of the field at B?

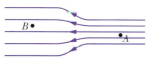

FIGURE 23-54 ■ Problem 59.

60. Sketch Sketch qualitatively the electric field lines both between and outside two concentric conducting spherical shells when a uniform positive charge q_A is on the inner shell and a uniform negative charge $-q_B$ is on the outer. Consider the cases $q_A > q_B$, $q_A = q_B$ and $q_A < q_B$.

61. Thin Circular Disk Sketch qualitatively the electric field lines for a thin, circular, uniformly charged disk of radius R. (*Hint:* Consider as limiting cases points very close to the disk, where the electric field is directed perpendicular to the surface, and points very far from it, where the electric field is like that of a point charge.)

62. Particles and Lines In Fig. 23-55, particles with charges $+1.0q$ and $-2.0q$ are fixed a distance d apart. Find the magnitude and direction of the net electric field at points (a) A, (b) B, and (c) C. (d) Sketch the electric field lines.

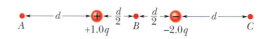

FIGURE 23-55 ■ Problem 62.

63. Three Point Charges Two In Fig. 23-56, three point charges are arranged in an equilateral triangle. (a) Sketch the field lines due to $+Q$ and $-Q$, and from them determine the direction of the force that acts on $+q$ because of the presence of the other two charges. (b) What is the magnitude of that net electric force on $+q$?

FIGURE 23-56 ■
Problem 63.

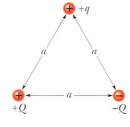

Additional Problems

64. Two Electric Charges—Electric Force Figure 23-57 shows seven arrangements of two electric charges. In each figure, a point labeled P is also identified. All of the charges are the same size, 20 nC, but they can be either positive or negative. The charges and point P all lie on a straight line. The distances between adjacent items, either between two charges or between a charge and point P, are all 5 cm. There are no other charges in this region. For this problem, we will place a +5 nC charge at point P.
Rank these arrangements from greatest to least on the basis of the magnitude of the electric force on the +5 nC charge when it is placed at point P. [Based on Ranking Task 126, O'Kuma, et. al., *Ranking Task Exercises* (Prentice Hall, Upper Saddle River, NJ, 2000).]

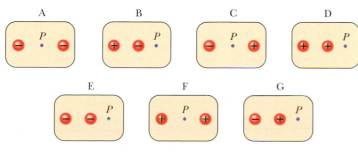

FIGURE 23-57 ■ Problem 64.

65. Fixed and Suspended Charges—Angle Figure 23-58 shows m_A, a stationary sphere with charge q_A. The charge m_B is suspended from the ceiling by a nonconducting string and has charge q_B. The masses m_A and m_B are conducting spheres of the same size. The charges q_A and q_B have the same sign. From the combinations below, rank the angle the string will form with the vertical from highest to lowest value. If any of the angles are the same, state that. Explain your reasoning.

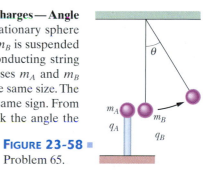

FIGURE 23-58 ■
Problem 65.

(a) $m_A = m$; $\quad q_A = q$
$\quad\;\; m_B = m$; $\quad q_B = q$

(b) $m_A = 2m$; $\quad q_A = q$
$\quad\;\; m_B = 2m$; $\quad q_B = 2q$

(c) $m_A = m$; $\quad q_A = 2q$
$\quad\;\; m_B = m$; $\quad q_B = 2q$

(d) $m_A = 2m$; $\quad q_A = q$
$\quad\;\; m_B = 2m$; $\quad q_B = q$

(e) $m_A = m$; $\quad q_A = 2q$
$\quad\;\; m_B = 2m$; $\quad q_B = 2q$

(f) $m_A = m$; $\quad q_A = q$
$\quad\;\; m_B = m$; $\quad q_B = 2q$

[Based on Ranking Task 138, O'Kuma, et. al., *Ranking Task Exercises* (Prentice Hall, Upper Saddle River, NJ, 2000).]

66. Electric Force on Same Charge Figure 23-59 shows a large region of space that has a uniform electric field in the x direction. At the point $(0, 0)$ m, the electric field is $(30$ N/C$)\hat{i}$. Rank the magnitude of the electric force from greatest to least on a 5 C charge when it is placed at each of the following points: A, $(0$ m, 0 m); B, $(0$ m, 3 m); C, $(-3$ m, 0 m); D, $(3$ m, 0 m); E, $(3$ m, 3 m); F, $(6$ m, 0 m). Explain your reasoning. [Based on Ranking Task 139, O'Kuma, et. al., *Ranking Task Exercises* (Prentice Hall, Upper Saddle River, NJ, 2000).]

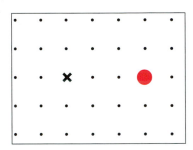

FIGURE 23-59 ■ Problem 66.

67. Dependence of *E* Figure 23-60 shows a fixed charge (specified by a circle) and a location (specified by the x). A test charge is placed at the x to measure the electric effect of the fixed charge. Complete the following two statements as quantitatively as you can. (For example, if the result is larger by a factor of three, don't say "increases"—say "triples" or "is multiplied by three.") Each statement is meant to be compared with the original situation. (The changes don't accumulate).

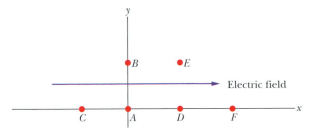

FIGURE 23-60 ■ Problem 67.

(a) If the test charge is replaced by one with half the amount of charge, then the electric field it experiences will ———.
(b) If the fixed charge is replaced by one with twice the amount of charge, then the electric field experiences by the test charge will ———.

68. What Is Going on at *P*? Figure 23-61 shows two charges of $-q$ arranged symmetrically about the y axis. Each produces an electric field at point P. (a) Are the magnitudes of the fields equal? Why or why not? (b) Does each electric field point toward or away from the charge producing it? Explain. (c) Is the magnitude of the net electric field equal to the sum of the magnitudes of the two field vectors (that is, equal to $2E$)? Why or why not? (d) Do the x-components of the two fields add or cancel? Explain. (e) Do the y-components of the two fields add or cancel? Explain. [Based on question 5.20, Arons, *Homework and Test Questions for Introductory Physics Teaching* (Wiley, New York, 1994).]

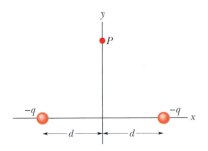

FIGURE 23-61 ■ Problem 68.

69. Field Lines Figure 23-62 shows the region in the neighborhood of a negatively charged conducting sphere and a large positively charged conducting plate extending far beyond the region shown. Someone claims that lines $A-F$ are possible field lines representing the electric field lying in the region between the two conductors. (a)

Examine each of the lines and indicate whether it is a correctly drawn field line. If a line is not correct, explain why. (b) Redraw the diagram with a pattern of field lines that is more correct.

70. Field Lines Two Figure 23-63 shows the electric field lines for three point charges that are positive and negative as indicated. (a) Show the *direction* of each of the electric field lines with an arrow, and (b) if the central charge is $+1.0$ μC what are the values of the outer charges?

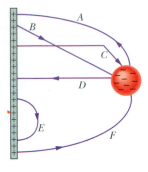

FIGURE 23-62 ■ Problem 69.

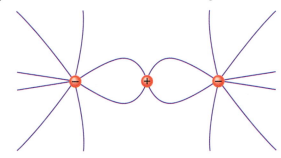

FIGURE 23-63 ■ Problem 70.

71. Field Lines Three Figure 23-64 shows the electric field lines for three point charges separated by a small distance. The two outer charges are identical and the one in the center is different. (a) Determine the ratio, q_A/q_B, of one of the outer charges to the inner one. (b) Determine the signs of q_A and q_B.

72. Functional Dependence and the Electric Field (a) Suppose you want to purchase a sweater in Maryland that has a list price of $40 for which you pay $2 in sales tax. Your friend bought the same sweater in

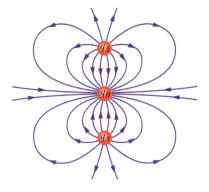

FIGURE 23-64 ■ Problem 71.

Maryland, but it had a list price of $80 for which she paid $4 in sales tax. How does the ratio of sales tax to price of the sweater compare for you and your friend [i.e., compare the ratios (sales tax)/(sweater price)]? What does that ratio tell us? As what is that ratio defined? (b) Suppose a charge exerts a repulsive force of 4 N on a test charge of 0.2 μC that is 2 cm from it. However, the charge exerts a repulsive force of 8 N on a test charge of 0.4 μC that is 2 cm from it. How does the ratio of the force on the test charge to the test charge itself compare in each case [i.e., compare (force felt by test charge)/(test charge)]? What does that ratio tell us? What is that ratio defined as? (c) Suppose a charge Q exerts a force F on a test charge q that is placed near it. By how much would the force exerted by Q increase if the test charge increased by a factor of α, where α can be any constant (i.e., $\alpha = -17$ or 5 or 7.812, etc.)? By how much would the ratio of the force on the test charge to the test charge itself increase if the test charge increased by a factor of α?

Explain. (d) When the value of one quantity depends on the value of a second quantity (and perhaps on others), we say that the first quantity is *a function of* the second. *How* the first quantity changes when the second changes is called the *functional dependence*. For example, if $t = As$, we say that t has a linear functional dependence on s. When s doubles, so does t. If s is divided by 10, so is t. As a second example, if we had $y = Bx^2$, we would say that y depends quadratically on x. If x doubles, y quadruples. If x is divided by 10, then y is divided by 100. (Try this with some numbers, picking whatever values of the constants A and B you would like.)

(i) What is the functional dependence on the sales tax paid on the price of the sweater in part (a)? Explain. Write an equation that relates the tax paid (t) to the cost of the sweater (s).

(ii) What is the functional dependence of the sales tax percentage rate on the price of the sweater in part (a)? Explain.

(iii) In part (c), what is the functional dependence of the force magnitude, F, on the amount of the test charge, $|q|$? Explain.

(iv) In part (c), what is the functional dependence of the electric field magnitude established by Q, E_Q, on the test charge, q? Explain.

73. E-Field Multiple Representations Figure 23-65a displays a grid with coordinates measured in meters. On the grid two charges are placed with their positions indicated as red circles. We call the charge at the position (1 m, 0 m) q_A, and the charge at the position (−1 m, 0 m) q_B. Figure 23-65b shows a set of possible vector directions. Below is a list of the components of possible E fields. For each of the following three cases:

I $q_A = 0$	$q_B = 8\pi\varepsilon_0$	E field at the point $(x, y) = (-1\text{ m}, 1\text{ m})$
II $q_A = 0$	$q_B = -8\pi\varepsilon_0$	E field at the point $(x, y) = (-1\text{ m}, -1\text{ m})$
III $q_A = 160\pi\varepsilon_0$	$q_B = -16\pi\varepsilon_0$	E field at the point $(x, y) = (0\text{ m}, -1\text{ m})$

specify an arrow corresponding to the directions of the E field from figure (b) and a set of components from the list on the right. Each of your answers should consist of a capital letter and a small letter.

(*Note:* The values of the charges in Coulombs are chosen to make the messy "$4\pi\varepsilon_0$" in Coulomb's law cancel. Don't put in numbers first!)

a. $\sqrt{8}$ N/C$\hat{\mathbf{i}}$

b. $(-\sqrt{8}$ N/C$)\hat{\mathbf{j}}$

c. $(-2$ N/C$)\hat{\mathbf{i}}$

d. $(2$ N/C$)\hat{\mathbf{j}}$

e. $(\frac{1}{2}$ N/C$)\hat{\mathbf{i}} + (\frac{1}{2}\sqrt{5}$ N/C$)\hat{\mathbf{j}}$

f. $(\frac{1}{2}$ N/C$)\hat{\mathbf{i}} - (\sqrt{5}$ N/C$)\hat{\mathbf{j}}$

g. None of the above

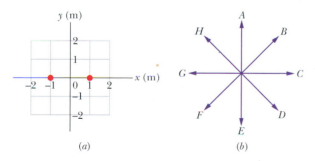

(a)   (b)

FIGURE 23-65 ▪ Problem 73.

74. The Size of an Oil Drop In the Millikan oil drop experiment, an atomizer (a sprayer with a fine nozzle) is used to introduce many tiny droplets of oil between two oppositely charged parallel metal plates. Some of the droplets pick up one or more excess electrons. The charge on the plates is adjusted so that the electric force on the excess electrons exactly balances the weight of the droplet. The idea is to look for a droplet that has the smallest electric force and assume that it has only one excess electron. This lets the observer measure the charge on the electron. Suppose we are using an electric field of 3×10^4 N/C. The charge on one electron is about 1.6×10^{-19} C. Estimate the radius of an oil drop whose weight could be balanced by the electric force of this field on one electron.

75. What's a Field? In this class, we repeatedly refer to an "electric field." Describe what an electric field is. Discuss how you would know a nonzero field was present and how you would measure it.

76. Charge from Field Lines Figure 23-66 shows some representative electric field lines associated with some charges. Both pictures show the same charges, but they are masked in different ways by imaginary closed surfaces drawn for the purpose of hiding the charges from your view.

(a) From the field lines in the two pictures, which of the following statements is most likely to be true?

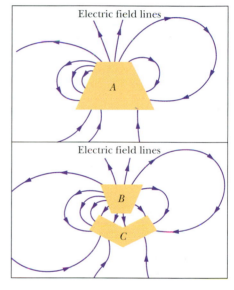

FIGURE 23-66 ▪ Problem 76.

A. There are no charges contained in A.

B. The charge contained in A is positive.

C. The charge contained in A is negative.

D. The total charge contained in A is zero.

E. None of the above can be true.

(b) From the field lines in the two pictures, which of the following statements is most likely to be true?

A. There are no charges contained in B.

B. The charge contained in B is positive.

C. The charge contained in B is negative.

D. The total charge contained in B is zero.

E. None of the above.

(c) From the field lines in the two pictures, which of the following statements is most likely to be true?

A. The charge contained in C is positive and greater in amount than the charge in B.

B. The charge contained in C is positive and smaller in amount than the charge in B.

C. The charged contained in C is negative and greater in amount than the charge in B.

D. The total charge contained in C is negative and smaller in amount than the charge in B.

E. None of the above.

77. Finding the E Field Figure 23-67 shows two charges placed on a coordinate grid. Each of the tic marks on the axes represents 1 m. The amount of the charge is represented by the solid circle is q_A and is at the position (2 m, 0 m), while the charge represented by the open circle is q_B and is at the position (0 m, 2 m). Below is a list of five sets of configurations (labeled a–e) specifying the value of the charges and the positions at which the E field is to be measured. For ease of calculation, these are represented in terms of the Coulomb contant $k = 1/4\pi\varepsilon_0$. On the right is a list of 12 possible electric fields represented as x- and y-components. For each of the five configurations, select the E-field components that represent the field found at that position.

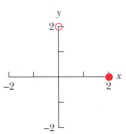

FIGURE 23-67 ■ Problem 77.

Configuration

	q_A	q_B	Position to Test the E Field
(a)	4/k	−4/k	(0 m, 0 m)
(b)	4/k	−4/k	(2 m, 2 m)
(c)	4/k	0	(0 m, 0 m)
(d)	0	−8/k	(−2 m, 2 m)
(e)	0	−8/k	(2 m, 0 m)

Possible E-Fields

1. $(1 \text{ N/C})\hat{i}$

2. $(-1 \text{ N/C})\hat{j}$

3. $(1 \text{ N/C})\hat{i} + (1 \text{ N/C})\hat{j}$

4. $(-1 \text{ N/C})\hat{i}$

5. $(-1 \text{ N/C})\hat{i} + (1 \text{ N/C})\hat{j}$

6. $(1 \text{ N/C})\hat{j}$

7. $\left(\frac{1}{\sqrt{2}} \text{ N/C}\right)\hat{i} - \left(\frac{1}{\sqrt{2}} \text{ N/C}\right)\hat{j}$

8. $\left(-\frac{1}{\sqrt{2}} \text{ N/C}\right)\hat{i} + \left(\frac{1}{\sqrt{2}} \text{ N/C}\right)\hat{j}$

9. $(2 \text{ N/C})\hat{i}$

10. $(2 \text{ N/C})\hat{j}$

11. $(-2 \text{ N/C})\hat{i}$

12. None of the above

78. Beads on a Ring Two charged beads are on the plastic ring in Fig. 23-68a. Bead 2, which is not shown, is fixed in place on the ring, which has radius $R = 60.0$ cm. Bead 1 is initially at the right side of the ring, at angle $\theta = 0°$. It is then moved to the left side, at angle $\theta = 180°$, through the first and second quadrants of the xy coordinate system. Figure 23-68b gives the x-component of the net electric field produced at the origin by the two beads as a function of θ. Similarly, Fig. 23-68c gives the y-component. (a) At what angle θ is bead 2 located? What are the charges of (b) bead 1 and (c) bead 2?

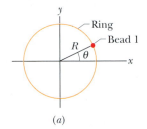

(a)

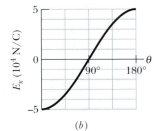

(b)

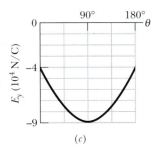

(c)

FIGURE 23-68 ■ Problem 78.

24 | Gauss' Law

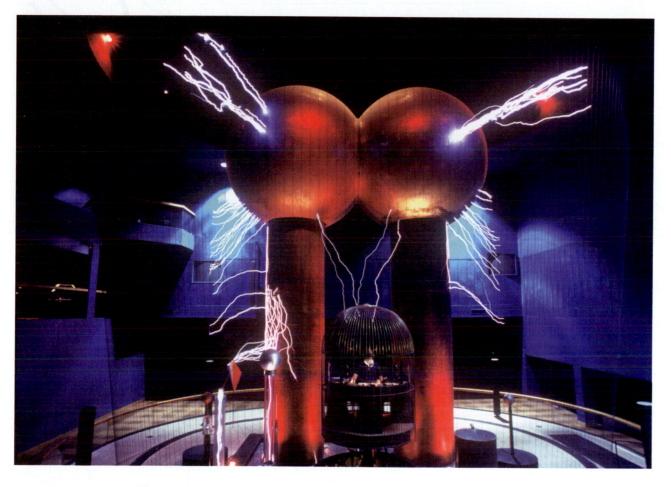

A demonstrator at the Boston Museum of Science is enclosed in a large conducting cage made of wire mesh. An electrical discharge from a giant Van de Graaff generator, like the one discussed in Chapter 25, is charging the metal cage to a dangerously high voltage. Yet the demonstrator cannot detect the fact that the cage is electrically charged even while touching the inside of the cage.

How can a closed conducting surface such as this metal cage or an automobile prevent someone from being harmed by lightning or other high-voltage sources?

The answer is in this chapter.

24-1 An Alternative to Coulomb's Law

We associate a vector electric field with a distribution of charges. The electric field has a vector at every location in space telling us what force a test charge q_t will experience at that location. In Sections 23-5 through 23-10 in the last chapter, we used Coulomb's law and the principle of superposition to calculate the electric field vectors at various points in space due to charges that were distributed in different ways. Although Coulomb's law can be used to calculate the electric force (and hence electric field) exerted on a test charge by any possible arrangement of charges we could imagine, this is usually a very difficult task. For example, even calculating the electric field outside the surface of a hollow, charged, conducting sphere would require us to do a triple integration.

In Chapter 23 we used Coulomb's law to find electric fields from charge distributions, but what if we want to turn our calculation around and determine a distribution of charges from an electric field pattern? Unless our distribution of charges is very simple, this reverse calculation is also difficult to perform using Coulomb's law. Thus Coulomb's law appears to be valid but difficult to use in many circumstances. In this chapter we introduce Gauss' law as another method for relating a known electric field to the charge distribution generating it and, conversely, for relating a known charge distribution to its associated electric field. Gauss' law in the integral form discussed in this chapter allows us to find electric fields easily for very symmetrical charge distributions.

To explore how we might find a general relationship between a collection of charges and their electric field, let's consider the electric field associated with the simplest possible charge distribution—a point charge (see Fig. 24-1). By applying Coulomb's law we have already found that the magnitude of the charge's electric field *decreases* as the inverse square of the distance r, as expressed in Eq. 23-8,

$$E = |\vec{E}| = k\frac{|q|}{r^2}.$$

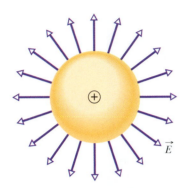

FIGURE 24-1 ■ If a single charge is located at the center of an imaginary sphere, Coulomb's law tells us the magnitudes of the electric field vectors are the same at all points on the surface of the sphere and the direction of each electric field vector is normal (perpendicular) to the surface. Only the field vectors that lie in the plane of the page are shown in this drawing.

However, if we construct an imaginary spherical surface around our source charge we find that the surface area of the sphere *increases* as the square of the distance of the spherical surface from the source charge. The equation for the surface area is given by $A = 4\pi r^2$. Thus, we see that the product of the electric field magnitude and the surface area of any imaginary spherical boundary is constant no matter how large or small the distance from the charge is, as shown in Eq. 24-1,

$$EA = k\frac{|q|}{r^2}(4\pi r^2) = \frac{1}{4\pi\varepsilon_0}\frac{|q|}{r^2}(4\pi r^2) = \frac{|q|}{\varepsilon_0}. \tag{24-1}$$

Here we use Eq. 22-7 to replace the electrostatic constant k with $1/4\pi\varepsilon_0$ where ε_0 is the electric (or permittivity) constant.

Equation 24-1 is remarkable for two reasons. First, as the electric field magnitude gets smaller, the area over which it can act gets larger by exactly the same factor. Second, the product of the electric field magnitude anywhere on a spherical surface and the area of the spherical surface is *proportional* to the amount of charge $|q|$ enclosed by that surface. Does this proportionality still exist when the closed surface takes on other shapes? These questions were addressed by German mathematician and physicist Carl Friedrich Gauss (1777–1855). We begin our study of Gauss' approach to relating charge distributions, electric fields, and closed surfaces to each other by defining a new quantity called electric flux.

24-2 Electric Flux

For the case of a single point charge at the center of an imaginary sphere, Eq. 24-1 tells us that the product of the electric field magnitude (at the surface of a sphere) and the surface area of the sphere are proportional to the charge. This product EA is known as the **electric flux** through the sphere. In our simple situation the directions of electric field vectors created by the point charge happen to be normal (that is, perpendicular) to the surface of our imaginary sphere at all points along its surface. What if we have a complex array of charges or decide to surround our charge with an imaginary enclosure with a different shape? In that case we need to break our surface into little elements of area and find the component of the electric field vector that is normal to each area element as depicted in Fig. 24-2. We took a similar approach in Section 15-10 in defining *volume flux* for fluids flowing in pipes and streams. If the definitions of volume flux and normal vector for an area are not familiar to you, we suggest you read this earlier section.

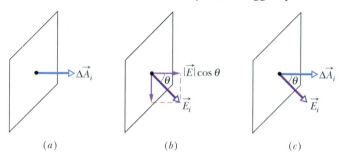

(a)   (b)   (c)

FIGURE 24-2 ■ (a) A small area vector element $\Delta \vec{A}$ is perpendicular to the plane of a square loop of area A with a magnitude of A. (b) The component of $\vec{E}$ perpendicular to the plane of the loop is $|\vec{E}| \cos \theta$, where θ is the angle between $\vec{E}$ and a normal to the plane. (c) The area vector $\Delta \vec{A}$ makes an angle θ with $\vec{E}$.

If we know the nature of the velocity vector field, $\vec{v}$, characterizing the motion of the fluid, we can use the definition of volume flux presented in Chapter 15 to calculate the amount of fluid flowing through any very small element, $\Delta \vec{A}_i$, of a larger surface area*. If we look at the ith element of a larger area, the *volume flux* element, Φ_i, for that small area is defined as the scalar or dot product of the normal vector representing an area element and the velocity vector at the location of the area element as shown in Eq. 15-33,

$$\Phi_i \equiv \vec{v}_i \cdot \Delta \vec{A}_i \qquad \text{(volume flux definition for a small area element).}$$

What is a normal vector? Recall that we defined the normal vector to a small flat area to allow us to represent both the magnitude and the orientation of an element of area. If the element of area is part of a closed surface completely surrounding a space, we define the normal vector to be pointing *out* of the surface (Fig. 24-3). The normal vector points at right angles, or normal, to the plane of the area and has a magnitude equal to the area (Fig. 24-4).

Although *electric flux* does not involve the flow of anything, we define it in a way mathematically analogous to volume flux introduced in Chapter 15. An **electric flux element** is defined as the dot product of the normal vector representing an area element and the electric field vector at the location of the area element as shown in Fig. 24-2 and in Eq. 24-2,

$$\Phi_i \equiv (E_i)(\Delta A_i) \cos \theta = \vec{E}_i \cdot \Delta \vec{A}_i \qquad \text{(electric flux definition for a small area),} \qquad (24\text{-}2)$$

where E_i and ΔA_i are magnitudes while θ is the angle between the two vectors. If a curved surface like the one in Fig. 24-3 is broken into small area elements, each of the $\Delta \vec{A}_i$ vectors can point in different directions.

───────────

*Our use of the symbol ΔA_i instead of just A_i is to signify that the areas are very small. In this context, the delta does not signify change.

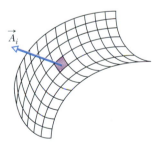

FIGURE 24-3 ■ In order to make net flux calculations, a curved surface area must be divided into N small area elements. Each element must be small enough so it is essentially flat and has electric field vectors that have the same magnitude and direction at every location on a given surface element. The ith area element and its normal vector are shown assuming that an outside piece of a closed surface is being shown here.

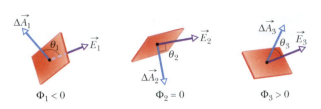

FIGURE 24-4 ■ Three small areas that subtend different angles with respect to various electric field vectors. The first flux element is negative, the second zero, and the third positive. Note that nothing is "flowing" in the case of electric flux to exist.

As is the case for volume flux, if our area is not small enough to be considered as flat or if the electric field vectors are not uniform over the area we choose, then we must break the area into smaller elements that are essentially flat (Fig. 24-4). We can then determine the net electric flux as the sum of individual flux elements. For N flux elements, this is given by

$$\Phi^{net} = \Phi_1 + \Phi_2 + \cdots + \Phi_N$$
$$= \vec{E}_1 \cdot \Delta\vec{A}_1 + \vec{E}_2 \cdot \Delta\vec{A}_2 + \cdots + \vec{E}_N \cdot \Delta\vec{A}_N \quad \text{(net electric flux)}, \quad (24\text{-}3)$$
$$= \sum_{n=1}^{N} \vec{E}_n \cdot \Delta\vec{A}_n$$

where $\vec{E}_1$, $\vec{E}_2$, $\vec{E}_3$, and so on represent the electric field vectors at the location of each of the N area elements. The flux associated with an electric field is a scalar, and its SI unit is the newton-meter-squared per coulomb or $[\text{N} \cdot \text{m}^2/\text{C}]$.

Some possible orientations for area elements and electric field vectors needed to calculate electric flux elements are shown in Fig. 24-4.

In everyday language the term flux is often used to represent flow or change. This is suggested by expressions such as "an influx of population" or "the economy is in a state of flux." These popular uses of the word flux can be deceptive when applied to electrostatic phenomena that we are dealing with in Chapters 22 through 25. Electric flux can be defined whenever an electric field exists, even when an electric field is static and not changing. Furthermore, even if a redistribution of charges causes an electric field to change over time, the changing flux associated with electric field is not related to the flow of anything.

> Instead of representing change or flow, **electric flux** at an area represents the summation over a surface of flux elements. Each flux element represents the product of an essentially flat area element on the surface and the component of the electric field vector that lies along the normal to that area element.

READING EXERCISE 24-1: The figure shows two situations in which the angle between a field vector and the normal vector representing the orientation of the area is $\theta = 60°$. Assume the magnitude of the area in each case is $\Delta A = 2 \times 10^{-4} \text{ m}^2$. (a) If the imaginary area element is placed at a location in a stream where the magnitude of the stream velocity is $v = 3$ m/s, what is the volume flux through the area? Is anything flowing through the area element? If so, what? (b) Suppose the imaginary area element is placed in an electric field where the magnitude of the field vector is $E = 3$ N/C. What is the electric flux through the area element? Is anything flowing through the area element? If so, what? ■

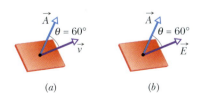

24-3 Net Flux at a Closed Surface

In the introductory section we posed the question of whether there is a proportionality between an enclosed charge distribution and the flux at a surface that encloses it. To answer this question we need to examine carefully the procedures for determining net electric flux at an imaginary surface that encloses charges. The word "enclose" is important here. In the discussion that follows, we will not be discussing calculations of electric flux at any arbitrary surface. We will limit our discussion to the electric flux at closed surfaces that are continuous and connected. That is, a **closed** surface must be without cuts or edges. Nothing can get into or out of such surfaces without passing through the surface itself.

In order to define the net electric flux at any closed surface, consider Fig. 24-5, which shows an arbitrary (irregularly shaped) imaginary surface immersed in a *nonuniform* electric field. For historical reasons, any imaginary closed surface used in the calculation of a net electric flux is called a **Gaussian surface.** Since the electric field vector might be different at each location on our Gaussian surface, we must divide the entire surface into small area elements and take the sum as shown in Eq. 24-3.

Let's consider the arbitrary closed surface shown in Fig. 24-5. The vectors $\Delta\vec{A}_i$ and $\vec{E}_i$ for each square have some angle θ_i between them. Figure 24-5 shows an enlarged view of three small squares (1, 2, and 3) on the Gaussian surface, and the angle θ_i between $\vec{E}_i$ and $\Delta\vec{A}_i$. Our net flux equation (Eq. 24-3) instructs us to visit each square on the Gaussian surface, to evaluate the scalar product $\vec{E}_i \cdot \Delta\vec{A}_i$ at the location of each, and to sum the results algebraically (that is, with signs included) for all the squares that make up the surface. The sign or a zero resulting from each scalar product determines whether the flux at a square is positive, negative, or zero. Squares like 1, in which $\vec{E}_1$ points inward, make a negative contribution to the sum. Squares like 2, in which $\vec{E}_2$ lies in the surface, make zero contribution. Squares like 3, in which $\vec{E}_3$ points outward, make a positive contribution. (Note that the particular signs for the flux elements discussed above are a consequence of the convention adopted on the previous page; the area vectors point outward for closed surfaces.)

The exact definition of the flux of the electric field at a surface is found by allowing the area of the squares shown in Fig. 24-5 to become smaller and smaller, approaching a differential limit dA. The normal vectors for each tiny surface area then approach a differential limit $d\vec{A}$. Thus, the electric flux at a closed surface is given by the integral of the electric field components parallel to the normal of each surface area element over the magnitude of each surface area element. In mathematical notation the equation for electric flux becomes

$$\Phi^{\text{net}} \equiv \lim_{\Delta\vec{A}\to 0} \sum_{i=1}^{N} \vec{E}_i \cdot \Delta\vec{A}_i$$

$$= \oint \vec{E} \cdot d\vec{A} \qquad \text{(net electric flux at a Gaussian surface).} \qquad (24\text{-}4)$$

The circle on the integral sign indicates that the integration is to be taken over the entire closed surface (Gaussian surface).

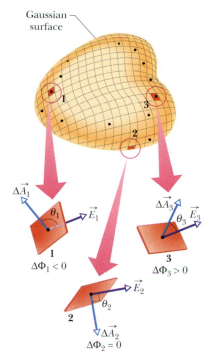

FIGURE 24-5 ■ A Gaussian surface of arbitrary shape is immersed in an electric field. The surface is divided into small area elements. The electric field vectors and the area vectors are shown for three representative area elements marked 1, 2, and 3. The other electric field vectors are not shown.

TOUCHSTONE EXAMPLE 24-1: Net Flux for a Uniform Field

Figure 24-6 shows a Gaussian surface in the form of a cylinder of radius R immersed in a uniform electric field $\vec{E}$, with the cylinder axis parallel to the field. What is the flux Φ^{net} of the electric field through this closed surface?

FIGURE 24-6 ■ A cylindrical Gaussian surface, closed by end caps, is immersed in a uniform electric field. The cylinder axis is parallel to the field direction.

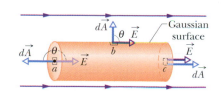

SOLUTION ■ The **Key Idea** here is that we can find the flux Φ through the surface by integrating the scalar product $\vec{E} \cdot d\vec{A}$ over the Gaussian surface. We can do this by writing the flux as the sum of three terms: integrals over the left disk cap a, the cylinder surface b, and the right disk cap c. Thus, from Eq. 24-4,

$$\Phi^{\text{net}} = \oint \vec{E} \cdot d\vec{A}$$

$$= \int_a \vec{E} \cdot d\vec{A} + \int_b \vec{E} \cdot d\vec{A} + \int_c \vec{E} \cdot d\vec{A}. \qquad (24\text{-}5)$$

For all points on the left cap, the angle θ between $\vec{E}$ and $d\vec{A}$ is 180°, and the magnitude E of the field is constant. Thus,

$$\int_a \vec{E} \cdot d\vec{A} = \int E(\cos 180°)\, dA = -E \int dA = -EA,$$

where $\int dA$ gives the cap's area, $A(= \pi R^2)$. Similarly, for the right cap, where $\theta = 0$ for all points,

$$\int_c \vec{E} \cdot d\vec{A} = \int E(\cos 0°)\, dA = E \int dA = +EA.$$

Finally, for the cylindrical surface, where the angle θ is 90° at all points,

$$\int_b \vec{E} \cdot d\vec{A} = \int E(\cos 90°)\, dA = 0.$$

Substituting these results into Eq. 24-5 leads us to

$$\Phi = -EA + EA = 0. \qquad \text{(Answer)}$$

This result is perhaps not surprising because the field lines that represent the electric field all pass entirely through the Gaussian surface, entering through the left end cap, leaving through the right end cap, and giving a net flux of zero.

TOUCHSTONE EXAMPLE 24-2: Flux for a Nonuniform Field

A *nonuniform* electric field given by $\vec{E} = (3.0 \text{ N/C} \cdot \text{m})x\hat{i} + (4.0 \text{ N/C})\hat{j}$ pierces the Gaussian cube shown in Fig. 24-7. What is the electric flux through the right face, the left face, and the top face?

SOLUTION ■ The **Key Idea** here is that we can find the flux Φ through the surface by integrating the scalar product $\vec{E} \cdot d\vec{A}$ over each face.

Right face: An area vector $\vec{A}$ is always perpendicular to its surface and always points away from the interior of a Gaussian surface. Thus, the vector $d\vec{A}$ for the right face of the cube

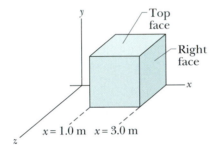

FIGURE 24-7 ■ A Gaussian cube with one edge on the x axis lies within a nonuniform electric field.

must point in the positive x direction. In unit vector notation, then,

$$d\vec{A} = dA\,\hat{i}.$$

From Eq. 24-4, the flux Φ_r through the right face is then

$$\Phi_r = \int \vec{E} \cdot d\vec{A} = \int [(3.0 \text{ N/C} \cdot \text{m})x\hat{i} + (4.0 \text{ N/C})\hat{j}] \cdot (dA\,\hat{i})$$

$$= \int [(3.0 \text{ N/C} \cdot \text{m})(x)(dA)\hat{i} \cdot \hat{i} + (4.0 \text{ N/C})(dA)\hat{j} \cdot \hat{i}]$$

$$= \int (3.0 \text{ N/C} \cdot \text{m})x\, dA + (0.0 \text{ N} \cdot \text{m}^2/\text{C}) = (3.0 \text{ N/C} \cdot \text{m}) \int x\, dA.$$

We are about to integrate over the right face, but we note that x has the same value everywhere on that face—namely, $x = 3.0$ m. This means we can substitute that constant value for x. Then

$$\Phi_r = (3.0 \text{ N/C} \cdot \text{m}) \int (3.0 \text{ m})\, dA = (9.0 \text{ N/C}) \int dA.$$

Now the integral merely gives us the area $A = 4.0 \text{ m}^2$ of the right face, so

$$\Phi_r = (9.0 \text{ N/C})(4.0 \text{ m}^2) = 36 \text{ N} \cdot \text{m}^2/\text{C}. \qquad \text{(Answer)}$$

Left face: The procedure for finding the flux through the left face is the same as that for the right face. However, two factors change. (1) The differential area vector $d\vec{A}$ points in the negative x direction and thus $d\vec{A} = -dA\,\hat{i}$. (2) The term x again appears in our integration, and it is again constant over the face being considered. However, on the left face, $x = 1.0$ m. With these two changes, we find that the flux Φ_l through the left face is

$$\Phi_l = -12 \text{ N} \cdot \text{m}^2/\text{C}. \qquad \text{(Answer)}$$

Top face: The differential area vector $d\vec{A}$ points in the positive y direction and thus $d\vec{A} = dA\,\hat{j}$. The flux Φ_t through the top face is then

$$\Phi_t = \int [(3.0 \text{ N/C} \cdot \text{m})x\hat{i} + (4.0 \text{ N/C})\hat{j}] \cdot (dA\,\hat{j})$$

$$= \int [(3.0 \text{ N/C} \cdot \text{m})(x\, dA)\hat{i} \cdot \hat{j} + (4.0 \text{ N/C})(dA)\hat{j} \cdot \hat{j}]$$

$$= (0.0 \text{ N} \cdot \text{m}^2/\text{C}) + \int (4.0 \text{ N/C})\, dA) = (4.0 \text{ N/C}) \int dA$$

$$= 16 \text{ N} \cdot \text{m}^2/\text{C}. \qquad \text{(Answer)}$$

24-4 Gauss' Law

Let's return for a moment to the consequence of Coulomb's law we presented in the first section, where we surrounded a single charge with a spherical Gaussian surface. We found that a flux-like quantity (namely, the product of the magnitude of the electric field at the sphere's surface multiplied by the area of the sphere's surface) is equal to a constant times the enclosed charge. The surprising thing is this is true no matter what the radius of the sphere is, because the amount by which the surface area of the sphere increases just compensates for the amount by which the electric field magnitude decreases. This suggests that the net flux through a Gaussian surface of any shape enclosing a single charge will be proportional to the amount of charge enclosed.

Visualizing Flux through a Gaussian Surface

Since the relationship between flux and charge enclosed by a Gaussian surface is hard to visualize in three dimensions, let's consider the special case of an infinitely long rod that has a uniform charge density. While infinitely long rods do not exist, our result will be valid providing the Gaussian surfaces are far from the ends of the rod. Imagine a Gaussian surface that has the shape of a coin and surrounds a small segment of the rod. This is shown in Fig. 24-8.

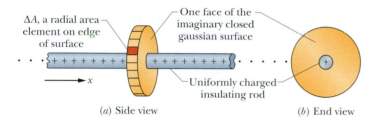

ΔA, a radial area element on edge of surface

One face of the imaginary closed gaussian surface

Uniformly charged insulating rod

(a) Side view

(b) End view

FIGURE 24-8 ■ (a) An infinitely long uniformly charged rod has a Gaussian surface that looks like a coin with front and back faces that are perpendicular to the rod and enclose a small charge. (b) An end view of the rod and Gaussian surface face can help us visualize flux at the surface's edges.

Because the charged rod is infinitely long it is symmetric about any point on it. As we showed in Section 23-5 (see Fig. 23-12), it turns out the electric field vectors created by a symmetric pair of charges point outward in a radial direction and have no components parallel to the line that the charges lie on (in this case, the line determined by the rod). We can also show that for a thin rod the field magnitude falls off as $1/r$ where r is the radial distance from the center of the rod. (Likewise, a similar negatively charged rod has electric field vectors pointing radially inward). The key factor in surrounding a piece of long rod with a coin-shaped closed surface is that all the flux at the surface will be at the edges and there will be no flux at the faces of the surface. For this reason, we can calculate and depict the "amount" of flux at elements of area on the edges of the surface by looking at an end view of the rod. This is true not only for coin-like closed surfaces that have circular faces but also for any shaped faces so long as the two faces are parallel to each other and perpendicular to the rod. End views depicting flux amounts as green rectangles are shown in Fig. 24-9 for three different imaginary Gaussian surfaces outlined in red.

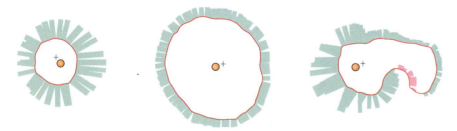

FIGURE 24-9 ■ Three imaginary Gaussian surfaces surround the same point charge. Here the red lines show only a two-dimensional cross section of three-dimensional surfaces. The contribution of electric flux at a series of small area elements is calculated and represented by rectangles. Note that whenever part of a surface is close to the charge, the flux elements are bigger but there are fewer of them. We can see visually that the net flux (which is proportional to the area occupied by all the outgoing flux (shown as green) minus the incoming flux (shown as pink) is approximately the same in the three cases.

A small bundle of enclosed charge yields the same net electric flux at a Gaussian surface no matter what the shape of the surface. By superposition, if there are two charges enclosed by a Gaussian surface, each charge contributes its proportional share to the net flux no matter where each of the charges is located, provided both are *inside* the Gaussian surface. This leads us to a statement of Gauss' law that describes a plausible general relationship between the net flux through a Gaussian surface of any shape and the total enclosed charge no matter how it is distributed.

> **GAUSS' LAW:** The net flux through any imaginary closed surface is directly proportional to the net charge enclosed by that surface.

Based on consideration of SI units, the constant of proportionality must be $1/\varepsilon_0$ where ε_0 is the permittivity constant, so that the mathematical expression of Gauss' law is

$$\Phi^{net} = \frac{q^{enc}}{\varepsilon_0} \qquad \text{(Gauss' law)}. \qquad (24\text{-}6)$$

By substituting the definition of electric flux at a Gaussian surface, $\Phi^{net} \equiv \oint \vec{E} \cdot d\vec{A}$, we can also write Gauss' law as

$$\Phi^{net} = \oint \vec{E} \cdot d\vec{A} = \frac{q^{enc}}{\varepsilon_0} \qquad \text{(Gauss' law)}. \qquad (24\text{-}7)$$

Here, the circle on the integral sign indicates that the surface over which we integrate must be "closed." The use of the permittivity constant for a vacuum, ε_0, in Eqs. 24-6 and 24-7 indicates this form of Gauss' law only holds when the net charge is located in air or some other medium that doesn't polarize easily. In Section 28-6, we modify Gauss' law to include situations in which so-called dielectric materials that can polarize, such as paper, oil, or water, are present. In Fig. 24-10 we show how the net flux can have the same value for two different charge distributions involving the same amount of enclosed charge.

Gauss' law is useful for finding both charge and flux. That is, if we can calculate the net flux through a closed surface, we can deduce the amount of charge enclosed. On the other hand, if we know the amount of charge enclosed, we can use Gauss' law to deduce the net flux through any surface that encloses the charge.

Interpreting Gauss' Law

One use of Gauss' law is to calculate how much net charge is contained inside any closed surface. To make the calculation, you need know only the net electric flux at the surface enclosing the collection of charges. This net flux is related to the strength of the normal components of the electric field at all locations on the surface.

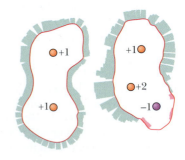

FIGURE 24-10 ■ Each Gaussian surface encloses a different charge distribution but encloses the same net charge. The electric flux calculated at the edges of the surface is represented by green rectangles (outward flux) or pink rectangles (inward flux). The total space covered by all of the green rectangles minus that occupied by the pink rectangles turns out to be the same for the two situations, which is compatible with the predictions of Gauss' law.

In Eqs. 24-6 and 24-7, the net charge q^{enc} is the algebraic sum of all the *enclosed* positive and negative charges, and it can be positive, negative, or zero. We include the sign, rather than just use the amount of enclosed charge, because the sign tells us something about the net flux at the Gaussian surface. Here we continue to use our convention that the normal area vectors representing the area elements of a closed surface point *outward*. If the net charge enclosed, q^{enc}, is positive, its electric field vectors point mostly outward too. This leads to a net flux that is *outward* and positive as shown in Fig. 24-11a. If q^{enc} is negative, the area vector still points outward but the electric field vector points inward. This leads to a net flux that is *inward* and negative, as shown in Fig. 24-11. Figure 24-11c shows how positive and negative charges inside a Gaussian surface can lead to zero net flux.

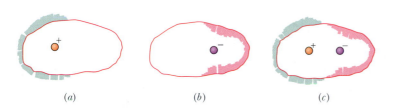

(a)　　　　(b)　　　　(c)

FIGURE 24-11 ■ Each of these Gaussian surfaces has the same shape. (a) One unit of enclosed positive charge causes a positive net outward flux shown in green. (b) One unit of enclosed negative charge causes a negative net inward flux shown in pink. Note that the amount of negative flux is the same as the amount of positive flux shown in the previous diagram. (c) If both the positive and negative charges are enclosed the net charge is zero and so is the net flux.

Charge outside a Gaussian surface, no matter how large or how close it may be, is not included in the term q^{enc} in Gauss' law. We expect this since there is no source of electric field inside the surface, and negative and positive flux elements will cancel each other, as shown in Fig. 24-12. The exact form or location of the charges inside the Gaussian surface is also of no concern; the only things that matter are the amount of the net charge enclosed and its sign. The quantity $\vec{E}$ on the left side of Eq. 24-7, however, is the electric field resulting from *all* charges, both those inside and those outside the Gaussian surface. This may seem to be inconsistent, but keep in mind the electric field due to a charge outside the Gaussian surface contributes zero net flux on the surface (as shown in Fig. 24-12). This is the case even though a charge outside the surface does contribute to the actual values of the electric field at each point on the surface.

FIGURE 24-12 ■ A charge element along a rod is located *outside* a Gaussian surface. When the electric flux is calculated at each area element using Coulomb's law, its outward values are represented by green rectangles and the inward flux by pink rectangles. The net flux is zero because the negative inward flux at the portion of the surface near the charge just cancels the positive outward flux at the location of the portions of the surface far away from the charge.

Let us apply these ideas to Fig. 24-13, which shows the electric field lines surrounding two point charges, equal in amount but opposite in sign. Four Gaussian surfaces are also shown, in cross section. Let us consider each in turn.

Surface S_1 (encloses only the positive charge): The electric field is dominated by the nearby positive charge and so points outward for the majority of the points on this surface. Thus, the flux of the electric field at this surface is positive, and so is the net charge within the surface, as Gauss' law requires. (That is, if Φ is positive, q^{enc} must be also.)

Surface S_2 (encloses only the negative charge): The electric field is dominated by the nearby negative charge and so points inward for the majority of the points on this surface. Thus, the flux of the electric field is negative and so is the enclosed charge, as Gauss' law requires.

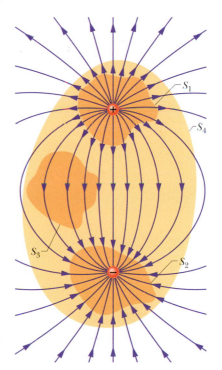

FIGURE 24-13 ■ An idealization showing two point charges of equal amount and opposite sign are shown with the field lines that depict their net electric field as if all lines lie in a plane. The cross sections of four Gaussian surfaces are shown. Surface S_1 encloses the positive charge, S_2 encloses the negative charge, and S_3 encloses no charge. Since S_4 surrounds both charges, it encloses no net charge.

Surface S_3 (encloses no charges): Since $q^{\text{enc}} = 0$ and there are comparable contributions to the electric field at points on the surface from both charges, the field on some parts of the surface will point out and on other parts it will point in. Gauss' law (Eq. 24-7) requires the net electric flux through this surface to be zero. That is reasonable because in calculating the net flux, the inward and outward flux elements cancel each other.

Surface S_4 (encloses both charges): This surface encloses no *net* charge, because equal amounts of positive and negative charge are enclosed. Gauss' law requires the net flux of the electric field at this surface be zero. That is reasonable because in this case the field vectors point outward for the portion of the surface nearest to the positive charge (yielding positive flux) and inward for the portion of the surface near the negative charge (yielding negative flux). In calculating the net flux, the positive and negative flux elements cancel each other, even though the field is nonzero along most of the surface.

What would happen if we were to bring an enormous charge Q up close to (but still outside of) surface S_4 in Fig. 24-13? The pattern of the electric field would certainly change, but the net flux for the four Gaussian surfaces would not change. We can understand this because the inward and outward flux elements associated with the added Q at any of the four surfaces would cancel each other, making no contribution to the net flux at any of them. The value of Q would not enter Gauss' law in any way, because Q lies outside all four of the Gaussian surfaces that we are considering.

READING EXERCISE 24-2: The figure shows three situations in which a Gaussian cube sits in an electric field. The arrows indicate the directions of the electric field vectors for the top, front, and right faces of each cube. The flux at the six sides of each cube is listed in the table below. In which situations do the cubes enclose (a) a positive net charge, (b) a negative net charge, and (c) zero net charge?

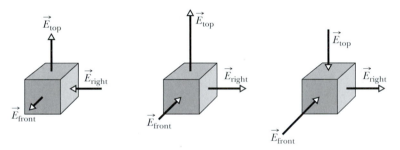

Flux [N · m²/C]

Face	Cube 1	Cube 2	Cube 3
Front	+2	−4	−7
Back	−3	−3	+8
Left	+7	+3	+2
Right	−4	+5	+5
Top	+5	+10	−6
Bottom	−7	−6	−5

24-5 Symmetry in Charge Distributions

Why go through all this trouble to develop a method of calculating electric fields that is equivalent to Coulomb's law? We suggested in the introduction to this chapter that it is because Gauss' law makes it possible to calculate the field for highly symmetric charge distributions. What we mean by symmetric charge distributions are arrangements of charges that can be rotated about an axis or reflected in a mirror and still look the same. Figure 24-14 shows several examples of symmetric objects.

Why do charge distributions need to be symmetric in order for Gauss' law to be helpful in finding an electric field? Because we can use symmetry arguments to find the direction of the electric field and surfaces along which it is constant. This allows us to choose an imaginary Gaussian surface over which the electric field is constant. Then we can take the dot product and turn the vectors into scalar magnitudes. Finally, we know the electric field magnitude is constant at the surface we are integrating over, so we can pull the electric field vector outside of the integral sign. By following the steps we outlined, in some cases Gauss' law can be reduced to

$$\varepsilon_0 \oint E \cos\theta \, dA = \varepsilon_0 E \oint \cos\theta \, dA = |q^{\text{enc}}|.$$

Better still is to be able to find a Gaussian surface over which both the electric field and the angle between the field and area vectors, θ, are constant over the entire area. In that case, both the electric field and the cosine functions can be moved outside the integral and Gauss' law reduces to:

$$(\varepsilon_0 E \cos\theta) \oint dA = |q^{\text{enc}}|.$$

This expression is very easy to evaluate because the integral of dA is simply the magnitude of the total area of the Gaussian surface, which we will denote as A. Hence, if we can find a Gaussian surface over which the field and angle θ are constant, Gauss' law allows us to calculate the electric field of an extended charge distribution without doing an integral. In those cases, Gauss' law tells us that the electric field magnitude is

$$E = \frac{|q^{\text{enc}}|}{\varepsilon_0 A \cos\theta} \qquad \text{(constant } E \text{ and } \theta\text{)}, \tag{24-8}$$

where A is the area of the Gaussian surface, θ is the angle between the field and each area vector, and q^{enc} is the net charge *enclosed by the Gaussian surface*. In some cases where the angle, θ, has one value for some parts of a surface and another value for other parts of a surface, we can handle the calculation by breaking the surface integral into parts.

A word of caution: There are only a few charge distributions with sufficient symmetry for Gauss' law to be useful. These include single point charges and spherically symmetric ones. Charge distributions that work with Gauss' law also include the infinitely long cylinder, with cylindrical symmetry, and that of a uniformly charged slab with infinitely long sides with planar symmetry. Fortunately, there are many physical situations for which these geometries are important. Hence, Gauss' law is an extraordinarily useful tool.

However, for many charge distributions, we cannot use Gauss' law to find the field because the flux integral on the left-hand side of the expression

$$\varepsilon_0 \oint \vec{E} \cdot d\vec{A} = q^{\text{enc}}$$

is too complicated to evaluate. In these cases, Gauss' law is still valid but not useful.

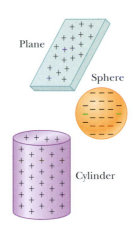

FIGURE 24-14 ■ Some symmetrically charged objects—a plane, a sphere, and a cylinder.

24-6 Application of Gauss' Law to Symmetric Charge Distributions

As we determined in the last section, Gauss' law is useful if we already know what the general shape of the vector electric field plot looks like. In some cases we can derive this knowledge from symmetry of the charge distribution without using equations or doing calculations. Only then can we choose an imaginary closed surface and use the mathematical form of Gauss' law to calculate the magnitude of the electric field at points on the surface. In this section we take this approach to determining the electric field for three highly symmetric charge distributions.

Spherical Symmetry for a Shell of Charge

Figure 24-15 shows a charged spherical shell of total positive charge q and radius R and two concentric spherical Gaussian surfaces, S_1 and S_2. (Note that we chose the shape of the Gaussian surface to mirror the symmetry of the charge distribution.) Because the charge distribution is spherically symmetric no matter how we rotate the spherical shell around its center, the shell looks the same. This means that the electric field must have a spherical symmetry too. Thus, it must have the same magnitude at every point on the spherical Gaussian surface S_2 and it must point in a radial direction. Further, since the area vector points radially outward at all points on S_2, the angle between the electric field $\vec{E}$ and the area $\vec{A}$ is constant. As a result of the spherical symmetry of the distributed charge, we know the electric field also points in a radial direction at all points on S_2. Hence, the angle θ is not only constant but it is also $0°$ at all points on the surface. Applying Gauss' law to surface S_2 then comes down to evaluating the expression for the electric field magnitude that we derived in Eq. 24-8 for constant E and θ,

$$E = \frac{|q^{enc}|}{\varepsilon_0 A \cos \theta}.$$

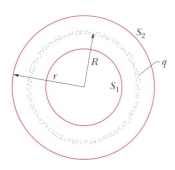

FIGURE 24-15 ◾ A thin, charged, spherical shell with total charge q, in cross section. Two Gaussian surfaces S_1 and S_2 are also shown in cross section. Surface S_2 encloses the shell, and S_1 encloses only the empty interior of the shell.

Note that $\cos \theta = \cos 0 = 1$ and the area of a sphere (the Gaussian sphere) of radius r is $4\pi r^2$. Hence, for any $r \geq R$, we find that

$$E = \frac{1}{4\pi\varepsilon_0}\frac{|q^{enc}|}{r^2} \qquad \text{(spherical shell, field at } r \geq R \text{).} \qquad (24\text{-}9)$$

What is surprising is that outside the shell the electric field is the same as if the shell of charge were replaced by a single point-like charge, q, provided that the single charge is placed where the center of the shell of charge was. Thus, if the charge on a shell is evenly distributed, a shell of total charge q would produce the same force on a small test charge placed anywhere *outside* the shell as a single point-like charge q would.

> A shell with a uniform charge distribution attracts or repels a charged particle that is outside the shell as if all the shell's charge were concentrated at the center of the shell.

This shell theorem is identical to the one developed by Isaac Newton for gravitation in Section 14-2.

What happens to the electric field inside the shell of charge? Applying Gauss' law to surface S_1, for which $r < R$, leads directly to

$$\vec{E} = 0 \, [N/C] \qquad \text{(spherical shell, field at } r < R \text{),} \qquad (24\text{-}10)$$

because this Gaussian surface encloses no charge. Thus, when a small test charge is enclosed by a shell of uniform charge distribution, the shell exerts no net electrostatic force on it.

> A shell of uniform charge exerts no electrostatic force on a charged particle that is located inside the shell.

A Spherically Symmetric Charge Distribution

Any spherically symmetric charge distribution, such as that of Fig. 24-16, can be constructed with a nest of concentric spherical shells. This is a good starting point for treating a wide variety of charged objects with nearly spherical distribution of charge such as nuclei and atoms. For purposes of applying the two shell theorems stated above, the volume charge density ρ, defined as the charge per unit volume, should have a single value for each shell but need not be the same from shell to shell. Thus, for the charge distribution as a whole, ρ, can vary only with r, the radial distance from the center of the sphere and not with direction. We can then examine the effect of the charge distribution "shell by shell."

In Fig. 24-16a the entire charge lies within a Gaussian surface with $r > R$. The charge produces an electric field on the Gaussian surface as if the charge were a point charge located at the center, and Eq. 24-9 holds.

Figure 24-16b shows a Gaussian surface with $r < R$. To find the electric field at points on this Gaussian surface, we consider two sets of charged shells—one set inside the Gaussian surface and one set outside. The charge lying *outside* the Gaussian surface does not set up a net electric field on the Gaussian surface. Gauss' law tells us that the charge *enclosed* by the surface sets up an electric field as if that enclosed charge were concentrated at the center. Letting q' represent that enclosed charge, we can then write the electric field magnitude as

$$|\vec{E}| = \frac{1}{4\pi\varepsilon_0} \frac{|q^{\text{enc}}|}{r^2} = \frac{1}{4\pi\varepsilon_0} \frac{|q'(r)|}{r^2} \qquad \text{(spherical distribution, field at } r < R\text{),} \quad (24\text{-}11)$$

where the term $q'(r)$ signifies that q' depends on r. (It is not the product of q' and r.)

Equation 24-11 is valid for any spherically symmetric charge distribution, even one that is not uniform. For example, Fig. 24-16 shows a situation in which the volume charge density is spherically symmetric but larger near the center of the sphere than further out. In other words, Eq. 24-11 is valid whenever $\rho = \rho(r)$ or ρ is a constant. But the equation is not useful unless we know how to use a knowledge of the volume charge density to determine the charge q' enclosed by a sphere of radius r.

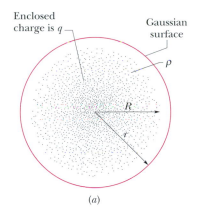

(a)

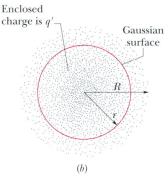

(b)

FIGURE 24-16 ■ Spherically symmetric distributions of charge of radius R, whose volume charge density ρ is a function only of distance from the center. The charged object is not a conductor, so the charge is assumed to be fixed in position. A cross-section of concentric spherical Gaussian surface with $r > R$ is shown in (a). A similar Gaussian surface with $r < R$ is shown in (b).

Spherical Symmetry for a Uniform Volume Charge Distribution

Consider the simple case where the charge is distributed uniformly through the volume of a sphere of radius R containing an excess charge q. In this case it is possible to find the magnitude of the electric field at any location inside the sphere in terms of the total charge in the sphere.

Whenever the total charge q enclosed within a sphere of radius R is distributed uniformly, we can use the definition of volume charge density (presented in Table 23-2) and the knowledge that the volume of a sphere of radius R is given by $\frac{4}{3}\pi R^3$ to write

$$\rho \equiv \frac{q}{V} = \frac{q}{\frac{4}{3}\pi R^3}. \qquad (24\text{-}12)$$

Since the charge density is a constant the amount of charge in a smaller sphere of radius r is proportional to its volume. Since its volume is $V' = \frac{4}{3}\pi r^3$ then $q' = \rho V' = \rho\left(\frac{4}{3}\pi r^3\right)$. Substituting Eq. 24-12 for ρ gives

$$q' = q\frac{r^3}{R^3}. \qquad (24\text{-}13)$$

Substituting this into Eq. 24-11 gives us the electric field magnitude in terms of the total charge on the sphere.

$$E = \left(\frac{|q|}{4\pi\varepsilon_0 R^3}\right)r \qquad \text{(uniform volume charge density for } r \le R\text{)}. \qquad (24\text{-}14)$$

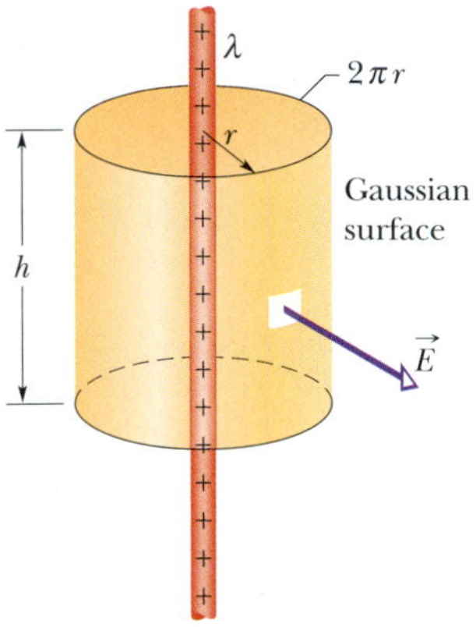

Cylindrical Symmetry for a Uniform Line Charge Distribution

Figure 24-17 shows a section of a very long thin cylindrical plastic rod with a uniform distribution of positive charge, so that linear charge density λ (as defined in Table 23-2) is constant. Let us find an expression for the magnitude of the electric field $\vec{E}$ outside of the rod at a distance r from its axis in terms of the linear charge density of the rod. In doing so, we assume that r is small compared to the length of the rod so that we can ignore the effect of the rod's ends.

We start by choosing a Gaussian surface that matches the cylindrical symmetry of the rod. So our imaginary surface is a circular cylinder of radius r and length h, coaxial with the rod. The Gaussian surface must be closed, so we include two end caps as part of the surface. We pick the end caps of the Gaussian surface so they are far from the end of the rod.

Imagine that, while you are not watching, someone rotates the plastic rod around its longitudinal axis or moves it a finite distance along the axis. When you look again at the rod, you will not be able to detect any change in either the appearance of the rod or the behavior of the electric field that surrounds it. Furthermore, when we experiment, we find that if the rod is flipped end for end we still detect no change in the rod's electric field. What does this tell us about the nature of the electric field? If the electric field has only a component that points radially inward or outward from the rod, then the field should be unaffected by the changes in orientation that we have discussed. If however, the field had any component tangent to the rod's surface, pointed toward or away from the rod, we would detect a change in the electric field as we rotated or flipped the rod. Hence, we conclude from these symmetry arguments that at every point on the cylindrical part of the Gaussian surface, the electric field must have the same magnitude $E = |\vec{E}|$ and must be directed radially outward (for a positively charged rod).

Since $2\pi r$ is the circumference of the cylinder and h is its height, the area A of the cylindrical surface is $2\pi rh$. The flux of $\vec{E}$ at this cylindrical surface is then

$$\Phi = EA\cos\theta = E(2\pi rh)\cos 0 = E(2\pi rh).$$

There is no flux at the end caps because $\vec{E}$, being radially directed, is parallel to the end caps at every point, so $\vec{E}$ is perpendicular to the normal and the dot product vanishes. Thus the flux through the cylindrical surface is equal to the net flux ($\Phi^{net} = \Phi$).

According to Gauss' law, shown in Eq. 24-6,

$$\Phi^{net} = \frac{q^{enc}}{\varepsilon_0}.$$

We can find the enclosed charge in terms of the linear charge density, defined as the charge per unit length. If the charge enclosed by the surface that encompasses a length h of the rod has a uniform density λ, then $q^{enc} = \lambda h$. Thus, the previous two equations reduce to $E(2\pi rh) = |\lambda|h/\varepsilon_0$, so that

$$E = \frac{|\lambda|}{2\pi\varepsilon_0 r} \qquad \text{(long line of uniformly distributed charge)}. \qquad (24\text{-}15)$$

FIGURE 24-17 ■ A Gaussian surface in the form of a closed cylinder surrounds a section of a very long, uniformly charged, cylindrical plastic rod.

This is the expression for the electric field magnitude due to a very long, straight line of uniformly distributed charge, at a point that is a radial distance r from the line. The direction of $\vec{E}$ is radially outward from the line of charge if the charge is positive, and radially inward if it is negative. Equation 24-15 also approximates the field of a *finite* line of charge, at points that are not too near the ends (compared with the distance from the line).

A Sheet of Uniform Charge

Figure 24-18 shows a portion of a thin, very large, sheet with a uniform (positive) surface charge density σ (as defined in Table 23-2). A large sheet of thin plastic wrap, uniformly charged on one side, can serve as a simple example of a nonconducting sheet. A large sheet of aluminum foil serves as an example of a conducting sheet. Let us find the electric field $\vec{E}$ a distance r from the uniformly charged sheet. Here we assume that we are far from the edges of the sheet and that the thickness of the sheet is much less than r.

Even though it doesn't have the same shape as a charged sheet, something called a Gaussian pillbox turns out to make a useful imaginary surface in this case. The pillbox is a closed cylinder with end caps of area A, arranged so that it is perpendicular to the sheet with each end cap located at the same distance from the sheet. This Gaussian pillbox is shown in Fig. 24-18a. Using symmetry (considerations like those used earlier in this section or those depicted in Fig. 23-12 and Fig. 23-13 in the previous chapter), $\vec{E}$ must be perpendicular to the sheet and hence to the end caps. Furthermore, since the charge is positive, $\vec{E}$ is directed *away* from the sheet, and thus the electric field vectors point in an outward direction from the two Gaussian end caps. Because the electric field vectors are perpendicular to the normal vectors on the curved surface, there is no flux at this portion of the Gaussian surface. Thus $\vec{E} \cdot d\vec{A}$ is simply EdA—the product of the magnitudes of $\vec{E}$ and $d\vec{A}$. In this case Gauss' law (Eq. 24-7) gives us

$$\oint \vec{E} \cdot d\vec{A} = q^{\text{enc}}/\varepsilon_0.$$

Since there are two caps on our pillbox we need to break the integral into two parts so in terms of the area and electric field magnitudes,

$$EA + EA = \int_{\substack{\text{end} \\ \text{caps}}} EdA = |q^{\text{enc}}|/\varepsilon_0.$$

Next we can find the amount of charge on the sheet enclosed by our Gaussian pillbox in terms of the surface charge density, σ, on the sheet. Since the surface charge is uniform and the surface charge density is defined as the ratio of the charge on a given surface to its area, we know that $\sigma = q^{\text{enc}}/A$. If we replace q^{enc} in the equation above with σA and solve it for the electric field magnitude we get

$$E = \frac{|\sigma|}{2\varepsilon_0} \qquad \text{(sheet of uniformly distributed charge).} \qquad (24\text{-}16)$$

The equation holds whether the sheet is conducting or nonconducting as long as the layer of charge on the sheet is thin.

Equation 24-16 tells us that *the electric field has the same value for all locations outside a large uniformly charged sheet and points in a direction that is perpendicular to the sheet.* This result is quite surprising! The fact that the net field is perpendicular to the sheet can be explained using symmetry arguments. But how can it be that as you get farther away from the charged sheet the electric field doesn't decrease? The answer lies in considering the influences of the charges as we move away from the sheet. When a test charge is placed very close to the sheet, the

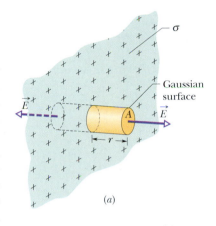

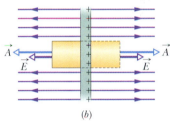

FIGURE 24-18 ■ Perspective view (*a*) and side view (*b*) of a portion of a very large, thin plastic sheet, uniformly charged with surface charge density σ. A closed cylindrical Gaussian surface passes through the sheet and is perpendicular to it.

influence on it by the charge closest to it dominates. If the test charge is moved far-ther from the sheet the influence of the nearest sheet charge gets weaker, but the normal components of the electric field vectors from neighboring sheet charges start to contribute and compensate for the loss of influence of the nearest sheet charge. If the test charge is moved even farther the influence of the nearest and nearby charges diminish but the components of additional surrounding charges come into play and so on.

Equation 24-16 agrees with what we would have found by integration of the elec-tric field components that are produced by individual charges. That would be a very time-consuming and challenging integration, and note how much more easily we ob-tain the result using Gauss' law. This is one reason for devoting a whole chapter to Gauss' law. For certain symmetric arrangements of charge, it is much easier to use it than to integrate field components.

READING EXERCISE 24-3: Consider an array of 9 charges evenly distributed on a square insulating sheet as shown in the diagram. Use symmetry arguments to explain why the electric field vector anywhere on a line normal to the central charge and passing through it has no component that is parallel to the sheet.

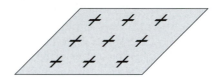

TOUCHSTONE EXAMPLE 24-3: $\vec{E}$ for Two Sheets of Charge

Figure 24-19a shows portions of two large, parallel, nonconducting sheets, each with a fixed uniform charge on one side. The amounts of the surface charge densities are $\sigma_{(+)} = 6.8\,\mu\text{C/m}^2$ for the posi-tively charged sheet and $\sigma_{(-)} = 4.3\,\mu\text{C/m}^2$ for the negatively charged sheet.

Find the electric field $\vec{E}$ (a) to the left of the sheets, (b) be-tween the sheets, and (c) to the right of the sheets.

SOLUTION ■ The **Key Idea** here is that with the charges fixed in place, we can find the electric field of the sheets in Fig. 24-19a by (1) finding the field of each sheet as if that sheet were iso-lated and (2) adding the vector fields of the isolated sheets via the superposition principle. (The vector addition is simple here since the fields lie along the same axis. We can add the fields algebraically because they are parallel to each other.) From Eq. 24-16, the mag-

nitude $E_{(+)}$ of the electric field due to the positive sheet at any point is

$$|\vec{E}_{(+)}| = \frac{|\sigma_{(+)}|}{2\varepsilon_0} = \frac{6.8 \times 10^{-6}\,\text{C/m}^2}{(2)(8.85 \times 10^{-12}\,\text{C}^2/\text{N}\cdot\text{m}^2)}$$

$$= 3.84 \times 10^5\,\text{N/C}.$$

Similarly, the magnitude $|\vec{E}_{(-)}|$ of the electric field at any point due to the negative sheet is

$$|\vec{E}_{(-)}| = \frac{|\sigma_{(-)}|}{2\varepsilon_0} = \frac{4.3 \times 10^{-6}\,\text{C/m}^2}{(2)(8.85 \times 10^{-12}\,\text{C}^2/\text{N}\cdot\text{m}^2)}$$

$$= 2.43 \times 10^5\,\text{N/C}.$$

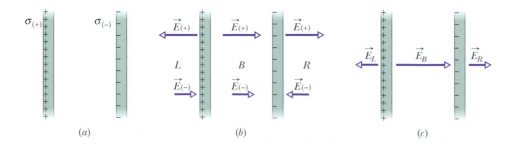

FIGURE 24-19 ■ (a) Two large, parallel insu-lating sheets, uniformly charged on one side. (b) The individual electric fields resulting from the two charged sheets. (c) The net field due to both charged sheets, found by superpo-sition.

Figure 24-19b shows the fields set up by the sheets to the left of the sheets (L), between them (B), and to their right (R).

The resultant fields in these three regions follow from the superposition principle. To the left of the sheets, the field magnitude is

$$|\vec{E}_L| = |\vec{E}_{(+)}| - |\vec{E}_{(-)}|$$
$$= 3.84 \times 10^5 \text{ N/C} - 2.43 \times 10^5 \text{ N/C}$$
$$= 1.4 \times 10^5 \text{ N/C.} \qquad \text{(Answer)}$$

Because $|E_{(+)}|$ is larger than $|E_{(-)}|$, the net electric field $\vec{E}_L$ in this region points to the left, as Fig. 24-19c shows. To the right of the

sheets, the electric field $\vec{E}_R$ has the same magnitude but points to the right, as Fig. 24-19c shows.

Between the sheets, the two fields add and we have

$$|\vec{E}_B| = |\vec{E}_{(+)}| + |\vec{E}_{(-)}|$$
$$= 3.84 \times 10^5 \text{ N/C} + 2.43 \times 10^5 \text{ N/C}$$
$$= 6.3 \times 10^5 \text{ N/C.} \qquad \text{(Answer)}$$

The electric field $\vec{E}_B$ points to the right.

24-7 Gauss' Law and Coulomb's Law

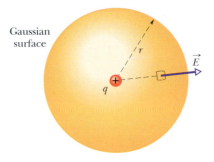

FIGURE 24-20 ■ A spherical Gaussian surface centered on a point charge q.

If Gauss' law and Coulomb's law are equivalent, we should be able to derive each from the other. Here we derive Coulomb's law from Gauss' law and some symmetry considerations.

Figure 24-20 shows a positive point charge q, around which we have drawn a concentric spherical Gaussian surface of radius r. Let us divide this surface into differential areas $d\vec{A}$. By definition, the area vector $d\vec{A}$ at any point is perpendicular to the surface and directed outward from the interior. From the symmetry of the situation, we know at any point the electric field $\vec{E}$ is also perpendicular to the surface and directed outward from the interior. Thus, since the angle θ between $\vec{E}$ and $d\vec{A}$ is zero, we can rewrite Gauss' law expressed in Eq. 24-7 as

$$\oint \vec{E} \cdot d\vec{A} = \oint E \, dA = q^{\text{enc}}/\varepsilon_0. \qquad (24\text{-}17)$$

Here $q^{\text{enc}} = q$. Although the magnitude of the vector $\vec{E}$ varies radially with the distance from q, it has the same value everywhere on the spherical surface. Since the integral in this equation is taken over that surface, the electric field magnitude ($E = |\vec{E}|$) is a constant in the integration and can be brought out in front of the integral sign. That gives us

$$\varepsilon_0 E \oint dA = |q^{\text{enc}}|. \qquad (24\text{-}18)$$

The integral is now merely the sum of the magnitudes of all the differential area elements $d\vec{A}$ on the sphere and thus is just the surface area, $4\pi r^2$. Substituting this, we have

$$\varepsilon_0 E (4\pi r^2) = |q^{\text{enc}}|,$$

or since $q = q^{\text{enc}}$
$$E = \frac{1}{4\pi\varepsilon_0} \frac{|q^{\text{enc}}|}{r^2} = k \frac{|q|}{r^2}. \qquad (24\text{-}19)$$

This is exactly the electric field due to a point charge (Eq. 23-8), which we found using Coulomb's law. Thus, we have shown that Gauss' law and Coulomb's law give us the same result for the electric field due to a single point-like charge. However, Gauss' law is also valid for complex arrays of charges. It can be shown using the principle of superposition that the information about electric fields obtained by using either Gauss' or Coulomb's law will yield the same results even for charge arrays. The difference between the two laws is this: It is easier to use Coulomb's law if we have an array of a few point-like charges, and it is easier to use Gauss' law if we have certain kinds of highly symmetric charge distributions like those discussed in Section 24-6. In still other situations, it is quite difficult to use either law.

READING EXERCISE 24-4: There is a certain net flux Φ^{net} at a Gaussian sphere of radius r enclosing an isolated charged particle. Suppose the enclosing Gaussian surface is changed to (a) a larger Gaussian sphere, (b) a Gaussian cube with edge length equal to r, and (c) a Gaussian cube with edge length equal to $2r$. In each case, is the net flux at the new Gaussian surface greater than, less than, or equal to Φ^{net}? ■

24-8 A Charged Isolated Conductor

Gauss' law permits us to prove an important theorem about isolated conductors:

> If excess charges are placed on an isolated conductor, that amount of charge will move entirely to the surface of the conductor. Once the charges stop moving, none of the excess charge will be found within the body of the conductor.

This might seem reasonable, considering charges with the same sign repel each other. You might imagine that by moving to the surface, the added charges are getting as far away from each other as they can. We turn to Gauss' law for verification of this speculation.

Figure 24-21a shows, in cross section, an isolated lump of copper hanging from an insulating thread and having an excess charge q. We place a Gaussian surface just inside the actual surface of the conductor.

Once the excess charges stop moving, the electric field inside this conductor must be zero. If this were not so, the field would exert forces on the conduction (free) electrons, which are always present in a conductor such as copper, and thus current would always exist within a conductor. (That is, charge would flow from place to place within the conductor.) Of course, there are no such perpetual currents in an isolated conductor, and so we know that the internal electric field is zero.

An internal electric field *does* appear as a conductor is being charged. However, the added charge quickly distributes itself in such a way that the net internal electric field—the vector sum of the electric fields due to all the charges, both inside and outside—is zero. The movement of charge then ceases because there are drag forces known as *resistance* in conductors that dissipate the charges' kinetic energies and eventually bring them to rest. Since the net field is zero, the net force on each charge is zero. So, once the charges are stopped by resistance in the conductor, they remain at rest. Some special materials can be "superconductors" at very low temperatures and allow charges to move without resistance. Therefore, these materials can support long-lasting currents.

If $\vec{E}$ is zero everywhere inside our copper conductor, it must be zero for all points on the Gaussian surface because that surface, though close to the surface of the conductor, is definitely inside the conductor. This means the flux at the Gaussian surface must be zero. Gauss' law then tells us the net charge inside the Gaussian surface must also be zero. Then because the excess charge is not inside the Gaussian surface, it must be outside that surface, which means it must lie on the actual surface of the conductor.

An Isolated Conductor with a Cavity

Figure 24-21b shows the same hanging conductor, but now with a cavity totally within the conductor. It is perhaps reasonable to suppose that when we scoop out the electrically neutral material to form the cavity we do not change the distribution of charge

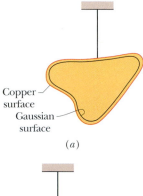

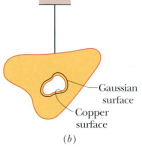

FIGURE 24-21 ■ (a) A lump of copper with a charge q hangs from an insulating thread. A Gaussian surface is placed within the metal, just inside the actual surface. (b) The lump of copper now has a cavity within it. A Gaussian surface lies within the metal, close to the cavity surface.

or the pattern of the electric field that exists in Fig. 24-21a. Again, we can turn to Gauss' law for a quantitative proof.

We draw a Gaussian surface surrounding the cavity, close to its surface but inside the conducting body. Because $\vec{E} = 0$ inside the conductor, there can be no flux at this new Gaussian surface. Therefore, from Gauss' law, that surface can enclose no net charge. We conclude there is no net charge on the cavity walls; all the excess charge remains on the outer surface of the conductor, as in Fig. 24-21a.

The Conductor Removed

Consider now an object that has the same shaped surface, but consists of only a conducting shell of charge. This is equivalent to enlarging the cavity of Fig. 24-21b until it consumes the entire conductor, leaving only the charges. The electric field would not change at all; it would remain zero inside the thin shell of charge and would remain unchanged for all external points. This reminds us that the electric field is set up by the charges and not by the conductor. The conductor simply provides an initial pathway for the charges to take up their positions.

The External Electric Field

You have seen that the excess charge on an isolated conductor moves entirely to the conductor's surface. However, unless the conductor is spherical, the charge does not distribute itself uniformly. Put another way, the surface charge density σ (charge per unit area) varies over the surface of any nonspherical conductor. Generally, this variation makes the determination of the electric field set up by the surface charges very difficult.

Suppose we know the surface charge density, σ, on a region of a conductor. Then it is easy to use Gauss' law to calculate the electric field just outside the surface of a conductor. To do this, we consider a section of the surface small enough to permit us to neglect any curvature and thus to take the section to be flat. We then imagine a tiny cylindrical Gaussian surface to be embedded in the section as in Fig. 24-22: One end cap is fully inside the conductor, the other is fully outside, and the cylinder is perpendicular to the conductor's surface.

The electric field $\vec{E}$ at and just outside the conductor's surface must also be perpendicular to that surface. If it were not, then it would have a component along the conductor's surface exerting forces on the surface charges, causing them to move. However, such motion would violate our implicit assumption that we are dealing with electrostatic equilibrium. Therefore, $\vec{E}$ is perpendicular to the conductor's surface.

We now sum the flux at the Gaussian surface. There is no flux at the internal end cap, because the electric field within the conductor is zero. There is no flux at the curved surface of the cylinder, because internally (in the conductor) there is no electric field and externally the electric field is parallel to the curved portion of the Gaussian surface. The only flux at the Gaussian surface is at the external end cap, where $\vec{E}$ is perpendicular to the plane of the cap. We assume the cap area A is small enough that the field magnitude $|\vec{E}|$ is constant over the cap. Then the amount of the flux at the cap is $|\vec{E}|A$, and that is the net amount of flux $|\Phi^{\text{net}}|$ at the Gaussian surface.

The charge q^{enc} enclosed by the Gaussian surface lies on the conductor's surface in an area A. If σ is the charge per unit area, then q^{enc} is equal to σA. When we substitute σA for q^{enc} and $|\vec{E}|A$ for $|\Phi^{\text{net}}|$, Gauss' law, $\varepsilon_0 |\Phi^{\text{net}}| = |q^{\text{enc}}|$, becomes $\varepsilon_0 |\vec{E}|A = |\sigma|A$, from which we find

$$|\vec{E}| = \frac{|\sigma|}{\varepsilon_0} \qquad \text{(conducting surface)}. \qquad (24\text{-}20)$$

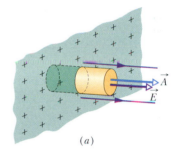

(a)

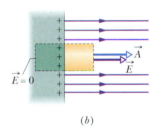

(b)

FIGURE 24-22 ■ Perspective view (a) and side view (b) of a tiny portion of a large, isolated conductor with excess positive charge on its surface. A (closed) cylindrical Gaussian surface, embedded perpendicularly in the conductor, encloses some of the charge. Electric field lines pierce the external end cap of the cylinder, but not the internal end cap. The external end cap has area A and area vector $\vec{A}$.

Thus, the magnitude of the electric field at a location just outside a conductor is proportional to the surface charge density at that location on the conductor. If the charge on the conductor is positive, the electric field is directed away from the conductor as in Fig. 24-22. It is directed toward the conductor if the charge is negative.

The difference between Eq. 24-20 and Eq. 24-16 ($|\vec{E}| = |\sigma|/2\varepsilon_0$) results from the fact that our conductor is no longer thin so that one of our Gaussian pillbox endcaps lies inside the conductor where the electric field is zero. Although the situation in Figs. 24-18 and 24-22 look similar, there is an important difference. There must be other charges in Fig. 24-22 that contribute to making the field zero inside the conductor. Even though these charges are outside the Gaussian surface and therefore do not contribute to the total flux, they change the values of the E field on the surface and therefore change the value we extract.

The field vectors in Fig. 24-22 point toward negative charges somewhere in the environment. If we bring those charges near the conductor, the charge density at any given location on the conductor's surface changes, and so does the magnitude of the electric field. However, the relation between the amount of the surface charge per unit area and the electric field magnitude is still given by Eq. 24-20,

$$E = \frac{|\sigma|}{\varepsilon_0}.$$

The Faraday Cage

The fact that an isolated conductor with a cavity has no electric field inside of it has led to the construction of a very valuable electrical device. Many research environments today involve the measurement of very low power electrical signals. This might occur when measuring the electrical signals from the neuron of a live mouse running a maze or while trying to measure the electrical properties of a microscopic device meant as part of a micro-miniaturized computer chip. In our modern world there are numerous electrical signals traveling through space, arising from everything from the 60 Hz power running in our walls to the radio signals from TV stations and cellular phones. These signals can interfere with sensitive electrical measurements.

To prevent these stray electric fields from ruining sensitive measurements, researchers often conduct their experiments inside a thin-walled metal cage known as a Faraday cage. Examples of Faraday cages are shown in the photo on the first page of this chapter as well as in Fig. 24-23. The Faraday cage in Fig. 24-23 is like the object shown in Fig. 24-21b except that now the "cavity" takes up almost the whole volume of the material. In addition, the thin metal shell in a Faraday cage is typically made of wire mesh. As long as the mesh is fairly fine, charge can spread out evenly on its surface. This type of cage can prevent even strong electrical signals from producing electric fields inside the cage. How? The external electric field induces charges on the surface of the Faraday cage to move so that the field they produce will precisely cancel the external field at points inside the surface. This rearrangement occurs naturally and is predictable by Gauss' law. This is why a demonstrator in a Faraday cage that is highly charged by a Van de Graaff generator can touch the inside of the cage and survive as shown in the opening photograph. The principle of the Faraday cage is also what makes it safe to be inside an automobile in a lightning storm. Even if lightning strikes your car, the effects inside the conductor are substantially reduced. This would not be the case if you were in a wooden crate, because the lightning could pass right through it. The crate could also catch on fire.

FIGURE 24-23 ■ A charged Faraday cage consisting of a sphere made of curved brass rods. Charges on the outside of the cage travel along conducting strings to the small balls causing them to be repelled from the cage. There is no charge inside the cage so the balls in the cage do not repel.

To external charge source

READING EXERCISE 24-5: Suppose a single positive charge is suddenly placed in the cavity shown in Fig. 24-21b. What has to happen in the conductor at the cavity walls to ensure that the electric field everywhere inside the conductor remains at zero? ■

TOUCHSTONE EXAMPLE 24-4: Spherical Metal Shell

Figure 24-24a shows a cross section of a spherical metal shell of inner radius R. A point charge of $-5.0\,\mu$C is located at a distance R/2 from the center of the shell. If the shell is electrically neutral, what are the (induced) charges on its inner and outer surfaces? Are those charges uniformly distributed? What is the field pattern inside and outside the shell?

SOLUTION ■ Figure 24-24b shows a cross section of a spherical Gaussian surface within the metal, just outside the inner wall of the

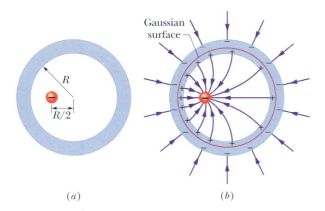

(a)   (b)

FIGURE 24-24 ■ (a) A negative point charge is located within a spherical metal shell that is electrically neutral. (b) As a result, positive charge is nonuniformly distributed on the inner wall of the shell, and an equal amount of negative charge is uniformly distributed on the outer wall. The electric field lines are shown.

shell. One **Key Idea** here is that the electric field must be zero inside the metal (and thus on the Gaussian surface inside the metal). This means that the electric flux through the Gaussian surface must also be zero. Gauss' law then tells us that the *net* charge enclosed by the Gaussian surface must be zero. With a point charge of $-5.0\,\mu$C within the shell, a charge of $+5.0\,\mu$C must lie on the inner wall of the shell.

If the point charge were centered, this positive charge would be uniformly distributed along the inner wall. However, since the point charge is off-center, the distribution of positive charge is skewed, as suggested by Fig. 24-24b, because the positive charge tends to collect on the section of the inner wall nearest the (negative) point charge.

A second **Key Idea** is that because the shell is electrically neutral, its inner wall can have a charge of $+5.0\,\mu$C only if electrons, with a total charge of $-5.0\,\mu$C, leave the inner wall and move to the outer wall. There they spread out uniformly, as is also suggested by Fig. 24-24b. This distribution of negative charge is uniform because the shell is spherical and because the skewed distribution of positive charge on the inner wall cannot produce an electric field in the shell to affect the distribution of charge on the outer wall.

The field lines inside and outside the shell are shown approximately in Fig. 24-24b. All the field lines intersect the shell and the point charge perpendicularly. Inside the shell the pattern of field lines is skewed owing to the skew of the positive charge distribution. Outside the shell the pattern is the same as if the point charge were centered and the shell were missing. In fact, this would be true no matter where inside the shell the point charge happened to be located.

Problems

SEC. 24-3 ■ NET FLUX AT A CLOSED SURFACE

1. Cube The cube in Fig. 24-25 has edge lengths of 1.40 m and is oriented as shown with its bottom face in the x-y plane at z = 0.00 m. Find the electric flux through the right face if the uniform electric field, in newtons per coulomb, is given by (a) $6.00\hat{i}$, (b) $-2.00\hat{j}$, and (c) $-3.00\hat{i} + 4.00\hat{k}$. (d) What is the total flux through the cube for each of these fields?

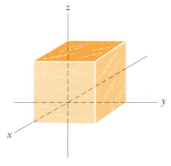

FIGURE 24-25 ■ Problems 1, 5, and 10.

2. Square Surface The square surface shown in Fig. 24-26 measures 3.2 mm on each side. It is immersed in a uniform electric field with magnitude $|\vec{E}| = 1800$ N/C. The field lines make an angle of 35° with a normal to the surface, as shown. Take that normal to be directed "outward," as though the surface were one face of a box. Calculate the electric flux through the surface.

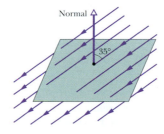

FIGURE 24-26 ■ Problem 2.

SEC. 24-4 ■ GAUSS' LAW

3. Charge at Center of Cube A point charge of 1.8 μC is at the center of a cubical Gaussian surface 55 cm on edge. What is the net electric flux through the surface?

4. Four Charges You have four point charges, $2q, q, -q$, and $-2q$. If possible describe how you would place a closed surface that encloses at least the charge $2q$ (and perhaps other charges) and through which the net electric flux is (a) 0 (b) $+3q/\varepsilon_0$, and (c) $-2q/\varepsilon_0$.

5. Flux Through Cube Find the net flux through the cube of Problem 1 and Fig. 24-25 if the electric field is given by (a) $\vec{E} = (3.00\,y\,[\text{N}/(\text{C}\cdot\text{m})])\hat{j}$

and (b) $\vec{E} = -(4.00 \text{ N/C})\hat{i} + (6.00 \text{ N/C} + 3.00 y [\text{N/(C·m)}]\hat{j}$. (c) In each case, how much charge is enclosed by the cube?

6. Butterfly Net In Fig. 24-27, a butterfly net is in a uniform electric field of magnitude $\vec{E}$. The rim, a circle of radius a, is aligned perpendicular to the field. Find the electric flux through the netting.

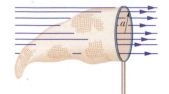

FIGURE 24-27 ■ Problem 6.

7. Earth's Atmosphere It is found experimentally that the electric field in a certain region of Earth's atmosphere is directed vertically down. At an altitude of 300 m the field has magnitude 60.0 N/C; at an altitude of 200 m, the magnitude is 100 N/C. Find the net amount of charge contained in a cube 100 m on edge, with horizontal faces at altitudes of 200 and 300 m. Neglect the curvature of Earth.

8. Shower When a shower is turned on in a closed bathroom, the splashing of the water on the bare tub can fill the room's air with negatively charged ions and produce an electric field in the air as great as 1000 N/C. Consider a bathroom with dimensions of 2.5 m × 3.0 m × 2.0 m. Along the ceiling, floor, and four walls, approximate the electric field in the air as being directed perpendicular to the surface and as having a uniform magnitude of 600 N/C. Also, treat those surfaces as forming a closed Gaussian surface around the room's air. What are (a) the volume charge density ρ and (b) the number of excess elementary charges e per cubic meter in the room's air?

9. Point Charge A point charge q is placed at one corner of a cube of edge a. What is the flux through each of the cube faces? (*Hint*: Use Gauss' law and symmetry arguments.)

10. Surface of Cube At each point on the surface of the cube shown in Fig 24-25, the electric field is along the y-axis. The length of each edge of the cube is 3.0 m. On the right surface of the cube, $\vec{E} = (-34 \text{ N/C})\hat{j}$, and on the left face of the cube $\vec{E} = (+20 \text{ N/C})\hat{j}$. Determine the net charge contained within the cube.

SEC. 24-6 ■ APPLICATION OF GAUSS' LAW TO SYMMETRIC CHARGE DISTRIBUTIONS

11. Conducting Sphere A conducting sphere of radius 10 cm has an unknown charge. If the electric field 15 cm from the center of the sphere has the magnitude 3.0×10^3 N/C and is directed radially inward, what is the net charge on the sphere?

12. Charge Causes Flux A point charge causes an electric flux of -750 N · m²/C to pass through a spherical Gaussian surface of 10.0 cm radius centered on the charge. (a) If the radius of the Gaussian surface were doubled, how much flux would pass through the surface? (b) What is the value of the point charge?

13. Rutherford In a 1911 paper, Ernest Rutherford said: "In order to form some idea of the forces required to deflect an α particle through a large angle, consider an atom [as] containing a point positive charge Ze at its center and surrounded by a distribution of negative electricity $-Ze$ uniformly distributed within a sphere of radius R. The electric field E . . . at a distance r from the center for a point inside the atom [is]

$$E = \frac{Ze}{4\pi\varepsilon_0}\left(\frac{1}{r^2} - \frac{r}{R^3}\right)."$$

Verify this equation.

14. Concentric Spheres Two charged concentric spheres have radii of 10.0 cm and 15.0 cm. The charge on the inner sphere is 4.00×10^{-8} C, and that on the outer sphere is 2.00×10^{-8} C. Find the electric field (a) at $r = 12.0$ cm and (b) at $r = 20.0$ cm.

15. Proton A proton with speed $v = 3.00 \times 10^5$ m/s orbits just outside a charged sphere of radius $r = 1.00$ cm. What is the charge on the sphere?

16. Charge at Center of Shell A point charge $+q$ is placed at the center of an electrically neutral, spherical conducting shell with inner radius a and outer radius b. What charge appears on (a) the inner surface of the shell and (b) the outer surface? What is the net electric field at a distance r from the center of the shell if (c) $r < a$, (d) $b > r > a$, and (e) $r > b$? Sketch field lines for those three regions. For $r > b$, what is the net electric field due to (f) the central point charge plus the inner surface charge and (g) the outer surface charge? A point charge $-q$ is now placed outside the shell. Does this point charge change the charge distribution on (h) the outer surface and (i) the inner surface? Sketch the field lines now. (j) Is there an electrostatic force on the second point charge? (k) Is there a net electrostatic force on the first point charge? (l) Does this situation violate Newton's Third Law?

17. Solid Nonconducting Sphere A solid nonconducting sphere of radius R has a nonuniform charge distribution of volume charge density $\rho = \rho_s r/R$, where ρ_s is a constant and r is the distance from the center of the sphere. Show (a) that the total charge on the sphere is $Q = \pi\rho_s R^3$ and (b) that

$$|\vec{E}| = k\frac{|Q|}{R^4}r^2$$

gives the magnitude of the electric field inside the sphere.

18. Hydrogen Atom A hydrogen atom can be considered as having a central point-like proton of positive charge $+e$ and an electron of negative charge $-e$ that is distributed about the proton according to the volume charge density $\rho = A \exp(-2r/a_1)$. Here A is a constant, $a_1 = 0.53 \times 10^{-10}$ m is the *Bohr radius*, and r is the distance from the center of the atom. (a) Using the fact that hydrogen is electrically neutral, find A. (b) Then find the electric field produced by the atom at the Bohr radius.

19. Sphere of Radius a In Fig 24-28 an insulating sphere, of radius a and charge $+q$ uniformly distributed throughout its volume, is concentric with a spherical conducting shell of inner radius b and outer radius c. This shell has a net charge of $-q$. Find expressions for the electric field, as a function of the radius r, (a) within the sphere $(r < a)$, (b) between the sphere and the shell $(a < r < b)$, (c) inside the shell $(b < r < c)$, and (d) outside the shell $(r > c)$. (e) What are the charges on the inner and outer surfaces of the shell?

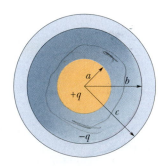

FIGURE 24-28 ■ Problem 19.

20. Uniform Volume Charge Density Figure 24-29a shows a spherical shell of charge with uniform volume charge density ρ. Plot E due to the shell for distances r from the center of the shell ranging from zero to 30 cm. Assume that $\rho = 1.0 \times 10^{-6}$ C/m³, $a = 10$ cm, and $b = 20$ cm.

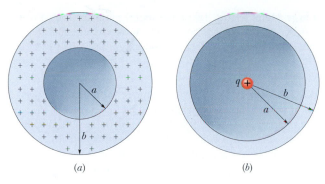

FIGURE 24-29 ■ Problems 20 and 21.

enclosed by a thin, nonconducting cylinder of outside radius 1.5 cm, coaxial with the wire. The cylinder is to have positive charge on its outside surface with a surface charge density σ such that the net external electric field is zero. Calculate the required σ.

27. Cylindrical Rod A very long conducting cylindrical rod of length L with a total charge $+q$ is surrounded by a conducting cylindrical shell (also of length L) with total charge $-2q$, as shown in Fig. 24-32. Use Gauss' law to find (a) the electric field at points outside the conducting shell, (b) the distribution of charge on the shell, and (c) the electric field in the region between the shell and rod. Neglect end effects.

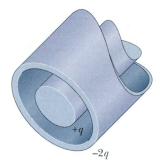

FIGURE 24-32 ■ Problem 27.

21. Nonconducting Spherical Shell In Fig. 24-29b, a nonconducting spherical shell, of inner radius a and outer radius b, has a positive volume charge density $\rho = A/r$ (within its thickness), where A is a constant and r is the distance from the center of the shell. In addition, a positive point charge q is located at that center. What value should A have if the electric field in the shell ($a \leq r \leq b$) is to be uniform? (*Hint:* The constant A depends on a but not on b.)

22. Show That A nonconducting sphere has a uniform volume charge density ρ. Let $\vec{r}$ be the vector from the center of the sphere to a general point P within the sphere. (a) Show that the electric field at P is given by $\vec{E} = \rho\vec{r}/3\varepsilon_0$. (Note that the result is independent of the radius of the sphere.) (b) A spherical cavity is hollowed out of the sphere, as shown in Fig. 24-30. Using superposition concepts, show that the electric field at all points within the cavity is uniform and equal to $\vec{E} = \rho\vec{a}/3\varepsilon_0$, where $\vec{a}$ is the position vector from the center of the sphere to the center of the cavity. (Note that this result is independent of the radius of the sphere and the radius of the cavity.)

FIGURE 24-30 ■ Problem 22.

23. Spherically Symmetrical A spherically symmetrical but nonuniform volume distribution of charge produces an electric field of magnitude $|\vec{E}| = Kr^4$, directed radially outward from the center of the sphere. Here r is the radial distance from that center, and K is a positive constant. What is the volume density ρ of the charge distribution as a function of r?

24. Long Metal Tube Figure 24-31 shows a section of a long, thin-walled metal tube of radius R, with a positive charge per unit length λ on its surface. Derive expressions for $|\vec{E}|$ in terms of the distance r from the tube axis, considering both (a) $r > R$ and (b) $r < R$. Plot your results for the range $r = 0$ to $r = 5.0$ cm, assuming that $\lambda = 2.0 \times 10^{-8}$ C/m and $R = 3.0$ cm. (*Hint:* Use cylinderical Gaussian surfaces, coaxial with the metal tube.)

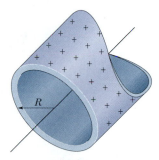

FIGURE 24-31 ■ Problem 24.

25. Infinite Line of Charge An infinite line of charge produces a field magnitude of 4.5×10^4 N/C at a distance of 2.0 m. Calculate the amount of linear charge density $|\lambda|$.

26. Long Straight Wire A long, straight wire has fixed negative charge with a linear charge density of -3.6 nC/m. The wire is to be

28. Solid Cylinder A long, noncon-ducting, solid cylinder of radius 4.0 cm has a nonuniform volume charge density ρ that is a function of the radial distance r from the axis of the cylinder, as given by $\rho = Ar^2$ with $A = 2.5$ μC/m⁵. What is the magnitude of the electric field at a radial distance of (a) 3.0 cm and (b) 5.0 cm from the axis of the cylinder?

29. Two Concentric Cylinders Two long, charged, concentric cylinders have radii of 3.0 and 6.0 cm. Assume the outer cylinder is hollow. The charge per unit length is 5.0×10^{-6} C/m on the inner cylinder and -7.0×10^{-6} C/m on the outer cylinder. Find the electric field at (a) $r = 4.0$ cm and (b) $r = 8.0$ cm, where r is the radial distance from the common central axis.

30. Geiger Counter Figure 24-33 shows a Geiger counter, a device used to detect ionizing radiation (radiation that causes ionization of atoms). The counter consists of a thin, positively charged central wire surrounded by a concentric, circular, conducting cylinder with an equal negative charge. Thus, a strong radial electric field is set up inside the cylinder. The cylinder contains a low-pressure inert gas. When a particle of radiation enters the device through the cylinder wall, it ionizes a few of the gas atoms. The resulting free electrons (labelled e) are drawn to the posi-tive wire. However, the electric field is so intense that, between

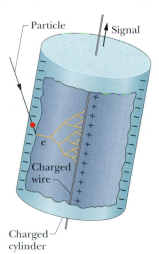

FIGURE 24-33 ■ Problem 30.

collisions with other gas atoms, the free electrons gain energy suffi-cient to ionize these atoms also. More free electrons are thereby cre-ated, and the process is repeated until the electrons reach the wire. The resulting "avalanche" of electrons is collected by the wire gener-ating a signal that is used to record the passage of the original parti-cle of radiation. Suppose that the radius of the central wire is 25 μm, the radius of the cylinder 1.4 cm, and the length of the tube 16 cm. If the electric field component E_r at the cylinder's inner wall is $+2.9 \times 10^4$ N/C, what is the total positive charge on the central wire?

31. Charge Is Distributed Uniformly Charge is distributed uni-formly throughout the volume of an infinitely long cylinder of radius R. (a) Show that, at a distance r from the cylinder axis (for $r < R$),

$$|\vec{E}| = \frac{|\rho|r}{2\varepsilon_0},$$

where $|\rho|$ is the amount of volume charge density. (b) Write an expression for $|\vec{E}|$ when $r > R$.

32. Parallel Sheets Figure 24-34 shows cross sections through two large, parallel, nonconducting sheets with identical distributions of positive charge with area charge density σ. What is $\vec{E}$ at points (a) above the sheets, (b) between them, and (c) below them?

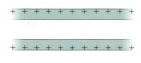

FIGURE 24-34 ■ Problem 32.

33. Square Metal Plate A square metal plate of edge length 8.0 cm and negligible thickness has a total charge of 6.0×10^{-6} C. (a) Estimate the magnitude E of the electric field just off the center of the plate (at, say, a distance of 0.50 mm) by assuming that the charge is spread uniformly over the two faces of the plate. (b) Estimate E at a distance of 30 m (large relative to the plate size) by assuming that the plate is a point charge.

34. Thin Metal Plates Two large, thin metal plates are parallel and close to each other. On their inner faces, the plates have excess surface charge of opposite signs. The amount of charge per unit area is given by $|\sigma| = 7.0 \times 10^{-22}$ C/m², with the negatively charged plate on the left. What are the magnitude and direction of the electric field $\vec{E}$ (a) to the left of the plates, (b) to the right of the plates, and (c) between the plates?

35. Ball on Thread In Fig. 24-35, a small, nonconducting ball of mass $m = 1.0$ mg and charge $q = 2.0 \times 10^{-8}$ C (distributed uniformly through its volume) hangs from an insulating thread that makes an angle $\theta = 30°$ with a vertical, uniformly charged nonconducting sheet (shown in cross section). Considering the gravitational force on the ball and assuming that the sheet extends far vertically and into and out of the page, calculate the surface charge density σ of the sheet.

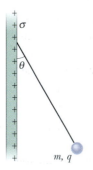

FIGURE 24-35 ■ Problem 35.

36. Large Metal Plates Two large metal plates of area 1.0 m² face each other. They are 5.0 cm apart and have equal but opposite charges on their inner surfaces. If the magnitude $|\vec{E}|$ of the electric field between the plates is 55 N/C, what is the amount of charge on each plate? Neglect edge effects.

37. An Electron Is Shot An electron is shot directly toward the center of a large metal plate that has excess negative charge with surface charge density -2.0×10^{-6} C/m². If the initial kinetic energy of the electron is 1.60×10^{-17} J and if the electron is to stop (owing to electrostatic repulsion from the plate) just as it reaches the plate, how far from the plate must it be shot?

38. Planar Slab A planar slab of thickness d has a uniform volume charge density ρ. Find the magnitude of the electric field at all points in space both (a) within and (b) outside the slab, in terms of x, the distance measured from the central plane of the slab.

SEC. 24-8 ■ A CHARGED ISOLATED CONDUCTOR

39. Photocopying Machine The electric field just above the surface of the charged drum of a photocopying machine has a magnitude $|\vec{E}|$ of 2.3×10^5 N/C. What is the surface charge density on the drum, assuming that the drum is a conductor?

40. Space Vehicles Space vehicles traveling through Earth's radiation belts can intercept a significant number of electrons. The resulting charge buildup can damage electronic components and disrupt operations. Suppose a spherical metallic satellite 1.3 m in diameter accumulates -2.4 μC of charge in one orbital revolution. (a) Find the resulting surface charge density. (b) Calculate the magnitude of the electric field just outside the surface of the satellite due to the surface charge.

41. Charged Sphere A uniformly charged conducting sphere of 1.2 m diameter has a surface charge density of 8.1 μC/m². (a) Find the net charge on the sphere. (b) What is the total electric flux leaving the surface of the sphere?

42. Arbitrary Shape Conductor An isolated conductor of arbitrary shape has a net charge of $+10 \times 10^{-6}$ C. Inside the conductor is a cavity within which is a point charge $q = +3.0 \times 10^{-6}$ C. What is the charge (a) on the cavity wall and (b) on the outer surface of the conductor?

Additional Problems

43. If/Can If the electric field in a region of space is zero, can you conclude there are no electric charges in that region? Explain.

44. If/Than If there are fewer electric field lines leaving a Gaussian surface than there are entering the surface, what can you conclude about the net charge enclosed by that surface?

45. Net Flux What is the net electric flux through each of the closed surfaces in Fig. 24-36 if the value of q is $+1.6 \times 10^{-19}$ C?

46. Net Flux Two What is the net electric flux through each of the closed surfaces in Fig. 24-37 if the value of q is 8.85×10^{-12} C? Explain the reasons for your answers.

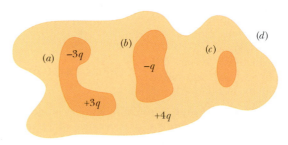

FIGURE 24-36 ■ Problem 45.

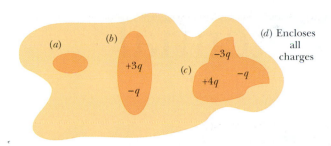

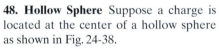

FIGURE 24-37 ▪ Problem 46.

47. Fair Weather During fair weather, an electric field of about 100 N/C points vertically downward into Earth's atmosphere. Assuming that this field arises from charge distributed in a spherically symmetric manner over the surface of Earth, determine the *net* charge of Earth and its atmosphere if the radius of Earth and its atmosphere is 6.37×10^6 m.

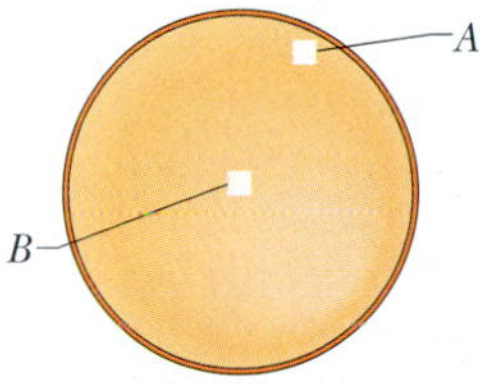

48. Hollow Sphere Suppose a charge is located at the center of a hollow sphere as shown in Fig. 24-38.

(a) Are the intersections of the field lines with the surface of the sphere uniformly distributed throughout? In other words, is the density of lines passing through the surface of the sphere uniform? Explain why or why not.

FIGURE 24-38 ▪ Problem 48.

(b) Consider surface elements A and B, which have exactly the same area. Is the number of field lines passing through surface element A greater than, less than, or equal to the number of field lines through surface element B? Explain.
(c) Is the flux through surface element A greater than, less than, or equal to the flux through surface element B? Explain.

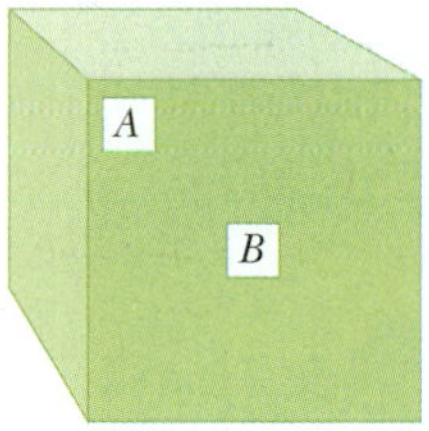

49. Center of Cube Suppose a charge is located at the center of the cube shown in Fig. 24-39.

(a) Are the intersections of the field lines with a side of the cube uniformly distributed across the side? In other words, is the density of lines passing through the box uniform? Explain why or why not.
(b) Is the number of field lines through surface element A greater than, less than, or equal to the number of field lines through surface element B? Explain.

FIGURE 24-39 ▪ Problem 49.

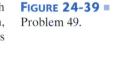

(c) Is the flux through surface element A greater than, less than, or equal to the flux through surface element B? Explain.

50. Using Gauss' Law Gauss' law is usually written as an equation in the form

$$\oint \vec{E} \cdot d\vec{A} = q^{\text{enc}}/\varepsilon_0.$$

(a) For this equation, specify what each term in this equation means and how it is to be calculated when doing some specific (but arbitrary—not a special case) calculation.

A long thin cylindrical shell like that shown in Fig. 24-40 has length L and radius R with $L \gg R$ and is uniformly covered with a charge Q. If we look for the field near the cylinder somewhere about the middle, we can treat the cylinder as if it were an infinitely long cylinder. Using this assumption, we can calculate the magnitude and direction of the field at a point a distance d from the *axis* of the cylinder (outside the cylindrical shell; i.e., $L \gg d > R$ but d not very close to R) using Gauss's law. Do so by explicitly following the steps below.

FIGURE 24-40 ▪ Problem 50.

(b) Select an appropriate Gaussian surface. Explain why you chose it.
(c) Carry out the integral on the left side of the equation, expressing it in terms of the unknown value of the magnitude of the E field.
(d) What is the relevant value of q for your surface?
(e) Use your results in (c) and (d) in the equation and solve for the magnitude of E.

51. Interpreting Gauss Gauss' law states

$$\oint_A \vec{E} \cdot d\vec{A} = q_A/\varepsilon_0,$$

where A is a surface and q_A is a charge.

(a) Which of the following statements are true about the surface A appearing in Gauss' law for the equation to hold? You may list any number of these statements including all or none.

  i. The surface A must be a closed surface (must cover a volume).

  ii. The surface A must contain all the charges in the problem.

  iii. The surface A must be a highly symmetrical surface like a sphere or a cylinder.

  iv. The surface A must be a conductor.

  v. The surface A is purely imaginary.

  vi. The normals to the surface A must all be in the same direction as the electric field on the surface.

(b) Which of the following statements are true about the charge q_A appearing in Gauss' law? You may list any number of these statements including all or none.

  i. The charge q_A must be all the charge lying on the Gaussian surface.

  ii. The charge q_A must be the charge lying within the Gaussian surface.

  iii. The charge q_A must be all the charge in the problem.

  iv. The charge q_A flows onto the Gaussian surface once the surface is established.

  v. The electric field E in the integral on the left of Gauss' law is due only to the charge q_A.

  vi. The electric field E in the integral on the left on Gauss' law is due to all charges in the problem.

25 | Electric Potential

At more than 115 years old, the Eiffel Tower is arguably the world's most famous landmark. Among other things, the tower is an engineering feat, a work of art, a scenic lookout, and a radio tower. Less well known is the tower's ability to protect people, trees, and other buildings from being struck by lightning that might emanate from thunderheads behind it.

How can the Eiffel Tower protect people from lightning?

The answer is in this chapter.

25-1 Introduction

In the last few chapters we have explored the nature of interaction forces between charged particles. We have developed the concept of electric field as a way to represent the forces a point charge would experience at any point in the space surrounding a collection of charges.

In certain situations it is difficult to understand the motions of charges in terms of an electric field. This difficulty is analogous to problems encountered in describing the motion of an object in the presence of gravitational forces. We developed the concepts of work and energy in Chapters 9 and 10 to deal with these problems. We will now investigate the application of the concepts of work and energy to situations in which the forces involved are electrostatic forces. In this chapter we develop the concept of electric potential—commonly referred to as voltage. We then explore some of its properties, including how charges are distributed on a metal conductor placed in an electric field.

Since the concept of potential or voltage is essential to an understanding of electric circuits, we will use the concept of electric potential in the next chapter to help us understand the role that batteries play in maintaining currents.

25-2 Electric Potential Energy

Newton's law for the gravitational force and Coulomb's law for the electrostatic force are mathematically similar. In Section 14-2 we saw that the gravitational force between two particle-like masses depends directly on the product of the masses and inversely with the square of the distance between them (Eq. 14-2). In like manner, the electrostatic forces between two point charges depend directly on the product of the charges and inversely with the square of the distance between them (Eq. 22-4). This similarity gives us a starting point in our search for additional useful concepts related to the interactions between charged objects. In this chapter, we consider whether some of the general features we have established for the gravitational force apply to the electrostatic force as well.

For example, the gravitational force is a *conservative force*. The work done by it is independent of the path along which an object moves. In experimental tests the work done by the electrostatic force has also been found to be *path independent*. If a charged particle moves from point i to point f while an electrostatic force is acting on it, the work W done by the force is the same for all paths between points i and f. Hence, we can infer that the electrostatic force is a conservative force as well.

Definition of Electric Potential Energy

In Chapter 10 we defined potential energy as the energy associated with the configuration of a system of objects that interact and hence exert forces on each other. We then proceeded to define gravitational potential energy as the negative of the amount of gravitational work objects in the system do on each other when their positions relative to one another change. From Eq. 10-5, $\Delta U \equiv -W^{\text{cons}}$ or $\Delta U^{\text{grav}} = -W^{\text{grav}}$ (Eq. 10-6). This general definition of work can be applied to a system of charges that interact by means of electrostatic forces.

Since electrostatic forces, like gravitational forces, are conservative, then it makes sense to assign an **electric potential energy change** ΔU to a system of interacting charges in a similar manner. If we cause or allow a system to change its configuration from an initial potential energy state U_1 to a different final state U_2, the internal electrostatic forces do a total amount of work W^{elec} on the particles in the system. As in

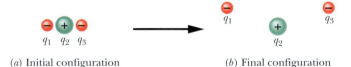

(a) Initial configuration   (b) Final configuration

FIGURE 25-1 ■ (a) A system of three charges is in an initial configuration in which the charges are separated and have an electric potential energy U_1 associated with them. (b) Since both of the negative charges will be attracted to the positive charge, they will coalesce into a final configuration with potential energy U_2. The net electrostatic work the charges do on each other is positive so the system loses potential energy. Thus $U_2 < U_1$ so that $\Delta U < 0$.

Chapter 10, we define the potential energy change ΔU as the negative of the work the system does on itself when it undergoes the reconfiguration. This can be expressed symbolically as

$$\Delta U = U_2 - U_1 \equiv W^{\text{ext}} = -W^{\text{elec}}. \tag{25-1}$$

Figure 25-1 shows a system of charges losing electrostatic potential energy as a result of a natural reconfiguration. Figure 25-2 shows the same system gaining electrostatic potential energy as an external agent (doing positive work) causes the system to reconfigure. This results in the system doing negative work on itself.

(a) Initial configuration   (b) Final configuration

FIGURE 25-2 ■ (a) A system of three charges is in an initial configuration in which the charges are close together and have an electric potential energy U_1 associated with them. (b) Since q_1 and q_3 are both attracted to q_2, it will take positive external work, W^{ext}, to pull the charges apart. The net electrostatic work the charges do on each other is negative so the system gains potential energy. Thus $U_2 > U_1$ so that $\Delta U > 0$.

As you may recall, we determined that only differences in gravitational potential energy were physically significant. In Chapter 10 the system of masses we considered consisted of the Earth and a single object near its surface. We chose a convenient height at which to set the gravitational potential energy to zero. For example, we may have defined an Earth–object system as having zero potential energy when an object is at floor level or at the level of a tabletop. In doing so, we set the absolute scale for gravitational potential energy differently in different situations. This is legitimate since only potential energy *differences* are meaningful.

Potential energy difference is also of primary importance in keeping track of electric potential energy. Typically, we *define the electric potential energy of a system of charges to be zero when the particles are all infinitely separated from each other*, just as we did in Chapter 14 with the general form of gravitational potential energy. Using this zero of electric potential energy makes sense because the charges making up such a system have no interaction forces in that configuration. Using a standard reference potential (instead of moving it around as we typically do for Earth–object systems) allows us to find unique values of U_1 and U_2. For example suppose several charged particles come together from initially infinite separations (state 1) to form a system of nearby particles (state 2). Then using the conventional reference configuration, the initial potential energy U_1 is zero. If W^{elec} represents the internal work done by the electrostatic forces between particles during the move in from infinity, then from Eq. 25-1,

$$\Delta U = U_2 - U_1 = U_2 = -W^{\text{elec}}. \tag{25-2}$$

Since U_1 is zero, the final potential energy U_2 of the system can simply be denoted as U. Then, in terms of symbols,

$$U \equiv -W^{\text{elec}} \qquad \text{(for initial potential energy = 0).} \qquad (25\text{-}3)$$

As usual, the use of the symbol "$\equiv$" signifies that the expression is a definition.

External Forces and Energy Conservation

Since opposite charges attract, they will come together naturally if they are free to move. In these cases the charges "fall together" and the potential energy of the system of charges will be reduced. Similarly, like charges that are free will move apart and their potential energy will also be reduced. However, we can raise the potential energy of a system of charges by using energy from another system. Two common examples of external agents that can raise the potential energy of a system of charges are the Van de Graaff generator and the battery. Van de Graaff generators (see Fig. 25-3) use mechanical energy to force charges of like sign onto metal conductors. Batteries use chemical potential energy (which is actually a combination of electric and quantum effects) to force charges onto an electrode having the same sign charges.

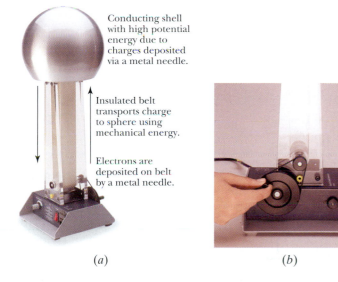

Conducting shell with high potential energy due to charges deposited via a metal needle.

Insulated belt transports charge to sphere using mechanical energy.

Electrons are deposited on belt by a metal needle.

(a)   (b)

FIGURE 25-3 ■ A Van de Graaff generator uses mechanical energy from either (a) a motor or (b) a hand crank to transport charge to a conducting sphere, raising its potential energy. (Photo courtesy of PASCO scientific.)

Suppose an *external force* outside of the system under consideration causes a test particle of charge q to move from an initial location to a final location in the presence of an unchanging electric field generated by the source charges in the system. As the test charge moves, our outside force does work W^{ext} on the charge. At the same time, the electric field does work W^{elec} on it. By the work-kinetic energy theorem, the change ΔK in the kinetic energy of the particle is

$$\Delta K = K_2 - K_1 = W^{\text{ext}} + W^{\text{elec}}.$$

But since $W^{\text{elec}} = -\Delta U$,

$$\Delta K + \Delta U = W^{\text{ext}}. \qquad (25\text{-}4)$$

Now suppose the particle is stationary before and after the move. Then K_2 and K_1 are both zero, and this reduces to

$$\Delta U = W^{\text{ext}} \qquad \text{(for no kinetic energy change).} \qquad (25\text{-}5)$$

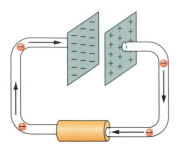

FIGURE 25-4 ■ A 1.5 V D-cell can act as an external agent that does the work needed to move electrons through a wire from a metal plate with excess positive charges to one with excess negative charges.

That is, the work, W^{ext}, done by our external force during the move is equal to the change in electric potential energy—provided there is no change in kinetic energy.

So in what direction will a positive or negative charge move if released? Will the charge move to raise or lower the potential energy of the system? The expression above can be used to determine this. For example, let the external force (perhaps the push or pull of your hand) do positive work. Recall from Section 9-4 the sign convention associated with work in general. If W^{ext} is positive, then ΔU must also be positive (by the equation above) and so we know that $U_2 > U_1$. In other words, the motion of a charge from a lower potential energy to a higher potential energy requires positive work to be done on the system. This motion would not happen if the particle were simply released. Spontaneous or naturally occuring motion is associated with reduced potential energy. This is very similar to the situations encountered in Section 9-8, where we considered an object under the influence of the Earth's attractive gravitational force. Often a battery does this work in a circuit, as in Fig. 25-4.

READING EXERCISE 25-1: Why is a configuration with charges separated by an infinite distance a good choice for our reference (zero) potential energy? Would a zero separation be equally good? Why or why not? ■

READING EXERCISE 25-2: In the figure, a proton moves from point 1 to point 2 in a uniform electric field directed as shown. (a) Does the electric field do positive or negative work on the proton? (b) Does the electric potential energy of the proton increase or decrease? (c) In this case we don't choose the potential energy to be zero at infinity. Why not? ■

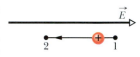

25-3 Electric Potential

When considering gravitational potential energy we dealt primarily with a system consisting of the Earth and a single object much smaller than the Earth. If the object were to fall toward the Earth the interaction forces between them would be equal in magnitude, but as the object moves toward the Earth, the Earth's motion would be negligibly small. Thus the change in the system's gravitational potential energy would simply be the change in potential energy of the falling object. Similarly, as we did in Section 23-2, we can consider systems in which a small "test" charge moves in the presence of an electric field but does not change the electric field significantly. In these systems, the electric potential energy of the system can be calculated as the negative of work done by the electric field on a single test charge as we bring it to a location of interest from infinity.

> In the next several chapters we will focus primarily on systems in which the change in potential energy of a single test charge moving in an electric field is for all practical purposes the same as the change in potential energy of the entire system of charges.

This situation applies if the only charge that moves is our test charge. In this case, the electric field generated by the fixed source charges remains the same, so that the change in the system's potential energy will be proportional to the magnitude of the test charge. (This will not be true if other charges move.)

Defining Electric Potential

Recall that we defined and used the concept of electric field as the *electric force per unit charge* so we could easily analyze the forces experienced by a charge of any sign or magnitude. It is advantageous to develop an analogous concept for the determination of the electric potential energy of a system associated with the change in location of a test charge of any reasonable sign or magnitude. We will do that now, defining **electric potential** as a potential energy *per unit charge*. Once we have chosen a reference configuration with zero energy, our electric potential (potential energy per unit charge) has a unique value at any point in space. For example, suppose we move a test particle of positive charge $q_t = 1.60 \times 10^{-19}$ C from a location at infinity where the electric potential energy is defined as zero to a location in an electric field where the particle has an electric potential energy of 2.40×10^{-17} J. Then the change in electric potential, ΔV, of the system associated with the change in location of a test charge can be calculated as

$$\Delta V = V_2 - V_1 = \frac{2.40 \times 10^{-17} \text{ J}}{1.60 \times 10^{-19} \text{ C}} - 0 = 150 \text{ J/C}.$$

Next, suppose we replace that test particle with one having twice as much positive charge, 3.20×10^{-19} C. We would find that, at the same point, the second particle has an electric potential energy of 4.80×10^{-17} J, twice that of the first particle. However, the potential energy per unit charge or electric potential would be the same, still 150 J/C.

Thus, the system potential energy per unit charge, which can be symbolized as U/q_t, is independent of the charge q_t of the test particle we happen to be considering (Fig. 25-5). It is *characteristic only of the electric field* that is present. The potential energy per unit charge at a point in an electric field is defined as the electric potential V (or simply the **potential**) at that point. Thus V is defined as

$$V \equiv \frac{U}{q}. \tag{25-6}$$

Note that potential energy and charge are both scalar quantities, so the electric potential is also a scalar, not a vector.

The *electric potential difference*, ΔV, associated with moving a charge q between any two points 1 and 2 in an electric field is equal to the difference between the potential energy per unit charge at the two points:

$$\Delta V = V_2 - V_1 = \frac{U_2}{q} - \frac{U_1}{q} = \frac{\Delta U}{q}. \tag{25-7}$$

Using $\Delta U = U_2 - U_1 = -W^{\text{elec}}$ (Eq. 25-1) to substitute the work done by electrostatic forces $-W^{\text{elec}}$ for ΔU in the equation above, we can define the potential difference between points 1 and 2 as

$$\Delta V = V_2 - V_1 \equiv -\frac{W^{\text{elec}}}{q} \qquad \text{(potential difference defined).} \tag{25-8}$$

That is, the potential difference between two points is the negative of the work done by the electrostatic force to move a unit charge from one point to the other. A potential difference can be positive, negative, or zero, depending on the signs and magnitudes of the charge q and the electrostatic work W^{elec}.

As we already mentioned, we have set $U_1 = 0$ infinitely far from any charges as our reference potential energy. So since $V \equiv U/q$ (Eq. 25-6), the electric potential

Charges fixed in an
insulating material

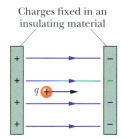

FIGURE 25-5 ■ A test charge moves in an electric field created by a stable configuration of source charges. If the test charge doesn't affect the electric field significantly as it changes location, the change in electric potential, ΔV, of the system (consisting of the source charges and the test charge) is due entirely to the work per unit charge done on the test charge by the electric field.

must also be zero there. Then using Eq. 25-8, we can define the electric potential V (measured relative to infinity) at any point in an electric field to be

$$V = -\frac{W_\infty^{\text{elec}}}{q} \qquad \text{(potential defined relative to infinity)}, \qquad (25\text{-}9)$$

where W_∞^{elec} is the work done by the electrostatic force on a charged particle as that particle moves in from infinity to point f. As was the case with potential difference ΔV, a potential V can be positive, negative, or zero, depending on the signs and magnitudes of q and W_∞^{elec}.

The SI unit for electric potential that follows from Eq. 25-9 is the joule per coulomb. This combination occurs so often that a special unit, the *volt* (abbreviated V) is used to represent it. Thus,

$$1 \text{ volt} \equiv 1 \text{ joule/coulomb}. \qquad (25\text{-}10)$$

Although the terms electric potential energy and electric potential are very similar, they are not the same thing. This is probably one of the reasons why it is so common to refer to electric potential as **voltage** after its unit—the volt.

This new unit called the volt allows us to adopt a more conventional unit for the electric field $\vec{E}$, which we have measured up to now in newtons per coulomb. With two unit conversions, we obtain

$$1 \text{ N/C} = \left[1\frac{\text{N}}{\text{C}} \right]\left[\frac{1 \text{ V}}{1 \text{ J/C}} \right]\left[\frac{1 \text{ J}}{1 \text{ N} \cdot \text{m}} \right] = 1 \text{ V/m}. \qquad (25\text{-}11)$$

The conversion factor in the second set of parentheses comes from Eq. 25-10, and that in the third set of parentheses is derived from the definition of the joule. From now on, we shall express values of the electric field in volts per meter rather than in newtons per coulomb.

The Electron Volt

Because we often have situations in which the charges involved are very small (a few times the charge of an electron), we define an energy unit that is a convenient one for energy measurements in the atomic and subatomic domain. One *electron-Volt* (eV) is the energy equal to the work required to move a single positive elementary charge e (the charge magnitude of the electron or the proton) through a potential difference of exactly one volt. Equation 25-8,

$$\Delta V = V_2 - V_1 = -\frac{W^{\text{elec}}}{q} = \frac{-W^{\text{elec}}}{e},$$

tells us that the magnitude of this work is $e \, \Delta V$, so

$$1 \text{ eV} = e(1 \text{ V})$$
$$= (1.60 \times 10^{-19} \text{ C})(1 \text{ J/C}) = 1.60 \times 10^{-19} \text{ J}, \qquad (25\text{-}12)$$

where the units for electron volt are joules because it is actually a unit of energy rather than electric potential.

READING EXERCISE 25-3: In the figure shown in Reading Exercise 25-2, we moved a proton from point i to point f in a uniform electric field directed as shown. (a) Does our external force do positive or negative work? (b) Does the proton move to a point of higher or lower potential? ■

TOUCHSTONE EXAMPLE 25-1: Electron Motion

An electron starts from rest at a point in space at which the electric potential is 9.0 V. If the only force acting on the electron is that associated with the electric potential, how fast will the electron be moving when it passes a second point in space where the electric potential is 10.0 V?

SOLUTION ■ First we need to convince ourselves that this problem describes a physical situation that is even possible. Equation 25-4 tells us that $\Delta K + \Delta U = W^{\text{ext}}$. Since $W^{\text{ext}} = 0$ here and since ΔK must be positive if the electron speeds up, this means that ΔU must be negative. But is it? After all, ΔV is positive here since $V_2 - V_1 = +1.0$ V. However, the charge of the electron is negative, so Eq. 25-7 tells us that:

$$\Delta U = q\,\Delta V = (-e)\Delta V = (-e)(+1.0\text{ V}) = -1.0\text{ eV},$$

so that $\qquad \Delta K = -\Delta U = -(-1.0\text{ eV}) = +1.0\text{ eV},$

which is positive.

The **Key Idea** here is that a negative charge *loses* potential energy and *gains* kinetic energy when it moves from a region of *lower* potential to a region of *higher* potential. This is just the opposite of what would happen to a positive charge!

Now that we know the electron has 1.0 eV of kinetic energy, we need to determine how fast it is going. The **Key Idea** here is that $1\text{ eV} = 1.60 \times 10^{-19}$ J (Eq. 25-12). Then

$$\Delta K = K_2 - K_1 = (\tfrac{1}{2})mv_2^2 - 0,$$

which gives us

$$
\begin{aligned}
v_2 &= \sqrt{2\,\Delta K/m} \\
&= \sqrt{2(1.0\text{ eV})(1.60 \times 10^{-19}\text{ J/eV}/(9.1 \times 10^{-31}\text{ kg})} \\
&= 5.9 \times 10^5 \text{ m/s}. \qquad\qquad \text{(Answer)}
\end{aligned}
$$

25-4 Equipotential Surfaces

We are interested in what our knowledge of electric potential can tell us about how small test charges might move. We can infer from the discussion above that charged particles will not spontaneously move from one point to another point of equal potential. This is quite analogous to movement of mass in a gravitational field. A skier on a flat surface with no kinetic energy will not spontaneously move from one part of the surface to another. On the other hand, if the skier is on a slope and is free to move, the skier will spontaneously start moving down the slope, from higher to lower potential energy. Thus, it would be useful to know where all the points of equal potential energy are in a given region of space. That way, we can easily infer the directions of the forces on each of the charges. An **equipotential surface** is defined as a surface having the same potential at all points on it. Topographical maps show equipotential surfaces (lines on a two-dimensional map) in regard to gravitational potential energy.

Let's consider the electric field associated with a source consisting of a single fixed point charge we designate as the source charge. What happens if we place a test charge at a distance r from the source charge and move it around? If we move the charge anywhere on the surface of a sphere of radius r, no electrostatic work is done on the test charge as it is always moving perpendicular to the electric field vectors. However, we cannot move our test charge from one distance from the source charge to another distance without the electric field doing work on it. This is illustrated in Fig. 25-6. Thus, any sphere centered on the source charge is an equipotential surface. If our source charge is positive, then the potential decreases as the distance from the source charge increases. We know this because from $\Delta U = -W^{\text{elec}}$ (Eq. 25-1) a charge naturally moves from high potential energy to low. Thus the equipotential surfaces

FIGURE 25-6 ■ All of the electric field vectors created by the presence of a single charge point radially outward in three dimensions. If a test charge moves around on a sphere that is centered on the charge (where the dashed circle shows a cross section of the sphere), no work is done on it by the electric field since all the electric field vectors on surface elements of the sphere are normal to the sphere. If the charge is moved from one radius to the other (black squiggly line) it has to move parallel to the field vectors some of the time, and work is done on it.

associated with a positive point charge consist of an infinite family of concentric spheres centered on the source charge. Each sphere has a different potential.

An equipotential surface can be either imaginary, such as a mathematical sphere, or a real, physical surface such as the outside of a wire. The set of all equipotential surfaces fills all of space, since every point in space has some value of electric potential associated with it. We could draw an equipotential surface through any one of these points, just like we can draw a field line through every point in space. However, in order to simplify illustrations and diagrams, we typically show just a few of the surfaces.

No work W^{elec} is done on a charged particle by an electric field when the particle moves between two points i and f on the same equipotential surface. This follows from Eq. 25-8,

$$\Delta V = V_2 - V_1 = -\frac{W^{\mathrm{elec}}}{q},$$

which tells us W^{elec} must be zero if $V_2 = V_1$. Because of the path independence of work (and thus of potential energy and potential), $W^{\mathrm{elec}} = 0$ for *any* path connecting points 1 and 2, regardless of whether that path lies entirely on the equipotential surface. In other words, if the charge moves away from the equipotential surface during the motion, the work done (positive or negative) is exactly canceled by the work done (negative or positive) in moving back onto the surface.

Figure 25-7 shows a *family* of equipotential surfaces associated with the electric field due to some distribution of charges. The work done by the electrostatic force on a charged particle as the particle moves from one end to the other of paths I and II is zero because each of these paths begins and ends on the same equipotential surface. The work done as the charged particle moves from one end to the other of paths III and IV is not zero but has the same value for both these paths because the initial and final potentials are identical for the two paths; that is, paths III and IV connect the same pair of equipotential surfaces.

As we already noted, the equipotential surfaces produced by a point charge or a spherically symmetrical charge distribution are a family of concentric spheres. For a uniform electric field it is not difficult to see that the equipotential surfaces are a family of planes perpendicular to the field lines.

The fact that the value of the potential is constant along an equipotential surface implies that the electric field must always be perpendicular to the equipotential surfaces. Why? Because, if $\vec{E}$ were *not* perpendicular to an equipotential surface, it would have a component lying along that surface. This component would then do work on a

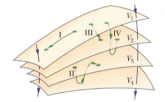

FIGURE 25-7 ■ Portions of four equipotential surfaces at electric potentials $V_1 = 100$ V, $V_2 = 80$ V, $V_3 = 60$ V, and $V_4 = 40$ V. Four paths along which a test charge may move are shown. Two electric field lines are also indicated.

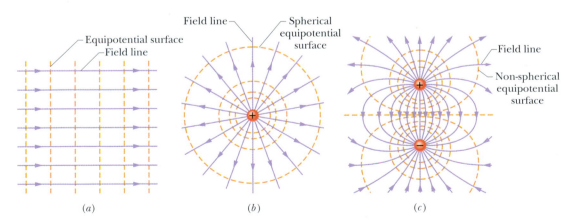

FIGURE 25-8 ■ Electric field lines (solid purple lines with arrows) and cross sections of equipotential surfaces (dashed gold lines) for (*a*) a uniform field with planar equipotential surfaces, (*b*) the field of a point charge with spherical equipotential surfaces, and (c) the field of an electric dipole with distorted equipotential surfaces that are not quite spherical.

charged particle as it moved along the surface. However, to prove that work cannot be done if the surface is truly an equipotential surface we use Eq. 25-8 once again,

$$\Delta V = V_2 - V_1 = -\frac{W^{\text{elec}}}{q}.$$

The only possible conclusion is that the electric field lines must be perpendicular to the surface everywhere along it.

    If electric field lines are perpendicular to an equipotential surface, then conversely the equipotential surface must be perpendicular to the field lines. Thus, equipotential surfaces are always perpendicular to the direction of the electric field $\vec{E}$, which is tangent to the field lines. Figure 25-8 shows electric field lines and cross sections of the equipotential surfaces for a uniform electric field and for the field associated with a point charge and with an electric dipole.

25-5 Calculating Potential from an *E*-Field

Can we calculate the potential difference between any two points 1 and 2 in an electric field if we know the electric field vector $\vec{E}$ all along any path connecting those points? We can if we can find the work done on a charge by the field as the charge moves from 1 to 2, and then use Eq. 25-8 again,

$$\Delta V = V_2 - V_1 = -\frac{W^{\text{elec}}}{q}.$$

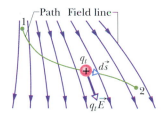

    For example, consider an arbitrary electric field, represented by the field lines in Fig. 25-9, and a positive test charge q_t moving along the path shown from point 1 to point 2. At any point on the path, an electrostatic force $q_t\vec{E}$ acts on the charge as it moves through an infinitesimally small differential displacement $d\vec{s}$. From Chapter 9, we know the differential work dW done on a particle by a force $\vec{F}$ during a displacement $d\vec{s}$ is

$$dW = \vec{F} \cdot d\vec{s}. \tag{25-13}$$

For the situation of Fig. 25-9, $\vec{F} = q_t\vec{E}$, and Eq. 25-13 becomes

$$dW^{\text{elec}} = q_t\vec{E} \cdot d\vec{s}. \tag{25-14}$$

FIGURE 25-9 ■ A test charge q_t moves from point 1 to point 2 along the path shown in a nonuniform electric field represented by curved electric field lines. During a displacement $d\vec{s}$, an electrostatic force $q_t\vec{E}$ acts on the test charge. This force points in the direction of the field line at the location of the test charge.

To find the total work W^{elec} done on the particle by the field as the particle moves from point 1 to point 2, we sum—via integration—the differential work done on the charge as it moves through all the differential displacements $d\vec{s}$ along the path:

$$W^{\text{elec}} = q_t\int_1^2 \vec{E} \cdot d\vec{s}. \tag{25-15}$$

If we substitute the total electrical work W^{elec} from Eq. 25-15 into Eq. 25-8, $\Delta V = V_2 - V_1 = -W^{\text{elec}}/q$, we find

$$V_2 - V_1 = -\int_1^2 \vec{E} \cdot d\vec{s}. \tag{25-16}$$

Thus, the potential difference $V_2 - V_1$ between any two points 1 and 2 in an electric field is equal to the negative of the *line integral* (meaning the integral along a particular path) of $\vec{E} \cdot d\vec{s}$ from 1 to 2. However, because the electrostatic force is conservative, all paths (whether easy or difficult to use) yield the same result. So, choose an easy-to-use path.

If the electric field is known throughout a certain region, Eq. 25-16 allows us to calculate the difference in potential between any two points in the field. If we choose the potential V_1 at point 1 to be zero, then Eq. 25-16 becomes

$$V = -\int_1^2 \vec{E} \cdot d\vec{s} \qquad \text{(for } V_1 = 0\text{)}, \qquad (25\text{-}17)$$

where we have dropped the subscript 2 on V_2. Equation 25-17 gives us the potential V at any point 2 in the electric field *relative to the zero potential* at point 1. If we let point 1 be at infinity, then Eq. 25-17 gives us the potential V at any point 2 relative to the zero potential at infinity.

READING EXERCISE 25-4: The figure shows a family of parallel equipotential surfaces (in cross section) and five paths along which we shall move an electron from one surface to another. (a) What is the direction of the electric field associated with the surfaces? (b) For each path, is the work we do positive, negative, or zero? (c) Rank the paths according to the work we do, greatest first.

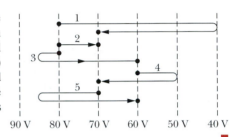

90 V   80 V   70 V   60 V   50 V   40 V

TOUCHSTONE EXAMPLE 25-2: Finding the Potential Difference

(a) Figure 25-10a shows two points 1 and 2 in a uniform electric field $\vec{E}$. The points lie on the same electric field line (not shown) and are separated by a distance d. Find the potential difference $V_2 - V_1$ by moving a positive test charge q_t from 1 to 2 along the path shown, which is parallel to the field direction.

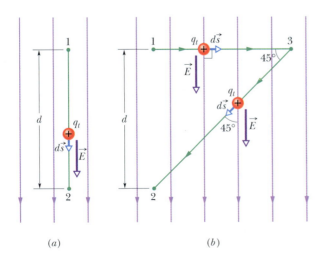

(a)   (b)

FIGURE 25-10 ■ (a) A test charge q_t moves in a straight line from point 1 to point 2, along the direction of a uniform electric field. (b) Charge q_t moves along path 1-3-2 in the same electric field.

SOLUTION ■ The **Key Idea** here is that we can find the potential difference between any two points in an electric field by integrating $\vec{E} \cdot d\vec{s}$ along a path connecting those two points according to Eq. 25-16. We do this by mentally moving a test charge q_t along that path, from initial point 1 to final point 2. As we move such a test charge along the path in Fig. 25-10a, its differential displacement $d\vec{s}$ always has the same direction as $\vec{E}$. Thus, the angle ϕ between $\vec{E}$ and $d\vec{s}$ is zero and the dot product in Eq. 25-16 is

$$\vec{E} \cdot d\vec{s} = |\vec{E}||d\vec{s}|\cos\phi = |\vec{E}||d\vec{s}|. \qquad (25\text{-}18)$$

Equations 25-16 and 25-18 then give us

$$V_2 - V_1 = -\int_1^2 \vec{E} \cdot d\vec{s} = -\int_1^2 |\vec{E}||d\vec{s}|. \qquad (25\text{-}19)$$

Since the field is uniform, E is constant over the path and can be moved outside the integral, giving us

$$V_2 - V_1 = -|\vec{E}|\int_1^2 |d\vec{s}| = -|\vec{E}||\vec{d}|,$$

in which the integral is simply the length d of the path. The minus sign in the result shows that the potential at point 2 in Fig. 25-10a is lower than the potential at point 1. This is a general result: The potential always decreases along a path that extends in the direction of the electric field lines.

(b) Now find the potential difference $V_2 - V_1$ by moving the positive test charge q_t from 1 to 2 along the path 1-3-2 shown in Fig. 25-10b.

SOLUTION ■ The **Key Idea** of (a) applies here too, except now we move the test charge along a path that consists of two lines: 1-3 and 3-2. At all points along line 1-3, the displacement $d\vec{s}$ of the test charge is perpendicular to $\vec{E}$. Thus, the angle ϕ between $\vec{E}$ and $d\vec{s}$ is 90°, and the dot product $\vec{E} \cdot d\vec{s}$ is 0. Equation 25-16 then tells us that points 1 and 3 are at the same potential: $V_3 - V_1 = 0$.

For line 3-2 we have $\phi = 45°$ and, from Eq. 25-16,

$$V_2 - V_1 = V_2 - V_3 = -\int_3^2 \vec{E} \cdot d\vec{s} = -\int_3^2 |\vec{E}|(\cos 45°)|d\vec{s}|$$

$$= -|\vec{E}|(\cos 45°)\int_3^2 |d\vec{s}|.$$

The integral in this equation is just the length of line 3-2; from Fig. 25-10b, that length is $d/\sin 45°$. Thus,

$$V_2 - V_1 = -|\vec{E}|(\cos 45°)\frac{|\vec{d}|}{\sin 45°} = -|\vec{E}||\vec{d}|. \qquad \text{(Answer)}$$

This is the same result we obtained in (a), as it must be; the potential difference between two points does not depend on the path connecting them. Moral: When you want to find the potential difference between two points by moving a test charge between them, you can save time and work by choosing a path that simplifies the use of Eq. 25-16.

25-6 Potential Due to a Point Charge

Imagine a single point charge in space. What would the value of the potential be at a distance of 3 m away from the charge? Consider a point P at a distance R from a fixed particle of positive charge q as in Fig. 25-11. To use Eq. 25-16,

$$V_2 - V_1 = -\int_1^2 \vec{E} \cdot d\vec{s},$$

we imagine that we move a positive test charge q_t from infinity to its final location at point P. We need to bring our test charge from infinity to a point P that is a distance R from the source charge. Because the path we choose will not change our final result, we are free to choose it. Mathematically, the simplest path between infinity and point P involves traveling along the same line that the electric field vectors lie along so no nonradial vector components of the electric field have to be considered.

We must then evaluate the dot product

$$\vec{E} \cdot d\vec{s} = |\vec{E}|\cos\phi|d\vec{s}| = (E)(ds)\cos\phi. \qquad (25\text{-}20)$$

The electric field $\vec{E}$ in Fig. 25-11 is directed radially outward from the fixed particle. So the differential displacement $d\vec{s}$ of the test particle along our chosen path is radially inward and has the opposite direction as $\vec{E}$. That means that the angle $\phi = 180°$ and $\cos\phi = -1$. Because the path is radial, let us write ds as dr. Then, substituting the limits ∞ and R, we can write Eq. 25-16,

$$V_2 - V_1 = -\int_1^2 \vec{E} \cdot d\vec{s},$$

as

$$V_2 - V_1 = -\int_\infty^R E_r \, dr, \qquad (25\text{-}21)$$

where E_r is the component of the electric field in the radial direction. Next we set $V_1 = 0$ (at ∞) and $V_2 = V$ (at R). Then, for the magnitude of the electric field at the site of the test charge, we substitute from Chapter 23:

$$E = k\frac{|q|}{r^2}. \qquad (25\text{-}22)$$

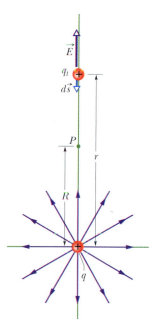

FIGURE 25-11 ■ The positive point charge q produces an electric field $\vec{E}$ and an electric potential ΔV at point P. We find the potential by moving a test charge q_t from its initial location at infinity to a point P. The test charge is shown at distance r from the point charge undergoing differential displacement $d\vec{s}$.

With these changes, Eq. 25-21 then gives us

$$V - 0 = -k|q| \int_\infty^R \frac{1}{r^2} dr = k|q| \left[\frac{1}{r} \right]_\infty^R$$

$$= k\frac{|q|}{R} \qquad \text{(for positive } q\text{).} \qquad (25\text{-}23)$$

We want to generalize finding the potential relative to infinity for any distance, not just distance R. So, switching from R to r, we have an expression for potential at a distance r from a source charge of

$$V = k\frac{|q|}{r} \qquad \text{(for positive } q\text{).}$$

Although we have derived this expression above for a positively charged particle, the derivation also holds for a negatively charged particle as well. However, if q in Fig. 25-11 were a negative charge, the electric field vectors would point in the same direction as the path (radially inward). Thus, the differential displacement $d\vec{s}$ of the test particle along our chosen path has the same direction as $\vec{E}$. That means the angle $\phi = 0°$ and so $\cos \phi = +1$. This introduces a negative sign that remains throughout the derivation and results in a negative final result for the potential. So, we conclude that *the sign of V is the same as the sign of q*. This gives us

$$V = k\frac{q}{r} \qquad \text{(relative to infinity for either sign of charge),} \qquad (25\text{-}24)$$

as the electric potential V relative to infinity due to a particle of charge q at any radial distance r from the particle.

> A positively charged particle produces a positive electric potential. A negatively charged particle produces a negative electric potential.

Figure 25-12 shows a computer-generated plot of Eq. 25-24 for a positively charged particle; the magnitude of V is plotted vertically. Note that the magnitude increases as $r \to 0$. In fact, according to the expression above, V is infinite at $r = 0$, although Fig. 25-12 shows a finite, smoothed-off value there.

Equation 25-24 also gives the electric potential *outside or on the external surface of* a spherically symmetric charge distribution. We can prove this by using an

FIGURE 25-12 ■ (a) A computer-generated plot of the electric potential $V(r)$ due to a positive point charge located at the origin of an x-y plane. The potentials at points in that plane are plotted vertically. (Curved lines have been added to help you visualize the plot.) The infinite value of V predicted by Eq. 25-24 for $r = 0$ is not plotted. (b) The same plot of electric potential is shown for a negative charge.

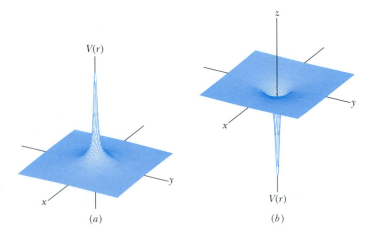

electrostatic analogy to the shell theorem we found so useful in our study of gravitation (Section 14-2). This theorem allows us to replace the actual spherical charge distribution with an equal charge concentrated at its center. Then the derivation leading to Eq. 25-24 follows, provided we do not consider a point within the actual distribution.

TOUCHSTONE EXAMPLE 25-3: Near a Proton

The nucleus of a hydrogen atom consists of a single proton, which can be treated as a particle (or point charge).

(a) With the electric potential equal to zero at infinite distance, what is the electric potential V due to the proton at a radial distance $r = 2.12 \times 10^{-10}$ m from it?

SOLUTION ■ The **Key Idea** here is that, because we can treat the proton as a particle, the electric potential V it produces at distance r is given by Eq. 25-24,

$$V = k\frac{q}{r}.$$

Here charge q is $e(=1.6 \times 10^{-19}$ C). Substituting this and the given value for r, we find

$$V = \frac{(8.99 \times 10^9 \text{ N} \cdot \text{m}^2/\text{C}^2)(1.60 \times 10^{-19} \text{ C})}{2.12 \times 10^{-10} \text{ m}}$$

$$= 6.78 \text{ V}. \qquad \text{(Answer)}$$

(b) What is the electric potential energy U in electron-volts of an electron at the given distance from the nucleus? (The potential energy is actually that of the electron–proton system—the hydrogen atom.)

SOLUTION ■ The **Key Idea** here is that when a particle of charge q is located at a point where the electric potential due to other charges is V, the electric potential energy U is given by Eq. 25-6 ($V = U/q$). Using the electron's charge $-e$, we find

$$U = qV = (-1.60 \times 10^{-19} \text{ C})(6.78 \text{ V})$$

$$= -1.0848 \times 10^{-18} \text{ J} = -6.78 \text{ eV}. \qquad \text{(Answer)}$$

(c) If the electron moves closer to the proton, does the electric potential energy increase or decrease?

SOLUTION ■ The **Key Ideas** of parts (a) and (b) apply here also. As the electron moves closer to the proton, the electric potential V due to the proton at the electron's position increases because r decreases). Thus, the value of V in part (b) increases. Because the electron is negatively charged, this means that the value of U becomes more negative. Hence, the potential energy U of the electron (that is, of the system or atom) decreases.

25-7 Potential and Potential Energy Due to a Group of Point Charges

Now let's consider what happens when there are lots of charges. First we will look at the case where we only move a small test charge while all the other charges remain fixed. In this case, the changes in the system's potential energy as the test charge moves lead us to the same definition of electric potential, V, as we already developed. At the end of this section we consider the situation in which many charges move and find that the total potential energy of the system changes. In this case, even though the system has a potential energy associated with it, we cannot define an electric potential.

We found in Chapter 23 that the electric field arising from a group of point charges satisfies a superposition principle. That is, the total electric field is the sum of the individual electric fields arising from each individual point charge. Since the potential V is the line integral of the electric field and the integral of a sum of terms is the sum of the integrals, the superposition principle also holds for electrostatic potential.

Hence, we use the principle of superposition to find the electric potential at a particular location due to a group of point charges. We calculate the potential resulting from the influence of each charge in the system one at a time, using Eq. 25-24 with the sign of the charge included. Then we sum the potentials. For n charges, the net potential (measured relative to a zero at infinity) is

$$V = \sum_{i=1}^{n} V_i = k \sum_{i=1}^{n} \frac{q_i}{r_i} \qquad \text{(potential due to } n \text{ point charges).} \qquad (25\text{-}25)$$

Here q_i is the value of the ith charge, and r_i is the radial distance of the given point from the ith charge. The sum in Eq. 25-25 is an *algebraic sum,* not a vector sum like the sum used to calculate the electric field resulting from a group of point charges. Herein lies an important computational advantage of potential over electric field: it is a lot easier to sum several scalar quantities than to sum several vector quantities whose directions and components must be considered.

In Section 25-2, we discussed the electric potential energy of a charged particle as an electrostatic force does work on it. In that section, we assumed that the charges that produced the force were fixed in place, so that neither the force nor the corresponding electric field could be influenced by the presence of the test charge. If we consider a system with charges that move when a test charge moves around, there is no logical way to determine a charge-independent electric potential for it. The electric field will keep changing due to the presence of the test charge. But we can take a broader view and find the electric potential energy of the entire *system* of charges due to the electric field produced *by* those same charges.

We can start simply by pushing two bodies that have charges of like sign into the same vicinity. For example, imagine that we have one excess electron on the conducting shell of the Van de Graaff generator shown in Fig. 25-3 and we want to put a second electron in place. Our second electron is sprayed on the insulated belt and the generator motor does work as it forces the second electron toward the conducting shell in the presence of the first one. The first electron is no doubt relocating and acting on the second electron during the forcing process. Nonetheless, we can keep track of the work the motor does. This work is stored as electric potential energy in the two-body system (provided the kinetic energy of the bodies does not change). As we bring up a third electron we can measure the work we have to bring it up to the shell in the presence of the other two electrons, which are relocating as a result of the interactions of all three electrons. The work needed to bring the third electron to the conducting shell adds to the work needed to bring up the second electron. The total work is stored as the potential energy of the three-body system. This process of doing more work and causing the excess electrons on the shell to relocate goes on until there are billions and billions of electrons on the conducting shell. If you later release the charges but touch the shell with a conductor attached to the ground, you can recover this stored energy, in whole or in part, as the kinetic energy of the charged bodies as they rush away from each other.

We define the electric potential energy *of a system of point charges* in terms of the final locations of all the charges as follows:

> The electric potential energy of a system of point charges that are not moving is equal to the work that must be done by an external agent to assemble the system one charge at a time.

We assume that the charges are stationary both in their initial infinitely distant positions and in their final assembled configuration. In equation form, the total electric potential energy of the system is given by the sum of the potential energies of all the possible pairs in the system so that

$$U = \sum_{\text{all pairs}} k \frac{q_i q_j}{|\vec{r}_i - \vec{r}_j|} \qquad \text{(system potential energy of } n \text{ point charges).} \qquad (25\text{-}26)$$

$i < j$

READING EXERCISE 25-5: So far in this chapter, we have discussed two ways to calculate the electric potential V. Describe how one would calculate the electric potential given information about the charge distribution (the magnitudes of the charges and where they are located). Describe how one would calculate the electric potential given information regarding the electric field $\vec{E}$. ∎

READING EXERCISE 25-6: The figure shows three arrangements of two protons. Rank the arrangements according to the net electric potential produced at point P by the protons, greatest first.

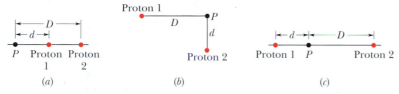

(a)   (b)   (c) ∎

TOUCHSTONE EXAMPLE 25-4: A Square of Charges

What is the electric potential at point P, located at the center of the square of point charges shown in Fig. 25-13a? The distance d is 1.3 m, and the charges are

$$q_1 = +12 \text{ nC}, \qquad q_3 = +31 \text{ nC},$$
$$q_2 = -24 \text{ nC}, \qquad q_4 = +17 \text{ nC}.$$

SOLUTION ∎ The **Key Idea** here is that the electric potential V at P is the algebraic sum of the electric potentials contributed

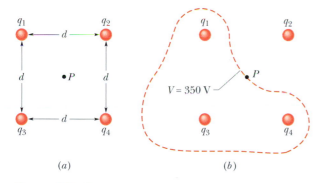

(a)   (b)

FIGURE 25-13 ∎ (a) Four point charges are held fixed at the corners of a square. (b) The closed curve is a cross section, in the plane of the figure, of the equipotential surface that contains point P. (The curve is only roughly drawn.)

by the four point charges. (Because electric potential is a scalar, the orientations of the point charges do not matter.) Thus, from Eq. 25-25, we have

$$V = \sum_{i=1}^{4} V_i = k \left(\frac{q_1}{r} + \frac{q_2}{r} + \frac{q_3}{r} + \frac{q_4}{r} \right).$$

The distance r is $d/\sqrt{2}$, which is 0.919 m, and the sum of the charges is

$$q_1 + q_2 + q_3 + q_4 = (12 - 24 + 31 + 17) \times 10^{-9} \text{ C}$$
$$= 36 \times 10^{-9} \text{ C}.$$

Thus, $\qquad V = \dfrac{(8.99 \times 10^9 \text{N} \cdot \text{m}^2/\text{C}^2)(36 \times 10^{-9} \text{ C})}{0.919 \text{ m}}$

$$\approx 350 \text{ V}. \qquad \text{(Answer)}$$

Close to any of the three positive charges in Fig. 25-13a, the potential has very large positive values. Close to the single negative charge, the potential has very large negative values. Therefore, there must be points within the square that have the same intermediate potential as that at point P. The curve in Fig. 25-13b shows the intersection of the plane of the figure with the equipotential surface that contains point P. Any point along that curve has the same potential as point P.

TOUCHSTONE EXAMPLE 25-5: A Dozen Electrons

(a) In Fig. 25-14a, 12 electrons (of charge $-e$) are equally spaced and fixed around a circle of radius R. Relative to $V = 0$ at infinity, what are the electric potential and electric field at the center C of the circle due to these electrons?

SOLUTION ∎ The **Key Idea** here is that the electric potential V at C is the algebraic sum of the electric potentials contributed

by all the electrons. (Because electric potential is a scalar, the orientations of the electrons do not matter.) Because the electrons all have the same negative charge $-e$ and are all the same distance R from C, Eq. 25-25 gives us

$$\Delta V = -12k\frac{e}{R}. \qquad \text{(Answer)} \qquad (25\text{-}27)$$

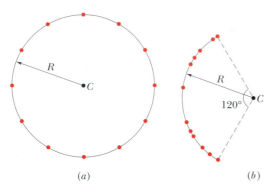

FIGURE 25-14 ■ (a) Twelve electrons uniformly spaced around a circle. (b) Those electrons are now nonuniformly spaced along an arc of the original circle.

For the electric field at C, the **Key Idea** is that electric field is a vector quantity and thus the orientation of the electrons *is* important. Because of the symmetry of the arrangement in Fig. 25-14a, the electric field vector at C due to any given electron is canceled by the field vector due to the electron that is diametrically opposite it. Thus, at C,

$$\vec{E} = 0. \qquad \text{(Answer)}$$

(b) If the electrons are moved along the circle until they are nonuniformly spaced over a 120° arc (Fig. 25-14b), what then is the potential at C? How does the electric field at C change (if at all)?

SOLUTION ■ The potential is still given by Eq. 25-27, because the distance between C and each electron is unchanged and orientation is irrelevant. The electric field is no longer zero, because the arrangement is no longer symmetric. There is now a net field that is directed toward the charge distribution.

25-8 Potential Due to an Electric Dipole

Electrically neutral matter is made of equal amounts of positive and negative charges. Electric forces pull in opposite directions on those charges. Thus, an electric field can cause a small separation of the positive and negative charges in matter (called polarization). In addition, many molecules distribute their electrons throughout their volume in a nonuniform way. This results in their having more positive charge on one end and one negative charge on the other end. For example, the water molecule shown in Fig. 23-22 has a nonuniform charge distribution.

A small separation produces an electric field very similar to that of a pair of equal and opposite charges separated by a small distance. If the charges were right on top of each other, their electric fields would cancel and they would appear neutral. But if they are a bit separated, their fields don't cancel perfectly, leaving a field pattern known as an electric dipole. The electric dipole fields produced by molecules play an essential role in a large number of processes in chemistry and biology, as well as in determining the electrical properties of matter such as color and transparency.

Now let us apply Eq. 25-25,

$$V = \sum_{i=1}^{n} V_i = \frac{1}{4\pi\varepsilon_0} \sum_{i=1}^{n} \frac{q_i}{r_i},$$

to an electric dipole to find the potential at an arbitrary point P in Fig. 25-15a. At P, the positive point charge (at distance $r_{(+)}$) sets up potential $V_{(+)}$ and the negative point charge (at distance $r_{(-)}$) sets up potential $V_{(-)}$. Then the net potential at P is given by Eq. 25-25 as

$$V = \sum_{i=1}^{2} V_i = V_{(+)} + V_{(-)} = \frac{1}{4\pi\varepsilon_0}\left(\frac{q}{r_{(+)}} + \frac{-q}{r_{(-)}}\right)$$

$$= \frac{q}{4\pi\varepsilon_0}\frac{r_{(-)} - r_{(+)}}{r_{(-)}r_{(+)}}.$$

(25-28)

Naturally occurring dipoles—such as those possessed by many molecules—are quite small, so we are usually interested only in points that are relatively far from the

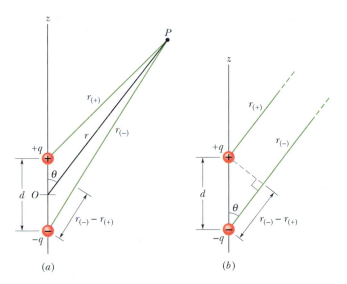

(a)   (b)

FIGURE 25-15 ■ (a) Point P is a distance r from the midpoint O of a dipole. The line OP makes an angle θ with the dipole axis. (b) If P is far from the dipole, the lines of lengths $r_{(+)}$ and $r_{(-)}$ are approximately parallel to the line of length r, and the dashed line is approximately perpendicular to the line of length $r_{(-)}$.

dipole, such that $r \gg d$, where d is the distance between the charges. Under those conditions, the approximations that follow from Fig. 25-15b are

$$r_{(-)} - r_{(+)} \approx d \cos \theta \quad \text{and} \quad r_{(-)}r_{(+)} \approx r^2.$$

If we substitute these quantities into Eq. 25-28, we can approximate V to be

$$V \approx \frac{q}{4\pi\varepsilon_0} \frac{d \cos \theta}{r^2} \quad \text{(for } r \gg d).$$

Here θ is measured from the dipole axis as shown in Fig. 25-15a. We can now write V as

$$V \approx k \frac{p \cos \theta}{r^2} \quad \text{(electric dipole for } r \gg d), \quad (25\text{-}29)$$

in which $p(=qd)$ is the magnitude of the electric dipole moment $\vec{p}$ defined in Section 23-7. The vector $\vec{p}$ is directed along the dipole axis, from the negative to the positive charge. (Thus, θ is measured from the direction of $\vec{p}$.)

Induced Dipole Moment

Many molecules such as water have *permanent* electric dipole moments. In other molecules (called nonpolar molecules) and in every isolated atom, the centers of the positive and negative charges coincide (Fig. 25-16a) and thus no dipole moment is set up. However, if we place an atom or a nonpolar molecule in an external electric field, the field affects the locations of the electrons relative to the nuclei and separates the

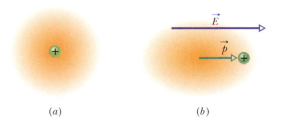

(a)   (b)

FIGURE 25-16 ■ (a) An atom, showing the positively charged nucleus (green) and a cloud of negatively charged electrons (gold shading). The centers of positive and negative charge coincide. (b) If the atom is placed in an external electric field $\vec{E}$, the electron orbits are distorted so that the centers of positive and negative charge no longer coincide. An induced dipole moment $\vec{p}$ appears. The distortion is exaggerated here by many orders of magnitude.

centers of positive and negative charge (Fig. 25-16b). Because the electrons are negatively charged, they tend to be shifted in a direction opposite the field. This shift sets up a dipole moment $\vec{p}$ pointing in the direction of the field. This dipole moment is said to be induced by the field, and the atom or molecule is then said to be polarized by the field (it has a positive side and a negative side). When the field is removed, the induced dipole moment and the polarization disappear.

READING EXERCISE 25-7: Suppose three points are set at equal (large) distances r from the center of the dipole in Fig. 25-15: Point a is on the dipole axis above the positive charge, point b is on the axis below the negative charge, and point c is on a perpendicular bisector through the line connecting the two charges. Rank the points according to the electric potential of the dipole there, greatest (most positive) first. ■

25-9 Potential Due to a Continuous Charge Distribution

When a charge distribution q is continuous (as on a uniformly charged thin rod or disk), we cannot use a summation to find the potential V at a point P. Instead, we must choose a differential element of charge dq. A differential element of charge is a very small bit of charge, small enough so we can treat it as if it were a point charge. We can then determine the potential dV at P due to dq, and then integrate over the entire charge distribution.

Let us again take the zero of potential to be at infinity. If we treat the element of charge dq as a point charge, then we can use Eq. 25-24,

$$V = k\frac{q}{r},$$

to express the potential dV at point P due to dq:

$$dV = k\frac{dq}{r} \qquad \text{(positive or negative } dq\text{).} \qquad (25\text{-}30)$$

Here r is the distance between P and dq. To find the total potential V at P, we integrate to sum the potentials due to all the charge elements:

$$V = \int dV = k\int \frac{dq}{r}. \qquad (25\text{-}31)$$

The integral must be taken over the entire charge distribution. Note that because the electric potential is a scalar, there are *no vector components* to consider in the equation above.

We now examine a continuous charge distribution, a line of charge.

Line of Charge

In Fig. 25-17a, a thin, nonconducting rod of length L has a positive charge of uniform linear density λ. Let us determine the electric potential V due to the rod at point P, a perpendicular distance d from the left end of the rod.

We consider a differential element dx of the rod as shown in Fig. 25-17b. This (or any other) element of the rod has a differential charge of

$$dq = \lambda\,dx. \qquad (25\text{-}32)$$

This element produces a potential dV at point P, which is a distance $r = (x^2 + d^2)^{1/2}$ from the element. Treating the element as a point charge, we can use Eq. 25-30,

$$dV = k\frac{dq}{r},$$

to write the potential dV as

$$dV = k\frac{dq}{r} = k\frac{\lambda\,dx}{(x^2 + d^2)^{1/2}}. \tag{25-33}$$

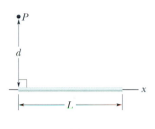

Since the charge on the rod is positive and we have taken $V = 0$ at infinity, we know dV in this expression must be positive.

We now find the total potential V (measured relative to a zero at infinity) produced by the rod at point P by integrating along the length of the rod, from $x = 0$ to $x = L$. We evaluate the integral using an integral table or a symbolic manipulation program like Mathcad or Maple. We then find

$$V = \int dV = \int_0^L k\frac{\lambda}{(x^2 + d^2)^{1/2}}\,dx$$

$$= k\lambda \int_0^L \frac{dx}{(x^2 + d^2)^{1/2}}$$

$$= k\lambda[\ln(x + (x^2 + d^2)^{1/2})]_0^L$$

$$= k\lambda[\ln(L + (L^2 + d^2)^{1/2}) - \ln d].$$

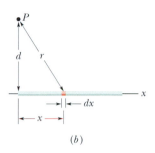

We can simplify this result by using the general relation $\ln A - \ln B = \ln(A/B)$. We then find

$$V = k\lambda \ln\left[\frac{L + (L^2 + d^2)^{1/2}}{d}\right]. \tag{25-34}$$

Because V is the sum of positive values of dV, it should be positive—but does this expression give a positive V? Since the argument of the logarithm is greater than one, the logarithm is a positive number and V is indeed positive.

FIGURE 25-17 ■ (a) A thin, uniformly charged rod produces an electric potential V at point P. (b) A differential element of charge produces a differential potential dV at P.

25-10 Calculating the Electric Field from the Potential

In Section 25-5, you saw how to find the potential at a point f if you know the electric field along a path from a reference point to point f. In this section, we propose to go the other way—that is, to find the electric field when we know the potential. As Fig. 25-8 shows, graphically finding the direction of the field is easy: If we know the potential V at all points near an assembly of charges, we can draw in a family of equipotential surfaces. The electric field lines, sketched perpendicular to those surfaces, reveal the direction of $\vec{E}$. What we are seeking here is the mathematical equivalent of this graphical procedure.

Figure 25-18 shows cross sections of a family of closely spaced equipotential surfaces, the potential difference between each pair of adjacent surfaces being dV. As the figure suggests, the field $\vec{E}$ at any point P is perpendicular to the equipotential surface through P.

Suppose a positive test charge q_t moves through a displacement $d\vec{s}$ from one equipotential surface to the adjacent surface. From Eq. 25-8, we can relate the change

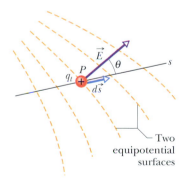

FIGURE 25-18 ■ A test charge q_t undergoes a displacement $d\vec{s}$ from one equipotential surface to another. (The separation between the surfaces has been exaggerated for clarity.) The displacement $d\vec{s}$ makes an angle θ with the direction of the electric field $\vec{E}$.

in electric potential to the work done by the electric field on our test charge

$$\Delta V = V_2 - V_1 = -\frac{W^{\text{elec}}}{q_t}.$$

Let's consider the potential difference associated with an infinitesimally small displacement denoted by $d\vec{s}$. We see that the electric field does an infinitesimal amount of work on the test charge during the move. Using Eq. 25-8, we can denote this as $-q_t dV$. From Eq. 25-14, $dW = q_t \vec{E} \cdot d\vec{s}$, and Fig. 25-18, we see that the infinitesimal work done by the force may also be written as $(q_t\vec{E}) \cdot d\vec{s}$ or $q_t |\vec{E}| (\cos \phi) |d\vec{s}|$, where ϕ is the angle between the electric field and displacement vectors as shown in Fig. 25-18. Equating these two expressions for the work yields

$$-q_t \, dV = q_t |\vec{E}|(\cos \phi) |d\vec{s}|, \tag{25-35}$$

or

$$\vec{E}(\cos \phi) = -\frac{dV}{ds}. \tag{25-36}$$

Since $E_s = |\vec{E}| \cos \phi$ is the component of $\vec{E}$ in the direction of $d\vec{s}$, the equation above becomes

$$E_s = -\frac{\partial V}{\partial s}. \tag{25-37}$$

We have added a subscript to the component of $\vec{E}$ and switched to the partial derivative symbols to emphasize that this expression involves only the variation of ΔV along a specified axis (here called the s axis) and only the component of $\vec{E}$ along that axis. In words, Eq. 25-37 is essentially the inverse of Eq. 25-16,

$$V_2 - V_1 = -\int_1^2 \vec{E} \cdot d\vec{s},$$

and states:

> The component of $\vec{E}$ in any direction is the negative of the rate of change of the electric potential with distance in that direction. Hence, $\vec{E}$ points in the direction of decreasing electric potential V.

If we take the s axis to be, in turn, the x, y, and z axes, we find that the x-, y-, and z-components of $\vec{E}$ at any point are

$$E_x = -\frac{\partial V}{\partial x}; \qquad E_y = -\frac{\partial V}{\partial y}; \qquad E_z = -\frac{\partial V}{\partial z}. \tag{25-38}$$

Thus, if we know V for all points in the region around a charge distribution—that is, if we know the function $V(x,y,z)$—we can find the components of $\vec{E}$, and thus $\vec{E}$ itself, at any point by taking partial derivatives. Each component of the electric field is simply the negative of the slope of the curve representing the electric potential vs. distance along each chosen axis.

For the simple situation in which the electric field $\vec{E}$ is uniform, the equipotential surfaces are a set of parallel planes that lie perpendicularly to the direction of the electric field. In addition, for a given potential difference, the distance between any two equipotential planes is the same. So, when the component of the electric field along the direction of $d\vec{s}$ is uniform, we can rewrite Eq. 25-37 ($E_s = -\partial V/\partial s$) in terms of the magnitude of the electric field $E = |\vec{E}|$ as

$$E = \left|\frac{\Delta V}{\Delta s}\right|,$$ (25-39)

where Δs is the component of displacement perpendicular to the equipotential surfaces. Equation 25-36 tells us that whenever the potential is constant along a surface so that $\Delta V = 0$, the electric field is zero. The component of the electric field is zero in any direction parallel to the equipotential surfaces. Thus, for a given potential difference ΔV, the magnitude of the electric field is given by the magnitude of the potential difference divided by the distance between any two equipotential surfaces.

READING EXERCISE 25-8: The figure shows three pairs of parallel plates with the same separation, and the electric potential of each plate. The electric field between the plates is uniform and perpendicular to the plates. (a) Rank the pairs according to the magnitude of the electric field between the plates, greatest first. (b) For which pair is the electric field pointing rightward? (c) If an electron is released midway between the third pair of plates, does it remain there, move rightward at constant speed, move leftward at constant speed, accelerate rightward, or accelerate leftward?

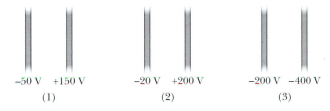

−50 V   +150 V   −20 V   +200 V   −200 V   −400 V

    (1)   (2)   (3) ■

READING EXERCISE 25-9: In what ways is the superposition principle for energy discussed above the same as, and different from, the superposition principle for electric field? ■

TOUCHSTONE EXAMPLE 25-6: Obtaining $\vec{E}$ from V

The electric potential at any point on the axis of a uniformly charged disk is given by

$$V = \frac{\sigma}{2\varepsilon_0}(\sqrt{z^2 + R^2} - z).$$

Starting with this expression, derive an expression for the electric field at any point on the axis of the disk.

SOLUTION ■ We want the electric field $\vec{E}$ as a function of distance z along the axis of the disk. For any value of z, the direction

of $\vec{E}$ must be along that axis because the disk has circular symmetry about that axis. Thus, we want the component E_z of $\vec{E}$ in the direction of z. Then the **Key Idea** is that this component is the negative of the rate of change of the electric potential with distance z. Thus, from the last of Eqs. 25-38, we can write

$$E_z = -\frac{\partial V}{\partial z} = -\frac{\sigma}{2\varepsilon_0}\frac{d}{dz}(\sqrt{z^2 + R^2} - z)$$

$$= \frac{\sigma}{2\varepsilon_0}\left(1 - \frac{z}{\sqrt{z^2 + R^2}}\right).$$ (Answer)

25-11 Potential of a Charged Isolated Conductor

In Section 24-8, we concluded $\vec{E} = 0$ for all points inside an electrically isolated conductor. We then used Gauss' law to prove that an excess charge placed on an isolated conductor lies entirely on its surface. (This is true even if the conductor has an empty internal cavity.) Here we use the first of these facts to prove an extension of the second:

An excess charge placed on an isolated conductor will distribute itself on the surface of that conductor so that all points of the conductor—whether on the surface or inside—come to the same potential. This is true even if the conductor has an internal cavity and even if that cavity contains a net charge.

This fact is rather obvious since any potential difference inside a conductor requires an electric field inside it. The nonzero electric field would, in turn, cause the free conduction electrons to redistribute themselves until the potential difference disappears.

The mathematical proof that an electrically isolated conductor is an equipotential region follows directly from Eq. 25-16,

$$V_2 - V_1 = -\int_1^2 \vec{E} \cdot d\vec{s}.$$

Since $\vec{E} = 0$ for all points within a conductor, it follows directly that $V_2 = V_1$ for all possible pairs of points i and f in the conductor.

A Spherical Shell with No External Electric Field

Figure 25-19a shows a plot of potential against radial distance r from the center for an isolated spherical conducting shell of 1.0 m radius, having a net excess charge of 1.0 μC. In the absence of an external field, we know by symmetry the surface charges will be uniformly distributed over the surface of the shell. For points outside the shell, we can calculate $V(r)$, the electric potential. Obviously this potential also has a spherical symmetry and can be given by Eq. 25-24,

$$V = k\frac{q}{r},$$

because the total charge on the shell, denoted as q, behaves for external points as if it were concentrated at the center of the shell. That equation holds right up to the surface of the shell. Now let us push a small test charge through the shell—assuming a small hole exists—to its center. No extra work is needed to do this because no net electric force acts on the test charge once it is inside the shell. Thus, the potential at all points inside the shell has the same value as on the surface, as shown in the Fig. 25-19a graph.

The Fig. 25-19b graph shows the variation of electric field with radial distance for the same shell. Note that $\vec{E} = 0$ everywhere inside the shell. The curves of Fig. 25-19b can be derived from the curve of Fig. 25-19a by differentiating with respect to r, using Eq. 25-37 (the derivative of a constant, recall, is zero). The curve of Fig. 25-19a can be derived from the curves of Fig. 25-19b by integrating

$$E_s = -\frac{\partial V}{\partial s}.$$

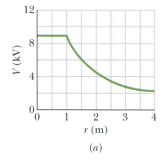

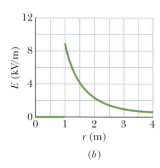

FIGURE 25-19 ■ (a) A plot of $V(r)$ both inside and outside a charged spherical shell of radius 1.0 m. (b) A plot of the electric field magnitude, $E(r)$, for the same shell.

The Charge Distribution on a Nonspherical Conductor

Consider a nonspherical charged conductor. Assume the conductor is electrically isolated and there is no external electric field in its vicinity. It turns out its surface charges do not distribute themselves uniformly. When compared to the uniform density of excess charge on a spherical conductor, the charges redistribute themselves so there is a higher charge density when the radius of curvature is convex and small and a lower charge density where the radius of curvature is concave and small (Fig. 25-20). Why? We can use the characteristics of equipotential surfaces to develop a qualitative explanation for this phenomenon.

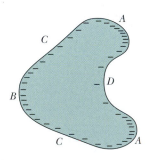

FIGURE 25-20 ■ The magnitude of the charge density on a conductor is greatest on a convex surface with a small radius of curvature (A) and least on a concave surface having small radius of curvature (D). The ranking of the magnitude of the charge density is $A > B > C > D$.

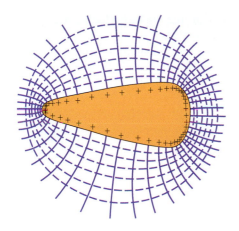

FIGURE 25-21 ■ The net positive charge on an odd-shaped isolated conductor distributes itself on the conductor's surface so the electric field generated by it is zero inside and normal to the surface elements of the conductor. This requires the equipotential surfaces (shown with dotted lines) to be closest together on the left where the conductor's convex radius of curvature is smallest. The electric field lines (shown with solid lines) and the excess charges also have the greatest density on the left where the curvature of the conductor's surface is smallest.

The explanation is as follows: There is no electric field inside the conductor, and the electric field at each point on the surface of the conductor must be normal (in other words perpendicular) to the surface. This requirement is obvious since any component of electric field parallel to the surface would cause free electrons to reconfigure themselves until all tangential components along the surface disappear. This also means the entire surface of our conductor is an equipotential surface no matter what its shape is. However, if we are far away from our charged conductor, the equipotential surfaces look more and more like those of a point charge. Thus, the family of equipotential surfaces that are each ΔV apart from the previous one become more and more spherical in shape. As the successive equipotential surfaces morph (change shape) slowly from that of our odd-shaped surface to that of a sphere, the parts of the equipotential surfaces near small-radius convex surface elements must be closer together than those elements having large radii of curvature. This is shown in Fig. 25-21. Now, equipotential surfaces more closely spaced occur where the electric field is the strongest and can do the most work on test charges, but the electric field is largest where the charge density that is its source is largest. The implication is that:

> On an isolated conductor the concentration of charges and hence the strength of the electric field is greater near sharp points where the curvature is large.

An Isolated Conductor in an External Electric Field

Suppose an *uncharged* isolated conductor is placed in an *external electric field*, as in Fig. 25-22. The electric field at the conductor's surface must have the same characteristics as it does when no external field is present. However, this doesn't mean its charges will be distributed in the same way as if no external electric field were present. All points of the conductor still come to a single potential regardless of whether the conductor is electrically neutral or has an excess charge. The free conduction electrons distribute themselves on the surface in such a way that the electric field they produce at interior points cancels the external electric field that would otherwise be there. Furthermore, the electron distribution causes the net electric field at all points on the surface to be normal to the surface. If the conductor in Fig. 25-22 could somehow be removed, leaving the surface charges frozen in place, the pattern of the electric field would remain absolutely unchanged for both exterior and interior points.

One common natural source of an external electric field that can affect isolated metal objects are excess negative charges at the bases of clouds contributing to the onset of thunderstorms. Such an external electric field can cause charge separation in conducting objects at the Earth's surface such as golf clubs and rock hammers. Since

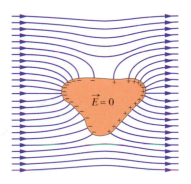

FIGURE 25-22 ■ An uncharged conductor is suspended in an external electric field. The free electrons in the conductor distribute themselves on the surface as shown, so as to reduce the net electric field inside the conductor to zero and make the field outside normal to each surface.

FIGURE 25-23 ■ This enhanced photograph shows the result of an overhead cloud system creating a strong electric field $\vec{E}$ near a woman's head. Many of the hair strands extended along the field, which was perpendicular to the equipotential surfaces and greatest where those surfaces were closest, near the top of her head.

these objects have points where the curvature is high, the surface charge density— and thus the external electric field, which is proportional to it—may reach very high values. The air around sharp points may become ionized, producing the corona discharge that golfers and mountaineers see on their tools when thunderstorms threaten. Such corona discharges, like hair that stands on end, are often the precursors of lightning strikes.

The cells and blood inside a human body contain salt water that acts as a conductor. The natural oil found on hair is also conductive. A person placed in a strong electric field can act like an uncharged conductor. For example, the woman shown in Fig. 25-23 was standing on a platform connected to the mountainside, and was at about the same potential as the mountainside. Overhead, a cloud system that had a high degree of charge separation with excess negative charges at its base moved in and created a strong electric field around her and the mountainside. Electrostatic forces due to this field drove some of the conduction electrons in the woman downward through her body, leaving her head and strands of her hair positively charged. The magnitude of this electric field was apparently large, but less than the value of about 3×10^6 V/m needed to cause electrical breakdown of the air molecules. (That value was exceeded when lightning struck the platform shortly after the picture was taken.)

As we just discussed, the surface charges on a nonspherical conductor concentrate in regions where the curvature is greatest. Thus, we expect the electric field to be greatest near the top of the woman's head—an equipotential surface. This suspicion is confirmed because the strands of her hair, containing excess positive charge, are pulled out most strongly where her head has the most curvature. Also, the strands of hair are extended along the direction of $\vec{E}$ perpendicular to her head. Since the magnitude of $\vec{E}$ was greatest just above her head, this is where the equipotential surfaces were most closely spaced. A sketch showing this close spacing is shown in Fig. 25-23.

The lesson here is simple. If an electric field causes the hairs on your head to stand up, you'd better run for shelter rather than pose for a snapshot.

What If Lightning Might Strike?

Speaking of lightning, what is the best way to protect yourself if lightning strikes? There are two ways to protect yourself using your knowledge of how conductors behave in electric fields. One is to enclose yourself in a relatively spherical conducting shell. The other is to use a lightning rod.

Using a Spherical Shell: If you enclose yourself inside a more or less spherical cavity, the electric field inside the cavity is guaranteed to be zero. A car (unless it is a convertible) is almost ideal (Fig. 25-24) because it protects the passengers from the effects of lightning for the same reason that the Faraday cage shown in Chapter 24 protects the demonstrator from the high voltage caused by the transfer of charge to the cage by a Van de Graaff generator.

Using a Lightning Rod: If you live in an area where thunderstorms are common, you can embed the base of a tall metal lightning rod in the ground. Recall that the bottoms of thunderclouds have an excess of negative charge that creates strong electric fields at the Earth's surface. What happens if a conducting rod, like the Eiffel Tower, has a sharp point *and* is taller than its immediate surroundings? A couple of factors come into play. First, the distance $|\Delta s|$ between the cloud bases and the top of a lightning rod is smaller than the distance to the ground, even though the electric field strength near the top of a tall rod is not really uniform. Equation 25-39 ($E = |\Delta V/\Delta s|$) tells us that the magnitude of the electric field between the cloud bases and the top of the rod is greater than that between the clouds and the ground. Second, as we discussed earlier in this section the magnitude of the electric field near a conductor that has a sharp point is quite strong compared to that on level

FIGURE 25-24 ■ A large spark jumps to a car's body and then exits by moving across the insulating left front tire, leaving the person inside unharmed because the electric potential difference remains zero inside the car.

ground. This means that the free electrons at the top of a lightning rod (such as the Eiffel Tower) will move toward the ground leaving a large accumulation of positive metal ions at the sharp point at the top of the tower. The tip of the rod will attract electrons from the atmosphere to it and down to the ground in a corona discharge process that can serve to prevent a major discharge or lightning strike in the vicinity of the tower. Lightning is shown hitting the Eiffel Tower in Fig. 25-25.

READING EXERCISE 25-10: The figure below shows the region in the neighborhood of a negatively charged conducting sphere and a large positively charged conducting plate extending far beyond the region shown. Someone claims lines *A* through *F* are possible field lines describing the electric field lying in the region between the two conductors. (a) Examine each of the lines and indicate whether it is a correctly drawn field line. If a line is not correct, explain why. (b) Redraw the diagram with a pattern of field lines that is more nearly correct. (Based on Arnold Arons' *Homework and Test Questions,* Wiley, New York, 1994.)

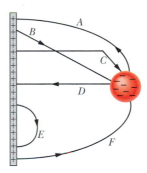

READING EXERCISE 25-11: Why are the equipotential surfaces shown in Fig. 25-23 closer together just above the woman's head than they are at the side of her head? ∎

FIGURE 25-25 ∎ In this historic 1902 postcard photo, bolts of lightning are shown converging at the top of the Eiffel Tower. The tower is acting as a "lightning rod" protecting people, trees, and other buildings from being struck by lightning.

Problems

SEC. 25-3 ∎ ELECTRIC POTENTIAL

1. Car Battery A particular 12 V car battery can send a total charge of ~~310 × 10³ C~~ through a circuit, from one terminal to the other. (a) How many coulombs of charge does this represent? (b) If this entire charge undergoes a potential difference of 12 V, how much energy is involved?

[handwritten: 84 ampers · Hours]

2. Ground and Cloud The electric potential difference between the ground and a cloud in a particular thunderstorm is 1.2×10^9 V. What is the magnitude of the change in the electric potential energy (in multiples of the electron-volt) of an electron that moves between the ground and the cloud?

3. Lightning Flash In a given lightning flash, the potential difference between a cloud and the ground is 1.0×10^9 V and the quantity of charge transferred is 30 C. (a) What is the decrease in energy of that transferred charge. (b) If all that energy could be used to accelerate a 1000 kg automobile from rest, what would be the automobile's final speed? (c) If the energy could be used to melt ice, how much ice would it melt at 0°C? The heat of fusion of ice is 3.33×10^5 J/kg.

SEC. 25-5 ∎ CALCULATING THE POTENTIAL FROM AN *E*-FIELD

4. From *A* to *B* When an electron moves from *A* to *B* along an electric field line in Fig. 25-26, the electric field does $3.94 \times$ 10^{-19} J of work on it. What are the electric potential differences (a) $V_B - V_A$, (b) $V_C - V_A$, and (c) $V_C - V_B$?

5. Infinite Sheet An infinite nonconducting sheet has a surface charge density $\sigma = 0.10\ \mu C/m^2$ on one side. How far apart are equipotential surfaces whose potentials differ by 50 V?

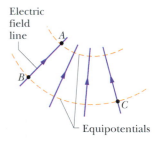

FIGURE 25-26 ∎ Problem 4.

6. Parallel Plates Two large, parallel, conducting plates are 12 cm apart and have charges of equal magnitude and opposite sign on their facing surfaces. An electrostatic force of 3.9×10^{-15} N acts on an electron placed anywhere between the two plates. (Neglect fringing.) (a) Find the electric field at the position of the electron. (b) What is the potential difference between the plates?

7. Geiger Counter A Geiger counter has a metal cylinder 2.00 cm in diameter along whose axis is stretched a wire 1.30×10^{-4} cm in diameter. If the potential difference between the wire and the cylinder is 850 V, what is the electric field at the surface of (a) the wire and (b) the cylinder? (*Hint:* Use the result of Problem 30 of Chapter 24.)

8. Field Inside The electric field inside a nonconducting sphere of radius R, with charge spread uniformly throughout its volume, is radially directed and has magnitude

$$E(r) = |\vec{E}(r)| = \frac{|q|r}{4\pi\varepsilon_0 R^3}.$$

Here q (positive or negative) is the total charge within the sphere, and r is the distance from the sphere's center. (a) Taking $V = 0$ at the center of the sphere, find the electric potential $V(r)$ inside the sphere. (b) What is the difference in electric potential between a point on the surface and the sphere's center? (c) If q is positive, which of those two points is at the higher potential?

9. Uniformly Distributed A charge q is distributed uniformly throughout a spherical volume of radius R. (a) Setting $V = 0$ at infinity, show that the potential at a distance r from the center, where $r < R$, is given by

$$V = \frac{q(3R^2 - r^2)}{8\pi\varepsilon_0 R^3}.$$

(*Hint:* See Section 24-6.) (b) Why does this result differ from that in (a) of Problem 8? (c) What is the potential difference between a point on the surface and the sphere's center? (d) Why doesn't this result differ from that of (b) of Problem 8?

10. Infinite Sheet Two Figure 25-27 shows, edge-on, an infinite nonconducting sheet with positive surface charge density σ on one side. (a) Use Eq. 25-16 and Eq. 24-16 to show that the electric potential of an infinite sheet of charge can be written $V = V_0 - (\sigma/2\varepsilon_0)z$, where V_0 is the electric potential at the surface of the sheet and z is the perpendicular distance from the sheet. (b) How much work is done by the electric field of the sheet as a small positive test charge q_0 is moved from an initial position on the sheet to a final position located a distance z from the sheet?

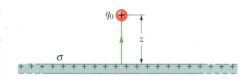

FIGURE 25-27 ■ Problem 10.

11. Thick Spherical Shell A thick spherical shell of charge Q and uniform volume charge density ρ is bounded by radii r_1 and r_2, where $r_2 > r_1$. With $V = 0$ at infinity, find the electric potential ΔV as a function of the distance r from the center of the distribution, considering the regions (a) $r > r_2$, (b) $r_2 > r > r_1$, and (c) $r < r_1$. (d) Do these solutions agree at $r = r_2$ and $r = r_1$? (*Hint:* See Section 24-6.)

SEC. 25-7 ■ POTENTIAL AND POTENTIAL ENERGY DUE TO A GROUP OF POINT CHARGES

12. Space Shuttle As a space shuttle moves through the dilute ionized gas of Earth's ionosphere, its potential is typically changed by -1.0 V during one revolution. By assuming that the shuttle is a sphere of radius 10 m, estimate the amount of charge it collects.

13. Diametrically Opposite Consider a point charge $q = 1.0$ μC, point A at distance $d_1 = 2.0$ m from q, and point B at distance $d_2 = 1.0$ m. (a) If these points are diametrically opposite each other, as in

Fig. 25-28a, what is the electric potential difference $V_A - V_B$? (b) What is that electric potential difference if points A and B are located as in Fig. 25-28b?

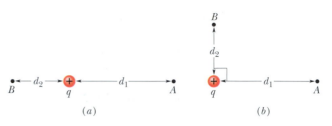

(a)   (b)

FIGURE 25-28 ■ Problem 13.

14. Field Lines and Equipotentials Figure 25-29 shows two charged particles on an axis. Sketch the electric field lines and the equipotential surfaces in the plane of the page for (a) $q_1 = +q$ and $q_2 = +2q$ and (b) $q_1 = +q$ and $q_2 = -3q$.

FIGURE 25-29 ■ Problems 14, 15, 16.

15. In Terms of d In Fig. 25-29, set $V = 0$ at infinity and let the particles have charges $q_1 = +q$ and $q_2 = -3q$. Then locate (in terms of the separation distance d) any point on the x axis (other than at infinity) at which the net potential due to the two particles is zero.

16. E-Field Is Zero Two particles, of charges q_1 and q_2, are separated by distance d in Fig. 25-29. The net electric field of the particles is zero at $x = d/4$. With $V = 0$ at infinity, locate (in terms of d) any point on the x axis (other than at infinity) at which the electric potential due to the two particles is zero.

17. Spherical Drop of Water A spherical drop of water carrying a charge of 30 pC has a potential of 500 V at its surface (with $V = 0$ at infinity). (a) What is the radius of the drop? (b) If two such drops of the same charge and radius combine to form a single spherical drop, what is the potential at the surface of the new drop?

18. Charge and Charge Density What are (a) the charge and (b) the charge density on the surface of a conducting sphere of radius 0.15 m whose potential is 200 V (with $V = 0$ at infinity)?

19. Field Near Earth An electric field of approximately 100 V/m is often observed near the surface of Earth. If this were the field over the entire surface, what would be the electric potential of a point on the surface? (Set $V = 0$ at infinity.)

20. Center of Rectangle In Fig. 25-30, point P is at the center of the rectangle. With $V = 0$ at infinity, what is the net electric potential at P due to the six charged particles?

21. Potential at P In Fig. 25-31, what is the net potential at point P due to the four point charges, if $V = 0$ at infinity?

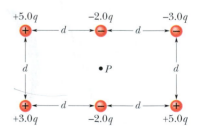

FIGURE 25-30 ■ Problem 20.

22. Potential Energy (a) What is the electric potential energy of two electrons separated by 2.00 nm? (b) If the separation increases, does the potential energy increase or decrease?

23. Work Required Derive an expression for the work required to set up the four-charge configuration of Fig. 25-32, assuming the charges are initially infinitely far apart.

24. Electric Potential Energy What is the electric potential energy of the charge configuration of Fig. 25-13a? Use the numerical values provided in Touchstone Example 25-4.

25. The Rectangle In the rectangle of Fig. 25-33, the sides have lengths 5.0 cm and 15 cm, $q_1 = -5.0 \mu C$, and $q_2 = +2.0 \mu C$. With $V = 0$ at infinity, what are the electric potentials (a) at corner A and (b) at corner B? (c) How much work is required to move a third charge $q_3 = +3.0 \mu C$ from B to A along a diagonal of the rectangle? (d) Does this work increase or decrease the electric energy of the three-charge system? Is more, less, or the same work required if q_3 is moved along paths that are (e) inside the rectangle but not on a diagonal and (f) outside the rectangle?

26. How Much Work In Fig. 25-34, how much work is required to bring the charge of $+5q$ in from infinity along the dashed line and place it as shown near the two fixed charges $+4q$ and $-2q$? Take distance $d = 1.40$ cm and charge $q = 1.6 \times 10^{-19}$ C.

27. A Particle of Positive Charge A particle of positive charge Q is fixed at point P. A second particle of mass m and negative charge $-q$ moves at constant speed in a circle of radius r_1, centered at P. Derive an expression for the work W that must be done by an external agent on the second particle to increase the radius of the circle of motion to r_2.

28. How Much Energy Calculate (a) the electric potential established by the nucleus of a hydrogen atom at the average distance ($r = 5.29 \times 10^{-11}$ m) of the atom's electron (take $V = 0$ at infinite distance), (b) the electric potential energy of the atom when the electron is at this radius, and (c) the kinetic energy of the electron, assuming it to be moving in a circular orbit of this radius centered on the nucleus. (d) How much energy is required to ionize the hydrogen atom (that is, to remove the electron from the nucleus so that the separation is effectively infinite)? Express all energies in electron-volts.

29. Fixed at Point P A particle of charge q is fixed at point P, and a second particle of mass m and the same charge q is initially held a distance r_1 from P. The second particle is then released. Determine

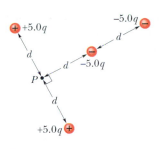

FIGURE 25-31 ■ Problem 21.

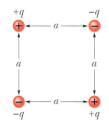

FIGURE 25-32 ■ Problem 23.

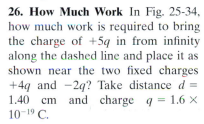

FIGURE 25-33 ■ Problem 25.

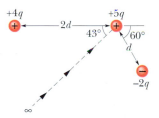

FIGURE 25-34 ■ Problem 26.

its speed when it is distance r_2 from P. Let $q = 3.1 \mu C$, $m = 20$ mg, $r_1 = 0.90$ mm, and $r_2 = 2.5$ mm.

30. Thin Plastic Ring A charge of -9.0 nC is uniformly distributed around a thin plastic ring of radius 1.5 m that lies in the yz plane with its center at the origin. A point charge of -6.0 pC is located on the x axis at $x = 3.0$ m. Calculate the work done on the point charge by an external force to move the point charge to the origin.

31. Tiny Metal Spheres Two tiny metal spheres A and B of mass $m_A = 5.00$ g and $m_B = 10.0$ g have equal positive charges $q = 5.00 \mu C$. The spheres are connected by a massless nonconducting string of length $d = 1.00$ m, which is much greater than the radii of the spheres. (a) What is the electric potential energy of the system? (b) Suppose you cut the string. At that instant, what is the acceleration of each sphere? (c) A long time after you cut the string, what is the speed of each sphere?

32. Conducting Shell on Support A thin, spherical, conducting shell of radius R is mounted on an isolating support and charged to a potential of $-V$. An electron is then fired from point P at distance r from the center of the shell ($r \gg R$) with initial speed v_1 and directly toward the shell's center. What value of v_1 is needed for the electron to just reach the shell before reversing direction?

33. Two Electrons Two electrons are fixed 2.0 cm apart. Another electron is shot from infinity and stops midway between the two. What is its initial speed?

34. Charged, Parallel Surfaces Two charged, parallel, flat conducting surfaces are spaced $d = 1.00$ cm apart and produce a potential difference $\Delta V = 625$ V between them. An electron is projected from one surface directly toward the second. What is the initial speed of the electron if it stops just at the second surface?

35. An Electron Is Projected An electron is projected with an initial speed of 3.2×10^5 m/s directly toward a proton that is fixed in place. If the electron is initially a great distance from the proton, at what distance from the proton is the speed of the electron instantaneously equal to twice the initial value?

SEC. 25-8 ■ POTENTIAL DUE TO AN ELECTRIC DIPOLE

36. Ammonia The ammonia molecule NH_3 has a permanent electric dipole moment equal to 1.47 D, where 1 D = 1 debye unit = 3.34×10^{-30} C · m. Calculate the electric potential due to an ammonia molecule at a point 52.0 nm away along the axis of the dipole. (Set $V = 0$ at infinity.)

37. Three Particles Figure 25-35 shows three charged particles located on a horizontal axis. For points (such as P) on the axis with $r \gg d$, show that the electric potential $V(r)$ is given by

$$V(r) = \frac{kq}{r}\left(1 + \frac{2d}{r}\right).$$

FIGURE 25-35 ■ Problem 37.

(*Hint:* The charge configuration can be viewed as the sum of an isolated charge and a dipole.)

SEC. 25-9 ■ POTENTIAL DUE TO A CONTINUOUS CHARGE DISTRIBUTION

38. Plastic Rod (a) Figure 25-36a shows a positively charged plastic rod of length L and uniform linear charge density λ. Setting $V = 0$ at infinity and considering Fig. 25-17 and Eq. 25-34, find the electric potential at point P without written calculation. (b) Figure 25-36b shows an identical rod, except that it is split in half and the right half is negatively charged; the left and right halves have the same magnitude λ of uniform linear charge density. With V still zero at infinity, what is the electric potential at point P in Fig. 25-36b?

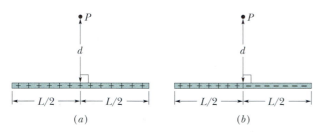

FIGURE 25-36 ■ Problem 38.

39. Nonlinear Charge Density The plastic rod shown in Fig. 25-37 has length L and a nonuniform linear charge density $\lambda = cx$, where c is a positive constant. With $V = 0$ at infinity, find the electric potential at point P_1 on the axis, at distance d from one end.

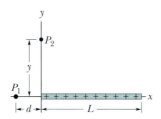

FIGURE 25-37 ■ Problems 39, 40, 44, 45.

40. Rod of Length L Figure 25-37 shows a plastic rod of length L and uniform positive charge Q lying on an x axis. With $V = 0$ at infinity, find the electric potential at point P_1 on the axis, at distance d from one end of the rod.

SEC. 25-10 ■ CALCULATING THE ELECTRIC FIELD FROM THE POTENTIAL

41. Points in the xy Plane The electric potential at points in an xy plane is given by $V = (2.0 \text{ V/m}^2)x^2 - (3.0 \text{ V/m}^2)y^2$. What are the magnitude and direction of the electric field at the point (3.0 m, 2.0 m)?

42. Parallel Metal Plates Two large parallel metal plates are 1.5 cm apart and have equal but opposite charges on their facing surfaces. Take the potential of the negative plate to be zero. If the potential halfway between the plates is then +5.0 V, what is the electric field in the region between the plates?

43. Show That (a) Using Eq. 25-31, show that the electric potential at a point on the central axis of a thin ring of charge of radius R and a distance z from the ring is

$$V = \frac{kq}{\sqrt{z^2 + R^2}}.$$

(b) From this result, derive an expression for the E-field magnitude

$|\vec{E}| = E$ at points on the ring's axis; compare your result with the calculation of E in Section 23-7

44. Why Not The plastic rod of length L in Fig. 25-37 has the non-uniform linear charge density $\lambda = cx$, where c is a positive constant. (a) With $V = 0$ at infinity, find the electric potential at point P_2 on the y axis, a distance y from one end of the rod. (b) From that result, find the electric field component E_y at P_2. (c) Why cannot the field component E_x at P_2 be found using the result of (a)?

45. Find Component (a) Use the result of Problem 39 to find the electric field component E_x at point P_1 in Fig. 25-37 (*Hint:* First substitute the variable x for the distance d in the result.) (b) Use symmetry to determine the electric field component E_y at P_1.

SEC. 25-11 ■ POTENTIAL OF A CHARGED ISOLATED CONDUCTOR

46. Hollow Metal Sphere An empty hollow metal sphere has a potential of +400 V with respect to ground (defined to be at $V = 0$) and has a charge of 5.0×10^{-9} C. Find the electric potential at the center of the sphere.

47. Excess Charge What is the excess charge on a conducting sphere of radius $r = 0.15$ m if the potential of the sphere is 1500 V and $V = 0$ at infinity?

48. Widely Separated Consider two widely separated conducting spheres, 1 and 2, the second having twice the diameter of the first. The smaller sphere initially has a positive charge q, and the larger one is initially uncharged. You now connect the spheres with a long thin wire. (a) How are the final potentials V_1 and V_2 of the spheres related? (b) What are the final charges q_1 and q_2 on the spheres, in terms of q? (c) What is the ratio of the final surface charge density of sphere 1 to that of sphere 2?

49. Two Metal Spheres Two metal spheres, each of radius 3.0 cm, have a center-to-center separation of 2.0 m. One has a charge of $+1.0 \times 10^{-8}$ C; the other has a charge of -3.0×10^{-8} C. Assume that the separation is large enough relative to the size of the spheres to permit us to consider the charge on each to be uniformly distributed (the spheres do not affect each other). With $V = 0$ at infinity, calculate (a) the potential at the point halfway between their centers and (b) the potential of each sphere.

50. Charged Metal Sphere A charged metal sphere of radius 15 cm has a net charge of 3.0×10^{-8} C. (a) What is the electric field at the sphere's surface? (b) If $V = 0$ at infinity, what is the electric potential at the sphere's surface? (c) At what distance from the sphere's surface has the electric potential decreased by 500 V?

51. Surface Charge Density (a) If Earth had a net surface charge density of 1.0 electron per square meter (a very artificial assumption), what would its potential be? (Set $V = 0$ at infinity.) (b) What would be the electric field due to the Earth just outside its surface?

52. Concentric Spheres Two thin, isolated, concentric conducting spheres of radii R_1 and R_2 (with $R_1 < R_2$) have charges q_1 and q_2. With $V = 0$ at infinity, derive expressions for the electric field magnitude $E(r)$ and the electric potential $V(r)$, where r is the distance from the center of the spheres. Plot $E(r)$ and $V(r)$ from $r = 0$ to $r = 4.0$ m for $R_1 = 0.50$ m, $R_2 = 1.0$ m, $q_1 = +2.0 \mu$C, and $q_2 = +1.0 \mu$C.

Additional Problems

53. Work Done Consider a charge $q = -2.0\ \mu C$ that moves from A to B or C to D along the paths shown in Fig. 25-38. This charge is moving in the presence of a uniform electric field of magnitude $E = 100$ N/C.

(a) What is the total work done on the charge if the distance between A and B is 0.62 m?

(b) What is the total work done on the charge if the distance between C and D is 0.58 m?

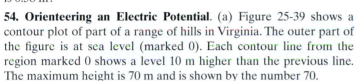

FIGURE 25-38 ■ Problem 53.

54. Orienteering an Electric Potential. (a) Figure 25-39 shows a contour plot of part of a range of hills in Virginia. The outer part of the figure is at sea level (marked 0). Each contour line from the region marked 0 shows a level 10 m higher than the previous line. The maximum height is 70 m and is shown by the number 70.

Answer the following questions by giving the pair of grid markers (a letter and a number) closest to the point being requested.

i. Where is there a steep cliff?
ii. Where is there a pass between two hills?
iii. Where is the easiest climb up the hill?

(b) Now suppose the figure represents a plot of the electric equipotentials for the surface of a glass plate, and the numbers now represent voltage. The maximum is 70 V and each contour line from the region marked 0 shows a level 10 V higher than the previous line.

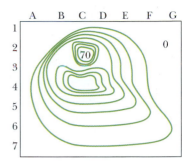

FIGURE 25-39 ■ Problem 54.

i. Where would a test charge placed on the glass feel the strongest electric force? In what direction would the force point?
ii. Is there a place on the glass where a charge could be placed so it feels no electric force? Where?

26 | Current and Resistance

When the zeppelin *Hindenburg* was built, it was the pride of Germany. Almost three football fields long, it was the largest flying machine ever built. Although the zeppelin was kept aloft by 16 cells of highly flammable hydrogen gas, it made many uneventful trans-Atlantic trips. However, on May 6, 1937, the *Hindenburg* burst into flames while landing at a U.S. naval air station in New Jersey during a rainstorm. While its handling ropes were being let down to a ground crew, ripples were sighted on the outer fabric near the rear of the ship. Seconds later, flames erupted from that region and 32 seconds after that the Hindenburg fell to the ground.

After so many successful flights of hydrogen zeppelins, why did this one burst into flames?

The answer is in this chapter.

26-1 Introduction

The interpretation of electrostatics experiments (described in Chapters 22 through 25) is that matter consists of two kinds of electrical charges, positive and negative. At least some negative charge can be moved from one object to another, leaving the first positively charged (with a deficit of negative charge) and the second negatively charged (with an excess of negative charge). Once the charges stopped moving we explored the electrostatic forces between them.

It turns out that the electrical devices we encounter most often in modern life such as computers, lights, and telephones are not purely electrostatic but involve moving charges which we will come to call *electric currents*. In addition, natural phenomena such as lightning, the flow of protons between the Earth's magnetic poles, and cosmic ray currents involve electric currents.

In this chapter we explore electric currents, or charge flow, with a primary focus on how current passes through conductors in electric circuits. We will see that the critical idea is to understand that a potential difference across a conductor causes a flow of charge (a current) through that conductor.

26-2 Batteries and Charge Flow

By the end of the 18th century, Alessandro Volta had discovered that when two metal plates were placed in contact with a moist piece of metal, they seemed to have electrical properties like those of rubbed amber and glass. To magnify this effect, Volta piled up pairs of unlike metals. When he grasped the plates (terminals) at each end of the pile with his hands, he claimed to feel electric charges move through his body on a continuous basis. Volta had invented the battery, and his experience with early batteries is an indication that there is a connection between electric charges, as discussed in Chapters 22–25, and the continuous flow of electricity created by batteries and other power sources. However, it is not obvious without further investigation that there is actually a connection between the sensation of electric flow that Volta experienced and the electric charges we believe exist based on the electrostatic observations discussed in Chapter 22.

In order to investigate this further, let's examine the results of several experiments involving a metal wire connected to oppositely charged conducting plates as shown in Fig. 26-1*a* and the same wire connected to a battery instead as shown in Fig. 26-1*b*.

Experiment 1 (Electrostatic Discharge): Suppose we use glass and amber rods that have been rubbed to transfer electrons to or from conducting plates. We can use a hanging amber or glass rod shown in Fig. 22-2 to verify that we have excess electric charge on each plate. Since it is easier to add excess electrons to a conductor than remove electrons from a conductor, the negatively charged plate will tend to have a

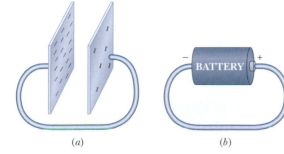

(a)   *(b)*

Figure 26-1 ■ (*a*) When a conducting wire connects two oppositely charged plates, charge flows from the negatively charged plate to the positively charged plate until both plates have the same number of excess electrons. As a result the wire becomes hot. (*b*) When a battery is placed between the ends of the wire instead, the wire also becomes hot, indicating that charge is also flowing.

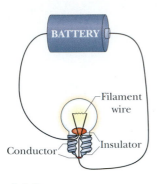

FIGURE 26-2 ■ If there is a complete conducting loop between the two terminals of a battery, a bulb will stay lit until the battery runs down.

greater magnitude of charge. Initially the negative electrons on the left plate repel each other and spread out but cannot leave the plate. Since a positive test charge placed between the plates will be repelled from the positive plate (on the right) and attracted to the negative plate (on the left), we know there is an electric field between the plates. So, Eq. 25-17 tells us there will be a potential difference between the plates.

If we connect the two plates with a piece of thread nothing happens. But when a conducting wire is connected across the two plates, (Fig. 26-1a) we observe that excess electrons on the left-hand plate will flow to the right-hand plate until both plates have the same number of excess electrons on them. This is not surprising since we expect the repulsive forces between the electrons on the left plate to push the charges through the wire while the attractive forces on the right plate pull on the charges. If we have enough excess charge on the plates, the wire will feel hot just after the discharge and then cool down again. If the wire has a properly connected small bulb in the middle of it, the bulb will light up briefly and then go out. We conclude from these observations that charge is flowing through the wire for a short time.

Experiment 2 (Battery Current): As we mentioned in Chapter 25, a battery is capable of doing work on electric charges and increasing their potential energy. So there must be a potential difference across its terminals. If we connect a piece of thread between the terminals of a battery nothing happens. On the other hand, if we connect a wire between the terminals of a battery, we observe that the wire gets very hot and stays that way for a long time as shown in Fig. 26-1b. If we also properly connect a bulb to the middle of the wire, as shown in Fig. 26-2, the bulb stays continuously lit until the battery eventually runs down. (In the next section we discuss how to connect a bulb to a battery properly, so that it lights.)

Because, at first, the electrostatic charging in Experiment 1 has the same result as the battery in Experiment 2, we infer that the underlying electric effects are the same in both cases. The hot wires and the lighting of bulbs lead us to conclude that charge is flowing through the wires. We call this flow of charge **electric current.**

READING EXERCISE 26-1: Although you were not provided with any details, what sensations might Volta have felt that led him to believe that electric charge was flowing through his body? ■

26-3 Batteries and Electric Current

There are some additional observations that help us understand the nature of electric current. Suppose we want to use a battery and perhaps a wire to light a flashlight bulb. By fiddling around we discover that many of the possible arrangements for lighting a bulb *do not work*. For example, none of the arrangements shown in Fig. 26-3 work.

To understand why these arrangements do not work, we need to examine a flashlight bulb much more carefully. The flashlight bulb consists of a piece of thin conducting "filament" wire encapsulated in glass that has no air inside. This wire glows and so gives off light when electric current passes through it. One end of the filament wire is in contact with a conductor that surrounds the bottom part of the bulb. The other end is connected to another conductor at the bulb's base. These conductors are separated by an insulator. A cutaway diagram of the bulb is shown in Fig. 26-4.

FIGURE 26-3 ■ Three of many arrangements of a battery and bulb and wire that do not cause the bulb to light.

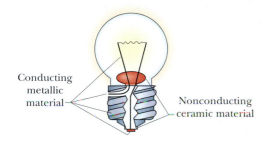

Conducting metallic material

Nonconducting ceramic material

FIGURE 26-4 ■ A cutaway diagram of a flashlight bulb.

After some more fiddling we discover that all of the arrangements of wires, bulb, and battery that cause the bulb to light have one thing in common. They all have a continuous, complete loop or **circuit** for current to pass from one terminal of a battery through conductors back to other terminal of the battery. In addition to the arrangement shown in Fig. 26-2, another of the many arrangements that forms a complete loop and causes a bulb to light is shown in Fig. 26-5.

When bulb filaments get old they sometimes break. In this case the circuit is incomplete and our "burned out" bulb does not light. Another requirement is that the battery must have a potential difference between its terminals. When a battery loses its potential difference after much use we refer to it as a "dead battery."

What Is Stored in a Battery?

It is commonly (and wrongly) believed that batteries store excess charge that can be "used up" in a circuit, and that a battery is "dead" when this excess charge is used up. The fact that people often refer to "charging" and "discharging" batteries is evidence of this belief. Careful observation tells us that this idea is wrong. The excess charge a fresh alkaline flashlight battery would have to store to keep a flashlight bulb lit as long as it does is more than 20 000 coulombs. This is a hundred million times the amount of charge we can typically place on a light metal-coated ball on a string. Yet, we observe no forces between such a charged ball and a fresh battery. There are also no forces between a charged ball and the wires carrying current in a circuit.

> Observations indicate that both batteries and any current-carrying wires connected to them are *electrically neutral*.

We conclude from these observations that batteries do not store charge. Batteries store energy. The energy in the battery is transformed to mechanical energy, light energy, and thermal energy as it pushes charges through wires and bulbs. Thus our observations support the idea that a battery acts as a pump that absorbs electrons at the negative terminal and releases higher potential energy electrons from the positive terminal. We discuss how chemical reactions can create a charge pump in more detail in Section 27-6.

If we connect two identical bulbs to the same battery as shown in Fig. 26-5, they shine with the same brightness. Based on this observation we conclude that the same current is passing through both bulbs.

> When wires and other conducting elements such as bulbs are placed between the battery terminals to make a continuous loop or circuit, the battery acts as a pump that pushes charge carriers already available in the wires around the loop. The battery is not a source of charge and electrical elements like bulbs do not use up charge.

Figure 26-6a shows a very simplified representation of a small segment of wire made up of electrically neutral atoms. In Fig. 26-6b the ends of the wire segment are

FIGURE 26-5 ■ When two identical bulbs in holders are connected in a row to a battery, they have the same brightness as each other. We conclude that the same current is passing through both bulbs. This indicates that the battery is not a source of excess charge used up by the bulbs.

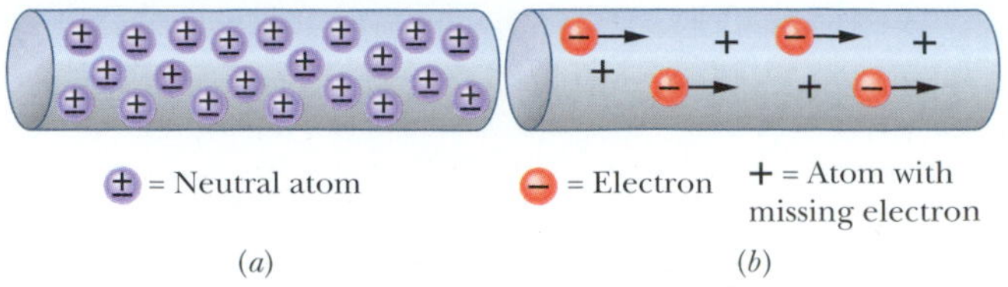

FIGURE 26-6 ▪ (*a*) A representation of many electrically neutral atoms in a wire. (*b*) A diagram that shows a potential difference across the ends of the wire so a very small fraction of the electrons surrounding atoms start moving and a few ions with missing conduction electrons are present. These ions have excess positive charge. *Note:* The neutral atoms are still present but are not shown.

⊕ = Neutral atom   ⊖ = Electron   + = Atom with missing electron

(*a*)   (*b*)

connected to a battery (not shown). A few of the conduction electrons in the metal start moving, but the stationary charges, consisting of neutral atoms and ions (with missing conduction electrons) still exist in the wire. The stationary ions neutralize the moving conduction electrons. To reduce clutter, Fig. 26-6*b* shows the stationary ions but not the neutral atoms. In other figures in this chapter we just show moving electrons and not the stationary ions. This type of depiction can give the false impression that there is excess charge in the wire. This is not so. Conducting wires are electrically neutral.

Defining Current Mathematically

Figure 26-2 shows a complete circuit with a battery (or other power source) that maintains a constant potential difference across its terminals. In this case, charge pushed through the circuit by the battery flows through a conducting wire, and then through the filament of a bulb, which is usually a very thin wire. In order to think more carefully about the current, we need to develop a mathematical definition for current.

Figure 26-7 shows a section of a conducting loop with different cross-sectional areas in which a current has been established. If net charge dq passes through a hypothetical plane (such as *a*) in time dt, then the current through that plane is defined as

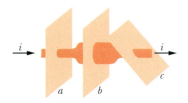

FIGURE 26-7 ▪ The current i or charge per unit time through the conductor has the same value at imaginary planes *a*, *b*, and *c* as long as the planes cut through the entire conductor at the points of intersection.

$$i \equiv \frac{dq}{dt} \qquad \text{(definition of current).} \qquad (26\text{-}1)$$

Regardless of the details of the geometry of the charge flow, we can find the net charge passing through any plane in a time interval extending from 0 to t by integration:

$$q = \int dq = \int_0^t i\, dt. \qquad (26\text{-}2)$$

Measurements of current through various locations in a single loop circuit show that the current is the same in all parts of a circuit where there are no junctions or alternate paths for the current to take. The current or rate of charge flow is the same passing through the imaginary planes *a*, *b*, and *c* shown in Fig. 26-7. Indeed, the current is the same for any plane that passes completely through the conducting elements in a continuous circuit with no branches, no matter what their locations or orientations. That is, a charge carrier must pass through plane *a* for every charge carrier that passes through plane *c*.

The unit for current is called the *ampere* (A), and it can be related to the coulomb by the expression

1 ampere = 1 A = 1 coulomb/second = 1C/s.

The Directions of Currents

How can we tell whether there are positive or negative charges moving when a current is established in electrically neutral conductors? When we place a conducting wire between the plates shown in Fig. 26-8, charge carriers flow until the plates are neutralized.

It is not possible for us to design an experiment based on macroscopic observations that will allow us to tell whether the charge carriers are positive or negative because the end result (neutralized plates) will be the same in either case. Early experimenters with electricity had no knowledge of atomic structure and could only use macroscopic observations of electrical effects to guide them. They assumed that charge carriers were positive. Even though we now know that negatively charged electrons are the charge carriers in conductors, for historical reasons we will stick with the assumption that the charge carriers are positive. This historical assumption makes it easier to use traditional references on electricity, and all the characteristics of circuits we will study on a macroscopic level will be exactly the same. Furthermore, this early assumption would have been correct if Benjamin Franklin had decided to designate the excess charges on rubber rods as positive and those on glass as negative instead of the other way around!

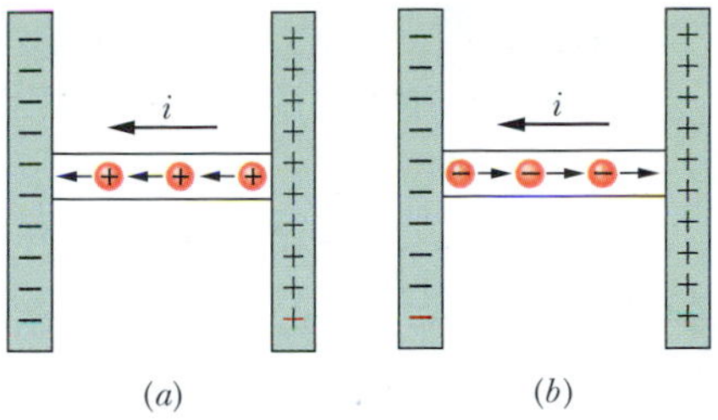

Although the charge carriers in conductors are negative, other currents, for example, protons streaming out of our Sun, create positive currents. Also charge carriers in fluids can be either positive ions (atoms with missing electrons) or negative electrons or ions (atoms with extra electrons). In fact, the movement of charge within most batteries is due to the migration of positive ions that undergo chemical reactions. Also, currents in biological systems are carried by sodium and potassium ions, which are positive charge carriers.

Current arrows show only a direction (or sense) of flow of charge carriers along the connected conductors as they bend and turn between battery terminals, not a fixed direction in space. Since current is actually a flux, which is a scalar quantity, *these current arrows do not represent vectors with magnitude and direction.*

> A current arrow, although not a mathematical vector, is drawn in the direction in which positive charge carriers would move through wires and circuit elements from a higher potential to a lower (more negative) potential, even though the actual charge carriers are usually negative and move in the opposite direction.

Charge Conservation at Junctions

So far we have only considered circuits like the one shown in Fig. 26-9a, ones for which there is only one path for charge carriers to follow. Such circuits are called **series** circuits. However, it is also common to find circuits or portions of circuits in which charge carriers encounter a junction where they can take either of two (or more) paths as shown in Fig. 26-9b. We call this type of circuit a **parallel** circuit. Although we introduce the terms "series" and "parallel" here, we will focus on the quantitative evaluation of series and parallel circuits in Chapter 27.

Figure 26-9b shows the moving charge carriers splitting up at a junction and then moving in parallel. If the bulbs are *identical*, how do the currents split at junction 1?

FIGURE 26-8 ■ No macroscopically-oriented experiment will allow us to detect whether the charge carriers in a conducting wire are (a) positive or (b) negative. So we define the current flow to be from right to left in both cases. Although stationary charges are not depicted, the conducting wires are neutral.

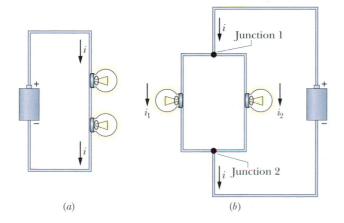

(a)   (b)

FIGURE 26-9 ■ We use a lightbulb as an example of a circuit element. (a) A series connection involves two or more circuit elements that are connected together so that the same current that passes through one element must pass through the other element. The potential differences across the elements is the sum of the drops across each element. (b) A parallel connection requires that one terminal of each two or more elements are connected together at one point and then the other terminal of each of the elements is connected together at another point. These points of connection are called junctions. Because of the connections at the junctions the potential difference across each element is the same when the parallel network is placed in a circuit.

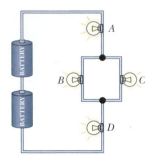

FIGURE 26-10 ■ To verify what we think would happen to current at the junctions, four identical lightbulbs and a battery are connected in a circuit having both series and parallel elements. Observations tell us that the brightness of bulb A is the same as that in bulb D. Bulbs B and C have the same brightness as each other but share the battery's current, so they are much dimmer than A and D. Note that even though they are not adjacent, bulbs A and D are in series.

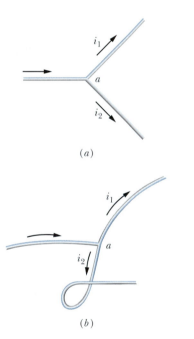

(a)

(b)

FIGURE 26-11 ■ The relation $i = i_1 + i_2$ is true at junction a no matter what the orientation of the three wire segments is.

What happens when they come back together at junction 2? Since the bulbs are identical, we expect that the current coming into junction 1 will divide equally so half takes the left path and half the right. If charge is conserved, when the currents combine at junction 2, they should add up to the original current. We can verify this by making a very simple observation with a couple of flashlight batteries in series and four bulbs as shown in Fig. 26-10. Bulbs A and D have the same brightness as each other. This fact indicates that the amount of current through bulb A is the same amount of current passing through bulb D and back through the battery. The fact that bulbs B and C have the same brightness as each other but are dimmer than A and D suggests that the current is splitting in half at the junction. We conclude

$$i = i_1 + i_2 = \frac{i}{2} + \frac{i}{2} \quad \text{(special case with identical parallel elements)}, \quad (26\text{-}3)$$

where i is the total current and i_1 and i_2 are the currents in the two branches.

If charge is conserved, the magnitudes of the currents in two parallel branches *must* add to yield the magnitude of the current in the original conductor even when the branches have different circuit elements. This statement, called Kirchhoff's current law, states that in general

$$i = i_1 + i_2 \quad \text{(three-way junction of Fig. 26-11a)}. \quad (26\text{-}4)$$

We treat Kirchhoff's circuit laws in more detail in Chapter 27.

Experiments indicate that bending or reorienting the wires in space does not change the validity of Eq. 26-4. The fact that current is not affected by wire orientation can be explained by the accumulation of static surface charges. These charges keep the electric field associated with a potential difference pointing along a wire, regardless of how the wire twists. The lack of influence of bending is depicted in Fig. 26-11.

READING EXERCISE 26-2: Explain why the word "circuit," as used in everyday speech, is an appropriate term for application to electrical situations. ■

READING EXERCISE 26-3: Suppose a battery sets up a flow of charges through wires and a bulb. (a) Will the overall circuit, consisting of the battery, bulb, and wires, remain electrically neutral, become positive, or become negative? Explain. (b) Will the wires remain electrically neutral, become positive, or become negative? (c) What ordinary observations would support your answers? ■

READING EXERCISE 26-4: Apply your understanding of the concept of flux, the nature of electrical charge, and the definition of current presented to explain why a flow of equal amounts of opposite charge in the same direction would not be considered a current. That is, explain why there must be a net flow of charge through a surface for there to be a current. ■

READING EXERCISE 26-5: The figure below shows a portion of a circuit. What are the magnitude and direction of the current i in the lower right-hand wire?

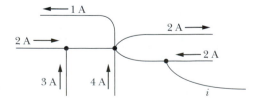

■

TOUCHSTONE EXAMPLE 26-1: Charged Fuel

If you've ever gone to a gas station to fill a gas can with fuel for your lawn mower, you may have noticed the sign that tells you to take the gas can out of your car and place it on the ground before you fill it with gasoline. Why is this important? As fuel is pumped from its underground storage tank, it can acquire a net electrical charge. If so, as you pump fuel into a container, the can will build up a net electrical charge if it is electrically isolated from its surroundings. If this charge builds up to a sufficient level, it can create a spark, igniting the fumes around your container with very unfortunate consequences.

Suppose the maximum safe charge that can be deposited on your 5.0 gal gas can is 1.0 μC.

(a) What is the maximum safe charge per liter that the fuel you are pumping can have?

SOLUTION ■ The **Key Idea** here is simply that the maximum safe "charge density" is

$$(1.0 \ \mu C)/(5.0 \ \text{gal}) = (0.20 \ \mu C/\text{gal})(264 \ \text{gal/m}^3)(1 \ \text{m}^3/1000 \ \text{L})$$
$$= 0.0528 \ \mu C/L. \qquad \text{(Answer)}$$

(b) If the pump delivers fuel at a rate of 8.0 gallons per minute, what is the maximum safe electrical current associated with the flow of the fuel into the can?

SOLUTION ■ The **Key Idea** here is that the fuel delivery rate is

$$(8.0 \ \text{gal/min})(1 \ \text{L}/0.264 \ \text{gal})(1 \ \text{min}/60 \ \text{s}) = 0.50505 \ \text{L/s}.$$

Since each liter of fuel can deliver no more than 0.0528 μC safely, the maximum safe electrical current is just $(0.0528 \ \mu C/L)$ $(0.50505 \ \text{L/s}) = 0.027 \ \mu C/s = 27 \ \text{nA}.$

26-4 Circuit Diagrams and Meters

As we move into the remaining sections in this chapter and the next, we will be drawing electric circuits with elements such as batteries, bulbs, wires, and switches. We will also be introducing new elements such as resistors and meters for measuring current and voltage.

Symbols for Basic Circuit Elements

Before proceeding with our study of current and resistance, we pause and introduce a few of the symbols scientists and engineers have created to represent circuit elements. Figure 26-12 shows the common symbols used to make the circuits we discuss in this chapter easier to draw.

Using these symbols, the circuit shown in Fig. 26-2 with a switch added can be represented as shown in Fig. 26-13.

Meters

Current and potential differences are very important properties of electrical circuits, and we have well-established convenient

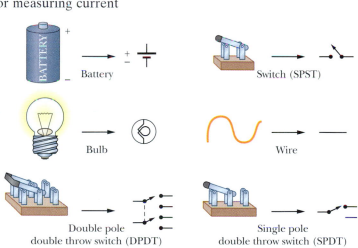

FIGURE 26-12 ■ Some circuit symbols.

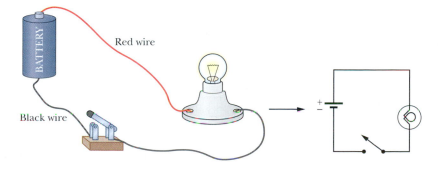

FIGURE 26-13 ■ A circuit sketch and corresponding diagram.

Current measurements

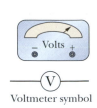

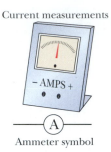

FIGURE 26-14 ■ An analog ammeter for measuring current and an analog voltmeter for measuring potential difference (or "voltage"), along with their circuit symbols.

Ammeter symbol   Voltmeter symbol

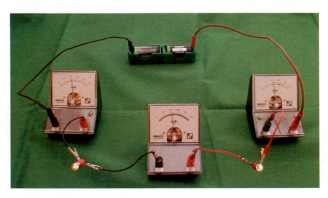

FIGURE 26-15 ■ Three analog ammeters measure the same current flowing through three locations in a series circuit consisting of two #14 flashlight bulbs.

ways to measure these quantities using meters. The device with which one measures current is called an **ammeter.** Potential difference is measured with a device called a **voltmeter.** An ammeter and voltmeter along with their circuit symbols are depicted in Fig. 26-14.

Since an ammeter measures current *through* a circuit (or a branch of a more complex circuit), it is placed in *series* with circuit elements. A voltmeter measures the potential difference between two locations (or points) in a circuit, so a voltmeter is placed across or in *parallel* with the two points of interest. This is shown in Fig. 26-15.

Often ammeters and voltmeters are combined in a device used to measure either potential difference or current. When the two or more meters are combined, the meter is typically called a **multimeter.** A digital multimeter is shown in Fig. 26-16. Many modern digital multimeters are also capable of measuring other quantities we will discuss, such as resistance and capacitance.

FIGURE 26-16 ■ The digital multimeter pictured can be configured to act as an ammeter to measure current *through* a given part of a circuit, a voltmeter to measure potential difference *across* any two points in a circuit, or the resistance of any circuit element.

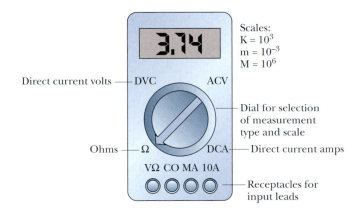

Scales:
$K = 10^3$
$m = 10^{-3}$
$M = 10^6$

Direct current volts — DVC   ACV

Dial for selection of measurement type and scale

Ohms — Ω DCA — Direct current amps

$V\Omega$ CO MA 10A

Receptacles for input leads

READING EXERCISE 26-6: In Fig. 26-17, the voltmeter is attached across the bulb and the ammeter is inserted into the circuit. Why are these devices connected this way? How would the ammeter reading change if it were inserted in the circuit before the bulb instead of after it? ■

26-5 Resistance and Ohm's Law

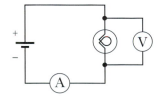

FIGURE 26-17 ■ A basic circuit for measuring the current flowing *through* a circuit element as a function of potential difference *across* it.

In professional applications of physics like designing electronic devices, we often need to know what effect adding more circuit elements will have on the flow of current. Given devices like ammeters and voltmeters, with which we can measure current and potential difference, we can do quantitative studies of the relationship between current and potential difference. For example, what will happen to the current in a circuit element, such as a bulb, that is part of a circuit if we add more batteries in series with our original battery? What will happen to the current in a conducting wire as voltage increases? The experimental setup for this investigation is shown in Fig. 26-17. The results are presented in Fig. 26-18 as graphs of applied potential difference ΔV and the resulting current i in two different circuit elements.

We can draw several interesting conclusions from looking at the two graphs in Fig. 26-18. First, we see in both graphs that as the potential difference increases, the amount of current through a given device increases. Second, it is not possible to tell how much current exists just by knowing the potential difference across a circuit element. For instance, when 1.0 V is placed across the lightbulb, the current through it is greater than the current in the Nichrome wire with the same potential difference

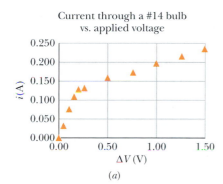
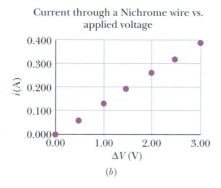

(a)   (b)

FIGURE 26-18 ■ Graph (*a*) shows ammeter data for current passing through a #14 lightbulb as a function of potential difference between the terminals of the lightbulb. Graph (*b*) shows ammeter data for the current through a length of cylindrical Nichrome wire as a function of potential difference between the ends of the wire.

across it. Third, for the length of Nichrome wire, the current is directly proportional to the potential difference, ΔV, across it. Thus, if we know the slope of the line, we can predict the current associated with any value of ΔV. Because of this direct proportionality, we refer to the Nichrome wire as a **linear** device. For the lightbulb, there is no convenient direct proportionality, so it is called a **nonlinear** device.

Definition of Resistance

In both the small bulb and the Nichrome wire, once we measure a specific potential difference, ΔV, across a circuit element and the corresponding current through it we have a measure of the *resistance* of the element to current but only at that ΔV. The **resistance** of a given circuit element is defined as the ratio of the potential difference across the element to the current through the element. When a small potential difference causes a relatively large current, the circuit element has a small resistance to flow of charge. Conversely, when the same potential difference produces a current that is small, we say the resistance is large. For example, in the data presented in Fig. 26-18, a potential difference of 1V across the bulb causes a current of 0.19A to flow, while the same potential difference across that Nichrome wire causes only 0.13A to flow. So we say that at the specific potential difference of 0.25V, the Nichrome wire has more resistance than the bulb.

We define resistance as the ratio of potential difference applied to the current that results:

$$R \equiv \frac{\Delta V}{i} \qquad \text{(definition of } R\text{)}. \qquad (26\text{-}5)$$

Here we use the notation ΔV to emphasize we are dealing with the *difference* in potential between two locations in a circuit, which changes the potential energy of the charges as they flow. When discussing circuits, potential difference is often referred to by an alternate name of **voltage.**

The SI unit for resistance that follows from Eq. 26-5 is the volt per ampere. This combination occurs so often that we give it a special name, the **ohm** (symbol Ω); that is,

$$1\,\text{ohm} = 1\,\Omega = 1\,\text{volt/ampere}$$
$$= 1\,\text{V/A.} \tag{26-6}$$

If we rewrite Eq. 26-5 as

$$i = \frac{\Delta V}{R},$$

it emphasizes the fact that the potential difference across a device with resistance R *produces* an electric current. The most common way to express the definition of resistance in Eq. 26-5 is

$$\Delta V = iR. \tag{26-7}$$

For a linear device like Nichrome wire we will get the same value for R no matter what potential difference we impress across the device. However, we must be *careful* in the case of a nonlinear device like a light bulb to specify at what potential difference we are measuring the current, i, in order to determine its resistance.

Ohm's Law

As we just pointed out, our Nichrome wire has the same resistance no matter what the value of the applied potential difference (as shown in Fig. 26-18*b*). Other conducting devices, such as lightbulbs, have resistances that change with the applied potential difference (as shown in Fig. 26-18*a*). Although both the Nichrome wire and the bulb contain metallic conductors, the wire in the bulb is so thin that its temperature rises noticeably as the potential difference increases, and the bulb's resistance increases.

In 1827, George Simm Ohm, a Bavarian, reported that he had observed a linear relationship between current and potential difference for metallic conductors kept at a fairly constant temperature. Because of this, linear devices such as the length of Nichrome wire are sometimes referred to as **ohmic.**

> A device is said to obey Ohm's law whenever the current through it is *always* directly proportional to the potential difference applied. That is, the device's resistance is constant in the $\Delta V = iR$ relation.

Many elements used in electric circuits, whether they are conductors like copper or semiconductors like pure silicon or silicon containing special impurities, obey Ohm's law within some range of values of potential difference. If the current in a resistive device is large enough to cause significant temperature changes in it, then Ohm's law often breaks down.

It is sometimes contended that $R = \Delta V/i$ (or $\Delta V = iR$) is a statement of Ohm's law. That is not true! This equation is the defining equation for resistance, and it applies to all conducting devices, whether they obey Ohm's law or not. If we measure the potential difference ΔV across and the current i through any device, even a bulb or other non-ohmic device, we can find its resistance *at that value of* ΔV as $R \equiv \Delta V/i$. The essence of Ohm's law, however, is a plot of i versus ΔV that is a straight line, so that the value of R is independent of the value of ΔV.

Resistors

A conductor whose function in a circuit is to obey Ohm's law so that it provides a specified resistance to the flow of charge independent of the potential difference impressed across it is called a **resistor** (see Fig. 26-19). Carbon resistors are the most standard sources of ohmic resistance used in electrical circuits for several reasons. Unlike a lightbulb, a resistor has a resistance that remains constant as current changes. Carbon resistors are inexpensive to manufacture, and they can be produced with a large range of resistances. The circuit diagram symbol for a resistor is shown in Fig. 26-20.

A typical carbon resistor contains graphite, a form of carbon, suspended in a hard glue binder. It usually is surrounded by a plastic case with a color code painted on it as shown in Fig. 26-21.

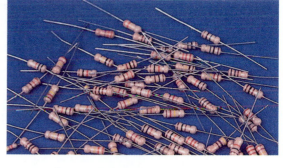

FIGURE 26-19 ■ An assortment of carbon resistors. The circular bands are color-coding marks that identify the value of the resistance.

READING EXERCISE 26-7: The following table gives the current i (in amperes) through three devices for several values of potential difference ΔV (in volts). From these data, determine which devices, if any, obey Ohm's law.

Device 1		Device 2		Device 3	
ΔV	i	ΔV	i	ΔV	i
2.00	4.50	2.00	1.50	2.00	6.50
3.00	6.75	3.00	2.50	3.00	8.75
4.00	9.00	4.00	3.00	4.00	11.00

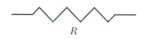

FIGURE 26-20 ■ Circuit diagram symbol for an ohmic resistor.

FIGURE 26-21 ■ Depiction of the four color bands on a color-coded resistor with $R = 47 \, \text{K}\Omega \pm 10\%$. See Table 26-1 for details

26-6 Resistance and Resistivity

Next we consider how the resistance of ohmic circuit elements such as metal wires or carbon resistors depends on their geometries. That is, how does the resistance of a short, broad object change if we stretch it so it is long and thin? To determine this, we fix our investigation on a single material. For example, we might experiment with copper wire. Relatively thick copper wire is commonly used in electric circuits because it has a very low resistance compared to other circuit elements. Thus, it can be used to connect circuit elements without adding much resistance to a circuit.

Observations

Consider a conducting wire with a potential difference across its ends as shown in Fig. 26-22. To start with, we will keep the thickness of the wire fixed and just decrease its length. If we apply a potential difference across the ends of the wire and use current and potential difference measurements, we can determine its resistance as a function of length. We find that its resistance is proportional to its length L. Thus, we can write

$$R = kL.$$

If instead we fix the length of the wire and decrease its thickness or cross-sectional area A, then the measured resistance of the wire increases as its cross-sectional area

TABLE 26-1
The Resistor Code[a]

Black	= 0	Blue	= 6
Brown	= 1	Violet	= 7
Red	= 2	Gray	= 8
Orange	= 3	White	= 9
Yellow	= 4	Silver	= ±10%
Green	= 5	Gold	= ±5%

[a]The value in ohms = $AB \times 10^C \pm D$. (AB means the A band digit placed beside the B band digit, not A times B). The colors on bands A, B, and C represent the digits shown in Table 26-1. The D band represents the "tolerance" of the resistor. No band denotes ±20%, a silver band denotes ±10%, and a gold band denotes ±5%. For example, a resistor with bands of Blue-Gray-Red-Silver has a value: $AB \times 10^C \pm D = 68 \times 10^2 \Omega \pm 10\%$ or $(6800 \pm 680)\Omega$, since $A = 6$, $B = 8$, $C = 2$, D = silver, (±10%).

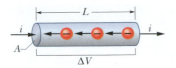

FIGURE 26-22 ■ A potential difference ΔV is applied between the ends of a conducting wire of length L and cross section A, establishing a current i. Although the stationary ions that neutralize the conduction electrons that make up the current are not shown, the wire is, as always, essentially neutral electrically.

decreases. In fact we get an inverse relationship so that

$$R = k'\frac{1}{A}.$$

To combine these two results, we write that R is proportional to L and inversely proportional to A with a new proportionality constant, ρ, which we define as the **resistivity** of the wire. Thus,

$$R = \rho\frac{L}{A}. \tag{26-8}$$

The results of these resistivity observations are important for two reasons. First, the fact that resistance varies inversely with cross-sectional area implies that current passes through the volume of the conductor, and not just along the surface. This knowledge will be useful as we continue to think about how charge moves through wires and other circuit elements.

Second, we know that every conducting material has a resistivity ρ. Is it the same for all materials? The answer is no. Is it the same if the length (or area) of a wire is changed? The answer is yes. What we observe is that if we apply the same potential difference between the ends of geometrically similar (same L and same A) rods of copper and of glass, very different currents result. This investigation reveals that resistivity varies with material. That is, it is a property of the *material* from which the object is fashioned.

We have just made an important distinction:

Resistance is a property of an object. Resistivity is a property of a material.

It is important to note that resistivity is analogous in many ways to the concept of density. Density depends only on the kind of material being used (such as lead or Styrofoam). The density can be used to calculate the mass of a certain volume of a substance. Similarly, resistivity depends only on the material being used in the wire and not on the length or cross-sectional area of the wire. If you know the resistivity of a material then the resistance of a given wire can be calculated using Eq. 26-8 once its length and cross-sectional area are known.

Variation of Resistivity with Temperature

The values of most physical properties vary with temperature, and resistivity is no exception. Figure 26-23, for example, shows the variation of this property for copper over a wide temperature range. The relation between temperature and resistivity for copper—and for metals in general—is fairly linear over the temperature range commonly found in circuits. For such linear relations we can write an approximation based on the results of measurements as

$$\rho - \rho_0 \approx \rho_0\alpha(T - T_0) \quad \text{(approx. temperature dependence of } \rho\text{)}. \tag{26-9}$$

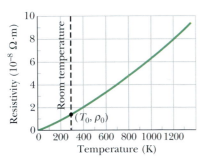

FIGURE 26-23 ■ The resistivity of copper as a function of temperature. The dot on the curve marks a convenient reference point ($T_0 = 293$ K and $\rho_0 = 1.69 \times 10^{-8}$ Ω·m).

Here T_0 is a selected reference temperature and ρ_0 is the resistivity at that temperature. Usually $T_0 = 293$ K (room temperature), for which $\rho_0 = 1.69 \times 10^{-8}$ Ω · m for copper. This approximate relationship is good enough for most engineering purposes.

Because temperature enters into this expression only as a difference, it does not matter whether you use the Celsius or Kelvin scale in that equation because the sizes of degrees on these scales are identical. The quantity α, called the **temperature coefficient of resistivity,** is chosen so that the equation gives good agreement with

Resistivities of Some Materials at Room Temperature (20°C)

Material	Resistivity, ρ ($\Omega \cdot$ m)	Temperature Coefficient of Resistivity, α (K^{-1})
Typical Metals		
Silver	1.62×10^{-8}	4.1×10^{-3}
Copper	1.69×10^{-8}	4.3×10^{-3}
Aluminum	2.75×10^{-8}	4.4×10^{-3}
Tungsten	5.25×10^{-8}	4.5×10^{-3}
Iron	9.68×10^{-8}	6.5×10^{-3}
Platinum	10.6×10^{-8}	3.9×10^{-3}
Manganin[a]	48.2×10^{-8}	0.002×10^{-3}
Typical Semiconductors		
Silicon, pure	2.5×10^3	-70×10^{-3}
Silicon, n-type[b]	8.7×10^{-4}	
Silicon, p-type[c]	2.8×10^{-3}	
Typical Insulators		
Glass	$10^{10} - 10^{14}$	
Fused quartz	$\sim 10^{16}$	

[a]An alloy specifically designed to have a small value of α.
[b]Pure silicon doped with phosphorus impurities to a charge carrier density of 10^{23} m^{-3}.
[c]Pure silicon doped with aluminum impurities to a charge carrier density of 10^{23} m^{-3}.

experimental values for temperatures in the chosen range. Some values of α for metals are listed in Table 26-2.

The *Hindenburg*

When the zeppelin *Hindenburg* was preparing to land on May 6th, 1937, the handling ropes were let down to the ground crew. Exposed to the rain, the ropes became wet (and thus were able to conduct a current). In this condition, the ropes "grounded" the metal framework of the zeppelin to which they were attached; that is, the wet ropes formed a conducting path between the framework and the ground, making the electric potential of the framework the same as the ground's. This should have also grounded the outer fabric of the zeppelin. The *Hindenburg*, however, was the first zeppelin to have its outer fabric painted with a sealant of large electrical resistivity. The fabric remained at the electric potential of the atmosphere at the zeppelin's altitude of about 43 m. Due to the rainstorm, that potential was large relative to the potential at ground level.

The handling of the ropes apparently ruptured one of the hydrogen cells and released hydrogen between that cell and the zeppelin's outer fabric, causing the reported rippling of the fabric. There was then a dangerous situation: the fabric was wet with conducting rainwater and was at a potential much different from the framework of the zeppelin. Apparently, charge flowed along the wet fabric and then sparked through the released hydrogen to reach the metal framework of the zeppelin, igniting the hydrogen in the process. The burning rapidly ignited the cells of hydrogen in the zeppelin and brought the ship down. If the sealant on the outer fabric of the *Hindenburg* had been of less resistivity (like that of other zeppelins), the *Hindenburg* disaster probably would not have occurred.

READING EXERCISE 26-8: Sketch a graph of i vs ΔV for a Nichrome wire like that in Fig. 26-22 but with the diameter of the wire cut in half. ∎

READING EXERCISE 26-9: In the section above, we cited the fact the resistance of a wire to current was inversely proportional to the cross-sectional area of the wire as evidence that the current passes through the volume of the wire rather than along the surface of the wire. (a) Justify this assertion. (b) What expression would you expect to replace Eq. 26-8 if the current was along the surface of the wire instead? ∎

READING EXERCISE 26-10: The figure shows three cylindrical copper conductors along with their face areas and lengths. Rank them according to the current through them, greatest first, when the same potential difference ΔV is placed across their lengths.

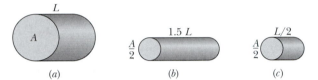

(a) (b) (c) ∎

26-7 Power in Electric Circuits

Batteries store a certain amount of chemical energy. This chemical energy is transformed to electrical and other forms of energy as current flows through various circuit elements. At times we are interested in the rate at which a battery's energy is used up by a circuit. Just as we did in Section 9-10 where power is defined as the rate at which work is done by a force, we also use the term power to describe the rate at which electrical energy is delivered to a circuit.

We start our consideration of power by examining the energy delivered to an electrical device that is connected to a battery by ideal wires. Figure 26-24 shows a circuit consisting of a battery B that is connected by wires to an unspecified conducting device. The device might be a resistor, a storage battery (a rechargeable battery), a motor, or some other electrical device. If the wires in the circuit are thick enough they are ideal because they have essentially no resistance. When current is present in a wire with no resistance the entire wire is at the same potential. In other words, there is no potential difference between one end of an ideal wire and the other end. In this case, a battery maintains a potential difference of magnitude ΔV across its own terminals, and thus across the terminals of the unspecified device, with a greater potential at terminal a of the device than at terminal b.

Since there is an external conducting path between the two terminals of the battery, and since the battery maintains a fixed potential difference, the battery produces a steady current i in the circuit. This current is directed from terminal a to terminal b. The amount of charge dq moving between those terminals in time interval dt is equal to $i\,dt$. This charge dq moves through a decrease in potential difference across the terminals of the device of magnitude ΔV, and thus its electric potential energy U decreases in magnitude by the amount

$$dU = -dq\,\Delta V = -i\,dt(\Delta V).$$

The principle of conservation of energy tells us that the decrease in electric potential energy from a to b is accompanied by a transfer of energy to some other form. Since $P = dW/dt$ (Eq. 9-48), the power P associated with that transfer is the rate at which the battery does work. Since $dW = -dU = i\,dt(\Delta V)$, we get

$$P = i\,\Delta V \qquad \text{(rate of electric energy transfer).} \tag{26-10}$$

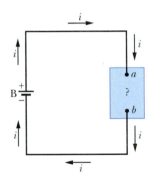

FIGURE 26-24 ∎ A battery B sets up a current i in a circuit containing an unspecified conducting device.

The wire coils within a toaster have appreciable resistance. When there is a current through them, electrical energy is transferred to thermal energy of the coils, increasing their temperature. The coils then emit infrared radiation and visible light that can toast bread.

Moreover, this power P is also the rate at which energy is transferred from the battery to the unspecified device. If that device is a motor connected to a mechanical load, the energy is transferred as work done on the load. If the device is a storage battery being charged, the energy is transferred to stored chemical energy in the storage battery. We know from observations that if the device is a resistor, the energy is transferred to internal thermal energy, tending to increase the resistor's temperature.

The unit of power following from the equation above is the volt-ampere (V · A). We can write it as

$$1\ \mathrm{V} \cdot \mathrm{A} = \left(1\frac{\mathrm{J}}{\mathrm{C}}\right)\left(1\frac{\mathrm{C}}{\mathrm{s}}\right) = 1\frac{\mathrm{J}}{\mathrm{s}} = 1\ \mathrm{W}.$$

The course of an electron moving through a resistor at constant speed is much like that of a stone falling through syrup at constant terminal speed. The average kinetic energy of the electron remains constant, and its lost electric potential energy appears as thermal energy in the resistor and its surroundings. On a microscopic scale this energy transfer is due to collisions between the electron and the molecules of the resistor, which leads to an increase in the temperature of the resistor lattice. The mechanical energy thus transferred to thermal energy is *lost* because the transfer cannot be reversed. This energy transfer due to atomic collisions is discussed in more detail in Sections 26-10 and 26-11.

For a resistor or some other device with resistance R, we can combine Eqs. 26-5 ($R = \Delta V/i$) and 26-10 to obtain, for the rate of electric energy loss (or dissipation) due to a resistance, either

$$P = i^2 R \qquad \text{(resistive dissipation)} \qquad (26\text{-}11)$$

or

$$P = \frac{(\Delta V)^2}{R} \qquad \text{(resistive dissipation).} \qquad (26\text{-}12)$$

Caution: We must be careful to distinguish these two new equations from Eq. 26-10: $P = i\,\Delta V$ applies to electric energy transfers of all kinds; $P = i^2 R$ and $P = (\Delta V)^2/R$ apply only to the transfer of electric potential energy to thermal energy in a device with resistance.

READING EXERCISE 26-11: A potential difference ΔV is connected across a device with resistance R, causing current i through the device. Rank the following variations according to the change in the rate at which electrical energy is converted to thermal energy due to the resistance, greatest change first: (a) ΔV is doubled with R unchanged, (b) i is doubled with R unchanged, (c) R is doubled with ΔV unchanged, (d) R is doubled with i unchanged. ■

TOUCHSTONE EXAMPLE 26-2: Heating Wire

You are given a length of uniform heating wire made of a nickel-chromium-iron alloy called Nichrome; it has a resistance R of 72 Ω. At what rate is energy dissipated in each of the following situations? (1) A potential difference of 120 V is applied across the full length of the wire. (2) The wire is cut in half, and a potential difference of 120 V is applied across the length of each half.

SOLUTION ■ The **Key Idea** is that a current in a resistive material produces a transfer of electrical energy to thermal energy; the rate of transfer (dissipation) is given by Eqs. 26-10 to 26-12. Because we know the potential ΔV and resistance R, we use

Eq. 26-12, which yields, for situation 1,

$$P = \frac{(\Delta V)^2}{R} = \frac{(120\ \mathrm{V})^2}{72\ \Omega} = 200\ \mathrm{W}. \qquad \text{(Answer)}$$

In situation 2, the resistance of each half of the wire is (72 Ω)/2, or 36 Ω. Thus, the dissipation rate for each half is

$$P' = \frac{(120\ \mathrm{V})^2}{36\ \Omega} = 400\ \mathrm{W},$$

and that for the two halves is

$$P = 2P' = 800 \text{ W}. \qquad \text{(Answer)}$$

This is four times the dissipation rate of the full length of wire.

Thus, you might conclude that you could buy a heating coil, cut it in half, and reconnect it to obtain four times the heat output. Why is this unwise? (What would happen to the amount of current in the coil?)

26-8 Current Density in a Conductor

We defined current so that it was a scalar—basically a "count" of the amount of charge crossing a surface per second with a sign to tell us in which direction the charge is crossing the surface—in the direction we choose as positive or opposite to it. Since in a current, charges are actually moving and have a velocity associated with them, there is a vector "hidden" in the concept of current. We can make it explicit by defining a new concept, the **current density.** If we have a volume that contains a set of moving charged particles, let the charge on each particle be e, let the density of the charges be n (number per unit volume), and let their average velocity be $\langle \vec{v} \rangle$. We then define the current density (or current percentage of cross-sectional area) as

$$\vec{J} \equiv ne\langle \vec{v} \rangle \qquad \text{(definition of current density)}. \qquad (26\text{-}13)$$

As is the case for the volume flux of water described in Eq. 15-33, the total amount of charge flowing through a given element of area can be defined as the dot product of the current density and an area element. If the area element is infinitesimal we can write the amount of current through it as $\vec{J} \cdot d\vec{A}$, where $d\vec{A}$ is the area vector of the element, perpendicular to the plane of the area element. The total *conventional current* through the surface of a cross section of wire is then

$$i = \int \vec{J} \cdot d\vec{A}. \qquad (26\text{-}14)$$

In most electrical conductors the charge carriers are negative. As we mentioned earlier, the term "conventional current" refers to the direction of flow of positive charge carriers. For a typical conductor such as copper, the electrons are moving in the opposite direction to the direction of the conventional current.

In Section 26-6 we concluded that in steady current through a conductor the charges must be flowing throughout the volume of the conductor. The key evidence for this is the inverse proportionality between resistance and the cross-sectional area of a conductor.* If we further assume that the direction of the current is parallel to $d\vec{A}$, then $\vec{J}$ is also uniform and parallel to $d\vec{A}$. In this case Eq. 26-14 can be rewritten in terms of the magnitudes of the current density and area.

$$|i| = \int J\,dA = J \int dA = JA,$$

so

$$J = \frac{|i|}{A}, \qquad (26\text{-}15)$$

---

*However, steady charge flow throughout a conductor is not true for the high-frequency alternating currents we treat in Chapter 33.

where A is the total area of the surface. From these equations, we see that the SI unit for current density is the ampere per square meter (A/m^2).

In Chapter 23 we represented an electric field with electric field lines. Figure 26-25 shows how current density can be represented with a similar set of lines, which we can call *streamlines*. The current, which is toward the right in Fig. 26-25, makes a transition from the wider conductor at the left to the narrower conductor at the right. Because charge is conserved during the transition, the current or rate at which the charges flow through the wire cannot change. However, the current density (or rate of charge flow per unit of cross-sectional area) does change—it is greater in the narrower conductor. The spacing of the streamlines suggests this increase in current density; streamlines that are closer together imply greater current density.

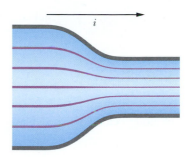

FIGURE 26-25 ▪ Streamlines representing current density in the flow of charge through a constricted conductor.

READING EXERCISE 26-12: The sketches below show several copper wires with the same potential difference across them. Rank the current density magnitude from largest to smallest.

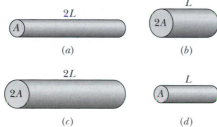

(a)   (b)

(c)   (d)   ▪

26-9 Resistivity and Current Density

Although i, ΔV, and R are the quantities that are directly measurable in electrical circuits, if we want to think more explicitly about what is happening in terms of the motion of charges it makes sense to reframe our Ohm's law relation in terms of the forces (or the electric field) and the current density. This gives us a generic relation that describes how forces affect the motion of charges without relying in any way on the properties of specific circuit elements in the way that Ohm's law does.

Recall that, for materials that obey Ohm's law, the resistance of a segment of a conductor R is related to the potential difference ΔV across it as well as the conventional current i passing through it. This relationship is given by

$$\Delta V = iR. \qquad \text{(Eq. 26-7)}$$

We can write this expression in an alternate form if we replace the potential difference ΔV with an expression involving the electric field $\vec{E}$. From Chapter 25, we know that the relationship between the electric field and the potential difference between two locations a and b is

$$V_a - V_b = \int_a^b \vec{E} \cdot d\vec{s}.$$

For a wire of length L with one end at location a and the other at location b, (Fig. 26-26), the electric field $\vec{E}$ set up within the wire is constant. As a result, the expression above can be expressed in terms of the electric field magnitude E and the length of the wire L as

$$V_a - V_b = \pm EL,$$

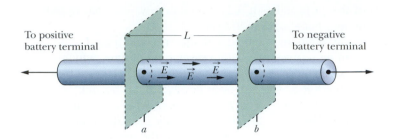

FIGURE 26-26 ■ A length L between points a and b along a current-carrying conductor.

where we use the plus sign if $\vec{E}$ and $\vec{ds}$ are in the same direction and the minus sign if $\vec{E}$ and $\vec{ds}$ point in opposite directions. Combining the expression above with

$$R = \frac{V_a - V_b}{i}$$

and ignoring signs gives us

$$R = \frac{EL}{i} = \frac{EL}{JA}.$$

The substitution for i comes from the relationship between current i, current density $\vec{J}$, and the cross section of the wire A. We compare this relation with that presented earlier when we introduced ρ as the resistivity of the material in Eq. 26-8:

$$R = \rho \frac{L}{A}.$$

By combining the previous two equations, we see that resistivity can be defined in terms of the magnitudes of the microscopic quantities $\vec{E}$ and $\vec{J}$ as

$$\rho \equiv \frac{E}{J} \qquad \text{(definition of } \rho\text{)}. \qquad (26\text{-}16)$$

If we combine the SI units of $\vec{E}$ and $\vec{J}$ we get, for the unit of ρ, the ohm-meter ($\Omega \cdot$ m):

$$\text{units of } \rho = \frac{\text{units of } E}{\text{units of } J} = \frac{\text{V/m}}{\text{A/m}^2} = \frac{\text{V}}{\text{A}}\,\text{m} = \Omega \cdot \text{m}.$$

(Do not confuse the *ohm-meter*, the unit of resistivity, with the *ohmmeter*, which is an instrument that measures resistance.)

Since $\vec{E}$ and $\vec{J}$ always point in the same direction, we can rewrite this expression in vector form as

$$\vec{E} = \rho \vec{J}. \qquad (26\text{-}17)$$

However, be aware that these two relations hold only for *isotropic* materials—materials whose electrical properties are the same in all directions (like the metals used to make wires).

26-10 A Microscopic View of Current and Resistance

Our macroscopic studies tell us that there is a current in a conductor whenever there is a potential difference across it. Whenever Ohm's law holds, the current is directly proportional to the potential difference that causes it. Let's consider a length L of thin conducting wire with a potential difference of ΔV between its ends. What happens microscopically to the charge carriers in this situation?

We already know that the conduction electrons in a metal serve as charge carriers, and that when there is a steady current, we can represent the density of electrons as n and the charge on each electron as e. What does Ohm's law tell us about the average velocity $\langle \vec{v} \rangle$ of these electrons? When Ohm's law holds so that $\Delta V = iR$ (Eq. 26-7), then according to Eq. 26-13, the current density is proportional to the average velocity of the charge carriers by definition,

$$\vec{J} \equiv ne\langle \vec{v} \rangle. \qquad \text{(Eq. 26-13)}$$

Since $\vec{E} = \rho \vec{J}$ (Eq. 26-17), we find that the electric field $\vec{E}$ across the wire (associated with potential difference ΔV across the wire) is also proportional to the average velocity of the charge carriers,

$$\vec{E} = \rho ne\langle \vec{v} \rangle. \qquad (26\text{-}18)$$

However, the electrostatic force on a charge carrier is given by $\vec{F}^{\text{elec}} = e\vec{E}$ (Eq. 23-4), so that

$$\vec{F}^{\text{elec}} = e\vec{E} = \rho ne^2\langle \vec{v} \rangle. \qquad (26\text{-}19)$$

This is a dramatic and interesting result. It tells us that the average velocity, $\langle \vec{v} \rangle$, of a charge carrier is proportional to the electrostatic force on it! However, if the electrostatic force is the only force acting on the electron, then Newton's Second Law tells us that the electron should accelerate and not maintain a constant average velocity. To maintain a constant velocity, the *net force* on the charge carrier must be zero. Thus, there must be a second force. This situation is very similar to that associated with air drag where an object falling in the presence of a gravitational force reaches a terminal velocity as a result of an air drag acting in the opposite direction. Using Eq. 6-24 we see that

$$\vec{F}^{\text{net}} = \vec{F}^{\text{elec}} + \vec{D} = \rho ne^2\langle \vec{v} \rangle + \vec{D} = 0$$

so that $\qquad\qquad \vec{D} = -e\vec{E} = -\rho ne^2\langle \vec{v} \rangle. \qquad (26\text{-}20)$

This leads us to conclude that there must be a drag force that is proportional to the average velocity of the charge carriers. The air drag force on a falling object is attributed to the action of many small air molecules hitting the falling object as it moves. Similarly, we can imagine that a charge carrier is being slowed down by hitting many stationary atoms and ions as it passes through the conductor. The interactions between charge carriers and the atoms in a conductor can only be described properly using quantum mechanics. Nonetheless, we attempt to picture the flow of charge past positive ions in Fig. 26-27.

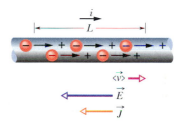

FIGURE 26-27 ■ Conduction electrons which are negative charge carriers drift at an average velocity $\langle \vec{v} \rangle$ in the opposite direction of the applied electric field $\vec{E}$. Their size is greatly exaggerated. By convention, the direction of the current density $\vec{J}$ and the sense of the arrow representing the flow of conventional current are drawn in that same direction.

What Is a Typical Average Charge Carrier Speed?

Solving $\vec{J} \equiv ne\langle \vec{v} \rangle$ (Eq. 26-13) for the average velocity and recalling Eq. 26-15 ($J = |i|/A$), we obtain the following expression for the average speed of the charge carrier,

$$|\langle \vec{v} \rangle| = \frac{|i|}{nAe} = \frac{J}{ne}. \qquad (26\text{-}21)$$

The product ne, whose SI unit is the coulomb per cubic meter (C/m³), is the *carrier charge density*.

At this point we can use Eq. 26-21 to find a typical value for the average speed for electrons flowing in a copper wire. Since copper has one conduction electron per atom we can use measurements for the density of copper atoms of $n = 8.5 \times 10^{28}$ atoms/m^3. Assume that our wire carries a current of 1.0 A and has a diameter of 2 mm so its cross-sectional area is 3×10^{-6} m^2. Then, according to Eq. 26-21, the average speed of the electrons is about

$$|\langle \vec{v} \rangle| = \frac{|i|}{nAe} = \frac{1 \text{ C/s}}{(8.5 \times 10^{28} \text{ atoms/m}^3)(3 \times 10^{-6} \text{ m}^2)(1.6 \times 10^{-19} \text{ C/atom})}$$
$$\approx 2.5 \times 10^{-5} \text{ m/s}.$$

This typical average speed is extremely small compared to very high speed random thermal motion of the electrons. It would take an electron about 11 hours to move across a 10 cm stretch of wire. Although the conduction electrons move along a wire very slowly like tired snails, there are so many of them that the current can actually be relatively large.

A Microscopic View of Resistivity

We can carry our microscopic analysis further, by relating the resistivity of a conductor to the properties of its charge carriers and the average time between electron collisions. If an electron of mass m is placed in an electric field of magnitude E, the electron will experience an acceleration given by Newton's Second Law:

$$\vec{a} = \frac{\vec{F}}{m} = \frac{e\vec{E}}{m}. \tag{26-22}$$

The nature of the collisions experienced by conduction electrons is such that, after a typical collision, each electron will—so to speak—completely lose its memory of its previous average velocity. Between collisions a conduction electron will have a mean free path λ like that derived in Section 20-5 for molecules traveling in a gas. However, it moves with a typical random speed $v^{\text{eff}} = \lambda/\tau$ where τ is the average time between collisions. Each electron will then start off fresh after every encounter, moving off in a random direction. In the average time τ between collisions, a typical electron will undergo an acceleration $\vec{a}$ in a direction opposite to that of the electric field as shown in Fig. 26-27. Thus, the average speed (often called the drift speed), the electron acquires in that direction is given by $|\langle \vec{v} \rangle| = a\tau$. Using Eq. 26-22 we get

$$|\langle \vec{v} \rangle| = a\tau = \frac{eE\tau}{m}. \tag{26-23}$$

Combining this result with $\vec{J} = ne\langle \vec{v} \rangle$ yields the average velocity of

$$\langle \vec{v} \rangle = \pm \frac{\vec{J}}{ne} = \pm \frac{e\vec{E}\tau}{m},$$

where we use the plus (+) sign for positive charge carriers and the minus (−) sign for negative charge carriers. We can combine the last two terms in the previous equation and solve for $\vec{E}$ to get

$$\vec{E} = \left(\frac{m}{e^2 n\tau} \right) \vec{J}.$$

This equation shows a proportionality between the electric field in a wire and the amount of current. Note that the magnitude of the electric field in a wire is in turn proportional to the potential difference across the wire. Thus, our microscopic picture of resistivity for metallic conductors is consistent with our macroscopic measurements, and it predicts a proportionality between potential difference and current.

Comparing the equation above with Eq. 26-17 ($\vec{E} = \rho\vec{J}$) leads to an expression for the resistivity in terms of the mass and charge of the carriers, the charge density n, and the average time between collisions

$$\rho = \frac{m}{e^2 n \tau}. \tag{26-24}$$

Conductivity

As well as referring to the resistivity of a material, we often speak of the conductivity σ of a material. This is simply the reciprocal of its resistivity, so

$$\sigma \equiv \frac{1}{\rho} \qquad \text{(definition of } \sigma\text{)}. \tag{26-25}$$

The SI unit of conductivity is the reciprocal ohm-meter $(\Omega \cdot m)^{-1}$. The unit name mhos per meter is sometimes used (mho is ohm backward). The definition of conductivity, σ, allows us to write Eq. 26-17($\vec{E} = \rho\vec{J}$) in the alternative form

$$\vec{J} = \sigma\vec{E}. \tag{26-26}$$

READING EXERCISE 26-13: The figure shows positive charge carriers moving leftward through a wire. Are the following leftward or rightward: (a) the conventional current i, (b) the current density $\vec{J}$, (c) the electric field $\vec{E}$ in the wire? *Hint*: You may want to review the discussion of conventional current in Section 26-8. ■

TOUCHSTONE EXAMPLE 26-3: Mean Free Time

What is the mean free time τ between collisions for the conduction electrons in copper?

SOLUTION ■ The **Key Idea** here is that the mean free time τ of copper is approximately constant, and in particular does not depend on any electric field that might be applied to a sample of the copper. Thus, we need not consider any particular value of applied electric field. However, because the resistivity ρ displayed by copper under an electric field depends on τ, we can find τ from Eq. 26-24 ($\rho = m/e^2 n\tau$). That equation gives us

$$\tau = \frac{m}{ne^2\rho}.$$

Taking the value of n, the number of conduction electrons per unit volume in copper, to be 8.5×10^{28} m^{-3}, and taking the value of ρ from Table 26-2, the denominator then becomes

$$(8.5 \times 10^{28} \text{ m}^{-3})(1.6 \times 10^{-19} \text{ C})^2(1.69 \times 10^{-8} \text{ } \Omega \cdot m)$$

$$= 3.67 \times 10^{-17} \text{ C}^2 \cdot \Omega/m^2$$

$$= 3.67 \times 10^{-17} \text{ kg/s},$$

where we converted units as

$$\frac{C^2 \cdot \Omega}{m^2} = \frac{C^2 \cdot V}{m^2 \cdot A} = \frac{C^2 \cdot J/C}{m^2 \cdot C/s} = \frac{kg \cdot m^2/s^2}{m^2/s} = \frac{kg}{s}.$$

Using these results and substituting for the electron mass m, we then have

$$\tau = \frac{9.1 \times 10^{-31} \text{ kg}}{3.67 \times 10^{-17} \text{ kg/s}} = 2.5 \times 10^{-14} \text{ s.} \qquad \text{(Answer)}$$

(b) The mean free path λ of the conduction electrons in a conductor is the average distance traveled by an electron between collisions. (This definition parallels that in Section 20-5 for the mean free path of molecules in a gas.) What is λ for the conduction electrons in copper?

SOLUTION ■ The **Key Idea** here is that the distance d any particle travels in a certain time t at a constant speed v is $d = vt$. To estimate v^{eff}, the speed at which the electrons typically move between collisions, we can think of the electrons as a "gas" of particles in thermal equilibrium with their surroundings inside the metal wire. Equation 20-21 then tells us that a typical electron has a kinetic energy related to the Kelvin temperature of its environment by $(\frac{1}{2})m\langle v^2 \rangle = (\frac{3}{2})k_B T$ (where k_B is the Boltzmann constant). Taking the electron's effective speed in a room temperature (300 K) environment to be $v^{rms} = \sqrt{\langle v^2 \rangle}$ gives

$$v^{eff} = \sqrt{3k_B T/m} = \sqrt{(3(1.38 \times 10^{-23} \text{ J/K})(300 \text{ K})/(9.11 \times 10^{-31} \text{ kg}))}$$

$$= 1.168 \times 10^5 \text{ m/s}$$

and

$$\lambda = v^{eff}\tau = (1.168 \times 10^5 \text{ m/s})(2.5 \times 10^{-14} \text{ s})$$

$$= 2.9 \times 10^{-9} \text{ m} = 2.9 \text{ nm}. \qquad \text{(Answer)}$$

This is about 10 times the distance between nearest-neighbor atoms in a copper lattice. While this is a reasonable sounding result, it turns out that the actual value of λ is about 10 times larger than this due to quantum effects.

26-11 Other Types of Conductors

In the last few chapters we have assumed that the conductors under consideration are metallic like copper or nichrome. As you can see from Table 26-2, one of the distinctive properties of metallic conductors is that they have positive temperature coefficients indicating that their resistivities *increase* with temperature. This property seems reasonable since the thermal energy in the metal lattice causes the atoms in the metal to vibrate more, which further impedes the flow of conduction electrons. In addition, Eq. 26-9 indicates that this increase of resistivity with temperature is approximately *linear*.

There are other types of conductors with resistivities that do not simply increase linearly with temperature. The most important of these are semiconductors, which lie at the heart of the microelectronic revolution. The resistivity of semiconductors decreases more or less linearly with temperature. Superconductors are another class of conductors that do not have the same temperature behavior as conductors. Although the resistivity of superconductors increases with temperature, it does so in a very nonlinear fashion.

Because of the importance of semiconductors and superconductors we describe some of their properties here. Both of these nonmetallic conductors have some amazing properties that we describe briefly in this section. However, in the next few chapters we return to the study—within the framework of classical physics—of *steady* currents of *conduction electrons* moving through *metallic conductors*.

Semiconductors

The basic element found in virtually all semiconductors is either silicon or germanium. Table 26-3 compares the properties of silicon—a typical semiconductor—and copper—a typical metallic conductor. We see that silicon has significantly fewer charge carriers, a much higher resistivity, and a temperature coefficient of resistivity that is both large and negative. Thus, although the resistivity of copper increases with temperature, that of pure silicon decreases.

Pure silicon has such a high resistivity that it is effectively an insulator and not of much direct use in microelectronic circuits. However, its resistivity can be greatly

TABLE 26-3
Some Electrical Properties of Copper and Silicon[a]

Property	Copper	Silicon
Type of material	Metal	Semiconductor
Charge carrier density, m^{-3}	9×10^{28}	1×10^{16}
Resistivity, $\Omega \cdot$ m	2×10^{-8}	3×10^3
Temperature coefficient of resistivity, K^{-1}	$+4 \times 10^{-3}$	-70×10^{-3}

[a]Rounded to one significant figure for easy comparison.

reduced in a controlled way by adding minute amounts of specific "impurity" atoms in a process called *doping*. Table 26-2 gives typical values of resistivity for silicon before and after doping with two different impurities, phosphorus and aluminum. Most semiconducting devices, such as transistors and junction diodes, are fabricated by the selective doping of different regions of the silicon with impurity atoms of different kinds.

A full explanation of the difference in resistivity between semiconductors and metallic conductors requires an understanding of quantum theory developed to explain atomic behavior. However, the difference has to do with the probability that electrons in a material can be made mobile. As we discuss in Section 22-6, in a metallic conductor some of the outermost electrons associated with an atom can move from one atom to the next without any additional energy. Thus, the electric field set up in the wire when a potential difference is applied drives current through a conductor.

In an insulator, considerable energy is required to free electrons so they can move through the material. Thermal energy cannot supply enough energy, and neither can any reasonable electric field applied to the insulator. Thus, no electrons are available to move through the insulator, and hence no current occurs even with an applied electric field. A semiconductor is like an insulator *except* that the energy required to free some electrons can be adjusted through doping. Doping can supply either electrons or positive charge carriers held very loosely within the material that are easy to get moving.*

In a semiconductor, the density of charge carriers is small but increases very rapidly with temperature as the increased thermal agitation makes more charge carriers available. This causes a *decrease* of resistivity with increasing temperature, as indicated by the negative temperature coefficient of resistivity for silicon in Table 26-3. The same increase in collision rate we noted for metals also occurs for semiconductors, but its effect is swamped by the rapid increase in the number of charge carriers.

Superconductors

In 1911, Dutch physicist Kamerlingh Onnes discovered that the resistivity of mercury absolutely disappears at temperatures below about 4 K (Fig. 26-28). This phenomenon of **superconductivity** is of vast potential importance in technology because it means charge can flow through a superconducting conductor without producing thermal energy losses. Currents created in a superconducting ring, for example, have persisted for several years without any measurable decrease; the electrons making up the current require a force and a source of energy at start-up time, but not thereafter.

Prior to 1986, the technological development of superconductivity was throttled by the cost of producing the extremely low temperatures that were required to achieve the effect. In 1986, however, new ceramic materials were discovered that become superconducting at considerably higher (and thus cheaper to produce) temperatures. Practical application of superconducting devices at room temperature may eventually become feasible.

Superconductivity is a much different phenomenon from conductivity. In fact, the best of the normal conductors, such as silver and copper, cannot become superconducting at any temperature, and the new ceramic superconductors are actually insulators when they are not at low enough temperatures to be in a superconducting state.

One explanation for superconductivity is that the electrons making up the current move in coordinated pairs. One of the electrons in a pair may electrically distort the molecular structure of the superconducting material as it moves through, creating a short-lived concentration of positive charge nearby. The other electron in the pair may then be attracted toward this positive charge. According to the theory, such coor-

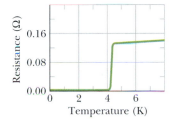

FIGURE 26-28 ■ The resistance of mercury drops to zero at a temperature of about 4 K.

A disk-shaped magnet is levitated above a superconducting material that has been cooled by liquid nitrogen. The goldfish is along for the ride.

---

*Explaining what positive charge carriers are and how they move is complex. For now just consider the charge carriers as negative (that is, electrons).

dination between electrons would prevent them from colliding with the molecules of the material and thus would eliminate electrical resistance. The theory worked well to explain the pre-1986, lower temperature superconductors, but new theories appear to be needed for the newer, higher temperature superconductors.

Problems

SEC. 26-3 ■ BATTERIES AND ELECTRIC CURRENT

1. Coulombs and Electrons A current of 5.0 A exists in a 10 Ω resistor for 4.0 min. How many (a) coulombs and (b) electrons pass through any cross section of the resistor in this time?

2. Charged Belt A charged belt, 50 cm wide, travels at 30 m/s between a source of charge and a sphere. The belt carries charge into the sphere at a rate corresponding to 100 μA. Compute the surface charge density on the belt.

3. Isolated Sphere An isolated conducting sphere has a 10 cm radius. One wire carries a current of 1.000 002 0 A into it. Another wire carries a current of 1.000 000 0 A out of it. How long would it take for the sphere to increase in potential by 1000 V?

SEC. 26-5 ■ RESISTANCE AND OHM'S LAW

4. Electrical Cable An electrical cable consists of 125 strands of fine wire, each having 2.65 μΩ resistance. The same potential difference is applied between the ends of all the strands and results in a total current of 0.750 A. (a) What is the current in each strand? (b) What is the applied potential difference? (c) What is the resistance of the cable?

5. Electrocution A human being can be electrocuted if a current as small as 50 mA passes near the heart. An electrician working with sweaty hands makes good contact with the two conductors he is holding, one in each hand. If his resistance is 2000 Ω, what might the fatal voltage be?

SEC. 26-6 ■ RESISTANCE AND RESISTIVITY

6. Trolley Car A steel trolley-car rail has a cross-sectional area of 56.0 cm². What is the resistance of 10.0 km of rail? The resistivity of the steel is 3.00×10^{-7} Ω · m.

7. Conducting Wire A conducting wire has a 1.0 mm diameter, a 2.0 m length, and a 50 mΩ resistance. What is the resistivity of the material?

8. A Wire A wire 4.00 m long and 6.00 mm in diameter has a resistance of 15.0 mΩ. A potential difference of 23.0 V is applied between the ends. (a) What is the current in the wire? (b) Calculate the resistivity of the wire material. Identify the material. (Use Table 26-2.)

9. A Coil A coil is formed by winding 250 turns of insulated 16-gauge copper wire (diameter = 1.3 mm) in a single layer on a cylindrical form of radius 12 cm. What is the resistance of the coil? Neglect the thickness of the insulation (Use Table 26-2.)

10. What Temperature (a) At what temperature would the resistance of a copper conductor be double its resistance at 20.0°C? (Use 20.0°C as the reference point in Eq. 26-9; compare your answer with Fig. 26-23.) (b) Does this same "doubling temperature" hold for all copper conductors regardless of shape or size?

11. Longer Wire A wire with a resistance of 6.0 Ω is drawn out through a die so that its new length is three times its original length. Find the resistance of the longer wire, assuming that the resistivity and density of the material are unchanged.

12. A Certain Wire A certain wire has a resistance R. What is the resistance of a second wire, made of the same material, that is half as long and has half the diameter?

13. Two Conductors Two conductors are made of the same material and have the same length. Conductor A is a solid wire of diameter 1.0 mm. Conductor B is a hollow tube of outside diameter 2.0 mm and inside diameter 1.0 mm. What is the resistance ratio R_A/R_B, measured between their ends?

14. Flashlight Bulb A common flashlight bulb is rated at 0.30 A and 2.9 V (the values of the current and voltage under operating conditions). If the resistance of the bulb filament at room temperature (20°C) is 1.1 Ω, what is the temperature of the filament when the bulb is on? The filament is made of tungsten.

15. Metal Rod When a metal rod is heated, not only its resistance but also its length and its cross-sectional area change. The relation $R = \rho L/A$ suggests that all three factors should be taken into account in measuring ρ at various temperatures. (a) If the temperature changes by 1.0 C°, what percentage changes in R, L, and A occur for a copper conductor? (b) The coefficient of linear expansion for copper is 1.7×10^{-5}/K. What conclusion do you draw?

16. Gauge Number If the gauge number of a wire is increased by 6, the diameter is halved; if a gauge number is increased by 1, the diameter decreases by the factor $2^{1/6}$ (see the table in Problem 32). Knowing this, and knowing that 1000 ft of 10-gauge copper wire has a resistance of approximately 1.00 Ω, estimate the resistance of 25 ft of 22-gauge copper wire.

SEC. 26-7 ■ POWER IN ELECTRIC CIRCUITS

17. X-Ray Tube A certain x-ray tube operates at a current of 7.0 mA and a potential difference of 80 kV. What is its power in watts?

18. A Student A student kept his 9.0 V, 7.0 W radio turned on at full volume from 9:00 P.M. until 2:00 A.M. How much charge went through it?

19. Space Heater A 120 V potential difference is applied to a space heater whose resistance is 14 Ω when hot. (a) At what rate is electric energy transferred to heat? (b) At 5.0¢/kW·h, what does it cost to operate the device for 5.0 h?

20. Thermal Energy Thermal energy is produced in a resistor at a rate of 100 W when the current is 3.00 A What is the resistance?

21. Energy Is Dissipated An unknown resistor is connected between the terminals of a 3.00 V battery. Energy is dissipated in the resistor at the rate of 0.540 W. The same resistor is then connected between the terminals of a 1.50 V battery. At what rate is energy now dissipated?

22. Space Heater Two A 120 V potential difference is applied to a space heater that dissipates 500 W during operation. (a) What is its resistance during operation? (b) At what rate do electrons flow through any cross section of the heater element?

23. Radiant Heater A 1250 W radiant heater is constructed to operate at 115 V. (a) What will be the current in the heater? (b) What is the resistance of the heating coil? (c) How much thermal energy is produced in 1.0 h by the heater?

24. Heating Element A heating element is made by maintaining a potential difference of 75.0 V across the length of a Nichrome wire that has a 2.60×10^{-6} m^2 cross section. Nichrome has a resistivity of 5.00×10^{-7} $\Omega \cdot$m. (a) If the element dissipates 5000 W, what is its length? (b) If a potential difference of 100 V is used to obtain the same dissipation rate, what should the length be?

25. Nichrome Heater A Nichrome heater dissipates 500 W when the applied potential difference is 110 V and the wire temperature is 800°C. What would be the dissipation rate if the wire temperature were held at 200°C by immersing the wire in a bath of cooling oil? The applied potential difference remains the same, and α for Nichrome at 800°C is 4.0×10^{-4}/K.

26. 100 W Lightbulb A 100 W lightbulb is plugged into a standard 120 V outlet. (a) How much does it cost per month to leave the light turned on continuously? Assume electric energy costs 12¢/kW·h. (b) What is the resistance of the bulb? (c) What is the current in the bulb? (d) Is the resistance different when the bulb is turned off?

27. Linear Accelerator A linear accelerator produces a pulsed beam of electrons. The pulse current is 0.50 A, and each pulse has a duration of 0.10 μs. (a) How many electrons are accelerated per pulse? (b) What is the average current for an accelerator operating at 500 pulses/s? (c) If the electrons are accelerated to an energy of 50 MeV, what are the average and peak powers of the accelerator?

28. Cylindrical Resistor A cylindrical resistor of radius 5.0 mm and length 2.0 cm is made of material that has a resistivity of 3.5×10^{-5} $\Omega \cdot$m. What is the potential difference when the energy dissipation rate in the resistor is 1.0 W?

29. Copper Wire A copper wire of cross-sectional area 2.0×10^{-6} m^2 and length 4.0 m has a current of 2.0 A uniformly distributed across that area. How much electric energy is transferred to thermal energy in 30 min?

SEC. 26-8 ■ CURRENT DENSITY IN A CONDUCTOR

30. Small But Measurable A small but measurable current of 1.2×10^{-10} A exists in a copper wire whose diameter is 2.5 mm. Assuming the current is uniform, calculate (a) the current density and (b) the average electron speed.

31. A Beam A beam contains 2.0×10^{8} doubly charged positive ions per cubic centimeter, all of which are moving north with a speed of 1.0×10^{5} m/s. (a) What are the magnitude and direction of the current density $\vec{J}$? (b) Can you calculate the total current i in this ion beam? If not what additional information is needed?

32. The U.S. Electric Code The (United States) National Electric Code, which sets maximum safe currents for insulated copper wires of various diameters, is given (in part) in the table. Plot the safe current density as a function of diameter. Which wire gauge has the maximum safe current density? ("Gauge" is a way of identifying wire diameters, and 1 mil = 10^{-3} in.)

Gauge	4	6	8	10	12	14	16	18
Diameter, mils	204	162	129	102	81	64	51	40
Safe current, A	70	50	35	25	20	15	6	3

33. A Fuse A fuse in an electric circuit is a wire that is designed to melt, and thereby open the circuit, if the current exceeds a predetermined value. Suppose that the material to be used in a fuse melts when the current density rises to 440 A/cm^2. What diameter of cylindrical wire should be used to make a fuse that will limit the current to 0.50 A?

34. Near Earth Near the Earth, the density of protons in the solar wind (a stream of particles from the Sun) is 8.70 cm^{-3}, and their speed is 470 km/s. (a) Find the current density of these protons. (b) If the Earth's magnetic field did not deflect them, the protons would strike the planet. What total current would the Earth then receive?

35. Steady Beam A steady beam of alpha particles ($q = +2e$) traveling with constant kinetic energy 20 MeV carries a current of 0.25 μA. (a) If the beam is directed perpendicular to a plane surface, how many alpha particles strike the surface in 3.0 s? (b) At any instant, how many alpha particles are there in a given 20 cm length of the beam? (c) Through what potential difference is it necessary to accelerate each alpha particle from rest to bring it to an energy of 20 MeV?

36. Current Density (a) The current density across a cylindrical conductor of radius R varies in magnitude according to the equation

$$J = J_0\left(1 - \frac{r}{R}\right),$$

where r is the distance from the central axis. Thus, the current density has a maximum magnitude of $J_0 = |\vec{J}_0|$ at that axis ($r = 0$) and decreases linearly to zero at the surface ($r = R$). Calculate the current in terms of J_0 and the conductor's cross-sectional area $A = \pi R^2$. (b) Suppose that, instead, the current density is a maximum J_0 at the cylinder's surface and decreases linearly to zero at the axis: $J = J_0 r/R$. Calculate the magnitude of the current. Why is the result different from that in (a)?

37. How Long How long does it take electrons to get from a car battery to the starting motor? Assume the current is 300 A and the electrons travel through a copper wire with cross-sectional area 0.21 cm^2 and length 0.85 m. (*Hint:* Assume one conduction electron per atom and take the number density of copper atoms to be 8.5×10^{28} atoms/m^2.)

38. Nichrome A wire of Nichrome (a nickel-chromium-iron alloy commonly used in heating elements) is 1.0 m long and 1.0 mm^2 in cross-sectional area. It carries a current of 4.0 A when a 2.0 V potential difference is applied between its ends. Calculate the conductivity σ of Nichrome.

39. When Applied When 115 V is applied across a wire that is 10 m long and has a 0.30 mm radius, the current density is 1.4×10^{4} A/m^2. Find the resistivity of the wire.

40. Truncated Right-Circular Cone A resistor has the shape of a truncated right-circular cone (Fig. 26-29). The end radii are a and b, and the altitude is L. If the taper is small, we may assume that the current density is uniform across any cross section. (a) Calculate the resistance of this object. (b) Show that your answer reduces to $\rho(L/A)$ for the special case of zero taper (that is, for $a = b$).

FIGURE 26-29 ■ Problem 40.

SEC. 26-10 ■ A MICROSCOPIC VIEW OF CURRENT AND RESISTANCE

41. Gas Discharge Tube A current is established in a gas discharge tube when a sufficiently high potential difference is applied across the two electrodes in the tube. The gas ionizes; electrons move toward the positive terminal and singly charged positive ions toward the negative terminal. (a) What is the magnitude of the current in a hydrogen discharge tube in which 3.1×10^{18} electrons and 1.1×10^{18} protons move past a cross-sectional area of the tube each second? (b) What is the direction of the current density $\vec{J}$?

42. A Block A block in the shape of a rectangular solid has a cross-sectional area of 3.50 cm^2 across its width, a front-to-rear length of 15.8 cm, and a resistance of 935 Ω. The material of which the block is made has 5.33×10^{22} conduction electrons/m^3. A potential difference of 35.8 V is maintained between its front and rear faces. (a) What is the current in the block? (b) If the current density is uniform, what is its value? (c) What is the average or drift speed of the conduction electrons? (d) What is the magnitude of the electric field in the block?

43. Earth's Lower Atmosphere Earth's lower atmosphere contains negative and positive ions that are produced by radioactive elements in the soil and cosmic rays from space. In a certain region, the atmospheric electric field strength is 120 V/m, directed vertically down. This field causes singly charged positive ions, at a density of 620/cm^3, to drift downward and singly charged negative ions, at a density of 550/cm^3, to drift upward (Fig. 26-30). The measured conductivity of the air in that region is $2.70 \times 10^{-14}(1/\Omega \cdot$ m$)$. Calculate (a) the average ion speed, assumed to be the same for positive and negative ions, and (b) the current density.

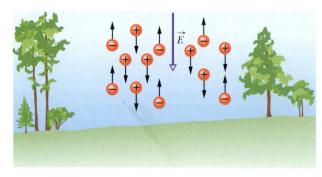

FIGURE 26-30 ■ Problem 43.

Additional Problems

44. Saving on Your Electric Bill Fluorescent bulbs deliver the same amount of light using much less power. If one kW-hr costs 12¢, estimate the amount of money you would save each month by replacing all the 75 W incandescent bulbs in your house by 10 W fluorescent ones than incandescent ones. *Be sure to clearly state your assumptions.*

45. Building a Water Heater The nickel-chromium alloy Nichrome has a resistivity of about 10^{-6} Ω-m. Suppose you want to build a small heater out of a coil of Nichrome wire and a 6 V battery in order to heat 30 ml of water from a temperature of 20 C to 40 C in 1 min. Assume the battery has negligible internal resistance.
(a) How much heat energy (in joules) do you need to do this?
(b) How much power (in watts) do you need to do it in the time indicated?
(c) What resistance should your Nichrome coil have in order to produce this much power in heat?
(d) Can you create a coil having these properties? (*Hint:* Can you find a plausible length and cross-sectional area for your wire that will give you the resistance you need?)
(e) If the internal resistance of the battery were 1/3 Ω, how would it affect your calculation? (Only explain what you would have to do; don't recalculate the size of your coil.)

46. A Confusing Thing One of the most confusing things about wiring circuits and figuring out what you've done is that many arrangements are electrically equivalent. Unless you have unusual powers of visualization it is often hard to recognize this. For example, three of the circuits shown in Fig. 26-31 are electrically equivalent and one is not. Answer questions (a) through (d) that follow.
(a) Which circuit is not like the others? Explain why it's different.
(b) Draw circuit diagrams for each of the arrangements and label each diagram as A, B, C, or D. (c) Examine your diagrams. Is it possible for neat circuit diagrams that look superficially different to represent the same set of electrical connections?

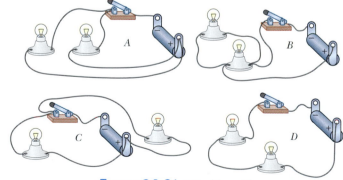

FIGURE 26-31 ■ Problem 46.

47. Draw the Circuit Diagram Draw a neat circuit diagram for each of the two circuits shown in Fig. 26-32 using the standard symbols for bulbs, batteries, and switches.

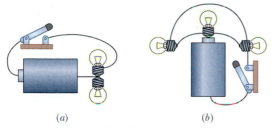

(a)   (b)

FIGURE 26-32 ■ Problem 47.

48. Charge Through Conductor The charge passing through a conductor increases over time as $q(t) = (1.6 \text{ C/s}^2)t^2 + (2.2 \text{ C/s})t$, where t is in seconds. (a) What equation describes the current in the circuit as a function of time? (b) What is the current in the conductor at $t = 0.0$ s and at $t = 2.0$ s?

49. Increases Over Time The charge passing through a conductor increases over time as $q(t) = (1.5 \text{ C/s}^3)t^3 - (4.5 \text{ C/s}^2)t^2 + (2 \text{ C/s})t$,

where t is in seconds. (a) What equation describes the current in the circuit as a function of time? (b) What is the current in the conductor at $t = 0.0$ s and at $t = 1.0$ s?

50. 1994 Honda Accord Consider a 1994 Honda Accord with a battery that is rated at 52 ampere-hours. This battery is supposed to be able to deliver 1 ampere of current to electrical devices in a car for at least 52 hours or 2 amperes for 26 hours, and so on. Suppose you leave the car lights turned on when you park the car and the car lights draw 20 amperes of current. How long will it be before your battery is dead?

51. The Resistance of a Pocket Calculator A typical AAA battery delivers a nearly constant voltage of 1.5 V and stores about 3 kJ of energy. From the time it takes you to use up the batteries in your calculator, estimate the resistance of your calculator. (If you don't have a calculator of this type, make a plausible estimate of how long it might take to use up the batteries. Give some reason for your estimate.)

27 | Circuits

The electric eel (*Electrophorus*) lurks in rivers of South America, killing the fish on which it preys with pulses of current. It does so by producing a potential difference of several hundred volts along its length; the resulting current in the surrounding water, from near the eel's head to the tail region, can be as much as one ampere. If you were to brush up against this eel while swimming, you might wonder (after recovering from the very painful stun):

How can the electric eel manage to produce a current that large without shocking itself?

The answer is in this chapter.

27-1 Electric Currents and Circuits

Knowing how to analyze circuits by predicting the currents through their elements and the potential differences across them is a valuable skill. Such knowledge enables engineers and scientists to design electrical devices and helps them make productive use of existing devices. Our goal in this chapter is to understand the behavior of relatively simple electric circuits by applying concepts such as current, potential difference, and resistors developed in the previous chapter. We will start by considering very simple ideal circuits and then go on to consider circuits with multiple loops and batteries such as those shown in Fig. 27-1. Toward the end of the chapter we will introduce the concept of emf or electromotive force associated with batteries and other power sources. In particular, we will consider how to extend our analysis to the behavior of circuits powered by nonideal batteries that have internal resistance.

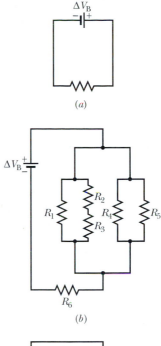

Ideal Circuits

As we so often do in developing physical ideas, we start by analyzing how a system behaves under ideal conditions. Only then do we introduce real-world complexities that require us to modify our methods of analysis. The ideal circuits we consider first have three characteristics:

1. **They are powered by ideal batteries.** As stated in Section 26-3, an ideal battery "maintains a constant potential difference across its terminals." This means there is a negligible amount of "electric friction" and the potential difference, ΔV_B, across the terminals of an ideal battery stays the same, regardless of the amount of charge flowing through it. But as the chemical potential energy of a real battery decreases, it develops some *internal resistance,* and the potential difference across its terminals decreases if its current increases.

2. **All circuit elements, other than the battery and connecting wires, are ohmic devices having a significant resistance.** As discussed in Section 26-5, an *ohmic device has a constant value of resistance, R, that is not a function of the amount of current passing through it.* Although lightbulbs and some other circuit elements are not ohmic, standard carbon resistors obey Ohm's law and have a constant resistance over a large current range. We make use of the fact that the potential difference across the terminals of an ohmic device is directly proportional to the current, i, flowing through it and is given by $\Delta V = iR$ (Eq. 26-7).

3. **Ideal conducting wires connect the battery to circuit elements.** Copper wiring is used in most circuits found in consumer devices, households, and industries. We can use Eq. 26-7 and data from Table 26-2 to determine that the resistance of a 30 cm length of common 22 gauge copper wire is about 0.1 Ω. If this wire was connected to a 10 Ω resistor, the additional resistance of the wire would add 1% to the overall resistance. In connecting larger resistors, the influence of the resistance of the wire is even smaller. Because the resistance in the wire is so small, the potential difference between the ends of even a relatively long continuous connecting wire is for all practical purposes negligible. In ideal circuits, we assume there is no potential drop across connecting wires.

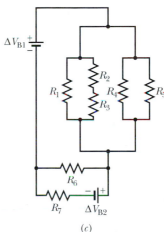

FIGURE 27-1 ■ Several types of ideal circuits we will learn to analyze in this chapter consist of ideal batteries, conducting wires with negligible resistance and ohmic resistors. (*a*) A single-loop circuit. (*b*) A single-battery, multiple-loop circuit. (*c*) A multiple-loop circuit with multiple batteries.

READING EXERCISE 27-1: Show that the resistance of a 30 cm ($\approx$ 12 inch) length of 22 gauge copper wire of diameter 0.024 cm has a resistance of about 0.1 Ω. *Hint:* You will need to use information from Table 26-2 along with Eq. 26-7. ■

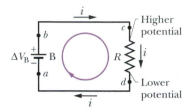

FIGURE 27-2 ■ A single-loop circuit in which a resistor R is connected across an ideal battery B with potential difference ΔV_B. The resulting current i is the same throughout the circuit.

27-2 Current and Potential Difference in Single-Loop Circuits

Suppose we want to design or operate an electrical device such as a CD player or refrigerator. The operation of the given device will require a certain minimum current or potential difference. How would we calculate the amount of current in a circuit or the potential difference between two points within the device? That is the topic of this section.

We start out our discussion of current in circuits by focusing on the part of the circuit outside of the battery. That is, we will focus on current that passes from one battery terminal, through the circuit, and back to the other terminal. At the end of the chapter we will review and extend our previous discussions about what goes on inside devices like batteries and generators.

Consider the simple *single-loop* circuit of Fig. 27-2 consisting of an ideal battery, a resistor, R, and two ideal connecting wires. Unless otherwise indicated, we assume that wires in circuits have negligible resistance. Their function, then, is merely to provide pathways along which charge carriers can move. Through use of stored chemical energy (a form of internal potential energy), the battery keeps one of its terminals (called the positive terminal and often labeled $+$) at a higher electric potential than the other terminal (called the negative terminal and labeled $-$).

The mobile negative charge carriers in the circuit wires move preferentially toward the positive terminal and away from the negative terminal. As a result, for the circuit shown in Fig. 27-2, we have a net flow of negative charge in a counterclockwise direction. In Chapter 26, we discussed the fact that a flow of negative electrons in one direction is macroscopically indistinguishable from a flow of positive charges in the other direction. For historical reasons we continue the practice established in that chapter of working with current as if the charge carriers are positive.

The direction of the conventional current in the circuit shown in Fig. 27-2 is noted with arrows that are labeled i. Unless otherwise noted, we will continue the practice of using conventional (positive) current in our analysis of electric circuits. We will reach the same conclusions about the fundamental behavior of circuits as we would if we had used electron currents.

To begin learning how to calculate currents in circuits, let's start with the ideal circuit depicted in Fig. 27-2. We have marked the points just before and after each element with the letters a, b, c, and d. Let's start at point a and proceed around the circuit in either direction, adding any changes in potential we encounter. Once we return to our starting point, we must also have returned to our starting potential. In words, the potential energy change per unit of charge traveling through the battery plus the potential energy change of the charge traveling through the wires and the resistors must be zero. This can be denoted as

$$\Delta V_{a\to b} + \Delta V_{b\to c} + \Delta V_{c\to d} + \Delta V_{d\to a} = \Delta V_{a\to a} = 0 \text{ V}.$$

For our simple circuit in Fig. 27-2 the charges gain potential while traveling from a to b due to the energy boost from the battery so that $\Delta V_{a\to b} = V_b - V_a = \Delta V_B$. The charges then flow freely from b to c through the first segment of the ideal conductor with no potential loss since the wire has a negligible resistance. Then the charges flow through the resistor, R. Finally, they flow back to point a, through another length of ideal wire.

$$\Delta V_B + \Delta V_{c\to d} = 0 \text{ V},$$

where ΔV_B represents a positive change in potential per unit charge as charges proceed from point a to point b by moving through the battery. Recall that if our ohmic resistor has a fixed value R, then we noted in Eq. 26-7 that $\Delta V = iR$ where i is the current passing through the circuit. However, Eq. 26-7 didn't specify whether the ΔV refers to $\Delta V_{c \to d}$ or $\Delta V_{d \to c} = -\Delta V_{c \to d}$. It is clear from the context that if we proceed through the loop from c to d, $\Delta V_{c \to d}$ must be negative so it will cancel the ΔV_{B}, which we know is positive. This tells us the following about the mathematics of finding the potential difference across a resistor:

$$\Delta V_{d \to c} = iR \quad \text{and} \quad \Delta V_{c \to d} = -iR.$$

In other words, charges lose potential as they travel through a resistor. This makes sense physically because resistors give off energy in the form of heat and light. So our battery acts as a pump to increase the potential energy of a charge and the charge loses potential energy in passing through a resistive device.

This can be summarized as the loop rule.

LOOP RULE: The algebraic sum of the changes in potential encountered in a complete traversal of any loop of a circuit must be zero.

This is often referred to as *Kirchhoff's loop rule* (or *Kirchhoff's voltage law*), after German physicist Gustav Robert Kirchhoff. This rule is analogous to what happens when you hike around a mountain. If you start from any point on a mountain and return to the same point after walking around it, the algebraic sum of the changes in elevation you encounter must be zero. Thus, you end up at the same gravitational potential as you had before you started. Although we developed this rule through consideration of a single-loop circuit, it also holds for any complete loop in a *multiloop* circuit, no matter how complicated.

In Fig. 27-2, we will start at point a, whose potential is V_a, and mentally walk clockwise around the circuit until we are back at a, keeping track of potential changes as we move. (Our starting point is at the low-potential terminal of the battery—the negative terminal.) The potential difference between the battery terminals is equal to ΔV_{B}. When we pass through the battery from the low to high-potential terminal, the change in potential is positive.

As we walk along the top wire to the top end of the resistor, there is no potential change because the wire has negligible resistance; it is at the same potential as the high-potential terminal of the battery. So too is the top end of the resistor. When we pass through the resistor in the direction of the current flow, the potential decreases by an amount equal to $-iR$. We know the potential decreases because we are moving from the higher potential terminal of the resistor to the lower potential terminal.

For a walk around a single-loop circuit of total resistance R in *the direction of the current* our loop rule gives us

$$\Delta V_{\mathrm{B}} - iR = 0 \text{ V}.$$

Solving this equation for i gives us

$$i = \frac{\Delta V_{\mathrm{B}}}{R} \qquad \text{(single-loop circuit).} \qquad (27\text{-}1)$$

If we apply the loop rule to a complete walk around a single-loop circuit of total resistance R *against the direction of current*, the rule gives us

$$-\Delta V_{\mathrm{B}} + iR = 0 \text{ V},$$

and we again find that

$$i = \frac{\Delta V_{\mathrm{B}}}{R} \qquad \text{(single-loop circuit).}$$

Thus, you may mentally circle a loop in either direction to apply the loop rule.

To prepare for circuits more complex than Fig. 27-2, let us summarize two rules for finding potential differences as we move around a chosen loop:

> **RESISTANCE RULE:** For a move through a resistor in the direction of the conventional current, the change in potential is $-iR$; in the opposite direction of current flow it is $+iR$.

> **POTENTIAL RULE:** For a move through a source of potential difference from low potential (for example, the negative terminal on a battery denoted a) to high potential (for example, the positive terminal on a battery denoted b) the change in potential is positive and given by $V_b - V_a = \Delta V_B$; in the opposite direction it is negative and given by $V_a - V_b = -\Delta V_B$.

What happens to the amount of current as it passes through a resistor? Is the current going into the resistor the same as the current coming out of the resistor? Or does a resistor (for example, a lightbulb) "use up" current? Recall that in Fig. 26-5 we depicted observations involving batteries and bulbs that clearly showed current is constant throughout a single loop circuit when resistors are connected in series. You can easily replicate these observations using fresh flashlight batteries, copper wires, and 1.5 V bulbs.

READING EXERCISE 27-2: It is asserted above that we can infer that the current flow into and out of a resistor is the same because three lightbulbs connected in series glow equally brightly. Suppose the resistors shown in Fig. 27-3a are lightbulbs. Describe the brightness of the third bulb relative to the first and second bulbs under the following assumptions: (a) All the current is used up by the first bulb; (b) most of the current is used up by the first bulb; (c) a small amount of the current was used up by the first bulb. ■

READING EXERCISE 27-3: The figure to the right shows the conventional current i in a single-loop circuit with a battery B and a resistor R (and wires of negligible resistance). At points a, b, and c, rank (a) the amount of the current and (b) the electric potential, greatest first.

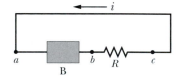

■

27-3 Series Resistance

We now turn our attention to more complicated single-loop circuits. Figure 27-3a shows three resistors connected in series to an ideal battery with potential difference ΔV_B between its terminals. Note that the three resistors are connected one after another between b and c, c and d, and d and a. Also an ideal battery maintains a potential difference across the series of resistors (between points a and b). If we apply the loop rule for charges moving in the direction of conventional current from point a at the negative terminal of the battery and proceeding through the loop until we encounter point a, again we get

$$\Delta V_{a \to b} + \Delta V_{b \to c} + \Delta V_{c \to d} + \Delta V_{d \to a} = 0 \; V. \tag{27-2}$$

Because we know that current is not used up by a resistor, we know the current flowing through the loop is the same everywhere, and so the current through each resistor must be the same. We also assume there is no potential difference along any segment of wire. If we consider the three resistors separately, applying the loop rule in the

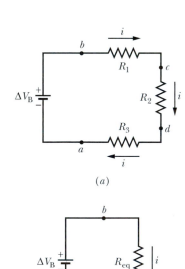

(a)

(b)

FIGURE 27-3 ■ (a) Three resistors are connected in series between points a and b. (b) An equivalent circuit, with the three resistors replaced with their equivalent resistance R_{eq}.

same manner (starting at the positive terminal of the battery and proceeding through the loop in the direction of conventional current) gives

$$\Delta V_B + (-iR_1) + (-iR_2) + (-iR_3) = 0 \text{ V}.$$

By rearranging terms in the equation above we get

$$\Delta V_B - i(R_1 + R_2 + R_3) = 0 \text{ V}, \tag{27-3}$$

and defining an equivalent resistance as $R_{eq} = R_1 + R_2 + R_3$ we find that Eq. 27-3 reduces to the same form as Eq. 27-1 with the equivalent resistance playing the role of the resistance in a circuit that has only one resistance. This is illustrated in Fig. 27-3.

Equating these two expressions tells us two things. First, the potential difference across the whole series of resistors is equal to the sum of the potential differences across the three resistors. Second, the potential difference across the whole series of resistors is equal to the potential difference across our ideal battery. Figure 27-3b shows the equivalent resistance, with a new resistor R_{eq}, that can replace the three resistors of Fig. 27-3a.

The result $R_{eq} = R_1 + R_2 + R_3$ is not surprising because it is compatible with the experimental findings we presented in Section 26-6: the resistance of a length of wire is directly proportional to its length (Eq. 26-8). Imagine three different carbon resistors like those depicted in the previous chapter (Fig. 26-21). Suppose these resistors are connected by ideal conductors (with almost no resistivity) having the same graphite material in their centers each with the same cross-sectional area. Giving the resistors different values of resistance would involve having the centers of the resistors be three different lengths. We would then expect the total resistance to be proportional to the sum of the three lengths of the resistors' graphite centers.

Obviously, we can extend our method of finding the equivalent resistance from 3 to N resistors by expanding Eq. 27-3 into the equation

$$R_{eq} = R_1 + R_2 + R_3 + \cdots + R_N = \sum_{j=1}^{N} R_j \quad (N \text{ resistors in series}). \tag{27-4}$$

Note that when resistors are in series, their equivalent resistance is always *greater* than that of any of the individual resistors. Also, the current moving through resistors wired in series can move along only a single route. If there are additional routes so the currents in different resistors are different, the resistors are not connected in series.

In general:

> If N resistors in series were covered by a box, the resistors could be replaced by a single equivalent resistor with a value. $R_{eq} = R_1 + R_2 + R_3 + \cdots + R_N$. Someone making measurements outside the box could not tell whether there is a single equivalent resistor or a series of individual resistors.

In short, we conclude that if we replace a series of resistors with a single equivalent resistor, the new circuit will have the same overall potential differences and currents as the original one (so long as we don't measure potential drops between the resistors wired in series).

More on Ammeters

Analog ammeters work by measuring the torque exerted by magnetic forces on a current-carrying wire. We discuss more about their operation in Chapter 29 on magnetic

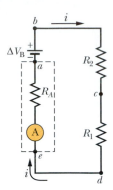

FIGURE 27-4 ■ This depicts how an ammeter can be inserted into a series circuit to measure the current. The third resistor represents the small resistance R_A of the ammeter itself.

fields. However, we continue our discussion of these devices from Chapter 26 and consider some important attributes the ammeter must have.

Recall from Chapter 26 that to measure the current in a wire, you are to break or cut the wire and insert the ammeter in series with an arm of the circuit so the current to be measured passes through the meter. (In Fig. 27-4, ammeter A is set up to measure current i).

When measuring the current in a circuit (or anything else for that matter) it is imperative that the measurement tool does not significantly change the quantity you are trying to measure. Hence, it is essential that the resistance R_A of the ammeter be very small compared to other resistances in the circuit. Otherwise, the presence of the meter will significantly change the current flow in the circuit, and measured current will be an inaccurate representation of the true current.

READING EXERCISE 27-4: In Fig. 27-3a, if $R_1 > R_2 > R_3$, rank the three resistances according to (a) the current through them and (b) the potential difference across them, greatest first. ■

READING EXERCISE 27-5: Consider an ammeter inserted into the circuit shown in Fig. 27-4. Compare the amount of current flowing through R_1 under the following three conditions: (a) without the ammeter inserted, (b) when the ammeter has a resistance much less than the equivalent resistance of $R_1 + R_2$, and (c) when the ammeter has a resistance equal to the equivalent resistance of $R_1 + R_2$. Explain your reasoning. Discuss the implications of your result on designing an ammeter. ■

27-4 Multiloop Circuits

Figure 27-5 shows a circuit containing more than one loop. There are two points (b and d) at which the current branches split off or come together. We call such branching points **junctions.** For the circuit shown in Fig. 27-5, we would say there are two junctions, at b and d, and there are three *branches* connecting these junctions. The branches are the left branch (bad), the right branch (bcd), and the central branch (bd).

What are the currents in the three branches? We arbitrarily label the currents, using a different subscript for each branch. Because current is not used up and there are no additional branching points, current i_1 has the same value everywhere in branch bad, i_2 has the same value everywhere in branch bcd, and i_3 is the current through branch bd. The directions of the currents are assigned arbitrarily.

Consider junction d for a moment: charge comes into that junction via incoming currents i_1 and i_3, and it leaves via outgoing current i_2. Because charged particles neither accumulate nor disperse at the junction, the total incoming charge must be equal to the total outgoing charge. Hence, through conservation of charge arguments, we conclude that the total current coming into junction d must equal the total current leaving junction d,

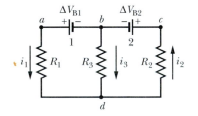

FIGURE 27-5 ■ A multiloop circuit consisting of three branches: left-hand branch bad, right-hand branch bcd, and central branch bd. The circuit has three loops we could choose to follow: left-hand loop $badb$, right-hand loop $bcdb$, and big loop $badcb$.

$$i_{in} = i_{out},$$

or
$$i_1 + i_3 = i_2. \qquad (27\text{-}5)$$

You can easily check that application of this condition to junction b leads to exactly the same equation. This expression for the current in branch 2 thus suggests a general principle:

> **JUNCTION RULE:** The sum of the currents entering any junction must be equal to the sum of the currents leaving that junction.

This rule is often called *Kirchhoff's junction rule* (or *Kirchhoff's current law*). It is simply a statement of the conservation of charge for a steady flow of charge—there is neither a buildup nor a depletion of charge at a junction. Thus, our basic tools for solving complex circuits are the *loop rule* (based on the conservation of energy) and the *junction rule* (based on the conservation of charge).

The relationship between i_1, i_2, and i_3 above is a single equation involving three unknowns. To solve the circuit completely (that is, to find all three currents), we need two more equations involving those same unknowns. We obtain them by applying the loop rule twice. In the circuit of Fig. 27-5, we have three loops from which to choose: the left-hand loop (*badb*), the right-hand loop (*bcdb*), and the big loop (*badcb*). Which two loops we choose turns out not to matter so long as we manage to pass through all the circuit elements at least once. For now, let's choose the left-hand loop and the right-hand loop.

If we traverse the left-hand loop in a counterclockwise direction from point *b*, the loop rule gives us

$$\Delta V_{B1} - i_1 R_1 + i_3 R_3 = 0 \text{ V}, \tag{27-6}$$

where ΔV_{B1} is the difference in potential between the terminals of battery 1. If we traverse the right-hand loop in a counterclockwise direction from point *b*, the loop rule gives us an equation involving battery 2,

$$-i_3 R_3 - i_2 R_2 - \Delta V_{B2} = 0 \text{ V}. \tag{27-7}$$

We now have three equations (Eqs. 27-5, 27-6, and 27-7) containing the three unknown currents, and they can be solved by a variety of mathematical techniques.

If we had applied the loop rule to the big loop, we would have obtained (moving counterclockwise from *b*) the equation

$$\Delta V_{B1} - i_1 R_1 - i_2 R_2 - \Delta V_{B2} = 0 \text{ V}.$$

This equation may look like fresh information, but in fact it is only the sum of Eqs. 27-6 and 27-7. (It would, however, yield the proper results when used with Eq. 27-5 and either 27-6 or 27-7.)

It is important to note that the assumed direction of the currents in a branch of the circuit do not have to be correct to get a correct solution. We must only keep track of the assumptions we have made. If in solving the resulting algebraic expressions we find that one of our currents turns out to have a negative value, then (because of the negative value) we know we made a wrong assumption about the direction of the current in that branch of the circuit.

In general, the total number of equations needed will be equal to the total number of independent loops in the circuit. The number of independent loops is simply the minimum number of loops needed to cover every branch in the circuit. Although some branches could be covered twice, every circuit element would be "covered" at least once. For example, we need at least two equations to cover all the loops in the circuit in Fig. 27-5 and at least three equations to cover all the loops in the more complex circuit in Fig. 27-6.

27-5 Parallel Resistance

Figure 27-6*a* shows three resistances connected by branching junctions. Resistances that are parts of separate loops like those in Fig. 27-6*a* are said to be connected *in parallel* to the battery. Resistors connected "in parallel" are directly wired together on one side and directly wired together on the other side, and a potential difference ΔV

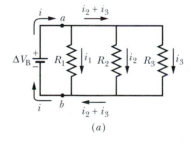

(*a*)

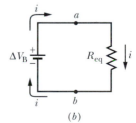

(*b*)

FIGURE 27-6 ■ (*a*) Three resistors connected in parallel across points *a* and *b*. (*b*) An equivalent circuit, with the three resistors replaced with their equivalent resistance R_{eq}.

is applied across the pair of connected sides. Thus, the resistances have the same potential difference ΔV across them, producing a current through each. Because we are assuming ideal wires, there is no potential difference across the wires. Therefore, the potential across the top branch of the circuit is constant everywhere equal to the potential at the positive pole of the battery, and the potential across the bottom branch of the circuit is constant everywhere equal to the potential at the negative pole of the battery. In general,

> When a potential difference ΔV is applied across resistances connected in parallel, each resistor has the same potential difference ΔV across it.

Notice that we have again labeled the currents in each of the branches i_1, i_2, and i_3. We have discussed the way in which the current into a junction is equal to the current out of the junction. We have not yet discussed in what proportions currents divide when there is a branch (a choice of path) in a circuit. Are all three currents i_1, i_2, and i_3 equal? If not, which of these currents is largest? The answer to this question becomes clear when we write out the expressions for current through each of the resistors in Fig. 27-6 using the potential rule for loops. For the case pictured here, we have

$$i_1 = \frac{\Delta V}{R_1}, \quad i_2 = \frac{\Delta V}{R_2}, \quad \text{and} \quad i_3 = \frac{\Delta V}{R_3}. \tag{27-8}$$

Since each resistor is connected so it has the same potential difference across it, it is straightforward to see how the sizes of the currents compare to each other. If the resistances are all equal, the current through each is the same. However, if the three resistances are not equal, more current flows through the smaller resistances. This outcome is consistent with what we might predict based solely on an understanding that a resistor is just a device that resists the flow of current.

If we want to simplify how we think about a circuit that has resistors wired in parallel (like that shown in Fig. 27-6a), we can treat the three resistors in parallel as if they have been replaced by a single equivalent resistor R_{eq}. Figure 27-6b shows the three parallel resistances replaced with an equivalent resistance R_{eq}. The applied potential difference ΔV_B is maintained by a battery. We can see from this figure that the potential difference across the equivalent resistance would have to be the same as the potential difference applied across each of the original resistors. Furthermore, the equivalent resistor would have to have the same total current ($i_1 + i_2 + i_3$) through it as the original three resistors.

> Resistances connected in parallel can be replaced with an equivalent resistance R_{eq}. If the equivalent resistance has the same potential difference applied across it, then the current through it will equal the sum of currents flowing through the original resistors.

To derive an expression for R_{eq} in Fig. 27-6b, we first write the current in each of the resistors in Fig. 27-6a as

$$i_1 = \frac{\Delta V}{R_1}, \quad i_2 = \frac{\Delta V}{R_2}, \quad \text{and} \quad i_3 = \frac{\Delta V}{R_3}, \tag{Eq. 27-8}$$

where ΔV is the potential difference between a and b. If we apply the junction rule at point a in Fig. 27-6a and then substitute these values, we find

$$i = i_1 + i_2 + i_3 = \Delta V \left(\frac{1}{R_1} + \frac{1}{R_2} + \frac{1}{R_3} \right). \tag{27-9}$$

If we instead consider the parallel combination with the equivalent resistance R_{eq} (Fig. 27-6b), we have

$$i = \frac{\Delta V}{R_{eq}} = \Delta V \left(\frac{1}{R_{eq}} \right). \tag{27-10}$$

Comparing the two equations above leads to

$$\frac{1}{R_{eq}} = \frac{1}{R_1} + \frac{1}{R_2} + \frac{1}{R_3}. \tag{27-11}$$

The result $1/R_{eq} = 1/R_1 + 1/R_2 + 1/R_3$ is not surprising because it is compatible with the experimental findings we presented in Section 26-6: the resistance of a length of wire is inversely proportional to its cross-sectional area (Eq. 26-8). To see this connection, imagine three different carbon resistors like those depicted in the last chapter (Fig. 26-21) connected in parallel. Then giving them different values of resistance would involve having the centers of the resistors have three different cross-sectional areas. Because the resistors are connected in parallel, we would then expect the total cross-sectional area to be the sum of the three cross-sectional areas of the resistors' graphite centers so $A_{eq} = A_1 + A_2 + A_3$. Since the cross-sectional area and resistance are inversely proportional, we get $1/R_{eq} = 1/R_1 + 1/R_2 + 1/R_3$.

Extending Eq. 27-11 to the case of n resistors, we have

$$\frac{1}{R_{eq}} = \sum_{j=1}^{n} \frac{1}{R_j} \qquad (n \text{ resistors in parallel}). \tag{27-12}$$

Since we often deal with the case of two resistors in parallel, it is worth it for us to consider this case a bit more. For the case of two resistors, the equivalent resistance is

$$\frac{1}{R_{eq}} = \frac{1}{R_1} + \frac{1}{R_2}.$$

With a bit of algebra, this becomes

$$R_{eq} = \frac{R_1 R_2}{R_1 + R_2} \qquad (2 \text{ resistors in parallel}). \tag{27-13}$$

If you accidentally took the equivalent resistance to be the sum divided by the product, you would notice at once that this result would be dimensionally incorrect.

Note that when two or more resistors are connected in parallel, the equivalent resistance is smaller than any of the combining resistances.

More on the Voltmeter

Recall from our discussion in Chapter 26 that a meter used to measure potential differences is called a *voltmeter*. To measure the potential difference between any two points in the circuit, the voltmeter terminals are connected across those points, without breaking or cutting the wire. In Fig. 27-7, voltmeter V is set up to measure the potential difference across a resistor R_1. The voltmeter is inserted in parallel to R_1 by connecting its terminals to points d and e in the circuit.

To prevent the voltmeter from affecting a measurement, it is essential that the resistance R_V of a voltmeter be *very large* compared to the resistance of the circuit element across which the voltmeter is connected. Otherwise, the meter becomes an important circuit element by drawing a significant current through itself. This change

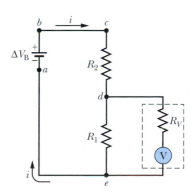

FIGURE 27-7 ■ A single-loop circuit, showing how to connect a voltmeter (V). The third resistor R_V represents the resistance of the voltmeter itself. We assume that R_V is very large compared to R_1 and R_2.

in current flow can alter the potential difference to be measured. On the other hand, even if the potential difference across the voltmeter is large, if a very small current flows through the voltmeter, the flow of current through R_1 will not change very much.

READING EXERCISE 27-6: A battery, with potential ΔV_B across it, is connected to a combination of two identical resistors and a current i flows through the battery. What is the potential difference across and the current through either resistor if the resistors are (a) in series, and (b) in parallel? ■

READING EXERCISE 27-7: Consider the voltmeter inserted into the circuit shown in Fig. 27-7. Describe what would happen if the voltmeter has a resistance $R_V \ll R_1$. How would this affect the potential difference measured across the resistor R_1? Describe what would happen if the voltmeter has a resistance $R_V \gg R_1$. How would this affect the potential difference measured across the resistor R_1? Which case would give the most "accurate" measure of the potential difference across the resistor when the voltmeter is not a part of the circuit? ■

READING EXERCISE 27-8: Suppose the resistors in Fig. 27-6a are all identical light-bulbs. Rank the brightness of the three bulbs. Compare the brightness of each of the bulbs to the brightness of one of the bulbs alone connected to the same battery. ■

TOUCHSTONE EXAMPLE 27-1: One Battery and Four Resistances

Figure 27-8a shows a multiloop circuit containing one ideal battery and four resistances with the following values:

$$R_1 = 20 \ \Omega, \quad R_2 = 20 \ \Omega, \quad \Delta V_B = 12 \ \text{V},$$

$$R_3 = 30 \ \Omega, \quad \text{and} \quad R_4 = 8.0 \ \Omega.$$

(a) What is the current through the battery?

SOLUTION ■ First note that the current through the battery must also be the current through R_1. Thus, one **Key Idea** here is that we might find that current by applying the loop rule to a loop that includes R_1 because the current would be included in the potential difference across R_1. Either the left-hand loop or the big loop will do. Noting that the potential difference arrow of the battery points upward so the current the battery supplies is clockwise, we might apply the loop rule to the left-hand loop, clockwise from point a. With i being the current through the battery, we would get

$$+\Delta V_B - iR_1 - iR_2 - iR_4 = 0 \ \text{V}. \quad \text{(incorrect)}$$

However, this equation is incorrect because it assumes that R_1, R_2, and R_4 all have the same current i. Resistances R_1 and R_4 do have the same current, because the current passing through R_4 must pass through the battery and then through R_1 with no change in value. However, that current splits at junction point b—only part passes through R_2, and the rest through R_3.

To distinguish the several currents in the circuit, we must label them individually as in Fig. 27-8b. Then, circling clockwise from a, we can write the loop rule for the left-hand loop as

$$+\Delta V_B - i_1R_1 - i_2R_2 - i_1R_4 = 0 \ \text{V}.$$

Unfortunately, this equation contains two unknowns, i_1 and i_2; we need at least one more equation to find them.

A second **Key Idea** is that an easier option is to simplify the circuit of Fig. 27-8b by finding equivalent resistances. Note carefully that R_1 and R_2 are *not* in series and thus cannot be replaced with an equivalent resistance. However, R_2 and R_3 are in parallel, so we can use either Eq. 27-12 or Eq. 27-13 to find their equivalent resistance R_{23}. From the latter,

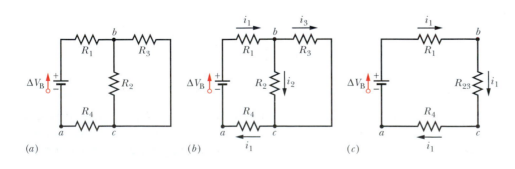

(a) (b) (c)

FIGURE 27-8 ■ (a) A multiloop circuit with an ideal battery of potential difference ΔV_B and four resistances. (b) Assumed currents through the resistances. (c) A simplification of the circuit, with resistances R_2 and R_3 replaced with their equivalent resistance R_{23}. The current through R_{23} is equal to that through R_1 and R_4.

$$R_{23} = \frac{R_2 R_3}{R_2 + R_3} = \frac{(20\ \Omega)(30\ \Omega)}{50\ \Omega} = 12\ \Omega.$$

We can now redraw the circuit as in Fig. 27-8c; note that the current through R_{23} must be i_1 because charge that moves through R_1 and R_4 must also move through R_{23}. For this simple one-loop circuit, the loop rule (applied clockwise from point a) yields

$$+\Delta V_B - i_1 R_1 - i_1 R_{23} - i_1 R_4 = 0\ \text{V}.$$

Substituting the given data, we find

$$12\ \text{V} - i_1(20\ \Omega) - i_1(12\ \Omega) - i_1(8.0\ \Omega) = 0\ \text{V},$$

which gives us

$$i_1 = \frac{12\ \text{V}}{40\ \Omega} = 0.30\ \text{A}. \qquad \text{(Answer)}$$

(b) What is the current i_2 through R_2?

SOLUTION ■ One **Key Idea** here is that we must work backward from the equivalent circuit of Fig. 27-8c, where R_{23} has replaced the parallel resistances R_2 and R_3. A second **Key Idea** is

that, because R_2 and R_3 are in parallel, they both have the same potential difference across them as their equivalent R_{23}. We know the current through R_{23} is $i_1 = 0.30$ A. Thus, we can use Eq. 26-5 $R = \Delta V/i$ to find the potential difference ΔV_{23} across R_{23}:

$$\Delta V_{23} = i_1 R_{23} = (0.30\ \text{A})(12\ \Omega) = 3.6\ \text{V}.$$

The potential difference across R_2 is thus 3.6 V, so the current i_2 in R_2 must be, by Eq. 26-5,

$$i_2 = \frac{\Delta V_2}{R_2} = \frac{3.6\ \text{V}}{20\ \Omega} = 0.18\ \text{A}. \qquad \text{(Answer)}$$

(c) What is the current i_3 through R_3?

SOLUTION ■ We can answer by using the same technique as in (b), or we can use this **Key Idea**: The junction rule tells us that at point b in Fig. 27-8b, the incoming current i_1 and the outgoing currents i_2 and i_3 are related by

$$i_1 = i_2 + i_3.$$

This gives us

$$i_3 = i_1 - i_2 = 0.30\ \text{A} - 0.18\ \text{A} = 0.12\ \text{A}. \qquad \text{(Answer)}$$

TOUCHSTONE EXAMPLE 27-2: Three Batteries and Five Resistances

Figure 27-9 shows a circuit with three ideal batteries in it. Two of these batteries labeled ΔV_{B2} are identical. The circuit elements have the following values:

$$\Delta V_{B1} = 3.0\ \text{V}, \quad \Delta V_{B2} = 6.0\ \text{V}, \quad R_1 = 2.0\ \Omega, \quad R_2 = 4.0\ \Omega.$$

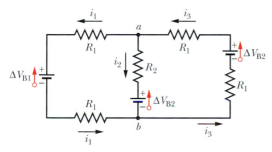

FIGURE 27-9 ■ A multiloop circuit with three ideal batteries and five resistances.

Find the amount and direction of the current in each of the three branches.

SOLUTION ■ It is not worthwhile to try to simplify this circuit, because no two resistors are in parallel, and the resistors that are in series (those in the right branch or those in the left branch) present no problem. So our **Key Idea** is to apply the junction and loop rules to this circuit.

Using arbitrarily chosen directions for the currents as shown in Fig. 27-9, we apply the junction rule at point a by writing

$$i_3 = i_1 + i_2. \qquad (27\text{-}14)$$

An application of the junction rule at junction b gives only the same equation, so we next apply the loop rule to any two of the three loops of the circuit. We first arbitrarily choose the left-hand loop, arbitrarily start at point a, and arbitrarily traverse the loop in the counterclockwise direction, obtaining

$$-i_1 R_1 - \Delta V_{B1} - i_1 R_1 + \Delta V_{B2} + i_2 R_2 = 0\ \text{V}.$$

Substituting the given data and simplifying yield

$$i_1(4.0\ \Omega) - i_2(4.0\ \Omega) = 3.0\ \text{V}. \qquad (27\text{-}15)$$

For our second application of the loop rule, we arbitrarily choose to traverse the right-hand loop clockwise from point a, finding

$$+i_3 R_1 - \Delta V_{B2} + i_3 R_1 + \Delta V_{B2} + i_2 R_2 = 0\ \text{V}.$$

Substituting the given data and simplifying yield

$$i_2(4.0\ \Omega) + i_3(4.0\ \Omega) = 0\ \text{V}. \qquad (27\text{-}16)$$

Using Eq. 27-14 to eliminate i_3 from Eq. 27-16 and simplifying give us

$$i_1(4.0\ \Omega) + i_2(8.0\ \Omega) = 0\ \text{V}. \qquad (27\text{-}17)$$

We now have a system of two equations (Eqs. 27-15 and 27-17) in two unknowns (i_1 and i_2) to solve either by hand (which is easy enough here) or with a math computer software package. (One solution technique is Cramer's rule, given in Appendix E.) We find

$$i_2 = -0.25 \text{ A}.$$

(The minus sign signals that our arbitrary choice of direction for i_2 in Fig. 27-9 is wrong; i_2 should point up through ΔV_{B2} and R_2.) Substituting $i_2 = -0.25$ A into Eq. 27-17 and solving for i_1 then give us

$$i_1 = 0.50 \text{ A}. \qquad \text{(Answer)}$$

With Eq. 27-14 we then find that

$$i_3 = i_1 + i_2 = 0.25 \text{ A}. \qquad \text{(Answer)}$$

The positive answers we obtained for i_1 and i_3 signal that our choices of directions for these currents are correct. We can now correct the direction for i_2 and write its amount as

$$i_2 = 0.25 \text{ A}. \qquad \text{(Answer)}$$

27-6 Batteries and Energy

So far we have discussed ideal batteries that can be characterized as maintaining a constant potential difference between their terminals no matter what current is flowing through them. Also, we have concentrated on analyzing what happens in the part of the circuit that lies outside the battery. In this section we consider more about what goes on inside batteries and how real, not so ideal, batteries behave.

The amazing thing about a battery is that positive charge carriers enter with a low potential energy and other carriers emerge from the battery at a higher potential. Energy transformations inside a battery enable charges to overcome the forces exerted on them by the electric field inside the battery. Positive carriers seem to move opposite to the battery's electric field, whereas negative charge carriers move with it. There must be some other force present inside an energy-providing device enabling charges to swim upstream against electrical forces. The outdated term given to this "force" is electromotive force. Its abbreviation, which we still use today, is emf. How is this "force" defined? Where does it come from in a typical battery?

We define the emf, $\mathscr{E}$, of a battery in terms of the work done per unit charge on charges flowing into it:

$$\mathscr{E} \equiv \frac{dW}{dq} \qquad \text{(definition of electromotive force).} \qquad (27\text{-}18)$$

In words, the battery emf is the work per unit charge it does to move charge from one terminal to the other. The SI unit for emf is the joule/coulomb. In Chapter 25 we defined one joule/coulomb as the *volt*. There must be some source of energy within a battery, enabling it to do work on the charges. The energy source may be chemical, as in a battery (or a fuel cell). Temperature differences may supply the energy, as in a thermopile; or the Sun may supply it, as in a solar cell. As you can see, the term electromotive force is very misleading since it is not a force at all, but has the same units as electrostatic potential (energy per unit charge). Furthermore emf is a scalar quantity and is not a vector quantity like a force is.

When a battery is connected to a circuit, it transfers energy to the charge carriers passing through it. Let's look at one example of how chemical action can do this. For this purpose we will consider the chemical reactions that take place inside one cell of a lead acid battery used in most automobiles. A lead acid battery consists of several cells wired together in series. Each cell has two metal plates surrounded by a liquid bath of chemicals. In a lead-acid cell, the negative plate is made of pure lead, and the positive plate is made of lead-oxide. These plates are immersed in sulfuric acid mixed with water. The acid dissociates in the water into hydronium ions (H_3O^+) and bisulfate ions (HSO_4^-). This is shown in Fig. 27-10. Both the lead and lead oxide can react

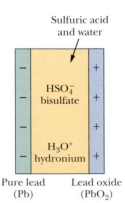

Sulfuric acid and water

HSO₄⁻ bisulfate

H₃O⁺ hydronium

Pure lead (Pb)   Lead oxide (PbO₂)

FIGURE 27-10 ■ The chemical constituents of the lead acid battery.

with the bisulfate ions as follows:

$$Pb + HSO_4^- + H_2O \rightarrow PbSO_4 + H_3O^+ + 2e^-$$

$$PbO_2 + HSO_4^- + 3H_3O^+ + 2e^- \rightarrow PbSO_4 + 4H_2O$$

The two electrons produced on the pure lead plate pile up on it. The second reaction removes the two electrons it needs from the lead oxide plate. Thus, each time the pair of reactions occur, electrons are added to the negative plate and removed from the positive plate. If the cell were not connected to a circuit, the reactions would stop when the charge difference gets so large that the energy needed to put more charges on the plates is greater than the energy released by the reactions. If the battery is connected to an external circuit, then as the charges flow through the circuit, they are removed from one plate and put back on the other; the process can keep going until all the sulfuric acid (HSO_4) is consumed.

Note that when we talk about a battery as a charge pump, this is somewhat misleading because the electrons removed by the chemical reaction at one battery terminal (plate) are not the same electrons released at the other terminal.

There are hundreds of different types of chemical batteries. The lead-acid battery action described here simply serves as an example of how chemical reactions can cause charge separation in a battery.

27-7 Internal Resistance and Power

In our evaluation of circuits up to this point, we have assumed the current passes through the battery (or other emf source) without encountering any resistance within it. We call such a battery or other emf device "ideal."

An **ideal emf device** is one that lacks any resistance to the movement of charge through it. The potential difference between the terminals of an *ideal* emf device is equal to the emf of the device. For example, an ideal battery with an emf of 12.0 V has a potential difference of 12.0 V between its terminals. Very fresh alkaline batteries are nearly ideal.

A **real emf device** has internal resistance to the movement of charge through it. For a real emf device (for example, a real battery), the only situation for which the potential difference between its terminals is equal to its emf is when the device is not connected to a circuit, and thus does not have current through it. However, when the device has current through it, the potential difference between its terminals differs from its emf.

Figure 27-11*a* shows circuit elements that describe the behavior of a real battery, with internal resistance r, wired to an external resistor of resistance R. The internal

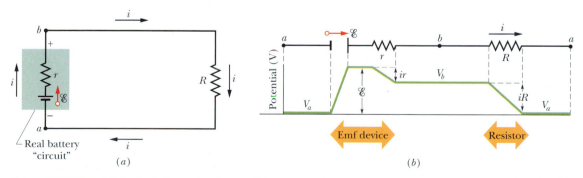

FIGURE 27-11 ■ (*a*) A single-loop circuit containing a real battery having internal resistance r and emf $\mathscr{E}$. (*b*) The same circuit, now spread out in a line. The potentials encountered in traversing the circuit clockwise from *a* are also shown. The potential V_a is arbitrarily assigned a value of zero, and other potentials in the circuit are graphed relative to V_a.

resistance of the battery is the electrical resistance of the conducting materials of the battery and thus is an unavoidable feature of any real battery. However, as an illustration, a real battery is depicted in Fig. 27-11b as if it could be separated into an ideal battery with potential difference $\mathscr{E}$ between its terminals and a resistor of resistance r. The order in which the symbols for these separated parts are drawn does not matter.

If we apply the potential (loop) rule, proceeding clockwise and beginning at point a, the *changes* in potential give us

$$\Delta V_{a \to b} + \Delta V_R = 0 \text{ V},$$

or
$$\mathscr{E} + \Delta V_{\text{internal resistance}} + \Delta V_R = 0 \text{ V}. \qquad (27\text{-}19)$$

It is customary to keep track of potential differences as if the charge carriers are positive. Thus, we go through both resistances in the direction of the *conventional* current (defined in the previous chapter as the direction of flow we would find if the charge carriers were positive instead of negative):

$$\mathscr{E} - ir - iR = 0 \text{ V}. \qquad (27\text{-}20)$$

Solving for the current, we find

$$i = \frac{\mathscr{E}}{R + r}. \qquad (27\text{-}21)$$

Note that this equation reduces to Eq. 27-1 if the battery is ideal so that $r = 0 \ \Omega$.

Figure 27-11b shows graphically the changes in electric potential around the circuit. (To better link Fig. 27-11b with the *closed circuit* in Fig. 27-11a, imagine curling the graph into a cylinder with point a at the left overlapping point a at the right.) Note how traversing the circuit is like walking up and down a (potential) mountain and returning to your starting point—you also return to the starting elevation.

In this book, if a battery is not described as real or if no internal resistance is indicated, you can assume for simplicity that it is ideal.

Implications of Internal Resistance in Real EMF Devices

To understand the implications of internal resistance in emf devices for real circuits, let's try to make our understanding a bit more quantitative. To start with, let's see how $\Delta V_B = \Delta V_{a \to b} = V_b - V_a$, the potential difference across the battery terminals in Fig. 27-11, is affected by the existence of an internal resistance in the battery. To calculate $V_b - V_a$, we start at point a and follow the shorter path around to b, which takes us clockwise through the battery. We then have

$$V_a + \mathscr{E} - ir = V_b,$$

or
$$V_b - V_a = \Delta V_B = \mathscr{E} - ir, \qquad (27\text{-}22)$$

where r is the internal resistance of the battery and $\mathscr{E}$ is the emf of the battery. This expression tells us the potential difference of the battery is equal to the emf minus the drop in potential associated with internal resistance.

Furthermore, if we refer back to Eq. 27-21,

$$i = \frac{\mathscr{E}}{R + r},$$

and substitute this expression for current (in the circuit shown in Fig. 27-11) into our expression for the potential difference across the battery terminals, we get

$$\Delta V_B = \mathscr{E} - \left(\frac{\mathscr{E}r}{R + r}\right).$$

With some algebra, we get the following generally applicable expression:

$$\Delta V_B = \mathscr{E}\frac{R}{R + r}. \tag{27-23}$$

For example, suppose that in Fig. 27-11, $\mathscr{E} = 12$ V, $R = 10\ \Omega$, and $r = 2.0\ \Omega$. Then the equation above tells us the potential across the battery's terminals is

$$\Delta V_B = (12\ \text{V})\frac{10\ \Omega}{10\ \Omega + 2.0\ \Omega} = 10\ \text{V}.$$

In "pumping" charge through itself, the battery (via electrochemical reactions) does work per unit charge of $\mathscr{E} = 12$ J/C, or 12 V. However, because of the internal resistance of the battery, it produces a potential difference of only 10 J/C, or 10 V, across its terminals.

If the internal resistance becomes large compared to the overall resistance in the circuit, the available potential difference of the battery, electrical generator, or other emf device will drop significantly. This drop in available potential difference results in a reduction in the amount of current in the circuit. This is especially important to consider when circuits are designed with a low resistance so they will carry a large current.

For example, consider the circuit shown in Fig. 27-6 (three resistors in parallel with a battery) and let $R = 3\ \Omega$ for each resistor. The equivalent resistance in the circuit is $R_{eq} = 1\ \Omega$. If the potential difference source is taken to be an ideal battery (internal resistance $r = 0$), the current in the circuit is

$$i = \frac{\Delta V_B}{R_{eq}} = \frac{12\ \text{V}}{1\ \Omega} = 12\ \text{A}.$$

The 12 amps are split evenly between each branch (because the resistances are all equal), so each resistor has 4 amps of current flowing through it.

However, if the potential difference source is a real battery with $\mathscr{E} = 12$ V and internal resistance $r = 2.0\ \Omega$, then the available potential difference from the battery is

$$\Delta V_B = (12\ \text{V})\frac{1\Omega}{1\ \Omega + 2.0\ \Omega} = 4\ \text{V}.$$

The total current in the circuit is then

$$i = \frac{\Delta V_B}{R_{eq}} = \frac{4\ \text{V}}{1\ \Omega} = 4\ \text{A}.$$

This current is still split between each of the branches of the circuit, so for the case of the real battery, the current flowing through each resistor is now only 4/3 amp. In comparison to the 4 amps produced by the ideal battery, one can see how the internal resistance of an emf device can play a significant role in the functioning of real circuits.

Power

When a battery or some other type of emf device does work on the charge carriers to establish a current i, it transfers energy from its source of energy (such as the chemical source in a battery) to the charge carriers. Because a real emf device has an internal resistance r, it also transfers energy to internal thermal energy via resistive dissipation, discussed in Chapter 26. Let us relate these transfers.

The net rate P of energy transfer from the emf device to the charge carriers is given by

$$P = i\,\Delta V, \tag{27-24}$$

where ΔV is the potential across the terminals of the emf device. (Note that this is the power associated with the transfer). If we apply this expression to the circuit shown in Fig. 27-11 (from Eq. 27-24 above), we can substitute $\Delta V_B = \mathscr{E} - ir$ into Eq. 27-24 to find

$$P = i(\mathscr{E} - ir) = i\mathscr{E} - i^2 r. \tag{27-25}$$

We see that the term $i^2 r$ in Eq. 27-25 is the rate P_r of energy transfer to thermal energy within the emf device:

$$P_r = i^2 r \quad \text{(internal dissipation rate)}. \tag{27-26}$$

Then the term $i\mathscr{E}$ in Eq. 27-25 must be the rate P_{emf} at which the emf device transfers energy to *both* the charge carriers and to internal thermal energy. Thus,

$$P_{\text{emf}} = i\mathscr{E} \quad \text{(power of emf device)}. \tag{27-27}$$

If a battery is being *recharged*, with a "wrong way" current through it, the energy transfer is then from the charge carriers to the battery—both to the battery's chemical energy and to the energy dissipated in the internal resistance r. The rate of change of the chemical energy is given by Eq. 27-27, the rate of dissipation is given by Eq. 27-26, and the rate at which the carriers supply energy is given by Eq. 27-24.

As is the case for mechanics, the accepted SI unit for electrical power is the watt. One watt is equal to one joule-sec.

TOUCHSTONE EXAMPLE 27-3: Two Real Batteries

Let's consider a circuit with two *nonideal* batteries that have internal resistances. Since the potential differences across the terminals of these batteries are not constant, we characterize each battery in terms of its emf ($\mathscr{E}_1$ or $\mathscr{E}_2$) and internal resistances (r_1 or r_2). The emfs and resistances in the circuit of Fig. 27-12a have the following values:

$$\mathscr{E}_1 = 4.4 \text{ V}, \quad \mathscr{E}_2 = 2.1 \text{ V}, \quad r_1 = 2.3\ \Omega, \quad r_2 = 1.8\ \Omega, \quad R = 5.5\ \Omega.$$

(a) What is the current i in the circuit?

SOLUTION ■ The **Key Idea** here is that we can get an expression involving the current i in this single-loop circuit by applying the loop rule. Although knowing the direction of i is not necessary, we can easily determine it from the emfs of the two batteries. Because $\mathscr{E}_1$ is greater than $\mathscr{E}_2$, battery 1 controls the direction of i, so that direction is clockwise. Let us then apply the loop rule by going counterclockwise—against the current—and starting at point a. We find

$$-\mathscr{E}_1 + ir_1 + iR + ir_2 + \mathscr{E}_2 = 0 \text{ V}.$$

Check that this equation also results if we apply the loop rule clockwise or start at some point other than a. Also, take the time to compare this equation term by term with Fig. 27-12b, which shows the potential changes graphically (with the potential at point a arbitrarily taken to be zero).

Solving the above loop equation for the current i, we obtain

$$i = \frac{\mathscr{E}_1 - \mathscr{E}_2}{R + r_1 + r_2} = \frac{4.4 \text{ V} - 2.1 \text{ V}}{5.5\ \Omega + 2.3\ \Omega + 1.8\ \Omega}$$

$$= 0.2396 \text{ A} \approx 240 \text{ mA}. \tag{Answer}$$

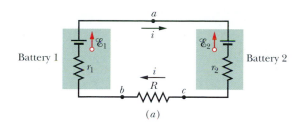

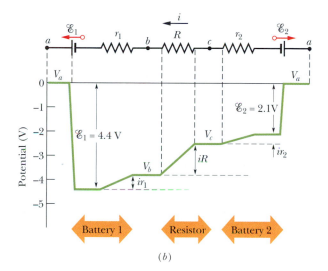

FIGURE 27-12 ■ (a) A single-loop circuit containing two real batteries and a resistor. The batteries oppose each other; that is, they tend to send current in opposite directions through the resistor. (b) A graph of the potentials encountered in traversing this circuit counterclockwise from point a, with the potential at a arbitrarily taken to be zero. (To better link the circuit with the graph, mentally cut the circuit at a and then unfold the left side of the circuit toward the left and the right side of the circuit toward the right.)

(b) What is the potential difference between the terminals of battery 1 in Fig. 27-12a?

SOLUTION ■ The **Key Idea** is to sum the potential differences between points a and b. Let us start at point b (effectively the negative terminal of battery 1) and travel clockwise through battery 1 to point a (effectively the positive terminal), keeping track of potential changes. We find that

$$V_b - ir_1 + \mathscr{E}_1 = V_a,$$

which gives us

$$V_a - V_b = -ir_1 + \mathscr{E}_1$$
$$= -(0.2396 \text{ A})(2.3 \text{ }\Omega) + 4.4 \text{ V}$$
$$= +3.84 \text{ V} \approx 3.8 \text{ V},$$

which is less than the emf of the battery. You can verify this result by starting at point b in Fig. 27-12a and traversing the circuit counterclockwise to point a.

TOUCHSTONE EXAMPLE 27-4: Electric Eel

Electric fish generate current with biological cells called *electroplaques*, which are physiological emf devices. The electroplaques in the South American eel shown in the photograph that opens this chapter are arranged in 140 rows, each row stretching horizontally along the body and each containing 5000 electroplaques. The arrangement is suggested in Fig. 27-13a; each electroplaque has an emf $\mathscr{E}$ of 0.15 V and an internal resistance r of 0.25 Ω. The water surrounding the eel completes a circuit between the two ends of the electroplaque array, one end at the animal's head and the other near its tail.

(a) If the water surrounding the eel has resistance $R_{water} = 800 \text{ }\Omega$, how much current can the eel produce in the water?

SOLUTION ■ The **Key Idea** here is that we can simplify the circuit of Fig. 27-13a by replacing combinations of emfs and internal

resistances with equivalent emfs and resistances. We first consider a single row. The total emf $\mathscr{E}_{row}$ along a row of 5000 electroplaques is the sum of the emfs:

$$\mathscr{E}_{row} = 5000\mathscr{E} = (5000)(0.15 \text{ V}) = 750 \text{ V}.$$

The total resistance R_{row} along a row is the sum of the internal resistances of the 5000 electroplaques:

$$R_{row} = 5000r = (5000)(0.25 \text{ }\Omega) = 1250 \text{ }\Omega.$$

We can now represent each of the 140 identical rows as having a single emf $\mathscr{E}_{row}$ and a single resistance R_{row}, as shown in Fig. 27-13b.

In Fig. 27-13b, the emf between point a and point b on any row is $\mathscr{E}_{row} = 750$ V. Because the rows are identical and because they are all connected together at the left in Fig. 27-13b, all points b in

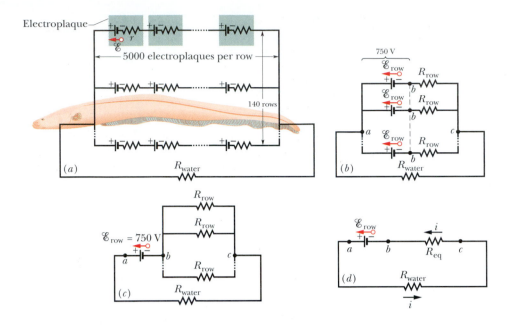

(a)

(b)

(c)

(d)

FIGURE 27-13 ■ (a) A model of the electric circuit of an eel in water. Each electroplaque of the eel has an emf $\mathscr{E}$ and internal resistance r. Along each of 140 rows extending from the head to the tail of the eel, there are 5000 electroplaques. The surrounding water has resistance R_{water}. (b) The emf $\mathscr{E}_{\text{row}}$ and resistance R_{row} of each row. (c) The emf between points a and b is $\mathscr{E}_{\text{row}}$. Between points b and c are 140 parallel resistances R_{row}. (d) The simplified circuit, with R_{eq} replacing the parallel combination.

that figure are at the same electric potential. Thus, we can consider them to be connected so that there is only a single point b. The emf between point a and this single point b is $\mathscr{E}_{\text{row}} = 750$ V, so we can draw the circuit as shown in Fig. 27-13c.

Between points b and c in Fig. 27-13c are 140 resistances $R_{\text{row}} = 1250\ \Omega$, all in parallel. The equivalent resistance R_{eq} of this combination is given by Eq. 27-12 as

$$\frac{1}{R_{\text{eq}}} = \sum_{j=1}^{140} \frac{1}{R_j} = 140\frac{1}{R_{\text{row}}},$$

or
$$R_{\text{eq}} = \frac{R_{\text{row}}}{140} = \frac{1250\ \Omega}{140} = 8.93\ \Omega.$$

Replacing the parallel combination with R_{eq}, we obtain the simplified circuit of Fig. 27-13d. Applying the loop rule to this circuit counterclockwise from point b, we have

$$\mathscr{E}_{\text{row}} - iR_{\text{water}} - iR_{\text{eq}} = 0\ \text{V}.$$

Solving for i and substituting the known data, we find

$$i = \frac{\mathscr{E}_{\text{row}}}{R_{\text{water}} + R_{\text{eq}}} = \frac{750\ \text{V}}{800\ \Omega + 8.93\ \Omega}$$

$$= 0.927\ \text{A} \approx 0.93\ \text{A}. \qquad \text{(Answer)}$$

If the head or tail of the eel is near a fish, much of this current could pass along a narrow path through the fish, stunning or killing it.

(b) How much current i_{row} travels through each row of Fig. 27-13a?

SOLUTION ■ The **Key Idea** here is that since the rows are identical, the current into and out of the eel is evenly divided among them:

$$i_{\text{row}} = \frac{i}{140} = \frac{0.927\ \text{A}}{140} = 6.6 \times 10^{-3}\ \text{A}. \qquad \text{(Answer)}$$

Thus, the current through each row is small, about two orders of magnitude smaller than the current through the water. This tends to spread the current through the eel's body, so that it need not stun or kill itself when it stuns or kills a fish.

Problems

SECS. 27-2 AND 27-3 ■ CURRENT AND POTENTIAL DIFFERENCE IN SINGLE LOOP CIRCUITS, SERIES RESISTANCE

1. Three Resistors In Fig. 27-14, take $R_1 = R_2 = R_3 = 10\ \Omega$. If the potential difference across the ideal battery is $\Delta V_B = 12$ V, find: (a) the equivalent resistance of the circuit and (b) the direction the current flows in the circuit. (c) Which point, A or B, is at higher potential?

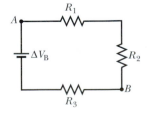

FIGURE 27-14 ■
Problems 1, 3, and 5.

2. Two Ideal Batteries Figure 27-15 shows two ideal batteries with $\Delta V_{B1} = 12$ V and $\Delta V_{B2} = 8$ V. (a) What is the direction of the current in the resistor? (b) Which battery is doing positive work? (c) Which point, A or B, is at the higher potential?

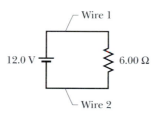

FIGURE 27-15 ■
Problem 2.

3. Total Current In Fig. 27-14, take $R_1 = 10 \ \Omega$, $R_2 = 15 \ \Omega$, and $R_3 = 20 \ \Omega$. If the potential difference across the ideal battery is $\Delta V_B = 15$ V, find: (a) the equivalent resistance of the circuit, (b) the current through each of the resistors, and (c) the total current in the circuit.

4. If Potential at P Is In Fig. 27-16, if the potential at point P is 100 V, what is the potential at point Q?

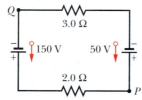

FIGURE 27-16 ■
Problem 4.

5. Voltages In Fig. 27-14, take $R_1 = 12 \ \Omega$, $R_2 = 15 \ \Omega$, and $R_3 = 25 \ \Omega$. If the potential difference across the ideal battery is $\Delta V_B = 15$ V, find the potential differences across each of the resistors.

6. Neglecting Wires Figure 27-17 shows a 6.00 Ω resistor connected to a 12.0 V battery by means of two copper wires. The wires each have length 20.0 cm and radius 1.00 mm. In such circuits we generally neglect the potential differences along wires and the transfer of energy to thermal energy in them. Check the validity of this neglect for the circuit of Fig. 27-17: What are the potential differences across (a) the resistor and (b) each of the two sections of wire? At what rate is energy lost to thermal energy in (c) the resistor and (d) each of the two sections of wire?

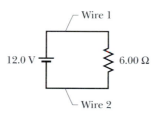

FIGURE 27-17 ■ Problem 6.

7. Single Loop The current in a single-loop circuit with one resistance R is 5.0 A. When an additional resistance of 2.0 Ω is inserted in series with R, the current drops to 4.0 A. What is R?

8. Ohmmeter A simple ohmmeter is made by connecting an ideal 1.50 V flashlight battery in series with a resistance R and an ammeter that reads from 0 to 1.00 mA, as shown in Fig. 27-18. Resistance R is adjusted so that when the clip leads are shorted together, the meter deflects to its full-scale value of 1.00 mA. What external resistance across the leads results in a deflection of (a) 10%, (b) 50%, and (c) 90% of full scale? (d) If the ammeter has a resistance of 20.0 Ω and the internal resistance of the battery is negligible, what is the value of R?

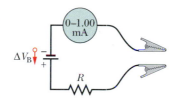

FIGURE 27-18 ■ Problem 8.

SECS. 27-4 AND 27-5 ■ MULTILOOP CIRCUITS AND PARALLEL RESISTANCE

9. Sizes and Directions What are the sizes and directions of the currents through resistors (a) R_2 and (b) R_3 in Fig. 27-19, where

each of the three resistances is 4.0 Ω?

10. Changes The resistances in Figs. 27-20a and b are all 6.0 Ω, and the batteries are ideal 12 V batteries. (a) When switch S in Fig. 27-20a is closed, what is the change in the electric potential difference ΔV_{R_1} across resistor 1, or does ΔV_{R_1} remain the same? (b) When switch S in Fig. 27-20b is closed what is the change in the electric potential difference ΔV_{R_1} across resistor 1, or does ΔV_{R_1} remain the same?

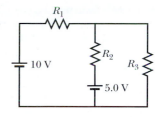

FIGURE 27-19 ■ Problem 9.

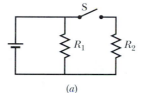

(a)

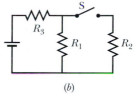

(b)

FIGURE 27-20 ■ Problem 10.

11. Equivalent (a) In Fig. 27-21, what is the equivalent resistance of the network shown? (b) What is the current in each resistor? Put $R_1 = 100 \ \Omega$, $R_2 = R_3 = 50 \ \Omega$, $R_4 = 75 \ \Omega$, and $\Delta V_B = 6.0$ V; assume the battery is ideal.

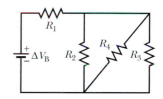

FIGURE 27-21 ■ Problem 11.

12. Plots Plot 1 in Fig. 27-22a gives the electric potential difference ΔV_{R_1} set up across R_1 versus the current i that can appear in resistor 1. Plots 2 and 3 are similar plots for resistors 2 and 3, respectively. Figure 27-22b shows a circuit with those three resistors and a 6.0 V battery. What is the current in resistor 2 in that circuit?

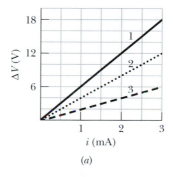

(a)

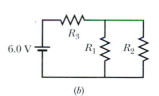

(b)

FIGURE 27-22 ■ Problem 12.

13. Equivalent Resistance Two In Fig. 27-23, $R = 10 \ \Omega$. What is the equivalent resistance between points A and B? (*Hint:* This circuit section might look simpler if you first assume that points A and B are connected to a battery.)

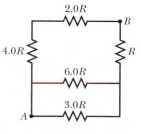

FIGURE 27-23 ■ Problem 13.

14. Three Switches Figure 27-24 shows a circuit containing three switches, labeled S_1, S_2, and S_3. Find the current at a for all possible combinations of switch settings. Put $\Delta V_B = 120$ V, $R_1 = 20.0$ Ω, and $R_2 = 10.0$ Ω. Assume that the battery has no resistance.

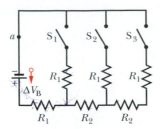

FIGURE 27-24 ▪
Problem 14.

15. Two Lightbulbs Two lightbulbs, one of resistance R_1 and the other of resistance R_2, are connected to a battery (a) in parallel and (b) in series. Which bulb is brighter in each case if $R_1 = R_2$? How is your answer different if $R_1 > R_2$?

16. Calculate Potential In Fig. 27-5, calculate the potential difference between points c and d by as many paths as possible. Assume that $\Delta V_{B1} = 4.0$ V, $\Delta V_{B2} = 1.0$ V, $R_1 = R_2 = 10$ Ω, and $R_3 = 5.0$ Ω.

17. Ammeter (a) In Fig. 27-25, determine what the ammeter will read, assuming $\Delta V_B = 5.0$ V (for the ideal battery), $R_1 = 2.0$ Ω, $R_2 = 4.0$ Ω, and $R_3 = 6.0$ Ω. (b) The ammeter and the source of emf are now physically interchanged. Show that the ammeter reading remains unchanged.

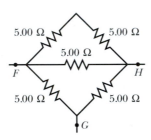

FIGURE 27-25 ▪
Problem 17.

18. Equivalent Resistance In Fig. 27-26, find the equivalent resistance between points (a) F and H and (b) F and G. (*Hint:* for each pair of points, imagine that a battery is connected across the pair).

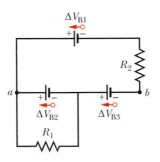

FIGURE 27-26 ▪
Problem 18.

19. Current in Each In Fig. 27-27 find the current in each resistor and the potential difference between points a and b. Put $\Delta V_{B1} = 6.0$ V, $\Delta V_{B2} = 5.0$ V, $\Delta V_{B3} = 4.0$ V, $R_1 = 100$ Ω, and $R_2 = 50$ Ω.

20. Two Resistors By using only two resistors—singly, in series, or in parallel—you are able to obtain resistances of 3.0, 4.0, 12, and 16 Ω. What are the two resistances?

21. Wire of Radius A copper wire of radius $a = 0.250$ mm has an aluminum jacket of outer radius $b = 0.380$ mm. (a) There is a current $i = 2.00$ A in the composite wire. Using Table 26-2, calculate the current in each material. (b) If a potential difference $V = 12.0$ V between the ends maintains the current, what is the length of the composite wire?

22. Between D and E In Fig. 27-28, find the equivalent resistance be-

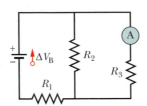

FIGURE 27-27 ▪
Problem 19.

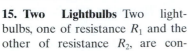

FIGURE 27-28 ▪
Problem 22.

tween points D and E. (*Hint:* Imagine that a battery is connected between points D and E.)

23. Four Resistors Four 18.0 Ω resistors are connected in parallel across a 25.0 V battery. What is the current through the battery?

24. Network Shown (a) In Fig. 27-29, what is the equivalent resistance of the network shown? (b) What is the current in each resistor? Put $R_1 = 100$ Ω, $R_2 = R_3 = 50$ Ω, $R_4 = 75$ Ω, and $\Delta V_B = 6.0$ V; assume the battery is ideal.

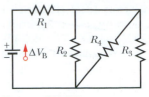

FIGURE 27-29 ▪
Problem 24.

25. Nine Copper Wires Nine copper wires of length l and diameter d are connected in parallel to form a single composite conductor of resistance R. What must be the diameter D of a single copper wire of length l if it is to have the same resistance?

26. Voltmeter A voltmeter (of resistance $R_{\Delta V}$) and an ammeter (of resistance R_A) are connected to measure a resistance R, as in Fig. 27-30a. The resistance is given by $R = \Delta V/i$, where ΔV is the voltmeter reading and i is the current in the resistance R. Some of the

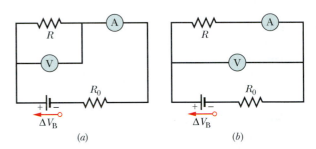

FIGURE 27-30 ▪ Problems 26 to 28.

current i' registered by the ammeter goes through the voltmeter, so that the ratio of the meter readings ($=\Delta V/i'$) gives only an *apparent* resistance reading R'. Show that R and R' are related by

$$\frac{1}{R} = \frac{1}{R'} - \frac{1}{R_{\Delta V}}.$$

Note that as $R_{\Delta V} \to \infty$, $R' \to R$. Ignore R_0 for now.

27. Ammeter and Voltmeter (See Problem 26.) If an ammeter and a voltmeter are used to measure resistance, they may also be connected as in Fig. 27-30b. Again the ratio of the meter readings gives only an apparent resistance R'. Show that now R' is related to R by

$$R = R' - R_A,$$

in which R_A is the ammeter resistance. Note that as $R_A \to 0$ Ω, $R' \to R$. Ignore R_0 for now.

28. What Will the Meters Read (See Problems 26 and 27.) In Fig. 27-30, the ammeter and voltmeter resistances are 3.00 Ω and 3.00 Ω, respectively. Take $\Delta V_B = 12.0$ V for the ideal battery and $R_0 = 100$ Ω. If $R = 85.0$ Ω, (a) what will the meters read for the two different connections (Figs. 27-30a and b)? (b) What apparent resistance R' will be computed in each case?

29. Given a Number You are given a number of 10 Ω resistors, each capable of dissipating only 1.0 W without being destroyed. What is the minimum number of such resistors that you need to combine in series or in parallel to make a 10 Ω resistance that is capable of dissipating at least 5.0 W?

30. Asymptote In Fig. 27-31a, resistor 3 is a variable resistor and the battery is an ideal 12 V battery. Figure 27-31b gives the current i through the battery as a function of R_3. The curve has an asymptote of 2.0 mA as $R_3 \to \infty$. What are (a) resistance R_1 and (b) resistance R_2?

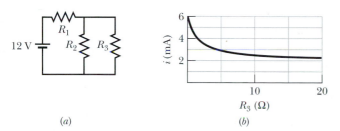

(a)

(b)

FIGURE 27-31 ■ Problem 30.

31. Box Figure 27-32 shows a section of a circuit. The electric potential difference between points A and B that connect the section to the rest of the circuit is $V_A - V_B$ = 78 V, and the cur-

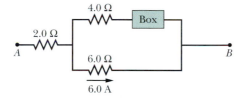

FIGURE 27-32 ■ Problem 31.

rent through the 6.0 Ω resistor is 6.0 A. Is the device represented by "Box" absorbing or providing energy to the circuit and at what rate?

32. Arrangement of N Resistors In Fig. 27-33, a resistor and an arrangement of n resistors in parallel are connected in series with an ideal battery. All the resistors have the same resistance. If one more identical resistor were added in parallel to the n resistors already in parallel, the current through the battery would change by 1.25%. What is the value of n?

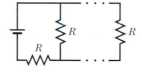

FIGURE 27-33 ■ Problem 32.

33. Rate of Energy Transfer In Fig. 27-34, where each resistance is 4.00 Ω, what are the sizes and directions of currents (a) i_1 and (b) i_2? At what rates is energy being transferred at (c) the 4.00 V battery and (d) the 12.0 V battery, and for each, is the battery supplying or absorbing energy?

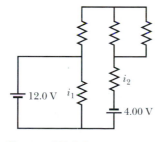

FIGURE 27-34 ■ Problem 33.

34. Both Batteries Are Ideal Both batteries in Fig. 27-35a are ideal. ΔV_{B1} of battery 1 has a fixed value but ΔV_{B2} of battery 2 can be varied between 1.0 V and 10 V. The plots in Fig. 27-35b give the currents through the two batteries as a function of ΔV_{B2}. You must decide which plot corresponds to which battery, but for both plots, a

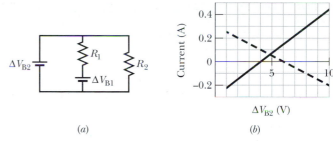

(a)

(b)

FIGURE 27-35 ■ Problem 34.

negative current occurs when the direction of the current through the battery is opposite the direction of that battery's potential difference. What are (a) ΔV_{B1} (b) resistance R_1, and (c) resistance R_2?

35. Work Done by Ideal Battery (a) How much work does an ideal battery with $\Delta V_B = 12.0$ V do on an electron that passes through the battery from the positive to the negative terminal? (b) If 3.4×10^{18} electrons pass through each second, what is the power of the battery?

36. Portion of a Circuit Figure 27-36 shows a portion of a circuit. The rest of the circuit draws current i at the connections A and B, as indicated. Take $\Delta V_{B1} = 10$ V, $\Delta V_{B2} = 15$ V, $R_1 = R_2 = 5.0$ Ω, $R_3 = R_4 = 8.0$ Ω, and $R_5 = 12$ Ω. For each of four values of i—0, 4.0, 8.0, and 12 A—find the current through each ideal battery and state whether the battery is charging or discharging. Also find the potential difference ΔV_{AB} between points A and B.

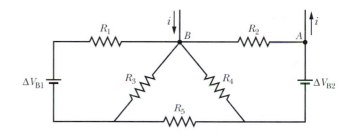

FIGURE 27-36 ■ Problem 36.

37. Adjusted Value In Fig. 27-37, R_s is to be adjusted in value by moving the sliding contact across it until points a and b are brought to the same potential. (One tests for this condition by momentarily connecting a sensitive ammeter between a and b; if these points are at the same potential, the ammeter will not deflect.) Show that when this adjustment is made, the following relation holds:

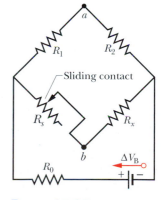

FIGURE 27-37 ■ Problem 37.

$$R_x = R_s \left(\frac{R_2}{R_1} \right).$$

An unknown resistance (R_x) can be measured in terms of a standard (R_s) using this device, which is called a Wheatstone bridge.

38. What Are the Currents In Fig. 27-38, what are currents (a) i_2, (b) i_4, (c) i_1, (d) i_3, and (e) i_5?

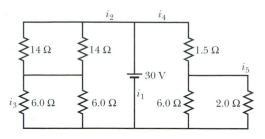

FIGURE 27-38 ■ Problem 38.

39. Sizes and Directions Two What are the sizes and directions of (a) current i_1 and (b) current i_2 in Fig. 27-39, where each resistance is 2.00 Ω? (Can you answer this making only mental calculations?) (c) At what rate is energy being transferred in the 5.00 V battery at the left, and is the energy being supplied or absorbed by the battery?

40. Size and Direction Three (a) What are the size and direction of current i_1 in Fig. 27-40, where each resistance is 2.0 Ω? What are the powers of (b) the 20 V battery, (c) the 10 V battery, and (d) the 5.0 V battery, and for each, is energy being supplied or absorbed?

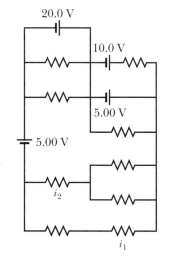

FIGURE 27-39 ■ Problem 39.

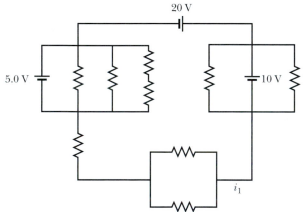

FIGURE 27-40 ■ Problem 40.

41. Size and Direction Four (a) What are the size and direction of current i_1 in Fig. 27-41? (b) How much energy is dissipated by all four resistors in 1.0 min?

SEC. 27-7 ■ INTERNAL RESISTANCE AND POWER

42. Chemical Energy A 5.0 A current is set up in a circuit for 6.0 min by a rechargeable battery with a 6.0 V emf. By how much is the chemical energy of the battery reduced?

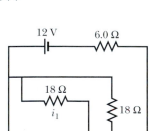

FIGURE 27-41 ■ Problem 41.

43. Flashlight Battery A standard flashlight battery can deliver about 2.0 W·h of energy before it runs down. (a) If a battery costs 80¢, what is the cost of operating a 100 W lamp for 8.0 h using batteries? (b) What is the cost if energy is provided at 12¢ per kilowatt-hour?

44. Power Supplied Power is supplied by a device of emf $\mathscr{E}$ to a transmission line with resistance R. Find the ratio of the power dissipated in the line for $\mathscr{E} = 110\ 000$ V to that dissipated for $\mathscr{E} = 110$ V, assuming the power supplied is the same for the two cases.

45. Car Battery A certain car battery with a 12 V emf has an initial charge of 120 A·h. Assuming that the potential across the terminals stays constant until the battery is completely discharged, for how long can it deliver energy at the rate of 100 W?

46. Energy Transferred A wire of resistance 5.0 Ω is connected to a battery whose emf $\mathscr{E}$ is 2.0 V and whose internal resistance is 1.0 Ω. In 2.0 min, (a) how much energy is transferred from chemical to electrical form? (b) How much energy appears in the wire as thermal energy? (c) Account for the difference between (a) and (b).

47. Assume the Batteries Assume that the batteries in Fig. 27-42 have negligible internal resistance. Find (a) the current in the circuit, (b) the power dissipated in each resistor, and (c) the power of each battery, stating whether energy is supplied by or absorbed by it.

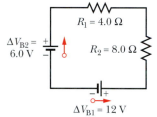

FIGURE 27-42 ■ Problem 47.

48. Both Batteries In Fig. 27-43a, both batteries have emf $\mathscr{E} = 1.20$ V and the external resistance R is a variable resistor. Figure 27-43b gives the electric potentials ΔV_T between the terminals of each battery as functions of R: Curve 1 corresponds to battery 1 and curve 2 corresponds to battery 2. What are the internal resistances of (a) battery 1 and (b) battery 2?

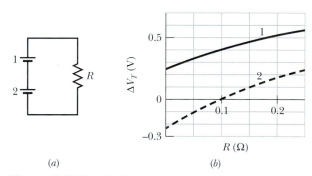

FIGURE 27-43 ■ Problem 48.

49. Find Internal Resistance The following table gives the electric potential difference ΔV_T across the terminals of a battery as a function of current i being drawn from the battery. (a) Write an equation that represents the relationship between the terminal potential difference ΔV_T and the current i. Enter the data into your graphing calculator and perform a linear regression fit of ΔV_T versus i. From the parameters of the fit, find (b) the battery's emf and (c) its internal resistance.

i (A):	50	75	100	125	150	175	200
ΔV_T (V):	10.7	9.0	7.7	6.0	4.8	3.0	1.7

50. Make Plots In Fig. 27-11a, put $\mathscr{E} = 2.0$ V and $r = 100$ Ω. Plot (a) the current and (b) the potential difference across R, as functions of R over the range 0 to 500 Ω. Make both plots on the same graph. (c) Make a third plot by multiplying together, for various values of R, the corresponding values on the two plotted curves. What is the physical significance of this third plot?

51. Energy Converted A car battery with a 12 V emf and an internal resistance of 0.040 Ω is being charged with a current of 50 A. (a) What is the potential difference across its terminals? (b) At what rate is energy being dissipated as thermal energy in the battery? (c) At what rate is electric energy being converted to chemical energy? (d) What are the answers to (a) and (b) when the battery is used to supply 50 A to the starter motor?

52. What Value of R (a) In Fig. 27-44, what value must R have if the current in the circuit is to be 1.0 mA? Take $\mathscr{E}_1 = 2.0$ V, $\mathscr{E}_2 = 3.0$ V, and $r_1 = r_2 = 3.0$ Ω. (b) What is the rate at which thermal energy appears in R?

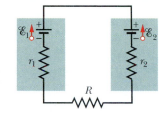

FIGURE 27-44 ■ Problem 52.

53. Circuit Section In Fig. 27-45, circuit section AB absorbs energy at a rate of 50 W when a current $i = 1.0$ A passes through it in the indicated direction. (a) What is the potential difference between A and B? (b) emf device X does not have internal resistance. What is its emf? (c) What is its *polarity* (the orientation of its positive and negative terminals)?

FIGURE 27-45 ■ Problem 53.

54. Lights of an Auto When the lights of an automobile are switched on, an ammeter in series with them reads 10 A and a voltmeter connected across them reads 12 V. See Fig. 27-46. When the electric starting motor is turned on, the ammeter reading drops to 8.0 A and the lights dim somewhat. If the internal resistance of the battery is 0.050 Ω and that of the ammeter is negligible, what are (a) the emf of the battery and (b) the current through the starting motor when the lights are on?

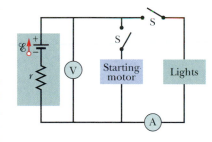

FIGURE 27-46 ■ Problem 54.

55. Same EMF Two batteries having the same emf $\mathscr{E}$ but different internal resistances r_1 and r_2 ($r_1 > r_2$) are connected in series to an external resistance R. (a) Find the value of R that makes the potential difference zero between the terminals of one battery. (b) Which battery is it?

56. Starting Motor The starting motor of an automobile is turning too slowly, and the mechanic has to decide whether to replace the motor, the cable, or the battery. The manufacturer's manual says that the 12 V battery should have no more than 0.020 Ω internal resistance, the motor no more than 0.200 Ω resistance, and the cable no more than 0.040 Ω resistance. The mechanic turns on the motor

and measures 11.4 V across the battery, 3.0 V across the cable, and a current of 50 A. Which part is defective?

57. Maximum Power (a) In Fig. 27-11a, show that the rate at which energy is dissipated in R as thermal energy is a maximum when $R = r$. (b) Show that this maximum power is $P = \mathscr{E}^2/4r$.

58. Solar Cell A solar cell generates a potential difference of 0.10 V when a 500 Ω resistor is connected across it, and a potential difference of 0.15 V when a 1000 Ω resistor is substituted. What are (a) the internal resistance and (b) the emf of the solar cell? (c) The area of the cell is 5.0 cm², and the rate per unit area at which it receives energy from light is 2.0 mW/cm². What is the efficiency of the cell for converting light energy to thermal energy in the 1000 Ω external resistor?

59. Maximum Energy Two batteries of emf $\mathscr{E}$ and internal resistance r are connected in parallel across a resistor R, as in Fig. 27-47a. (a) For what value of R is the rate of electrical energy dissipation by the resistor a maximum? (b) What is the maximum energy dissipation rate?

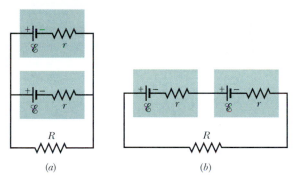

FIGURE 27-47 ■ Problems 59 and 60.

60. Either Parallel or Series You are given two batteries of emf $\mathscr{E}$ and internal resistance r. They may be connected either in parallel (Fig. 27-47a) or in series (Fig. 27-47b) and are to be used to establish a current in a resistor R. (a) Derive expressions for the current in R for both arrangements. Which will yield the larger current (b) when $R > r$ and (c) when $R < r$?

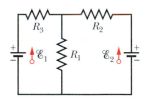

FIGURE 27-48 ■ Problem 61.

61. Batteries Are Ideal In Fig. 27-48, $\mathscr{E}_1 = 3.00$ V, $\mathscr{E}_2 = 1.00$ V, $R_1 = 5.00$ Ω, $R_2 = 2.00$ Ω, $R_3 = 4.00$ Ω, and both batteries are ideal. What is the rate at which energy is dissipated in (a) R_1, (b) R_2, and (c) R_3? What is the power of (d) battery 1 and (e) battery 2?

62. For What Value of R In the circuit of Fig. 27-49, for what value of R will the ideal battery transfer energy to the resistors (a) at a rate of 60.0 W, (b) at the maximum possible rate, and (c) at the minimum possible rate? (d) What are those rates?

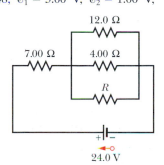

FIGURE 27-49 ■ Problem 62.

63. Calculate Current (a) Calculate the current through each ideal battery in Fig. 27-50. Since the batteries are ideal $\mathscr{E} = \Delta V_B$ in each case. Assume that $R_1 = 1.0$ Ω, $R_2 = 2.0$ Ω, $\mathscr{E}_1 = 2.0$ V and $\mathscr{E}_2 = \mathscr{E}_3 = 4.0$ V. (b) Calculate $V_a - V_b$.

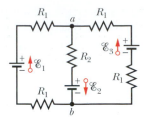

FIGURE 27-50 ■ Problem 63.

64. Constant Value In the circuit of Fig. 27-51, $\mathscr{E}$ has a constant value but R can be varied. Find the value of R that results in the maximum heating in that resistor. The battery is ideal.

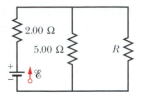

FIGURE 27-51 ■ Problem 64.

Additional Problems

65. True or False For the circuit in Fig. 27-52, indicate whether the statements are true or false. If a statement is false, give a correct statement.

(a) Some of the current is used up when the bulb is lit; the current in wire B is smaller than the current in wire A.

(b) A current probe will have the same readings if connected to read the current in wire A or wire B. The current flows from the battery, through wire A, through the bulb, and then back to the battery through wire B.

(c) The current flows toward the bulb in both wires A and B.

(d) The (positive) current flows from the battery, through wire A, and then back to the battery through wire B.

(e) If wire A is left connected but wire B is disconnected, the bulb will still light.

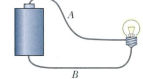

FIGURE 27-52 ■ Problem 65.

66. Use the Model (a) Use our model for electric current to rank the networks shown in Fig. 27-53 in order by resistance. Explain your reasoning. (b) If a battery were connected to each of the circuits, in which case would the current through the battery be the largest? The smallest? Explain your reasoning.

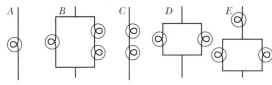

FIGURE 27-53 ■ Problem 66.

67. Examine the Circuits Examine the circuits shown in Fig. 27-54 and indicate whether you think each of the following two statements are true or false. Please explain your reasoning.

(a) Circuits 1 and 2 are different. The brightness of the two bulbs in circuit 1 are the same, but in circuit 2 the bulb closest to the battery in brighter than the bulb that is further away.

(b) Circuit diagrams only show electrical connections, so the drawings in circuits 1 and 2 are electrically equivalent and the brightness of the two bulbs is the same in both circuits 1 and 2.

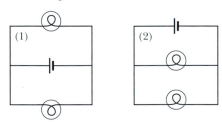

FIGURE 27-54 ■ Problem 67.

68. Which Diagram (a) Identify which of the nice, neat circuit diagrams (A, B, C, or D) in Fig. 27-55c corresponds to the messy circuit drawing in Fig. 27-55a. Explain the reasons for your answer.

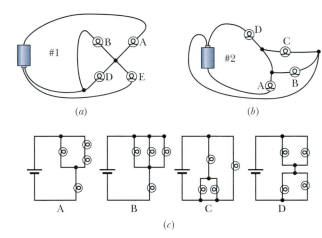

FIGURE 27-55 ■ Problem 68.

(b) Which neat circuit diagram corresponds to the messy circuit drawing in Fig. 27-55b. Explain the reasons for your answer.

69. At Which Point (a) For the circuit in Fig. 27-56, at which point A, B, C, D or E is the voltage the lowest? Explain. (b) At which point is the potential energy of a positive charge the highest? Explain. (c) At which point is the current the largest? Explain.

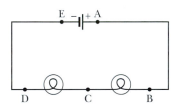

FIGURE 27-56 ■ Problem 69.

70. Bulbs 1 Through 6 (a) For the circuit shown in Fig. 27-57, rank bulbs 1 through 6 in order of descending brightness. Explain the reasoning for your ranking. (b) Now assume that the filament of lightbulb 6 breaks. Again rank the bulbs in order of descending brightness. Explain the reasoning for your ranking.

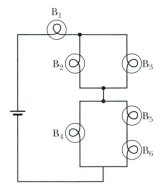

FIGURE 27-57 ■ Problem 70.

71. The Circuit Diagram The circuit diagram in Fig. 27-58 shows two unlabeled resistors attached to identical bulbs. Explain how you would interpret the brightness of bulbs *A* and *B* to decide which resistor is larger.

72. Three Circuits Which of the three circuits shown in Fig. 27-59, if any, are electrically identical? Which are different? Explain your answers.

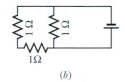

FIGURE 27-58 ■
Problem 71.

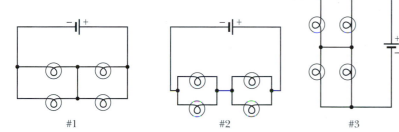

FIGURE 27-59 ■ Problem 72.

73. An Unscrewed Bulb Examine the circuit shown in Fig. 27-60. (a) Rank the bulbs according to brightness and explain your reasoning. (b) How will the brightness of bulbs 1 and 3 change if bulb 4 is unscrewed? Explain. (c) How will the brightness of bulbs 1, 3, 5, and 6 change if a conducting wire is connected between points *A* and *F*? Explain.

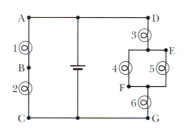

FIGURE 27-60 ■ Problem 73.

74. Examine the Circuit Examine the circuit shown in Fig. 27-61. (a) Assume that the switch is *open*. State which bulbs or combination of bulbs are in series, and in parallel. (b) Assume that the switch is *closed*. State whether the bulbs in the circuit are arranged in series or parallel.

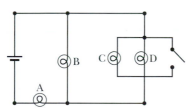

FIGURE 27-61 ■ Problem 74.

75. Examine the Circuit Two Examine the circuit shown in Fig. 27-62. (a) Assume that the switch is *open*. Rank the bulbs according to brightness and explain your reasoning. (b) Assume that the switch is *closed*. Rank the bulbs according to brightness and explain your reasoning.

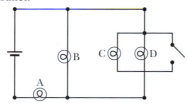

FIGURE 27-62 ■ Problem 75.

76. More Current If the batteries in Fig. 27-63 are identical, which circuit draws more current? Circuit *A*? Circuit *B*? Neither? Show your calculations and reasoning.

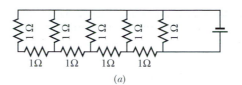

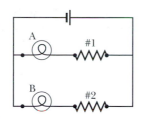

FIGURE 27-63 ■ Problem 76.

77. Which Are Connected In the circuits shown in Fig. 27-64, state which resistors are connected in series with which other resistors, which are connected in parallel with which other resistors, and which are neither in series nor parallel.

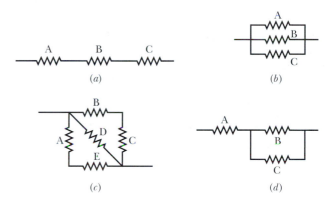

FIGURE 27-64 ■ Problem 77.

78. Lots of Batteries and a Bulb Figure 27-65 shows identical batteries connected in different arrangements to the same lightbulb. Assume the batteries have negligible internal resistances. The positive terminal of each battery is marked with a plus. Rank these arrangements on the basis of bulb brightness from the highest to the lowest. Please explain your reasoning.

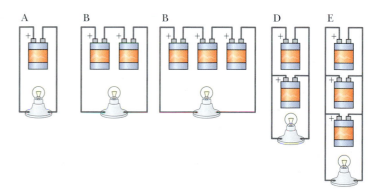

FIGURE 27-65 ■ Problem 78.

79. Constant Current Source We have studied batteries that provide a fixed voltage across their terminals. In that case, we had to examine our circuit and use our physical principles in order to calculate the current through the battery. In neuroscience, it is sometimes useful to use a constant current source (CCS), which instead provides a fixed amount of current through itself. In this case, we have to use our physical principles in order to calculate the voltage drop across the source.

Suppose we have a constant current source (denoted CSS) that always provides a current of $i_c = 10^{-6}$ amps. For the three circuits shown in Fig. 27-66, find the voltage drop across the current source. Each resistor has a resistance $R = 2000 \, \Omega$. (If you prefer, you may leave your answer in terms of the symbols i_c and R.)

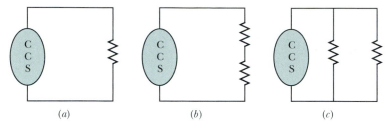

(a)   (b)   (c)

FIGURE 27-66 ▪ Problem 79.

80. Tracking Around a Circuit The circuit shown in Fig. 27-67 contains an ideal battery and three resistors. The battery has an emf of 1.5 V, $R_1 = 2 \, \Omega$, $R_2 = 3 \, \Omega$, and $R_3 = 5 \, \Omega$. Also shown in Fig. 27-67 is a graph tracking some quantity around the circuit. Make three

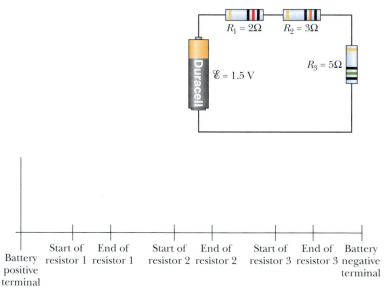

FIGURE 27-67 ▪ Problem 80.

copies of this graph. On the first, plot the voltage a test charge would experience as it moved through the circuit. On the second, plot the electric field a test charge would experience as it moved through the circuit. On the third, plot the current one would measure crossing a plane perpendicular to the wire of the circuit as one goes through the circuit.

81. Modeling a Nerve Membrane (From a homework set in a graduate course in synaptic physiology) As a result of a complex set of biochemical reactions, the cell membrane of a nerve cell pumps ions (Na^+ and K^+) back and forth across itself, thereby maintaining an electrostatic potential difference from the inside to the outside of the membrane. Modifications on the conditions can result in changes in those potentials.

Part of the process can be modeled by treating the membrane as if it were a simple electric circuit consisting of batteries, resistors, and a switch. A simple model of the membrane of a nerve cell is shown in Fig. 27-68. It consists of two batteries (ion pumps) with voltages $\Delta V_1 = 100$ mV and $V_2 = 50$ mV. The resistance to flow across the membrane is represented by two resistors with resistances $R_1 = 10 \, K \, \Omega$ and $R_2 = 90 \, K \, \Omega$. The variability is represented by a switch, S_1.

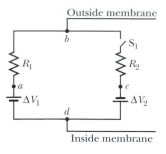

FIGURE 27-68 ▪ Problem 81.

Four points on the circuit are labeled by the letters a–d. The point b represents the outside of the membrane and the point d the inside of the membrane.

(a) What is the voltage difference across the membrane (i.e., between d and b) when the switch is open?
(b) What is the current flowing around the loop when the switch is closed?
(c) What is the voltage drop across the resistor R_1 when the switch is open? Closed?
(d) What is the voltage drop across the resistor R_2 when the switch is open? Closed?
(e) What is the potential difference across the membrane (i.e., between d and b) when the switch is closed?
(f) If the locations of resistances R_1 and R_2 were reversed, would the voltages across the cell membrane be different?

82. Find the Five Currents Consider the circuit in Fig. 27-69. (a) Apply the junction rule to junctions d and a and the loop rule to the three loops to produce five simultaneous, linearly independent equations. (b) Represent the five linear equations by the matrix equation $[A][B] = [C]$, where

$$[B] = \begin{bmatrix} i_1 \\ i_2 \\ i_3 \\ i_4 \\ i_5 \end{bmatrix}.$$

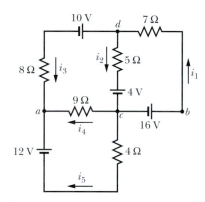

FIGURE 27-69 ▪ Problem 82 and 83.

What are the matrices $[A]$ and $[C]$? (c) Have the calculator perform $[A]^{-1}[C]$ to find the values of i_1, i_2, i_3, i_4, and i_5.

83. Knowing the Currents For the same situation as in Problem 82 and having already solved for the five unknown currents, do the following. (a) Find the electric potential difference across the 9 Ω resistor. (b) Find the rate at which work is being done on the 7 Ω resistor. (c) Find the rate at which the 12 V battery is doing work on the circuit. (d) Find the rate at which the 4 V battery is doing work on the circuit. (e) Of the points in the circuit labeled a and c, which is at the higher electric potential?

28 | Capacitance

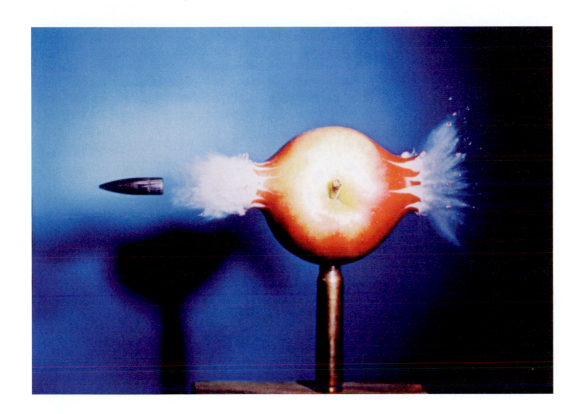

In 1964 Harold "Doc" Edgerton of MIT, who was renowned for his ability to take high-quality stop-action photos, captured this image of a bullet penetrating an apple. This stop-action photo was made by leaving the camera shutter open and tripping a high-speed electronic flash device at just the right time to illuminate the apple and bullet. Since the bullet was moving at 900 m/s, Edgerton used a flash with a duration of only 0.3 μs. This meant that the bullet only moved 0.3 mm during the flash. If we were doing ordinary photography we would probably illuminate the apple with a 100 W lightbulb using an exposure time of 1/20 s to provide about 5 J of energy for illumination. But providing the necessary 5 J of electrical energy for the illumination in a time period of 0.3 μs requires 15 MW of power.

How is it possible to provide the energy needed to stop a bullet's action when it is only illuminated for a tiny fraction of a second?

The answer is in this chapter.

28-1 The Uses of Capacitors

In Chapter 26 we discussed transferring excess charge to a pair of metal plates as shown in Fig. 26-1. The pair of metal plates is an example of the basic component of a **capacitor.** A capacitor can be constructed using any two conductors separated by an insulator. If we connect each conductor making up a capacitor to one of the terminals of a source of potential difference such as a battery, one conductor acquires a net positive charge while the other conductor acquires the same amount of net negative charge. The conductors can be any shape. Figure 28-1 shows some possible capacitor

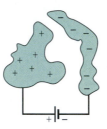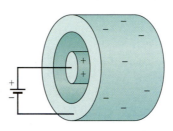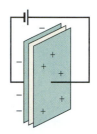

Amorphous capacitor (blobs) with air as an insulator

Cylindrical capacitor with air as an insulator

Parallel plate capacitor with paper and air as an insulator

FIGURE 28-1 ■ Three capacitors of different sizes and shapes have been connected to a battery. They each consist of a pair of conductors separated by an insulator. In each case the battery removes electrons from one of the two conductors, leaving it with excess positive charge and forces the same number of electrons to the opposite conductor.

geometries. No matter what shape or size a capacitor's conductors are, we often casually refer to the conductors as "plates."

There are many reasons for constructing and studying capacitors: they are useful circuit elements and they can store energy.

FIGURE 28-2 ■ An assortment of capacitors commonly found in electrical circuits. The structures of these devices are hidden.

Capacitors in Electrical Circuits

Since a capacitor consists of conductors separated by an insulator, no current can flow *through* it. So at first glance, it doesn't seem to make sense to use a capacitor as a circuit element. Surprisingly, capacitors have very interesting and useful properties in circuits with changing currents through their other components. For example, variable capacitors are vital elements that enable us to tune radio and television receivers. They are found in most household electrical devices. Capacitors are used to control the frequency of the flashing lights used for warning signals at construction sites. The coaxial cables used to carry high-frequency microwave and radio signals are cylindrical capacitors. Microscopic capacitors are used in communications and computers to shape the timing and strength of time-varying signal transmissions. Figure 28-2 shows some of the many sizes and shapes of capacitors commonly found in electric circuits.

FIGURE 28-3 ■ When a battery is connected across the terminals of a capacitor, the capacitor stores electrical energy.

Capacitors as Energy Storage Devices

Just as you can store potential energy by pulling a bowstring, stretching a spring, compressing a gas, or lifting a book, you can also store electrical energy in the electric field found inside a "charged" capacitor as shown in Figs. 28-3 and 28-4. For example, energy storage in microscopic capacitors enables them to function as memory devices in modern digital computers and in the charge-coupled devices (CCDs) used in video cameras. Energy stored in capacitors can also be used to keep computer circuits running smoothly during brief power outages. A much larger capacitor lies at the heart of a battery-powered photoflash unit. This capacitor accumulates electrical energy relatively slowly during the time between flashes, building up an electric field as it does so.

The electric field across the capacitor plates stores energy that can be released rapidly to create an intensive flash of light. (It is important to note that because capacitors are storehouses for electrical energy, some electrical devices can give you a nasty shock if you open them and accidentally touch both terminals of a capacitor—even when the device is turned off.)

28-2 Capacitance

Figure 28-5 shows a capacitor made from a conventional arrangement of a pair of metal plates. A device consisting of two parallel conducting plates of area A separated by a distance d is called a *parallel-plate capacitor*. The circuit symbol we use to represent a capacitor ($-||-$) is based on the structure of a parallel-plate capacitor but is used for capacitors of all shapes. For the purpose of defining capacitance in a simple manner, we will consider an ideal capacitor as two flat parallel conductors (or **plates**) with a perfect insulator between its plates. This perfect insulator allows absolutely no current to pass between them. For simplicity, at first we choose to consider the situation where there is no matter (such as air, glass, or plastic) between the capacitor plates. We just have a vacuum between the plates. We further assume we will charge our capacitor with an ideal battery. Recall that an ideal battery has no internal resistance, so its emf and the potential difference across its terminals are always the same. In Section 28-6 and those following we will relax some of these idealized restrictions.

FIGURE 28-4 ■ When a "charged" capacitor is disconnected from its battery and wired in series with a bulb, the energy stored in it can light the bulb for a short period of time.

Equal and Opposite Excess Charge on Plates

When capacitor plates of any shape are connected to a battery or some other voltage source, electrons flow from the negative terminal of the battery through the connecting wire and onto one plate of the capacitor. Meanwhile, the positive terminal of the battery attracts electrons from the other plate. These electrons are pulled through the wires of the circuit, away from the capacitor plate, and leave behind an excess of positive metal atoms with missing electrons. During this process, we cannot find an electric field outside of the capacitor so the overall capacitor seems to be electrically neutral. Hence, we must conclude that at any given time one plate has net or excess charge of $+q$ while the other has a net charge of $-q$. The chemical reactions taking place in the battery are complex, so the electrons pulled off one plate are not necessarily the same ones being pushed through the wires of the circuit onto the other plate. However, the battery does deposit one electron on the negative plate for every one it pulls off the positive plate. We will call this process *charge separation*. Sometimes the process is called *charging*.

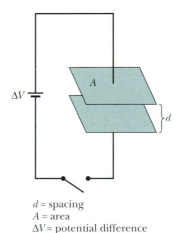

d = spacing
A = area
ΔV = potential difference

> To understand how a capacitor works, it is important to note that charge separation occurs as a result of charge flow in the wires of the circuit. Charges are not transferred from one plate to the other inside an ideal capacitor.

FIGURE 28-5 ■ A parallel-plate capacitor with identical plates of area A and spacing d is connected to a battery with potential difference ΔV. The plates have equal and opposite excess charges of amount $|q|$ on their facing surfaces.

Why Do Capacitor Plates Stop Accumulating Charge?

Observations show that the battery eventually stops pulling electrons off the positively charged plate and depositing electrons on the negatively charged plate. This is because as electrons build up on the negative plate, they oppose the battery's action and start repelling the flow of additional electrons. Similarly, it becomes harder and harder for the battery to pull electrons off the positive plate as the atoms carrying positive net charge pull back on them. When enough charge has accumulated on the

Electric field lines

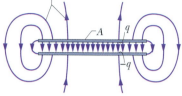

FIGURE 28-6 ■ As the field lines show, the electric field due to the charged plates is uniform in the central region between the plates. The field is not uniform at the edges of the plates, as indicated by the "fringing" of the field lines there.

plates, the force exerted on an electron by the battery and the oppositely directed forces exerted on it by the other charges on a plate cancel each other. No more electrons can flow from one plate to the other. We can use a high-quality voltmeter to measure the potential difference across a capacitor just disconnected from a battery. This measurement shows that *charge separation stops when the potential difference across a capacitor is the same as the potential difference across the battery*.

Factors Affecting Charge Separation Capacity

By convention we refer to the *charge on a capacitor* as $|q|$, the absolute value of the net charge on each plate. Although we refer to a capacitor with charges q and $-q$ on its plates as "charged," a capacitor is electrically neutral so we are actually describing its charge separation created by a voltage source. What factors might affect the capacity for charge separation in a parallel-plate capacitor? We can use our knowledge of electrostatics to explore the effects of several factors. In particular, we will explore how we expect charge to depend on the potential difference across the battery terminals and on geometric factors such as the area of the plates and their spacing (Fig. 28-6):

1. **Potential Difference, ΔV:** For a given capacitor of any shape, we would expect the charge separation to be larger when the potential difference the battery places across the capacitor plates is larger. How much larger? Consider a group of n charges distributed on the plates of a capacitor. Since the plates are conductors, each one is an equipotential surface. According to Eq. 25-25 we can find the electric potential at a given point on a plate relative to infinity. We just need to know the locations of the group of n charges distributed on the capacitor plates. The potential is given by

$$V = k \sum_{i=1}^{n} \frac{q_i}{r_i} \qquad \text{(Eq. 25-25)}$$

where r_i represents the radial distance between the point where the potential is being calculated and the location of the ith charge. By examining this equation we can see that if the potential is to be doubled, there needs to be twice as much charge at each location on the capacitor plates. We expect the amount of the charge separation on a capacitor to be proportional to the potential difference across its plates. We predict

$$|q| \propto |\Delta V|.$$

As you will see in the next subsection, the constant of proportionality between the amount of excess charge on each plate and the potential difference across the plates for a given capacitor is known as its *capacitance*. We will deal more formally with the definition of capacitance and its units in the next section.

2. **Influence of Plate Area, A:** Consider a parallel-plate capacitor. For a given potential difference and plate spacing, d, how do we expect the charge separation capacity to depend on the area of the plates? When the plates have a large area, the electrons the battery is trying to push on the negative plate have more room to spread out. Likewise, the unneutralized atoms left behind when electrons are pulled off the positive plate can be distributed further apart. We expect that as the area of the plates increases, it will be easier to remove or deposit electrons on them.

A simple experiment can be done to show that the charge separation capacity is in fact directly proportional to area. In this experiment, two sheets of aluminum

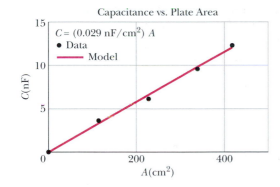

Capacitance vs. Plate Area

$C = (0.029 \text{ nF/cm}^2) \, A$

FIGURE 28-7 ■ Two rectangular pieces of aluminum foil are wedged between the insulating pages of a book. A multimeter is used to measure the capacitance of the system. The result shows capacitance increasing in direct proportion to the area of the conducting aluminum plates.

foil are placed opposite each other and separated by the insulating pages of a book. A multimeter like that described in Section 26-4 can be used to measure capacitance. Measurements are taken for different areas of foil. The results are shown in Fig. 28-7. We derive this relationship theoretically in the next section.

3. **Influence of Plate Spacing, *d*:** Once again we consider a parallel-plate capacitor. For a given potential difference and plate area, *A*, how do we expect the charge separation capacity to depend on the spacing between the plates? When the plates have a small spacing, the excess positive charges on one plate are quite close to the excess negative charges on the other plate. Since opposite charges attract each other, these charges pull on each other across the insulating gap even though they cannot cross the gap. This attraction helps to counterbalance the repulsion between the like charges on each plate. As the spacing between plates becomes smaller, we expect the overall capacity for the charge separation caused by the battery action to become larger.

A simple experiment can be done to show that the charge separation capacity does in fact increase as the spacing between plates decreases. In this experiment, two sheets of aluminum foil are placed opposite each other and separated by the insulating pages of a book. A multimeter is used to measure capacitance as different numbers of pages are inserted between the foil plates. The results are shown in Fig. 28-8. This graph shows that the capacitance of the foil plate system is inversely proportional to the spacing, *d*, between the plates. We derive this relationship theoretically in the next section.

Defining Capacitance

As we just discussed in the last subsection, the amount of the excess charge on each plate of a capacitor, $|q|$, and the size of the potential difference, $|\Delta V|$, across it should be proportional to each other, so

$$|q| = C|\Delta V|. \tag{28-1}$$

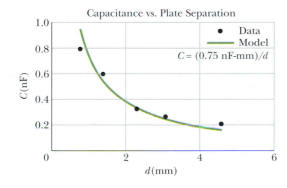

Capacitance vs. Plate Separation

$C = (0.75 \text{ nF-mm})/d$

FIGURE 28-8 ■ Two rectangular pieces of aluminum foil are wedged between the insulating pages of a book. The capacitance of the system is measured as a function of the spacing between the plates. The result shows that capacitance is inversely proportional to the spacing.

The proportionality constant C is defined as the **capacitance.** The capacitance is a measure of how much excess charge must be put on each of the plates to produce a certain potential difference between them: the greater the capacitance, the larger the charge separation created by a given potential difference.

For a parallel-plate capacitor, experimental results have shown us its capacitance depends directly on the plate areas and inversely on the spacing between plates. We will see in Sections 28-6 and 28-7 that capacitance will also depend on the nature of the insulating material inserted between the plates. Capacitors having different shapes will not have the same simple relationships between plate area and spacing. In the next section, we will use the definition of electric potential and Gauss' law to identify the theoretical geometric factors for several different types of capacitors including parallel-plate, cylindrical, and spherical capacitors.

Capacitance Units

The SI unit of capacitance following from this expression is the coulomb per volt. This unit occurs so often that it is given a special name, the *farad* (F):

$$1 \text{ farad} = 1 \text{ F} = 1 \text{ coulomb per volt} = 1 \text{ C/V}. \tag{28-2}$$

As you will see, the farad is a very large unit. Fractions of the farad, such as the microfarad (1 μF $= 10^{-6}$ F) and the picofarad (1 pF $= 10^{-12}$ F), are more convenient units in practice. A summary of units and their common notations is shown in Table 28-1.

TABLE 28-1
Units of Capacitance

microfarad: 10^{-6} F = 1 μF

nanofarad: 10^{-9} F = 1 nF = 1000 $\mu\mu$F

picofarad: 10^{-12} F = 1 pF = 1 $\mu\mu$F

READING EXERCISE 28-1: Does the capacitance C of a capacitor increase, decrease, or remain the same (a) when the excess charge of amount $|q|$ on its plates is doubled and (b) when the potential difference ΔV_c across it is tripled? ∎

28-3 Calculating the Capacitance

Our task here is to calculate the capacitance of a capacitor once we know its geometry. Because we will consider a number of different geometries, it seems wise to develop a general plan to simplify the work. In brief, our plan is as follows:

1. Assume a charge of amount $|q|$ on each of the "plates."

2. Calculate the electric field $\vec{E}$ between the plates in terms of this amount of charge, using Gauss' law.

3. Knowing $\vec{E}$, calculate the potential difference ΔV between the plates from

$$V_2 - V_1 = -\int_1^2 \vec{E} \cdot d\vec{s}. \tag{Eq. 25-16}$$

4. Calculate C from $|q| = C|\Delta V|$ (Eq. 28-1).

Before we start, we can simplify the calculation of both the electric field and the potential difference by making certain assumptions. We discuss each in turn.

Calculating the Electric Field

To relate the electric field $\vec{E}$ between the plates of a capacitor to the amount of excess charge $|q|$ on either plate, we shall use Gauss' law:

$$\varepsilon_0 \oint \vec{E} \cdot d\vec{A} = q^{\text{net}} = q. \tag{28-3}$$

Here q is the net charge enclosed by a Gaussian surface, and $\oint \vec{E} \cdot d\vec{A}$ is the net electric flux through that surface. In all cases we shall consider, the Gaussian surface will be such whenever electric flux passes through it, $\vec{E}$ will have a uniform magnitude $E = |\vec{E}|$, and the vectors $\vec{E}$ and $d\vec{A}$ will be parallel. This equation will then reduce to

$$|q| = \varepsilon_0 EA \qquad \text{(special case of Eq. 28-3)}, \qquad (28\text{-}4)$$

in which A is the area of the part of the Gaussian surface through which flux passes. For convenience, we shall always draw the Gaussian surface in such a way it completely encloses the charge on the positive plate; see Fig. 28-9 for an example.

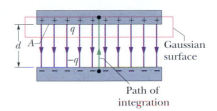

FIGURE 28-9 ■ A charged parallel-plate capacitor. A Gaussian surface encloses the charge on the positive plate. The integration of Eq. 28-6 is taken along a path extending directly from the negative plate to the positive plate.

Calculating the Potential Difference

In the notation of Chapter 25 (Eq. 25-16), the potential difference between the plates of a capacitor is related to the field $\vec{E}$ by

$$V_2 - V_1 = -\int_1^2 \vec{E} \cdot d\vec{s}, \qquad (28\text{-}5)$$

in which the integral is to be evaluated along any path starting on one plate and ending on the other. We shall always choose a path following an electric field line, from the negative plate to the positive plate. For this path, the vectors $\vec{E}$ and $d\vec{s}$ will have opposite directions, so the dot product $\vec{E} \cdot d\vec{s}$ will be equal to $-|\vec{E}||d\vec{s}|$. The right side of this equation will then be positive. Letting ΔV represent the difference, $V_2 - V_1$, we can then recast the relationship as

$$\Delta V = -\int_-^+ |\vec{E}||d\vec{s}| \qquad \text{(special case of Eq. 28-5)}, \qquad (28\text{-}6)$$

in which the " $-$ " and " $+$ " remind us that our path of integration starts on the negative plate and ends on the positive plate.

We are now ready to apply $|q| = \varepsilon_0 EA$ (Eq. 28-4) and $\Delta V = -\int_-^+ |\vec{E}||d\vec{s}|$ (Eq. 28-6) to some particular cases.

A Parallel-Plate Capacitor

We assume, as Fig. 28-9 suggests, that the plates of our parallel-plate capacitor are so large and so close together we can neglect the fringing of the electric field at the edges of the plates, taking $\vec{E}$ to be constant throughout the region between the plates. This configuration was used in old-time radios. As we will see in Chapter 33, the frequency of an oscillating circuit depends on the capacitance. In old radios (those built before the time that tiny transistors became ubiquitous), the dial was connected to a set of nested metal plates. When the dial was turned, some of the plates rotated while others stayed fixed. By turning the dial, the overlap of the plates changed, changing the capacitance and thereby the frequency of the signal selected.

We draw a Gaussian surface enclosing just the excess charge q on the positive plate, as in Fig. 28-9. Recall from above that

$$|q| = \varepsilon_0 EA, \qquad (28\text{-}7)$$

where A is the area of each of the plates.

Equation 28-6 yields

$$\Delta V = \int_-^+ |\vec{E}||d\vec{s}| = |\vec{E}| \int_0^d ds = Ed. \qquad (28\text{-}8)$$

Here, $|\vec{E}| = E$ can be placed outside the integral because it is a constant; the second integral then is simply the plate separation d.

Combining these two expressions with the relation $|q| = C|\Delta V|$ (Eq. 28-1), we find

$$C = \frac{\varepsilon_0 A}{d} \quad \text{(parallel-plate capacitor)}. \tag{28-9}$$

This theoretical relationship matches the results of the experiments we presented in the last section. The capacitance does indeed depend only on geometrical factors—namely, the plate area A and the plate separation d. Note that C increases as we increase the plate area A or decrease the separation d.

As an aside, we point out that this expression suggests one of our reasons for writing the electrostatic constant in Coulomb's law in the form $1/4\pi\varepsilon_0$. If we had not done so, the expression for the capacitance of a parallel-plate capacitor above—which is used more often in engineering practice than Coulomb's law—would have been less simple in form. We note further that it permits us to express the permittivity constant ε_0 in a unit more appropriate for use in problems involving capacitors; namely,

$$\varepsilon_0 = 8.85 \times 10^{-12} \text{ F/m} = 8.85 \text{ pF/m}. \tag{28-10}$$

We have previously expressed this constant as

$$\varepsilon_0 = 8.85 \times 10^{-12} \text{ C}^2/\text{N} \cdot \text{m}^2. \tag{28-11}$$

A Cylindrical Capacitor

Fig. 28-10 shows, in cross section, a cylindrical capacitor of length L formed by two coaxial cylinders of radii a and b. We assume $L \gg b$ so we can neglect the fringing of the electric field occurring at the ends of the cylinders. Each plate contains an amount of excess charge $|q|$. This configuration is important because coaxial cables are used in the communications industry for the long distance transmission of electrical signals (Fig 28-11).

The electric field inside the cylinder is highly symmetrical, so we can use Gauss's law to determine its values. As a Gaussian surface, we choose a cylinder

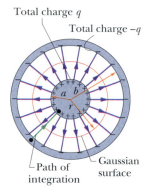

FIGURE 28-10 ■ A cross section of a long cylindrical capacitor, showing a cylindrical Gaussian surface of radius r (that encloses the positive "plate") and the radial path of integration along which Eq. 28-6 is to be applied. If we visualize the central conductor as the cross section of a sphere rather than that of a long cylindrical wire then this figure also illustrates a spherical capacitor.

FIGURE 28-11 ■ Coaxial cables and connectors are used for long-distance transmission of television and radio signals. The cable consists of a central conducting wire surrounded by a layer of insulation and then a cylindrical conductor. All three elements are centered on the same axis. Coaxial cables are good examples of cylindrical capacitors.

of length L and radius r, closed by end caps and placed as is shown in Fig. 28-10. Then

$$|q| = \varepsilon_0 |\vec{E}| A = \varepsilon_0 |\vec{E}| (2\pi r L),$$

in which $2\pi r L$ is the area of the curved part of the Gaussian surface. There is no flux through the end caps. Solving for $|\vec{E}|$ yields

$$|\vec{E}| = \frac{|q|}{2\pi \varepsilon_0 L r}. \qquad (28\text{-}12)$$

Substitution of this result into our general expression for potential difference yields

$$\Delta V = \int_-^+ \vec{E} \cdot d\vec{s} = -\frac{q}{2\pi \varepsilon_0 L} \int_b^a \frac{dr}{r} = \frac{q}{2\pi \varepsilon_0 L} \ln\left(\frac{b}{a}\right), \qquad (28\text{-}13)$$

where here $ds = -dr$ (we integrated radially inward). From the relation $C = |q/\Delta V|$, we then have

$$C = 2\pi \varepsilon_0 \frac{L}{\ln(b/a)} \qquad \text{(cylindrical capacitor).} \qquad (28\text{-}14)$$

We see that the capacitance of a cylindrical capacitor, like that of a parallel-plate capacitor, depends only on geometrical factors, in this case L, b, and a.

A Spherical Capacitor

Fig. 28-10 can also serve as a central cross section of a capacitor consisting of two concentric spherical shells, of radii a and b. As a Gaussian surface we draw a sphere of radius r concentric with the two shells; then

$$|q| = \varepsilon_0 EA = \varepsilon_0 E(4\pi r^2),$$

in which $4\pi r^2$ is the area of the spherical Gaussian surface. We solve this equation for $|\vec{E}|$, obtaining

$$E = |\vec{E}| = k\frac{|q|}{r^2} = \frac{1}{4\pi \varepsilon_0} \frac{|q|}{r^2}, \qquad (28\text{-}15)$$

which we recognize as the expression for the electric field due to a uniform spherical charge distribution from Chapter 24.

If we substitute this expression into Eq. 28-6, we find

$$\Delta V = \int_-^+ \vec{E} \cdot d\vec{s} = -\frac{|q|}{4\pi \varepsilon_0} \int_b^a \frac{dr}{r^2} = \frac{|q|}{4\pi \varepsilon_0}\left(\frac{1}{a} - \frac{1}{b}\right) = \frac{|q|}{4\pi \varepsilon_0} \frac{b-a}{ab}, \qquad (28\text{-}16)$$

where again we have substituted $-dr$ for ds. If we now substitute this into $|q| = C|\Delta V|$ (Eq. 28-1) and solve for C, we find

$$C = 4\pi \varepsilon_0 \frac{ab}{b-a} \qquad \text{(spherical capacitor).} \qquad (28\text{-}17)$$

An Isolated Sphere

We can assign a capacitance to a *single* isolated spherical conductor of radius R by assuming that the "missing plate" is a conducting sphere of infinite radius. After all, the field lines leaving the surface of a positively charged isolated conductor must end somewhere; the walls of the room in which the conductor is housed can serve effectively as our sphere of infinite radius.

To find the capacitance of the isolated conductor, we first rewrite the expression for a spherical capacitor above as

$$C = 4\pi\varepsilon_0 \frac{a}{1 - a/b}.$$

If we then let $b \rightarrow \infty$ and substitute R for a, we find

$$C = 4\pi\varepsilon_0 R \quad \text{(isolated sphere).} \tag{28-18}$$

Note that this formula and the others we have derived for capacitance (Eqs. 28-9, 28-14, and 28-17) involve the constant ε_0 multiplied by a quantity having the dimensions of a length.

READING EXERCISE 28-2: Consider capacitors charged by and then removed from the same battery. Does the charge on the capacitor plates increase, decrease, or remain the same in each of the following situations? (a) The plate separation of a parallel-plate capacitor is increased. (b) The radius of the inner cylinder of a cylindrical capacitor is increased. (c) The radius of the outer spherical shell of a spherical capacitor is increased. ∎

READING EXERCISE 28-3: Consider capacitors charged by identical batteries. If the capacitors stay connected to the batteries, does the amount of excess charge on the capacitor plates increase, decrease, or remain the same in each of the following situations? (a) The plate separation of a parallel-plate capacitor is increased. (b) The radius of the inner cylinder of a cylindrical capacitor is increased. (c) The radius of the outer spherical shell of a spherical capacitor is increased. ∎

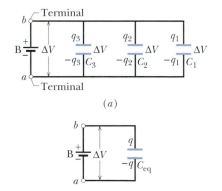

(a)

(b)

FIGURE 28-12 ∎ (a) Three capacitors connected in parallel to battery B. The battery maintains a positive potential difference $\Delta V = V_b - V_a$ across its terminals and thus across each fully charged capacitor. (b) The equivalent capacitor, with capacitance C_{eq}, replaces the parallel combination.

28-4 Capacitors in Parallel and in Series

When there is a combination of capacitors in a circuit, we can often replace that combination with an **equivalent capacitor**—that is, a single capacitor having the same behavior as the actual combination of capacitors. With such a replacement, we can simplify circuits This is similar to the approach we took with resistors in Chapter 27. In addition, circuits often have what is termed *stray capacitance* due to the presence of conductors and insulators in other types of circuit elements. Knowing how the effective capacitance of such elements might combine with each other and other capacitors in the vicinity is vital to the design of high-performance circuits. In this section we discuss the behavior of two basic types of capacitor combinations—parallel and series.

Capacitors in Parallel

Figure 28-12a shows an electric circuit in which three capacitors are connected *in parallel* with battery B. This description has little to do with where the capacitor plates appear in the diagram. Rather, "in parallel" means that one plate of each capacitor is wired directly to one plate of the other capacitors. The opposite plates of the capacitors are also wired to each other. When the parallel combination is connected to a battery, the battery's potential difference ΔV_B is applied across all three capacitors as shown in Fig. 28-12a.

We can anticipate how the parallel combination will behave by considering the special case in which all three capacitors are parallel-plate capacitors with the same spacing. What happens in this case is that the effective area of the plates of the combined network of capacitors is equal to the sum of the three areas. Using Eq. 28-9 we see

$$C_{eq} = \frac{\varepsilon_0 A}{d} = \frac{\varepsilon_0(A_1 + A_2 + A_3)}{d} = \frac{\varepsilon_0 A_1}{d} + \frac{\varepsilon_0 A_2}{d} + \frac{\varepsilon_0 A_3}{d} = C_1 + C_2 + C_3.$$

Even if the three capacitors are of different types with each having a different geometry, we expect the effective area of the combination will be increased. The proof of the pudding is in the experiment. It turns out that a multimeter set to measure capacitance can be used to verify

$$C_{eq} = C_1 + C_2 + C_3,$$

for parallel combinations of three capacitors of all sorts of different types. Since the potential difference across a parallel combination of capacitors connected to a voltage source is the same, we can use the expression $|q| = C|\Delta V|$ (Eq. 28-1) to show that if $C_{eq} = C_1 + C_2 + C_3$, then

$$\frac{|q_{eq}|}{|\Delta V|} = \frac{|q_1|}{|\Delta V|} + \frac{|q_2|}{|\Delta V|} + \frac{|q_3|}{|\Delta V|},$$

so that

$$|q_{eq}| = |q_1| + |q_2| + |q_3|.$$

In general,

> When a potential difference ΔV is applied across several capacitors connected in parallel, that potential difference ΔV is applied across each capacitor. The total amount of the excess charge $|q|$ found on each plate of the equivalent capacitor is equal to the sum of the excess charge amounts found on each of the capacitors.

When we analyze a circuit of capacitors in parallel, we can simplify it with this mental replacement:

> Capacitors connected in parallel can be replaced with an equivalent capacitor that has the same total charge $|q|$ and the same potential difference ΔV as the actual capacitors.

We can easily extend our method for finding the equivalent capacitance for three capacitors to any number of capacitors. For n capacitors wired in parallel,

$$C_{eq} = \sum_{j=1}^{n} C_j \quad (n \text{ capacitors in parallel}). \tag{28-19}$$

To find the equivalent capacitance of a parallel combination, we simply add the individual capacitances.

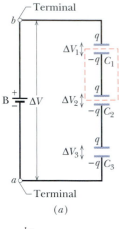

Terminal

Terminal

(a)

(b)

FIGURE 28-13 ■ (a) Three capacitors connected in series to battery B. The battery maintains a positive potential difference ΔV between the top and bottom plates of the series combination. (b) The equivalent capacitor, with capacitance C_{eq}, replaces the series combination.

Capacitors in Series

Figure 28-13a shows three capacitors connected *in series* to battery B. This description has little to do with where the capacitors are located on the drawing. Rather, "in series" means the capacitors are wired serially, one after the other, so a battery can set up a potential difference ΔV across the two ends of the series as shown in Fig. 28-13a.

Let's consider what goes on with the charges on the capacitor plates by following a *chain reaction* of events, in which the charging of each capacitor causes the charging of the next capacitor. We start with capacitor 3 and work upward to capacitor 1. When the battery is first connected to the series of capacitors, it produces a net charge $-q$ on the bottom plate of capacitor 3. That charge then repels negative charge from the top plate of capacitor 3 (leaving it with a net or excess charge $+q$). The repelled negative charge moves to the bottom plate of capacitor 2 (giving it charge $-q$). That excess negative charge on the bottom plate of capacitor 2 then repels negative charge from the top plate of capacitor 2 (leaving it with charge $+q$) to the bottom plate of capacitor 1 (giving it a net charge $-q$). Finally the excess charge on the bottom plate of capacitor 1 helps move negative charge from the top plate of capacitor 1 to the battery, leaving that top plate with net charge $+q$. We see then that the potential differences existing across the capacitors in the series produce identical amounts of excess charge $|q|$ on their plates.

Since the amounts of excess charge on each pair of plates in a series connection are the same, we can use Eq. 28-1, $|q| = C|\Delta V|$, to summarize our reasoning in equation form:

$$|q_1| = |q_2| = |q_3| = |q|,$$

and so

$$|\Delta V_1| = \frac{|q|}{C_1}, \quad |\Delta V_2| = \frac{|q|}{C_2}, \quad \text{and} \quad |\Delta V_3| = \frac{|q|}{C_3}.$$

The total potential difference ΔV due to the battery is the sum of these three potential differences. Thus,

$$|\Delta V| = |\Delta V_1| + |\Delta V_2| + |\Delta V_3|,$$

so that

$$\frac{|q|}{C_{eq}} = \frac{|q|}{C_1} + \frac{|q|}{C_2} + \frac{|q|}{C_3}.$$

The equivalent capacitance is then

$$C_{eq} = \frac{|q|}{|\Delta V|} \quad \text{and also} \quad \frac{1}{C_{eq}} = \frac{1}{C_1} + \frac{1}{C_2} + \frac{1}{C_3}.$$

> When a potential difference of size $|\Delta V|$ is applied across several capacitors connected in series, each of the capacitors has the same amount of excess charge $|q|$ on its plates. The sum of the potential differences across the entire network of capacitors is equal to the size of the applied potential difference $|\Delta V|$.

Here is an important point about capacitors in series: When charge is shifted from one capacitor to another in a series of capacitors, it can move along only one route, such as from capacitor 3 to capacitor 2 in Fig. 28-13a. If there are additional routes, the capacitors are not in series. Hence, when we analyze a circuit of capacitors in series, we can simplify it with this mental replacement:

TABLE 28-2
Series and Parallel Resistors and Capacitors

Resistors		Capacitors	
Series	**Parallel**	**Series**	**Parallel**
$R_{eq} = \sum_{j=1}^{n} R_j$	$\dfrac{1}{R_{eq}} = \sum_{j=1}^{n} \dfrac{1}{R_j}$	$\dfrac{1}{C_{eq}} = \sum_{j=1}^{n} \dfrac{1}{C_j}$	$C_{eq} = \sum_{j=1}^{n} C_j$
Eq. 27-4	Eq. 27-12	Eq. 28-20	Eq. 28-19
1. Same current through all resistors	1. Same potential difference across all resistors	1. Same excess charge on all capacitors	1. Same potential difference across all capacitors
2. Potential differences across each resistor add	2. Currents through each resistor add	2. Potential differences across each capacitor add	2. Excess charges on attached plates add

Capacitors connected in series can be replaced with an equivalent capacitor having the same amount of excess charge $|q|$ on each plate and the same size of potential difference $|\Delta V|$ as the size of the total potential differences across the individual capacitors.

We can easily extend our method of determining the equivalent capacitance of a set of capacitors wired in series from three capacitors to n capacitors by using the expression

$$\frac{1}{C_{eq}} = \sum_{j=1}^{n} \frac{1}{C_j} \qquad (n \text{ capacitors in series}). \qquad (28\text{-}20)$$

Using this expression, you can show that the equivalent of a series of capacitances is always less than the least capacitance in the series. This can also be predicted qualitatively since the effective insulated separation between the top and bottom plate increases since $d = d_1 + d_2 + d_3$. According to Eq. 28-9, capacitance is inversely proportional to plate separation.

Table 28-2 summarizes the equivalence relations for resistors and capacitors in series and in parallel. It also presents the information about potential differences and charges on the combinations we determined by thinking about the physics of how the charges move and distribute themselves in these different geometrical configurations.

READING EXERCISE 28-4: A battery with a potential difference ΔV is used to store an amount of excess charge $|q|$ on each of two identical capacitors and is then disconnected. The two capacitors are then connected to each other. What is the potential difference across each capacitor and the amount of excess charge on each capacitor plate when the capacitors are wired (a) in parallel and (b) in series? ■

TOUCHSTONE EXAMPLE 28-1: Equivalent Capacitance

(a) Find the equivalent capacitance for the combination of capacitances shown in Fig. 28-14a, across which potential difference ΔV is applied. Assume

$$C_1 = 12.0 \ \mu F, \quad C_2 = 5.30 \ \mu F, \quad \text{and} \quad C_3 = 4.50 \ \mu F.$$

SOLUTION ■ The **Key Idea** here is that any capacitors connected in series can be replaced with their equivalent capacitor, and

any capacitors connected in parallel can be replaced with their equivalent capacitor. Therefore, we should first check whether any of the capacitors in Fig. 28-14a are in parallel or series.

Capacitors 1 and 3 are connected one after the other, but are they in series? No. The potential ΔV that is applied to the capacitors forces excess charge on the bottom plate of capacitor 3. That charge causes charge to shift from the top plate of capacitor 3. However, note that the shifting charge can move to the bottom

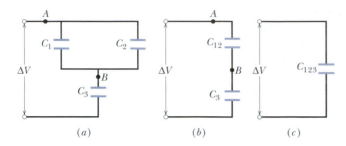

FIGURE 28-14 ■ (*a*) Three capacitors. (*b*) C_1 and C_2, a parallel combination, are replaced by C_{12}. (*c*) C_{12} and C_3, a series combination, are replaced by the equivalent capacitance C_{123}.

plates of both capacitor 1 and capacitor 2. Because there is more than one route for the shifting charge, capacitor 3 is *not* in series with capacitor 1 (or capacitor 2).

Are capacitor 1 and capacitor 2 in parallel? Yes. Their top plates are directly wired together and their bottom plates are directly wired together, and electric potential is applied between the top-plate pair and the bottom-plate pair. Thus, capacitor 1 and capacitor 2 are in parallel, and Eq. 28-19 tells us that their equivalent capacitance C_{12} is

$$C_{12} = C_1 + C_2 = 12.0 \ \mu\text{F} + 5.30 \ \mu\text{F} = 17.3 \ \mu\text{F}.$$

In Fig. 28-14*b*, we have replaced capacitors 1 and 2 with their equivalent capacitor, call it capacitor 12 (say "one two"). (The connections at points A and B are exactly the same in Figs. 28-14*a* and *b*.)

Is capacitor 12 in series with capacitor 3? Again applying the test for series capacitances, we see that the charge that shifts from the top plate of capacitor 3 must entirely go to the bottom plate of capacitor 12. Thus, capacitor 12 and capacitor 3 are in series, and we can replace them with their equivalent C_{123}, as shown in Fig. 28-14*c*.

From Eq. 28-20, we have

$$\frac{1}{C_{123}} = \frac{1}{C_{12}} + \frac{1}{C_3} = \frac{1}{17.3 \ \mu\text{F}} + \frac{1}{4.50 \ \mu\text{F}} = 0.280 \ \mu\text{F}^{-1},$$

from which

$$C_{123} = \frac{1}{0.280 \ \mu\text{F}^{-1}} = 3.57 \ \mu\text{F}. \qquad \text{(Answer)}$$

(b) The potential difference that is applied to the input terminals in Fig. 28-14*a* is $V = 12.5$ V. What is the excess charge on each plate of C_1?

SOLUTION ■ One **Key Idea** here is that, to get the excess charge q_1 on each plate of capacitor 1, we now have to work backward to that capacitor, starting with the equivalent capacitor 123. Since the given potential difference $\Delta V = 12.5$ V is applied across the actual combination of three capacitors in Fig. 28-14*a*, it is also applied across capacitor 123 in Fig. 28-14*c*. Thus, Eq. 28-1 ($|q| = C|\Delta V|$) gives us

$$|q_{123}| = C_{123}|\Delta V| = (3.57 \ \mu\text{F})(12.5 \text{ V}) = 44.6 \ \mu\text{C}.$$

A second **Key Idea** is that the series capacitors 12 and 3 in Fig. 28-1*b* have the same charge as their equivalent capacitor 123. Thus, capacitor 12 has charge $q_{12} = q_{123} = 44.6 \ \mu\text{C}$. From Eq. 28-1, the potential difference across capacitor 12 must be

$$|\Delta V_{12}| = \frac{|q_{12}|}{C_{12}} = \frac{44.6 \ \mu\text{C}}{17.3 \ \mu\text{F}} = 2.58 \text{ V}.$$

A third **Key Idea** is that the parallel capacitors 1 and 2 both have the same potential difference as their equivalent capacitor 12. Thus, capacitor 1 has the potential difference $\Delta V_1 = \Delta V_{12} = 2.58$ V. Thus, from Eq. 28-1, the excess charge on each plate of capacitor 1 must be

$$|q_1| = C_1|\Delta V_1| = (12.0 \ \mu\text{F})(2.58 \text{ V})$$

$$= 31.0 \ \mu\text{C}. \qquad \text{(Answer)}$$

28-5 Energy Stored in an Electric Field

Work must be done by an external agent to charge a capacitor. Starting with an uncharged capacitor, for example, imagine—using "magic tweezers"—that you remove electrons from one plate and deposit them one at a time to the other plate. The electric field building up in the space between the plates has a direction that tends to oppose further separation of charge. As excess charge accumulates on the capacitor plates, you have to do increasingly larger amounts of work to transfer additional electrons. In practice, this work is done not by "magic tweezers" but by a battery, at the expense of its store of chemical energy.

We visualize the work required to charge a capacitor as being stored in the form of **electric potential energy** U in the electric field between the plates. You can recover this energy at will, by discharging the capacitor in a circuit, just as you can recover the potential energy stored in a stretched bow by releasing the bowstring to transfer the energy to the kinetic energy of an arrow. Another example is carrying rocks up a hill against gravity. Energy is stored because of the hill's height and can be recovered by letting the rocks fall down again. In a capacitor, we can recover the stored energy by connecting wires to the ends.

Suppose that at a given instant, a charge $|q'|$ has been moved from one plate of a capacitor, through the wires in the circuit, to the other plate. The amount of the potential difference $|\Delta V'|$ between the plates at that instant will be $|q'|/C$. If an extra increment of charge $|dq'|$ is then removed from one plate and deposited on the other, the amount of the increment of work required will be (from Chapter 25)

$$|dW| = |\Delta V'||dq'| = \frac{|q'|}{C}|dq'|.$$

The work required to bring the total capacitor charge separation up to a final value $|q|$ is

$$W = \int dW = \frac{1}{C}\int_0^q q'|dq'| = \frac{|q|^2}{2C}.$$

This work is stored as potential energy U in the capacitor, and since $q^2 = |q|^2$

$$U = \frac{q^2}{2C} \qquad \text{(potential energy).} \tag{28-21}$$

From $|q| = C|\Delta V|$, we can also write this as

$$U = \tfrac{1}{2}C(\Delta V)^2 = \tfrac{1}{2}q\Delta V \qquad \text{(potential energy).} \tag{28-22}$$

These relations hold no matter what the geometry of the capacitor is.

To gain some physical insight into energy storage, consider two parallel-plate capacitors identical except that capacitor 1 has twice the plate separation of capacitor 2. Then capacitor 1 has twice the volume between its plates and also, from Eq. 28-9, half the capacitance of capacitor 2. Equation 28-4 tells us that if both capacitors have the same amount of charge $|q|$, the electric fields between their plates are identical. Equation 28-21 tells us capacitor 1 has twice the stored potential energy of capacitor 2. Of two otherwise identical capacitors with the same charge and same electric field, the one with twice the volume between its plates has twice the stored potential energy. Arguments like this tend to verify our earlier assumption:

> The potential energy of a charged capacitor may be viewed as being stored in the electric field between its plates.

A High-Speed Electronic Flash Unit

The ability of a capacitor to store potential energy is the basis of *high-speed electronic flash* devices, like those used in stop-action photography. In an electronic flash unit, a battery charges a capacitor relatively slowly to a high potential difference, storing a large amount of energy in the capacitor. The battery maintains only a modest potential difference; an electronic circuit repeatedly uses that potential difference to greatly increase the potential difference of the capacitor. The power, or rate of energy transfer, during this process is also modest.

When a high-speed flash unit fires, the capacitor releases its stored energy by sending a burst of electric current through a Xenon gas discharge tube that gives off a brief flash of white light. As an example, when a 200 μF capacitor in a high-speed flash unit is charged to 300 V, Eq. 28-22 gives the energy stored in the capacitor as

$$U = \tfrac{1}{2}C(\Delta V)^2 = \tfrac{1}{2}(200 \times 10^{-6}\ \text{F})(300\ \text{V})^2 = 9\ \text{J}.$$

As mentioned in the puzzler at the beginning of this chapter, this should be more than enough energy to provide the illumination needed to take a photograph with ordinary film. Suppose the flashtube in the high-speed flash unit Edgerton used to take the photo of the bullet passing through the apple has a very rapid discharge rate. If the Xenon tube takes only one-third of a microsecond to discharge, then the power associated with the discharge is

$$P = \frac{U}{t} = \frac{9 \text{ J}}{0.33 \times 10^{-6} \text{ s}} = 27 \times 10^6 \text{ W} = 27 \text{ MW}.$$

Energy Density

In a parallel-plate capacitor, neglecting fringing, the electric field has the same value at all points between the plates. The **energy density** u—that is, the potential energy per unit volume between the plates—should also be uniform. We can find u by dividing the total potential energy by the volume Ad of the space between the plates. Using Eq. 28-22, we obtain

$$u = \frac{U}{Ad} = \frac{C(\Delta V)^2}{2Ad}.$$

With Eq. 28-9 ($C = \varepsilon_0 A/d$), this result becomes

$$u = \tfrac{1}{2}\varepsilon_0 \left(\frac{\Delta V}{d}\right)^2.$$

However, from Eq. 25-39, $\Delta V/d$ equals the electric field magnitude $|\vec{E}| = E$, so

$$u = \tfrac{1}{2}\varepsilon_0 E^2 \quad \text{(energy density).} \tag{28-23}$$

Although we derived this result for the special case of a parallel-plate capacitor, it holds generally, whatever may be the source of the electric field. If an electric field $\vec{E}$ exists at any point in space, we can think of that point as a site of electric potential energy whose amount per unit volume is given by Eq. 28-23.

TOUCHSTONE EXAMPLE 28-2: Redistributing Charge

(a) Capacitor 1, with $C_1 = 3.55 \ \mu\text{F}$, is charged to a potential difference $\Delta V_0 = 6.30 \text{ V}$, using a 6.30 V battery. The battery is then removed and the capacitor is connected as in Fig. 28-15 to an uncharged capacitor 2, with $C_2 = 8.95 \ \mu\text{F}$. When switch S is closed, charge flows between the capacitors until they have the same potential difference ΔV. Find ΔV.

SOLUTION ■ The situation here differs from Touchstone Example 28-1 because an applied electric potential is *not* maintained

FIGURE 28-15 ■ A potential difference ΔV_0 is applied to capacitor 1 and the charging battery is removed. Switch S is then closed so that the charge on capacitor 1 is shared with capacitor 2.

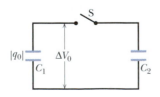

across a combination of capacitors by a battery or some other source. Here, just after switch S is closed, the only applied electric potential is that of capacitor 1 on capacitor 2, and that potential is decreasing. Thus, although the capacitors in Fig. 28-15 are connected end to end, in this situation they are not *in series;* and although they are drawn parallel, in this situation they are not *in parallel.*

To find the final electric potential (when the system comes to equilibrium and charge stops flowing), we use this **Key Idea**: After the switch is closed, the original excess charge $|q_0|$ on each plate of capacitor 1 is redistributed (shared) between capacitor 1 and capacitor 2. When equilibrium is reached, we can relate the original charge $|q_0|$ with the final charges $|q_1|$ and q_2 by writing

$$|q_0| = |q_1| + |q_2|.$$

Applying the relation $|q| = C|\Delta V|$ (Eq. 28-1) to each term of this equation yields

$$C_1|\Delta V_0| = C_1|\Delta V| + C_2|\Delta V|,$$

from which

$$|\Delta V| = |\Delta V_0|\frac{C_1}{C_1 + C_2} = \frac{(6.30 \text{ V})(3.55 \text{ }\mu\text{F})}{3.55 \text{ }\mu\text{F} + 8.95 \text{ }\mu\text{F}}$$

$$= 1.79 \text{ V}. \qquad \text{(Answer)}$$

When the capacitors reach this steady value of electric potential difference, the charge flow stops.

(b) How much energy is stored in the original capacitor when it is first charged up?

SOLUTION ▪ The **Key Idea** here is that the potential energy stored in a capacitor, given by Eq. 28-22, is just

$$U = (\tfrac{1}{2})C(\Delta V)^2$$

$$= (\tfrac{1}{2})(3.55 \text{ }\mu\text{F})(6.30 \text{ V})^2$$

$$= 70.4 \text{ }\mu\text{J}. \qquad \text{(Answer)}$$

(c) How much energy is stored in the two capacitors after they are connected together?

SOLUTION ▪ The **Key Idea** here is that the potential energy stored in *each* capacitor, given by Eq. 28-22, so that

$$U^{\text{total}} = U_1 + U_2 = (\tfrac{1}{2})C_1(\Delta V_1)^2 + (\tfrac{1}{2})C_2(\Delta V_2)^2$$

$$= (\tfrac{1}{2})((3.55 \text{ }\mu\text{F}) + (8.95 \text{ }\mu\text{F}))(1.79 \text{ V})^2$$

$$= 20.0 \text{ }\mu\text{J}. \qquad \text{(Answer)}$$

But how can this be? Before the second capacitor was placed across the first one, there was over 70 μJ of energy stored in the system. What happened to the 50 μJ of energy that seems to have vanished when the second capacitor was charged from the first one? You might argue that the "lost" energy must have been dissipated as heat in the resistance of the wires connecting the two capacitors. But suppose we used superconducting wires with zero resistance? Then where does the missing energy go? The answer, as you will learn in Chapters 33 and 34, is that the charge would oscillate back and forth between the two capacitors until the 50 μJ of "excess" energy was radiated away in the form of electromagnetic waves.

28-6 Capacitor with a Dielectric

If you fill the space between the plates of a capacitor with a *dielectric,* which is usually an insulating material such as mineral oil or plastic, what happens to the capacitance? Michael Faraday—to whom the whole concept of capacitance is largely due and for whom the SI unit of capacitance is named—first looked into this matter in 1837. Using simple equipment much like that shown in Fig. 28-16, he found that the capacitance *increased* by a numerical factor κ, which he called the dielectric constant of the insulating material. Table 28-3 shows some dielectric materials and their dielectric constants. The dielectric constant of a vacuum is unity by definition. Because air is mostly empty space, its measured dielectric constant is only slightly greater than unity.

Another effect of the introduction of a dielectric is to limit the potential difference that can be applied between the plates to a certain value ΔV^{max}, called the *breakdown potential.* If this value is substantially exceeded, the dielectric material will break down and form a conducting path between the plates. That is, when the capacitor is filled with a dielectric, the charge separation you can maintain with a given potential difference increases. Every dielectric material has a characteristic *dielectric strength,* which is the maximum value of the electric field that it can tolerate without breakdown. A few such values are listed in Table 28-3.

As we discussed in connection with Eq. 28-18, the capacitance of any capacitor can be written in the form

$$C = \varepsilon_0 L, \qquad (28\text{-}24)$$

in which L has the dimensions of a length. For example, $L = A/d$ for a parallel-plate capacitor. Faraday's discovery was, with a dielectric *completely* filling the space between the plates, Eq. 28-24 becomes

$$C = \kappa\varepsilon_0 L = \kappa C_{\text{air}}, \qquad (28\text{-}25)$$

where C_{air} is the value of the capacitance with only air between the plates.

FIGURE 28-16 ▪ The simple electrostatic apparatus used by Faraday. An assembled apparatus (second from left) forms a spherical capacitor consisting of a central brass ball and a concentric brass shell. Faraday placed dielectric materials in the space between the ball and the shell.

TABLE 28-3
Some Properties of Dielectrics[a]

Material	Dielectric Constant κ	Dielectric Strength (kV/mm)
Air (1 atm)	1.00054	3
Polystyrene	2.6	24
Paper	3.5	16
Transformer oil	4.5	
Pyrex	4.7	14
Ruby mica	5.4	
Porcelain	6.5	
Tantalum oxide	11.6	
Silicon	12	
Germanium	16	
Ethanol	25	
Water (20°C)[b]	80.4	
Water (25°C)[b]	78.5	
Titania ceramic	130	
Strontium titanate	310	8

For a vacuum, κ = unity.

[a]Measured at room temperature, except for the water.
[b]Note that water is not an insulating material. It is listed because it has dielectric properties.

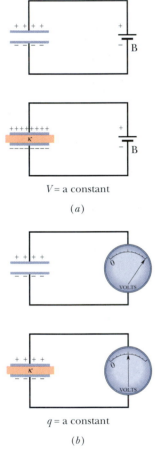

V = a constant

(a)

q = a constant

(b)

FIGURE 28-17 ■ (a) If the potential difference between the plates of a capacitor is maintained, as by battery B, the effect of a dielectric is to increase the excess charge on each plate. (b) If the charge on the capacitor plates is maintained, as in this case, the effect of a dielectric is to reduce the potential difference between the plates. The scale shown is that of a *potentiometer*, a device used to measure potential difference (here, between the plates). A capacitor cannot discharge through a potentiometer.

Figure 28-17 provides some insight into Faraday's experiments. In Fig. 28-17a the battery ensures that the potential difference ΔV between the plates will remain constant. When a dielectric slab is inserted between the plates, the excess amount of charge $|q|$ on the plates increases by a factor of κ, where κ is always greater than 1; the additional charge is delivered to the capacitor plates by the battery. In Fig. 28-17b there is no battery and therefore the amount of excess charge $|q|$ must remain constant when the dielectric slab is inserted; then the potential difference ΔV between the plates decreases by a factor of κ. Both these observations are consistent (through the relation $|q| = C|\Delta V|$) with the increase in capacitance caused by the dielectric.

Comparison of Eqs. 28-24 and 28-25 suggests that the effect of a dielectric can be summed up in more general terms:

> In a region completely filled by a dielectric material of dielectric constant κ, all electrostatic equations containing the permittivity constant ε_0 are to be modified by replacing ε_0 with $\kappa\varepsilon_0$.

A point charge inside a dielectric produces an electric field that, by Coulomb's law, has the magnitude

$$|\vec{E}| = \frac{1}{4\pi\kappa\varepsilon_0}\frac{|q|}{r^2}. \qquad (28-26)$$

Also, the expression for the electric field just outside an isolated conductor immersed in a dielectric (see Eq. 24-20) becomes

$$|\vec{E}| = \frac{|\sigma|}{\kappa\varepsilon_0}. \qquad (28-27)$$

Both these equations show that *for a fixed distribution of charges, the effect of a dielectric is to weaken the magnitude of the electric field that would otherwise be present.* In addition, the amount of energy stored is reduced because work must be done by the field to pull in the dielectric.

TOUCHSTONE EXAMPLE 28-3: A Dielectric's Energetics

A parallel-plate capacitor whose capacitance C is 13.5 pF is charged by a battery to a potential difference $\Delta V = 12.5$ V between its plates. The charging battery is now disconnected and a porcelain slab ($\kappa = 6.50$) is slipped between the plates. What is the potential energy of the device, both before and after the slab is put into place?

SOLUTION ■ The **Key Idea** here is that we can relate the potential energy U of the capacitor to the capacitance C and either the potential ΔV (with Eq. 28-22) or the capacitor charge $|q|$ (with Eq. 28-21):

$$U_1 = \tfrac{1}{2}C\Delta V^2 = \frac{q^2}{2C}.$$

Because we are given the initial potential $\Delta V(=12.5\text{V})$, we use Eq. 28-22 to find the initial stored energy:

$$U_1 = \tfrac{1}{2}CV^2 = \tfrac{1}{2}(13.5 \times 10^{-12}\,\text{F})(12.5\,\text{V})^2$$

$$= 1.055 \times 10^{-9}\,\text{J} = 1055\,\text{pJ} \approx 1100\,\text{pJ}. \quad \text{(Answer)}$$

To find the final potential energy U_2 (after the slab is introduced), we need another **Key Idea**: Because the battery has been disconnected, the amount of excess charge on each capacitor plate cannot change when the dielectric is inserted. However, the potential *does* change. Thus, we must now use Eq. 28-21 (based on q) to write the final potential energy U_2, but now that the slab is within the capacitor, the capacitance is κC. We then have

$$U_2 = \frac{q^2}{2\kappa C} = \frac{U_1}{\kappa} = \frac{1055\,\text{pJ}}{6.50} = 162\,\text{pJ} \approx 160\,\text{pJ}. \quad \text{(Answer)}$$

When the slab is introduced, the potential energy decreases by a factor of κ.

The "missing" energy, in principle, would be apparent to the person who introduced the slab. The capacitor would exert a tiny tug on the slab and would do work on it, in amount

$$W = U_1 - U_2 = (1055 - 162)\,\text{pJ} = 893\,\text{pJ}.$$

If the slab were allowed to slide between the plates with no restraint and if there were no friction, the slab would oscillate back and forth between the plates with a (constant) mechanical energy of 893 pJ, and this system energy would transfer back and forth between kinetic energy of the moving slab and potential energy stored in the electric field.

28-7 Dielectrics: An Atomic View

What happens, in atomic and molecular terms, when we put a dielectric in an electric field? There are two possibilities, depending on the nature of the molecules:

1. *Polar dielectrics.* The molecules of some dielectrics, like water, have permanent electric dipole moments. In such materials (called *polar dielectrics*), the electric dipoles tend to line up with an external electric field as in Fig. 28-18. Because the molecules are continuously jostling each other as a result of their random thermal motion, this alignment is not complete, but it becomes more complete as the magnitude of the applied field is increased (or as the temperature, and thus the jostling, is decreased). The alignment of the electric dipoles produces an electric field directed opposite the applied field and smaller in magnitude.

2. *Nonpolar dielectrics.* Regardless of whether they have permanent electric dipole moments, molecules acquire dipole moments by induction when placed in an external electric field. In Section 25-9 (see Fig. 25-16), we saw that this occurs because the external field tends to "stretch" the molecules, slightly separating the centers of negative and positive charge.

Figure 28-19a shows a nonpolar dielectric slab with no external electric field applied. An electric field $\vec{E}_0$ is present due to the excess charges shown on the capacitor plates in Fig. 28-19a. The result is a slight separation of the centers of the positive and negative charge distributions within the slab, producing positive charge on one face of the slab (due to the positive ends of dipoles there) and negative charge on the opposite

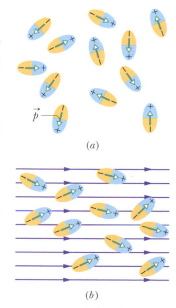

FIGURE 28-18 ■ (a) Molecules with a permanent electric dipole moment, showing their random orientation in the absence of an external electric field. (b) An electric field is applied, producing partial alignment of the dipoles. Thermal agitation prevents complete alignment.

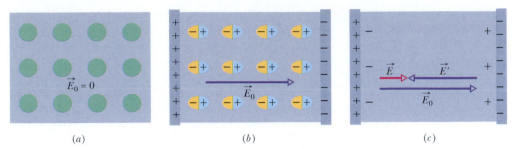

(a)   (b)   (c)

FIGURE 28-19 ▪ (a) A nonpolar dielectric slab. The circles represent the electrically neutral atoms within the slab. (b) An electric field is applied via charged capacitor plates; the field slightly stretches the atoms, separating the centers of positive and negative charge. (c) The separation produces surface charges on the slab faces. These charges set up a field $\vec{E}'$, which opposes the applied field $\vec{E}_0$. The resultant field $\vec{E}$ inside the dielectric (the vector sum of $\vec{E}_0$ and $\vec{E}'$) has the same direction as $\vec{E}_0$ but smaller magnitude.

face (due to the negative ends of dipoles there). The slab as a whole remains electrically neutral and—within the slab—there is no excess charge in any volume element.

Figure 28-19c shows that the induced surface charges on the faces produce an electric field $\vec{E}'$, in the direction opposite the applied electric field $\vec{E}_0$. The resultant field $\vec{E}$ inside the dielectric (the vector sum of fields $\vec{E}_0$ and $\vec{E}'$) has the direction of $\vec{E}_0$ but is smaller in magnitude.

Both the field $\vec{E}'$ produced by the surface charges in Fig. 28-19c and the electric field produced by the permanent electric dipoles in Fig. 28-18 act in the same way—they oppose the applied field $\vec{E}$. (Inside the material, the $\vec{E}$ field fluctuates wildly, depending on whether you are close to one side of a molecule or another. The effects we are looking at are the average effects of the molecules.) Thus, the effect of both polar and nonpolar dielectrics is to weaken any applied field within them, as between the plates of a capacitor. As a result, a given charge separation can be maintained at a lower potential difference, ΔV, with a dielectric than with a vacuum. This means that a capacitor with a dielectric added has a higher capacitance.

We can now see why the dielectric porcelain slab in Touchstone Example 28-3 is pulled into the capacitor: As it enters the space between the plates, the excess surface charge appearing on each slab face has a sign that is opposite to that of the excess charge on the nearby capacitor plate. Thus, slab and plates attract each other.

28-8 Dielectrics and Gauss' Law

In our discussion of Gauss' law in Chapter 24, we assumed that the charges existed in a vacuum. Here we shall see how to modify and generalize that law if dielectric materials, such as those listed in Table 28-3, are present. Figure 28-20 shows a parallel-plate capacitor of plate area A, both with and without a dielectric. We assume the amount of excess charge $|q|$ on the plates is the same in both situations. Note the field between the plates induces charge buildup on the faces of the dielectric by one of the methods discussed in Section 28-7.

For the situation of Fig. 28-20a, without a dielectric, we can find the electric field $\vec{E}_0$ between the plates as we did in Fig. 28-9: We enclose the excess charge q on the top plate with a Gaussian surface and then apply Gauss' law. Letting $E_0 = |\vec{E}_0|$ represent the magnitude of the field, we find

$$\left| \varepsilon_0 \oint \vec{E} \cdot d\vec{A} \right| = \varepsilon_0 E_0 A = |q^{\text{net}}| = |q|, \tag{28-28}$$

or

$$E_0 = \frac{|q|}{\varepsilon_0 A}. \tag{28-29}$$

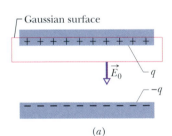

(a)

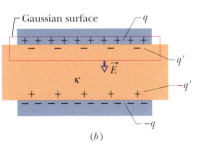

(b)

FIGURE 28-20 ▪ A parallel-plate capacitor (a) without and (b) with a dielectric slab inserted. The excess charge q on the plates is assumed to be the same in both cases.

In Fig. 28-20b, with the dielectric in place, we can find the electric field between the plates (and within the dielectric) by using the same Gaussian surface. However, now the surface encloses two types of charge: it still encloses a net charge q on the top plate but it now also encloses the induced charge q' on the top face of the dielectric. The excess charge on each conducting plate is said to be *free charge* because it can move through the circuit if we change the electric potential of the plate. The induced charge on the surfaces of the dielectric is bound charge. It's stuck to the molecules of an insulator. It can only be displaced from its original position by microscopic amounts and cannot move from the surface.

The amount of net charge enclosed by the Gaussian surface in Fig. 28-20b is $|q + q'|$, so Gauss' law now gives

$$\left| \varepsilon_0 \oint \vec{E} \cdot d\vec{A} \right| = \varepsilon_0 E A = |q + q'|, \tag{28-30}$$

or

$$E = \frac{|q + q'|}{\varepsilon_0 A}. \tag{28-31}$$

Since q' and q have different signs, this means that the effect of the dielectric is to weaken the original field E_0 by a factor of κ, so we may write

$$E = \frac{E_0}{\kappa} = \frac{|q|}{\kappa \varepsilon_0 A}. \tag{28-32}$$

Comparison of Eqs. 28-31 and 28-32 shows

$$|q^{\text{net}}| = |q + q'| = \frac{|q|}{\kappa}. \tag{28-33}$$

Equation 28-33 shows correctly that the amount of induced surface charge is less than that of the excess free charge and is zero if no dielectric is present (then, $\kappa = 1$ in Eq. 28-33).

By substituting for $|q + q'|$ from Eq. 28-33 in Eq. 28-30, we can write Gauss' law in the form

$$\varepsilon_0 \oint \kappa \vec{E} \cdot d\vec{A} = q \qquad \text{(Gauss' law with dielectric),} \tag{28-34}$$

where q is the net free charge on the plate of interest. Here we drop the absolute value sign to account for the fact that the excess charge on a plate of interest, q, can be either positive or negative.

This important equation, although derived for a parallel-plate capacitor, is true generally and is the most general form in which Gauss' law can be written. Note the following:

1. The flux integral now involves $\kappa \vec{E}$, not just $\vec{E}$. (The vector $\varepsilon_0 \kappa \vec{E}$ is sometimes called the electric displacement $\vec{D}$, so Eq. 28-34 can be written in the form $\oint \vec{D} \cdot d\vec{A} = q$).

2. The amount of excess charge $|q|$ enclosed by the Gaussian surface is now taken to be the free charge only. The induced surface charge is deliberately ignored on the right side of Eq. 28-34, having been taken fully into account by introducing the dielectric constant κ on the left side.

3. Equation 28-34 differs from Eq. 24-7, our original statement of Gauss' law, only in that ε_0 in the latter equation has been replaced by $\kappa \varepsilon_0$. We keep κ inside the integral of Eq. 28-34 to allow for cases in which κ is not constant over the entire Gaussian surface.

Gauss's law still holds when charged molecules are present, but it's hard to use, since we don't know where those molecular charges are. We only know their average effect, which is summarized by the measured constant κ. Here, we saw how to create a form of Gauss's law including the effect of the molecules automatically, and this allows us to work only with the charges we control directly—the "free" charges.

TOUCHSTONE EXAMPLE 28-4: Adding a Dielectric

Figure 28-21 shows a parallel-plate capacitor of plate area A and plate separation d. A potential difference ΔV_0 is applied between the plates. The battery is then disconnected, and a dielectric slab of thickness b and dielectric constant κ is placed between the plates as shown. Assume

$$A = 115 \text{ cm}^2, \quad d = 1.24 \text{ cm}, \quad \Delta V_0 = 85.5 \text{ V},$$

$$b = 0.780 \text{ cm}, \quad \text{and} \quad \kappa = 2.61.$$

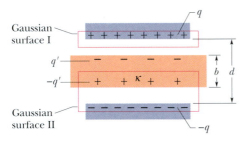

FIGURE 28-21 ■ A parallel-plate capacitor containing a dielectric slab that only partially fills the space between the plates.

(a) What is the capacitance C_0 before the dielectric slab is inserted?

SOLUTION ■ From Eq. 28-9 we have

$$C_0 = \frac{\varepsilon_0 A}{d} = \frac{(8.85 \times 10^{-12} \text{ F/m})(115 \times 10^{-4} \text{ m}^2)}{1.24 \times 10^{-2} \text{ m}}$$

$$= 8.21 \times 10^{-12} \text{ F} = 8.21 \text{ pF}. \quad \text{(Answer)}$$

(b) What is the amount of free excess charge that appears on each plate?

SOLUTION ■ From Eq. 28-1,

$$|q| = C_0 |\Delta V_0| = (8.21 \times 10^{-12} \text{ F})(85.5 \text{ V})$$

$$= 7.02 \times 10^{-10} \text{ C} = 702 \text{ pC}. \quad \text{(Answer)}$$

Because the charging battery was disconnected before the slab was introduced, the free charge remains unchanged as the slab is put into place.

(c) What is the magnitude of the electric field E_0 in the gaps between the plates and the dielectric slab?

SOLUTION ■ A **Key Idea** here is to apply Gauss' law, in the form of Eq. 28-34, to Gaussian surface I in Fig. 28-21—that surface

passes through the gap, and so it encloses *only* the free charge on the upper capacitor plate. Because the area vector $d\vec{A}$ and the field vector $\vec{E}_0$ are both directed downward, the dot product in Eq. 28-34 becomes

$$\vec{E}_0 \cdot d\vec{A} = |\vec{E}_0| dA \cos 0° = E_0 dA.$$

Equation 28-34 then becomes

$$\varepsilon_0 \kappa E_0 \oint dA = q.$$

The integration now simply gives the surface area A of the plate. Thus, we obtain

$$\varepsilon_0 \kappa |\vec{E}_0| \oint dA = q,$$

or

$$E_0 = \frac{q}{\varepsilon_0 \kappa A}.$$

One more **Key Idea** is needed before we evaluate E_0; that is, we must put $\kappa = 1$ here because Gaussian surface I does not pass through the dielectric. Since the charge q on the upper plate is positive, we have

$$E_0 = \frac{q}{\varepsilon_0 \kappa A} = \frac{7.02 \times 10^{-10} \text{ C}}{(8.85 \times 10^{-12} \text{ F/m})(1)(115 \times 10^{-4} \text{ m}^2)}$$

$$= 6900 \text{ V/m} = 6.90 \text{ kV/m}. \quad \text{(Answer)}$$

Note that the value of E_0 does not change when the slab is introduced because the amount of charge enclosed by Gaussian surface I in Fig. 28-21 does not change.

(d) What is the magnitude of the electric field E_1 in the dielectric slab?

SOLUTION ■ The **Key Idea** here is to apply Eq. 28-34 to Gaussian surface II in Fig. 28-21. That surface encloses free charge $-q$ and induced charge $-q'$, but we ignore the latter when we use Eq. 28-34. We find

$$\varepsilon_0 \oint \kappa \vec{E}_1 \cdot d\vec{A} = -\varepsilon_0 \kappa E_1 A = -q. \quad (28\text{-}35)$$

(The first minus sign in this equation comes from the dot product $\vec{E}_1 \cdot d\vec{A}$, because now the field vector $\vec{E}_1$ is directed downward and the area vector $d\vec{A}'$ is directed upward.) Equation 28-35 gives us

$$E_1 = \frac{q}{\varepsilon_0 \kappa A} = \frac{E_0}{\kappa} = \frac{6.90 \text{ kV/m}}{2.61} = 2.64 \text{ kV/m}. \quad \text{(Answer)}$$

(e) What is the potential difference ΔV between the plates after the slab has been introduced?

SOLUTION ▪ The **Key Idea** here is to find ΔV by integrating along a straight-line path extending directly from the bottom plate to the top plate. Within the dielectric, the path length is b and the electric field is E_1. Within the two gaps above and below the dielectric, the total path length is $d - b$ and the electric field is E_0. Equation 28-6 then yields

$$\Delta V = \int_-^+ |\vec{E}||d\vec{s}| = E_0(d - b) + E_1 b$$

$$= (6900 \text{ V/m})(0.0124 \text{ m} - 0.00780 \text{ m})$$

$$+ (2640 \text{ V/m})(0.00780 \text{ m})$$

$$= 52.3 \text{ V.} \qquad \text{(Answer)}$$

This is less than the original potential difference of 85.5 V.

(f) What is the capacitance with the slab in place?

SOLUTION ▪ The **Key Idea** now is that the capacitance C is related to the free charge q and the potential difference ΔV via Eq. 28-1, just as when a dielectric is not in place. Taking q from (b) and ΔV from (e), we have

$$C = \frac{|q|}{|\Delta V|} = \frac{7.02 \times 10^{-10} \text{ C}}{52.3 \text{ V}}$$

$$= 1.34 \times 10^{-11} \text{ F} = 13.4 \text{ pF.} \qquad \text{(Answer)}$$

This is greater than the original capacitance of 8.21 pF.

28-9 *RC* Circuits

In preceding sections we dealt only with circuits in which the currents did not vary with time. Here we begin a discussion of time-varying currents.

Charging a Capacitor

The capacitor of capacitance C in Fig. 28-22 is initially uncharged. To charge it, we close switch S on point a. This completes an *RC series circuit* consisting of the capacitor, an ideal battery of emf ε, and a resistance R. Since an ideal battery has no internal resistance, its emf is the same as the potential difference across the battery, ΔV_B.

From Section 28-2, we already know that as soon as the circuit is complete, charge begins to flow (current exists) between a capacitor plate and a battery terminal on each side of the capacitor. This current increases the amount of excess charge on the plates, q and the size of the potential difference $|\Delta V_C| = |q|/C$ across the capacitor. When that potential difference across the capacitor equals the potential difference across the battery (which here is equal to the emf of the battery, ΔV_B), the current is zero. From Eq. 28-1 ($|q| = C|\Delta V_C|$), the *equilibrium* (final) amount of excess *charge* on each plate of the fully charged capacitor is equal to $C|\Delta V_B|$.

Here we want to examine the charging process. In particular we want to know how the amount of excess charge $|q(t)|$ on each capacitor plate, the potential difference $\Delta V_C(t)$ across the capacitor, and the current $i(t)$ in the circuit vary with time during the charging process. We begin by applying the loop rule to the circuit, traversing it clockwise from the negative terminal of the battery. We find

$$\Delta V_B - iR - \frac{q}{C} = 0, \qquad (28\text{-}36)$$

where q represents the excess charge on the top plate of the capacitor, which is positive in this case.

The last term on the left side represents the potential difference across the capacitor. The term is negative because the capacitor's top plate, which is connected to the battery's positive terminal, is at a higher potential than the lower plate. Thus, there is a drop in potential as we move down through the capacitor.

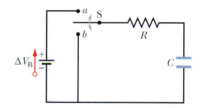

FIGURE 28-22 ▪ When switch S is closed on a, the capacitor is *charged* through the resistor. When the switch is afterward closed on b, the capacitor *discharges* through the resistor.

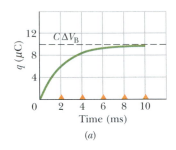

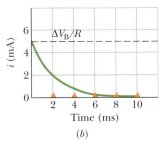

FIGURE 28-23 ■ (*a*) A plot of Eq. 28-39, which shows the buildup of excess charge on the capacitor plates of Fig. 28-22. (*b*) A plot of Eq. 28-40. The charging current in the circuit of Fig. 28-22 declines as the capacitor becomes more fully charged. The curves are plotted for $R = 2000 \ \Omega$, $C = 1 \ \mu$F, and $\Delta V_B = 10$ V. The small triangles represent successive intervals of one time constant τ.

We cannot immediately solve Eq. 28-36 because it contains two variables, i and q. However, those variables are not independent but are related by

$$i = \frac{dq}{dt}. \qquad (28\text{-}37)$$

Substituting this for i and rearranging, we find

$$R\frac{dq}{dt} + \frac{q}{C} = \Delta V_B \qquad \text{(charging equation)}. \qquad (28\text{-}38)$$

This differential equation describes the time variation of the excess positive charge q on the top plate of the capacitor shown in Fig. 28-23. To solve it, we need to find the function $q(t)$ that satisfies this equation and also satisfies the condition the capacitor be initially uncharged: $q = 0$ C at $t = 0$ s.

The solution to Eq. 28-38 is

$$q = C\Delta V_B(1 - e^{-t/RC}) \qquad \text{(charging a capacitor)}. \qquad (28\text{-}39)$$

(Here e is the exponential base, 2.718 . . . , and not the elementary charge.) You can verify by substitution that Eq. 28-39 is indeed a solution to Eq. 28-38. We can see that this expression does indeed satisfy our required initial condition, because at $t = 0$ the term $e^{-t/RC}$ is unity, so the equation gives $q = 0$. Note also that as t goes to ∞ (that is, a long time later), the term $e^{-t/RC}$ goes to zero; so the equation gives the proper value for the full (equilibrium) excess charge on the positive plate of the capacitor — namely, $q = C\Delta V_B$. A plot of $q(t)$ for the charging process is given in Fig. 28-23*a*.

The derivative of $q(t)$ is the positive current $i(t)$ charging the capacitor:

$$i = \frac{dq}{dt} = \left(\frac{\Delta V_B}{R}\right)e^{-t/RC} \qquad \text{(charging a capacitor)}. \qquad (28\text{-}40)$$

A plot of $i(t)$ for the charging process is given in Fig. 28-23*b*. Note that the current has the initial value $\Delta V_B/R$ and it decreases to zero as the capacitor becomes fully charged.

> A capacitor being charged initially acts like ordinary connecting wire relative to the charging current. A long time later, it acts like a broken wire.

By combining $|q| = C|\Delta V_C|$ (Eq. 28-1) and $q = C\Delta V_B(1 - e^{-t/RC})$ (Eq. 28-39), we find the potential difference $\Delta V_C(t)$ across the capacitor during the charging process is

$$|\Delta V_C| = \frac{q}{C} = |\Delta V_B(1 - e^{-t/RC})| \qquad \text{(charging a capacitor)}. \qquad (28\text{-}41)$$

This tells us $\Delta V_C = 0$ at $t = 0$ and $\Delta V_C = \Delta V_B$ when the capacitor is fully charged as the time approaches infinity ($t \rightarrow \infty$).

The Time Constant

The product RC appearing in the equations above has the dimensions of time (both because the argument of an exponential must be dimensionless and because, in fact, $1.0 \ \Omega \times 1.0$ F $= 1.0$ s). RC is called the **capacitive time constant** of the circuit and is represented with the symbol τ.

$$\tau = RC \qquad \text{(time constant)}. \qquad (28\text{-}42)$$

From the expression for the excess charge as a function of time on one plate of a charging capacitor $q = C\Delta V_B(1 - e^{-t/RC})$ (Eq. 28-39), we can now see that at time $t = \tau\,(=RC)$, the excess charge on the top plate of the initially uncharged capacitor of Fig. 28-22 has increased from zero to

$$q = C\Delta V_B(1 - e^{-1}) = 0.63C\Delta V_B. \qquad (28\text{-}43)$$

In words, after the first time constant, τ, the amount of excess charge has increased from zero to 63% of its final value, $C\Delta V_B$. In Fig. 28-22, the small triangles along the time axes mark successive intervals of one time constant during the charging of the capacitor. The charging times for RC circuits are often stated in terms of τ. The greater τ is, the greater is the charging time.

Discharging a Capacitor

Assume that now the capacitor of Fig. 28-22 is fully charged to a potential ΔV_0 equal to the potential difference, ΔV_B, of the battery. At a new time $t = 0$, switch S is thrown from a to b so the capacitor can *discharge* through resistance R. How do the excess charge $q(t)$ on the top plate of the capacitor and the current $i(t)$ through the discharge loop of capacitor and resistance now vary with time?

The differential equation describing $q(t)$ in this case is similar to the one we worked with for the case of charging Eq. 28-38, except now there is no battery in the discharge loop and so $\Delta V_B = 0$. Thus,

$$R\frac{dq}{dt} + \frac{q}{C} = 0 \qquad \text{(discharging equation)}, \qquad (28\text{-}44)$$

where the current term, dq/dt, and the voltage across the capacitor, q/C, can be positive or negative. The solution to this differential equation is

$$q = q_0e^{-t/RC} \qquad \text{(discharging a capacitor)}, \qquad (28\text{-}45)$$

where $|q_0|(=C|\Delta V_0|)$ is the initial amount of excess charge on the capacitor plates. You can verify by substitution that Eq. 28-45 is indeed a solution of Eq. 28-44.

Equation 28-45 tells us that the amount of excess charge on each capacitor plate decreases exponentially with time, at a rate set by the capacitive time constant $\tau = RC$. At time $t = \tau$, the capacitor's excess charge has been reduced to $|q_0|e^{-1}$, or about 37% of the initial value. That is, the amount of excess charge on the plates has decreased by 63%. Note that a greater τ means a greater discharge time.

Differentiating Eq. 28-45 gives us the current $i(t)$:

$$i = \frac{dq}{dt} = -\left(\frac{q_0}{RC}\right)e^{-t/RC} \qquad \text{(discharging a capacitor)}. \qquad (28\text{-}46)$$

This tells us the current also decreases exponentially with time, at a rate set by τ. The initial current i_0 is equal to q_0/RC. Note that you can find i_0 by simply applying the loop rule to the circuit at $t = 0$ the moment when the capacitor's initial potential ΔV_0 is connected across the resistance R. So the current must be

$$i_0 = \frac{\Delta V_0}{R} = \frac{(q_0/C)}{R} = \frac{q_0}{RC}.$$

The minus sign in the discharging capacitor expression (Eq. 28-46) can be ignored; it merely means the amount of excess charge on the plate is decreasing.

READING EXERCISE 28-5: The table gives four sets of values for the circuit elements in Fig. 28-22. Rank the sets according to (a) the initial current (as the switch is closed on *a*) and (b) the time required for the current to decrease to half its initial value, greatest first.

	1	2	3	4
ΔV_B (V)	12.0	12.0	10.0	10.0
R (Ω)	2.0	3.0	10.0	5.0
C (μF)	3.0	2.0	0.5	2.0

■

TOUCHSTONE EXAMPLE 28-5: Discharging a Capacitor

A capacitor of capacitance C is discharging through a resistor of resistance R.

(a) In terms of the time constant $\tau = RC$, when will the excess charge on each plate of the capacitor be half its initial value?

SOLUTION ■ The **Key Idea** here is that the excess charge on each plate of the capacitor varies according to Eq. 28-45,

$$q = q_0 e^{-t/RC},$$

in which q_0 is the initial charge. We are asked to find the time t at which $q = \frac{1}{2}q_0$ or at which

$$\tfrac{1}{2}q_0 = q_0 e^{-t/RC}. \qquad (28\text{-}47)$$

After canceling q_0, we realize that the time t we seek is "buried" inside an exponential function. To expose the symbol t in Eq. 28-47, we take the natural logarithms of both sides of the equation. (The natural logarithm is the inverse function of the exponential function.) We find

$$\ln\tfrac{1}{2} = \ln(e^{-t/RC}) = -\frac{t}{RC},$$

or $\qquad t = (-\ln\tfrac{1}{2})RC = 0.69RC = 0.69\tau.$ (Answer)

(b) When will the energy stored in the capacitor be half its initial value?

SOLUTION ■ There are two **Key Idea**s here. First, the energy U stored in a capacitor is related to the charge $|q|$ on the each plate according to Eq. 28-21 ($U = q^2/2C$). Second, that charge is decreasing according to Eq. 28-45. Combining these two ideas gives us

$$U = \frac{q^2}{2C} = \frac{q_0^2}{2C} e^{-2t/RC} = U_0 e^{-2t/RC},$$

in which U_0 is the initial stored energy. We are asked to find the time at which $U = \frac{1}{2}U_0$, or at which

$$\tfrac{1}{2}U_0 = U_0 e^{-2t/RC}.$$

Canceling U_0 and taking the natural logarithms of both sides, we obtain

$$\ln\tfrac{1}{2} = -\frac{2t}{RC},$$

or $\qquad t = -RC\frac{\ln\tfrac{1}{2}}{2} = 0.35RC = 0.35\tau.$ (Answer)

It takes longer (0.69τ versus 0.35τ) for the *charge* to fall to half its initial value than for the *stored energy* to fall to half its initial value. Does this result surprise you?

Problems

SEC. 28-2 ■ CAPACITANCE

1. Electrometer An electrometer is a device used to measure static charge—an unknown excess charge is placed on the plates of the meter's capacitor, and the potential difference is measured. What minimum charge can be measured by an electrometer with a capacitance of 50 pF and a voltage sensitivity of 0.15 V?

2. Two Metal Objects The two metal objects in Fig. 28-24 have net

FIGURE 28-24 ■
Problem 2.

(or excess) charges of +70 pC and −70 pC, which result in a 20 V potential difference between them. (a) What is the capacitance of the system? (b) If the excess charges are changed to +200 pC and −200 pC, what does the capacitance become? (c) What does the potential difference become?

3. Initially Uncharged The capacitor in Fig. 28-25 has a capacitance of

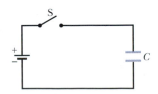

FIGURE 28-25 ■
Problem 3.

25 μF and is initially uncharged. The battery provides a potential difference of 120 V. After switch S is closed, how much charge will pass through it?

SEC. 28-3 ■ CALCULATING THE CAPACITANCE

4. Show That If we solve Eq. 28-9 for ε_0 we see that its SI unit is the farad per meter. Show that this unit is equivalent to that obtained earlier for ε_0—namely, the coulomb squared per newton-meter squared ($C^2/N \cdot m^2$).

5. Circular Plates A parallel-plate capacitor has circular plates of 8.2 cm radius and 1.3 mm separation. (a) Calculate the capacitance. (b) What excess charge will appear on each of the plates if a potential difference of 120 V is applied?

6. Two Flat Metal Plates You have two flat metal plates, each of area 1.00 m^2, with which to construct a parallel-plate capacitor. If the capacitance of the device is to be 1.00 F, what must be the separation between the plates? Could this capacitor actually be constructed?

7. Spherical Drop of Mercury A spherical drop of mercury of radius R has a capacitance given by $C = 4\pi\varepsilon_0 R$. If two such drops combine to form a single larger drop what is its capacitance?

8. Spherical Capacitor The plates of a spherical capacitor have radii 38.0 mm and 40.0 mm. (a) Calculate the capacitance. (b) What must be the plate area of a parallel-plate capacitor with the same plate separation and capacitance?

9. Two Spherical Shells Suppose that the two spherical shells of a spherical capacitor have approximately equal radii. Under these conditions the device approximates a parallel-plate capacitor with $b - a = d$. Show that Eq. 28-17 does indeed reduce to Eq. 28-9 in this case.

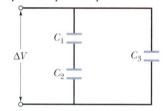

SEC. 28-4 ■ CAPACITORS IN PARALLEL AND IN SERIES

10. Equivalent In Fig. 28-26, find the equivalent capacitance of the combination. Assume that $C_1 = 10.0$ μF, $C_2 = 5.00$ μF, and $C_3 = 4.00$ μF.

FIGURE 28-26 ■
Problems 10 and 30.

11. How Many How many 1.00 μF capacitors must be connected in parallel to store an excess charge of 1.00 C with a potential of 110 V across the capacitors?

12. Each Uncharged Each of the uncharged capacitors in Fig. 28-27 has a capacitance of 25.0 μF. A potential difference of 4200 V is established when the switch is closed. How many coulombs of charge then pass through meter A?

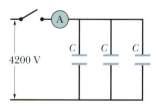

FIGURE 28-27 ■ Problem 12.

13. Combo In Fig. 28-28 find the equivalent capacitance of the combination. Assume that $C_1 = 10.0$ μF, $C_2 = 5.00$ μF, and $C_3 = 4.00$ μF.

14. Breaks Down In Fig. 28-28 suppose that capacitor 3 breaks down electrically, becoming equivalent to

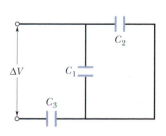

FIGURE 28-28 ■
Problems 13, 14, and 28.

a conducting path. What *changes* in (a) the amount of excess charge and (b) the potential difference occur for capacitor 1? Assume that $\Delta V = 100$ V.

15. Two in Series Figure 28-29 shows two capacitors in series; the center section of length b is movable vertically. Show that the equivalent capacitance of this series combination is independent of the position of the center section and is given by $C = \varepsilon_0 A/(a - b)$, where A is the plate area.

16. Battery Potential In Fig. 28-30, the battery has a potential difference of 10 V and the five capacitors each have a capacitance of 10 μF. What is the excess charge on (a) capacitor 1 and (b) capacitor 2?

17. Parallel with Second 100 pF capacitor is charged to a potential difference of 50 V, and the charging battery is disconnected. The capacitor is then connected in parallel with a second (initially uncharged) capacitor. If the potential difference across the first capacitor drops to 35 V, what is the capacitance of this second capacitor?

FIGURE 28-29 ■
Problem 15.

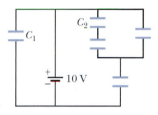

FIGURE 28-30 ■
Problem 16.

18. Charge Stored In Fig. 28-31, the battery has a potential difference of 20 V. Find (a) the equivalent capacitance of all the capacitors and (b) the excess charge stored by that equivalent capacitance. Find the potential across and charge on (c) capacitor 1, (d) capacitor 2, and (e) capacitor 3.

19. Opposite Polarity In Fig. 28-32, the capacitances are $C_1 = 1.0$ μF and $C_2 = 3.0$ μF and both capacitors are charged to a potential difference of $\Delta V = 100$ V but with opposite polarity as shown. Switches S_1, and S_2 are now closed. (a) What is now the potential difference between points a and b? What are now the amounts of excess charge on capacitors (b) 1 and (c) 2?

20. Battery Supplies In Fig. 28-33, battery B supplies 12 V. Find the excess charge on each capacitor (a) first when only switch S_1

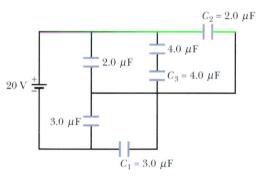

FIGURE 28-31 ■ Problem 18.

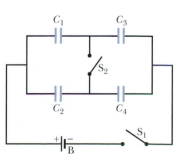

FIGURE 28-32 ■
Problem 19.

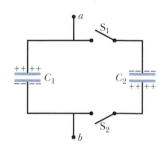

FIGURE 28-33 ■ Problem 20.

switch S_2 is also closed. Take $C_1 = 1.0$ μF, $C_2 = 2.0$ μF, $C_3 = 3.0$ μF, and $C_4 = 4.0$ μF.

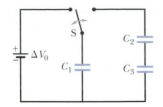

21. Switch Is Thrown When switch S is thrown to the left in Fig. 28-34, the plates of capacitor 1 acquire a potential difference ΔV_0. Capacitors 2 and 3 are initially uncharged. The switch is now thrown to the right. What are the final amounts of excess charge $|q_1|, |q_2|$, and $|q_3|$ on the capacitors?

FIGURE 28-34 ■
Problem 21.

SEC. 28-5 ■ ENERGY STORED IN AN ELECTRIC FIELD

22. Air How much energy is stored in one cubic meter of air due to the "fair weather" electric field of magnitude 150 V/m?

23. Capacitance Required What capacitance is required to store an energy of 10 kW · h at a potential difference of 1000 V?

24. Air-Filled Capacitor A parallel-plate air-filled capacitor having area 40 cm² and plate spacing 1.0 mm is charged to a potential difference of 600 V. Find (a) the capacitance, (b) the amount of excess charge on each plate, (c) the stored energy, (d) the electric field between the plates, and (e) the energy density between the plates.

25. Two Capacitors Two capacitors, of 2.0 and 4.0 μF capacitance, are connected in parallel across a 300 V potential difference. Calculate the total energy stored in the capacitors.

26. Connected Bank A parallel-connected bank of 5.00 μF capacitors is used to store electric energy. What does it cost to charge the 2000 capacitors of the bank to 50,000 V assuming 12.0¢/kW · h?

27. One Capacitor One capacitor is charged until its stored energy is 4.0 J. A second uncharged capacitor is then connected to it in parallel. (a) If the charge distributes equally, what is now the total energy stored in the electric fields? (b) Where did the excess energy go?

28. Find In Fig. 28-28 find (a) the excess charge, (b) the potential difference, and (c) the stored energy for each capacitor. Assume the numerical values of Problem 13, with $\Delta V = 100$ V.

29. Plates of Area A A parallel-plate capacitor has plates of area A and separation d and is charged to a potential difference ΔV. The charging battery is then disconnected, and the plates are pulled apart until their separation is $2d$. Derive expressions in terms of A, d, and ΔV for (a) the new potential difference; (b) the initial and final stored energies, U_i and U_f and (c) the work required to separate the plates.

30. Find the Charge In Fig. 28-26, find (a) the excess charge, (b) the potential difference, and (c) the stored energy for each capacitor. Assume the numerical values of Problem 10, with $\Delta V = 100$ V.

31. Cylindrical Capacitor A cylindrical capacitor has radii a and b as in Fig. 28-10. Show that half the stored electric potential energy lies within a cylinder whose radius is $r = \sqrt{ab}$.

32. Metal Sphere A charged isolated metal sphere of diameter 10 cm has a potential of 8000 V relative to $V = 0$ at infinity. Calculate the energy density in the electric field near the surface of the sphere.

33. Force of Magnitude (a) Show that the plates of a parallel-plate capacitor attract each other with a force of magnitude given by $F = q^2/2\varepsilon_0 A$. Do so by calculating the work needed to increase the

plate separation from x to $x + dx$, with the excess charge $|q|$ remaining constant. (b) Next show that the magnitude of the force per unit area (the *electrostatic stress*) acting on either capacitor plate is given by $\frac{1}{2}\varepsilon_0 E^2$. (Actually, this is the force per unit area on *any* conductor of *any* shape with an electric field $\vec{E}$ at its surface.)

SEC. 28-6 ■ CAPACITOR WITH A DIELECTRIC

34. Wax An air-filled parallel-plate capacitor has a capacitance of 1.3 pF. The separation of the plates is doubled and wax is inserted between them. The new capacitance is 2.6 pF. Find the dielectric constant of the wax.

35. Convert It Given a 7.4 pF air-filled capacitor, you are asked to convert it to a capacitor that can store up to 7.4 μJ with a maximum potential difference of 652 V. What dielectric in Table 28-3 should you use to fill the gap in the air capacitor if you do not allow for a margin of error?

36. Separation A parallel-plate air-filled capacitor has a capacitance of 50 pF. (a) If each of its plates has an area of 0.35 m², what is the separation? (b) If the region between the plates is now filled with material having $\kappa = 5.6$, what is the capacitance?

37. Coaxial Cable A coaxial cable used in a transmission line has an inner radius of 0.10 mm and an outer radius of 0.60 mm. Calculate the capacitance per meter for the cable. Assume that the space between the conductors is filled with polystyrene.

38. Construct a Capacitor You are asked to construct a capacitor having a capacitance near 1 nF and a breakdown potential in excess of 10 000 V. You think of using the sides of a tall Pyrex drinking glass as a dielectric, lining the inside and outside curved surfaces with aluminum foil to act as the plates. The glass is 15 cm tall with an inner radius of 3.6 cm and an outer radius of 3.8 cm. What are the (a) capacitance and (b) breakdown potential of this capacitor?

39. Certain Substance A certain substance has a dielectric constant of 2.8 and a dielectric strength of 18 MV/m. If it is used as the dielectric material in a parallel-plate capacitor, what minimum area should the plates of the capacitor have to obtain a capacitance of 7.0×10^{-2} μF and to ensure that the capacitor will be able to withstand a potential difference of 4.0 kV?

40. Two Dielectrics A parallel-plate capacitor of plate area A is filled with two dieletrics as in Fig. 28-35a. Show that the capacitance is

$$C = \frac{\varepsilon_0 A}{d} \frac{\kappa_1 + \kappa_2}{2}.$$

Check this formula for limiting cases. (*Hint*: Can you justify this arrangement as being two capacitors in parallel?)

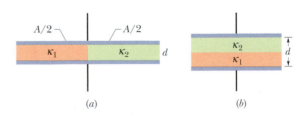

FIGURE 28-35 ■ Problems 40 and 41.

41. Limiting Cases A parallel-plate capacitor of plate area A is filled with two dielectrics as in Fig. 28-35b. Show that the capacitance is

$$C = \frac{2\varepsilon_0 A}{d} \frac{\kappa_1 \kappa_2}{\kappa_1 + \kappa_2}.$$

Check this formula for limiting cases. (*Hint:* Can you justify this arrangement as being two capacitors in series?)

42. What is Capacitance What is the capacitance of the capacitor, of plate area A, shown in Fig. 28-36? (*Hint:* See Problems 40 and 41.)

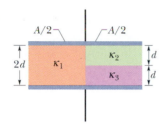

FIGURE 28-36 ■ Problem 42.

SEC. 28-8 ■ DIELECTRICS AND GAUSS' LAW

43. Mica A parallel-plate capacitor has a capacitance of 100 pF, a plate area of 100 cm², and a mica dielectric ($\kappa = 5.4$) completely filling the space between the plates. At 50 V potential difference, calculate (a) the electric field magnitude E in the mica, (b) the amount of excess free charge on each plate, and (c) the amount of induced surface charge on the mica.

44. Electric Field Two parallel plates of area 100 cm² are given excess charges of equal amounts 8.9×10^{-7} C but opposite signs. The electric field within the dielectric material filling the space between the plates is 1.4×10^6 V/m. (a) Calculate the dielectric constant of the material. (b) Determine the amount of bound charge induced on each dielectric surface.

45. Concentric Conducting Shells The space between two concentric conducting spherical shells of radii b and a (where $b > a$) is filled with a substance of dielectric constant κ. A potential difference ΔV exists between the inner and outer shells. Determine (a) the capacitance of the device, (b) the excess free charge q on the inner shell, and (c) the charge q' induced along the surface of the inner shell.

SEC. 28-9 ■ *RC* CIRCUITS

46. Initial Charge A capacitor with initial excess charge of amount $|q_0|$ is discharged through a resistor. In terms of the time constant τ, how long is required for the capacitor to lose (a) the first one-third of its charge and (b) two-thirds of its charge?

47. How Many Time Constants How many time constants must elapse for an initially uncharged capacitor in an *RC* series circuit to be charged to 99.0% of its equilibrium charge?

48. Leaky Capacitor The potential difference between the plates of a leaky (meaning that charges leak directly across the "insulated" space between the plates) 2.0 μF capacitor drops to one-fourth its initial value in 2.0 s. What is the equivalent resistance between the capacitor plates?

49. Time Constant A 15.0 kΩ resistor and a capacitor are connected in series and then a 12.0 V potential difference is suddenly applied across them. The potential difference across the capacitor rises to 5.00 V in 1.30 μs. (a) Calculate the time constant of the circuit. (b) Find the capacitance of the capacitor.

50. Flashing Lamp Figure 28-37 shows the circuit of a flashing lamp, like those attached to barrels at highway construction sites. The fluorescent lamp L (of negligible capacitance) is connected in parallel across the capacitor C of an *RC* circuit. There is a current through the lamp only when the potential difference across it reaches the breakdown voltage V_L; in this event, the capacitor discharges completely through the lamp and the lamp flashes briefly.

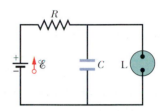

FIGURE 28-37 ■ Problem 50.

Suppose that two flashes per second are needed. For a lamp with breakdown voltage $\Delta V_L = 72.0$ V, wired to a 95.0 V ideal battery and a 0.150 μF capacitor, what should be the resistance R?

51. Initial Potential Difference A capacitor with an initial potential difference of 100 V is discharged through a resistor when a switch between them is closed at $t = 0$. At $t = 10.0$ s, the potential difference across the capacitor is 1.00 V. (a) What is the time constant of the circuit? (b) What is the potential difference across the capacitor at $t = 17.0$ s?

52. Electronic Arcade Game A controller on an electronics arcade games consists of a variable resistor connected across the plates of a 0.220 μF capacitor. The capacitor is charged to 5.00 V, then discharged through the resistor. The time for the potential difference across the plates to decrease to 0.800 V is measured by a clock inside the game. If the range of discharge times that can be handled effectively is from 10.0 μs to 6.00 ms, what should be the resistance range of the resistor?

53. Initial Stored Energy A 1.0 μF capacitor with an initial stored energy of 0.50 J is discharged through a 1.0 MΩ resistor. (a) What is the initial amount of excess charge on the capacitor plates? (b) What is the current through the resistor when the discharge starts? (c) Determine ΔV_C, the potential difference across the capacitor, and ΔV_R, the potential difference across the resistor, as functions of time. (d) Express the production rate of thermal energy in the resistor as a function of time.

Additional Problems

54. Capacitance (a) What is the physical definition and description of a capacitor? (b) What is the mathematical definition of capacitance? (c) Based on the physical description of a capacitor, why would you expect it to hold more excess charge on each of its conducting surfaces when the voltage difference between the two pieces of conductor increases?

55. Net Charge What is the net charge on a capacitor in a circuit? Is it ever possible for the amount of excess charge on one conductor to be different from the amount of excess charge on the other conductor? Explain.

56. Attraction and Repulsion Consider the attraction and repulsion of different types of charge. (a) Explain why you expect to find that the amount of excess charge a battery can pump onto a parallel-plate capacitor will double if the area of each plate doubles. (b) Explain why you expect to find that the amount of excess charge a battery can pump onto a parallel-plate capacitor will be cut in half if the distance between each plate doubles.

57. Three Parallel-Plate Capacitors Suppose you have three parallel-plate capacitors as follows:

Capacitor 1: Area A, spacing d
Capacitor 2: Area A, spacing $2d$
Capacitor 3: Area $2A$, spacing d

The three graph lines (labeled a, b, and c) in Fig. 28-38 represent data for the amounts of excess of charge on the plates of each capacitor as a function of the potential difference across it. Which capacitor {1, 2, or 3} belongs to which line {a, b, and c}? Explain your reasoning carefully.

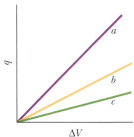

FIGURE 28-38 ■
Problem 57.

58. Capacitors in Series Give as clear an explanation as possible as to why it is physically reasonable to expect that two identical parallel-plate capacitors that are placed in series ought to have half the capacitance as one capacitor. *Hints:* What happens to the effective spacing between the first plate of capacitor 1 and the second plate of capacitor 2 when they are wired in series? What does the fact that like charges repel and opposites attract have to do with anything?

59. Capacitors in Parallel Give as clear an explanation as possible as to why it is physically reasonable to expect that two identical parallel-plate capacitors placed in parallel ought to have twice the capacitance as one capacitor. *Hints:* What happens to the effective area of capacitors wired in parallel? What does the fact that like charges repel and opposites attract have to do with anything?

60. *Charge Ratios on Capacitors* (Adapted from a TYC WS Project ranking task by D. Takahashi). Eight capacitor circuits are

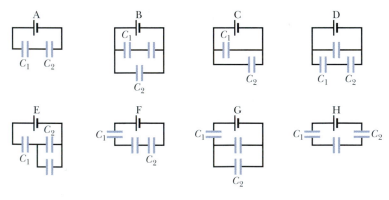

FIGURE 28-39 ■ Problem 60.

shown in Fig. 28-39. All of the capacitors are identical and all are fully charged. The batteries are also identical. In each circuit, one capacitor is labeled C_1 and another is labeled C_2. Assuming $|q_1|$ denotes the amount of excess charge on C_1, $|q_2|$ denotes the amount of excess charge on C_2, and the value of the ratio is denoted $|q_1/q_2|$, rank the circuit in which the value of the ratio $|q_1/q_2|$ is largest *first*, and rank the circuit in which the value of the ratio is the smallest *last*. If two or more circuits result in identical values for the ratio, give these circuits equal ranking. Express your ranking symbolically. (For example, suppose the ratio was highest for D and G and lowest for A and E with the in-between ratios being equal, then the symbolic ranking would be

$$D = G > B = C = H = F > A = E$$

(*Beware:* This is only a sample, not a correct answer!)

61. Physicists Claim Physicists claim that charge <u>never</u> flows *through* an ideal capacitor. Yet when an uncharged capacitor is first placed in series with a resistor and a battery, current flows through the battery and the resistor. Explain how this is possible.

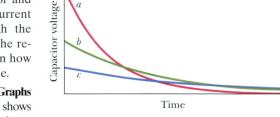

FIGURE 28-40 ■ Problem 62.

62. Voltage Graphs Figure 28–40 shows plots of voltage across the capacitor as a function of time for three different capacitors that have each been separately discharged through the same resistor. Rank the plots according to the capacitances, the greatest first. Explain the reasons for your rankings.

63. A Cell Membrane The inner and outer surfaces of a cell membrane carry excess negative and positive charge, respectively. Because of these charges, a potential difference of about 70 mV exists across the membrane. The thickness of the membrane is 8 nm.

(a) If the membrane were empty (filled with air), what would be the magnitude of the electric field inside the membrane?
(b) If the dielectric constant of the membrane were $\kappa = 3$ what would the field be inside the membrane?
(c) Cells can carry ions across a membrane *against the field* ("uphill") using a variety of active transport mechanisms. One mechanism does so by using up some of the cell's stored energy converting ATP to ADP. How much work does it take to carry one sodium ion (charge $= +e$) across the membrane against the field? Calculate your answer in eV, joules, and kcal/mole (the last for 1 mole of sodium ions).

29 | Magnetic Fields

Ocean water contains huge quantities of the light atomic nuclei found in "heavy water" needed to produce fusion power. If we could produce a cost-effective fusion reactor, the world's power problems could be solved. We have known this for over 50 years and still not produced fusion power. Why? A key problem is that it takes a temperature of at least 100 million degrees Celsius to force two light nuclei to fuse together. At this temperature, any material we tried to squeeze together to fuse would be so hot that it would vaporize any material it touches. The torus-shaped chamber of the large Tokomak reactor in this photo was built in an attempt to contain fusion reactions.

How can this Tokomak contain matter at 100 million degrees Celsius?

The answer is in this chapter.

29-1 A New Kind of Force?

In Chapter 14 we studied gravitational interaction forces that we experience on an everyday basis. Gravitational forces are so weak that it takes a source the size of a planet or star to produce a noticeable effect. This made the study of the effects of gravity near the Earth's surface relatively simple. In most cases, we treat the gravitational force on an object as a constant.

Then, in Chapter 22, we studied the electrostatic force—a long-range force that is much stronger than the gravitational force. If you run a comb through your hair, a bit of paper near the comb hops up and sticks to the comb. The electrostatic force exerted on the paper by the comb is somewhat larger than the gravitational force that the whole Earth exerts on the paper.

Are there any other long-range (or, action-at-a-distance) forces, or are we done? If you think about your personal experiences, you probably have had the opportunity to play with small disk-shaped refrigerator magnets or pairs of bar magnets. On a larger scale, **electromagnets** are used for sorting scrap metal (Fig. 29-1) and many other things. Magnets are fun because they behave in such an unusual way. You can use one magnet to chase a second magnet around a table without even touching it. But if you come at the magnet from a slightly different direction, it will suddenly seem to change what it's doing and will be pulled toward the other. A refrigerator magnet will seem to leap to the door of the refrigerator, being drawn to it from a distance. Clearly a long-range force is at work here. But is it a new kind of force? Or is it merely a form of gravitational or electrical force?

FIGURE 29-1 ■ A large electromagnet is used to collect and transport scrap metal at a steel mill.

29-2 Probing Magnetic Interactions

We know from our everyday experiences with small bar magnets that we can *feel* a force on one bar magnet as it interacts with another. This means we can use a bar magnet as a test object for investigating the nature of magnetic interactions. In order to answer the question of whether magnetic interactions are really gravitational or electrostatic forces, let's investigate what happens when a small bar magnet or disk-shaped refrigerator magnet experiences a significant force.

Is the Magnetic Force a Type of Gravitational Force?

The force on our test magnet near the Earth's surface is clearly *in addition* to the gravitational attraction of the Earth. The fact that a refrigerator magnet can stick to the refrigerator and not fall means that it is experiencing a force that is stronger than the gravitational force exerted on it by the entire Earth.

What happens if we replace our test magnet with another *nonmagnetic* object of equal mass and the same shape? We find that the magnetic force disappears. Hence, we must surmise that the force we detected with the bar magnet is not a gravitational force associated with the presence of another object. It is too strong and exists only for certain probe objects. Furthermore, we know from playing with magnets that the force can be attractive or repulsive. As we know, this is not true for the gravitational force.

Is the Magnetic Force a Type of Electrostatic Force?

Could the magnetic force be the electrostatic force we have learned about? After all, the magnetic force, like the electrostatic force, is sometimes attractive and sometimes repulsive. To test this idea, we replace our test magnet with a test charge (such as a tiny Styrofoam ball charged by a rubber rod) at the former location of our test magnet. Again, we find that our new probe (the charge) is only weakly attracted—as is any charged object to a neutral object. So, we must also surmise that *the force the bar magnet detects is not a type of electrostatic force.*

The Magnetic Force and a Moving Charge

We have just described observations that show that forces between magnets are fundamentally different from either electrostatic or gravitational forces. So it appears that we have a new action-at-a-distance force to learn about. This force can be either attractive or repulsive. We can detect this force with a magnet, and so we will refer to it as a **magnetic force.**

Having completed our investigations of electric force in earlier chapters, we now take the electric charge we had been using as a probe and move it rapidly away from a magnet. When we do this, we find something strange. When we *move* the charge, we do detect a force!

> **OBSERVATION:** A magnet exerts a force on a moving charged object, but not on a stationary charged object.

Furthermore, when we try moving the charge at different velocities, we find that the larger the magnitude of the velocity, the larger is the force exerted on the charge. Is the same true for *uncharged,* nonmagnetic masses? Experimentation shows the same is *not* true for uncharged masses. No magnetic force is detected when an uncharged, nonmagnetic mass is used as a probe—regardless of whether the probe is moving or stationary.

In the early 19th century, both Oersted and Ampère discovered that magnets interact with moving charges. In fact, these two scientists showed that current-carrying wires both exert forces on and feel forces from bar magnets. Their observations provide us with important information in our quest to understand the magnetic force. We have found that magnetic forces are not just exerted on other magnets. Magnetic forces are also exerted on a nonmagnetic small charged particle in rough proportion to the degree to which the particle is *both* charged and moving. What is the simplest relationship between magnetic force, charge, and velocity that is consistent with our observations? Mathematically stated, it is a proportional relationship given by

$$|\vec{F}^{\,\text{mag}}| \propto |q|\,v,$$

where $|q|$ represents the amount of electric charge on the particle and v is the particle's speed.

Is this relationship correct? Well, if it is, we should see a doubling of the force when we double the velocity of the charged particle we are using as a probe. Experimentally, this does turn out to be the case. Furthermore, we also find that doubling the charge on the probe doubles the force detected. Hence, the linear relationship expressed above is a good start toward a more precise mathematical description of the magnetic force on a moving charged particle. We will return to experimentation as a means for developing a precise expression for the magnetic force in just a moment.

29-3 Defining a Magnetic Field $\vec{B}$

When we play with two bar magnets, we quickly see that the magnetic force can be attractive or repulsive. Furthermore, if we observe more carefully, we find that the strength of the force decreases as the distance between the two magnets is increased. These observations are distinctly reminiscent of our observations of the electrostatic force between two charges. So our first guess in developing a model of the magnetic force might turn out to be somewhat similar to our model of the electrostatic force.

In order to develop a model of magnetism that parallels our model of electrostatics, we should have two different kinds of "magnetic charges." These conceptual objects are referred to as **magnetic monopoles.** We can model our bar magnet as containing a south and a north pole with at least some separation between them. If we assume that like poles repel and unlike poles attract, then this model allows us to correctly predict all our observations. Playing with bar magnets informs us that poles of the same kind repel one another and poles of different kinds attract one another. This is just as we found for electric charges. However, careful observation of the interaction between bar magnets shows that their behavior is similar to that of electric dipoles. Recall that an electric dipole consists of two charges of opposite sign with a small spacing between them. If two electric dipoles that are placed with all their charges lying on the same line are brought together, they will attract. Why? Because a negative charge from the end of one dipole will be closest to the positive charge of the other dipole. However, if we turn one of the electric dipoles around so the dipoles are anti-aligned, then the two like charges will be closest together. Now the dipoles will repel.

Two bar magnets when aligned and then anti-aligned will behave just like electric dipoles. For this reason, we often refer to magnets as *magnetic dipoles*. That is, one end appears to be one kind of magnetic charge and the other end appears to be the other kind of magnetic charge. By convention, we can assign names to the poles of a bar magnet as follows. If we suspend a bar magnet by a string placed halfway between its ends and take other magnetic sources away from its vicinity, one pole of the magnet will point more or less north and the other more or less south. We can call the north-pointing end the north pole of the magnet and the other end the south pole of the magnet.

This idea that a bar magnet is a magnetic dipole with a north charge at one end and a south charge at the other end provides us with a start in describing magnetic interactions. However, to continue with the analogy between the magnetism and electricity, we would like to isolate a magnetic charge. After all, we can separate a negative charge from a positive charge. So we need to be able to separate the north pole of a bar magnet from the south pole of a bar magnet. To do this, we take our bar magnet and cut it in half. But, when we do this we find a surprising thing. The result of breaking the bar magnet in half is simply that we have two weaker half-sized bar magnets. Each one still behaves as a dipole with both a north and a south pole. If we again try to break the magnet in half, we find we have a still smaller magnet, but still with a north and south pole (Fig. 29-2). In fact, if we break the magnet down into subatomic parts, we find that even the electrons, protons, and neutrons within atoms behave as magnetic dipoles (that is, *very* little bar magnets).

As it turns out, the magnetic effect of a bar magnet arises from the combination of the effects of the little bar magnets in the electrons in iron, nickel, and cobalt aligning with each other and producing a strong effect. Each electron's magnet is small, but when you turn them in the same direction and add them all up, the total effect is strong—the full magnetic effect of the bar magnet. So, in short, although the existence of separate magnetic charges (or magnetic monopoles) have been predicted by some physicists, they have never actually been found.

Does the fact that we cannot find an isolated magnetic monopole mean that we must abandon our effort to find parallels between magnetic and electrostatic forces? Not at all. In Chapter 23, we found that the concept of an electric field was quite useful. With so many different possible sources of significant electrostatic forces, it was helpful to think about the force field associated with a given charge (the source of electrostatic force)—without having to decide on what object the force will be exerted on. That is, we wanted to separate the discussion of the source of the

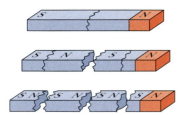

FIGURE 29-2 ■ Whenever a magnet is broken into pieces between its poles, the pieces behave like smaller, weaker magnets.

from the discussion of the object the force is exerted on. So we defined the electric field $\vec{E}$ as

$$\vec{E} = \frac{\vec{F}^{elec}}{q}. \qquad \text{(Eq. 23-4)}$$

We determined the electric field $\vec{E}$ at a point by putting a test particle of charge q at rest at that point and measuring the electrostatic force $\vec{F}^{elec}$ acting on the particle. We saw that *electric charges* set up an electric field that can then affect other electric charges.

Perhaps the same idea could be useful to us in describing magnetic forces. If we could develop a parallel concept of a magnetic field, we could separate the issue of sources of magnetic forces from discussions of the objects that magnetic forces are exerted on. This would be helpful since the concept of a magnetic monopole is so problematic. If a magnetic monopole were available, we could define the magnetic field $\vec{B}$ in a way similar to that used for electric fields. However, because such particles have not been found, we must using another method to define a magnetic field $\vec{B}$.

For nonmagnetic particles, we have already observed that the magnetic force is proportional to the charge and the magnitude of the velocity of the particle being acted on (the probe). We can use this information and define the magnetic field in terms of the force $\vec{F}^{mag}$ exerted on a moving, electrically charged test particle. The magnitude of the force seems to depend on the direction of the particle's velocity $\vec{v}$ as well. We will examine this effect in more detail in the next section, but for now we define the magnetic field $\vec{B}$ in terms of the *maximum* force magnitude we measure after trying all different directions for $\vec{v}$. So we can express the *magnitude* of the magnetic field $\vec{B}$ in terms of this maximum force magnitude as:

$$B = \frac{F_{max}^{mag}}{|q|v}, \qquad \text{(29-1)}$$

where q is the particle's charge and v is its speed.

Having defined the magnitude of the magnetic field is a big step forward. It is a concept that will turn out to be extremely useful. Right now, it is helpful because we have not identified the source of the force exerted on our probe. But, having defined the magnitude of the magnetic field in this way, we can at least say that we know that there is a vector magnetic field in the region of space we have been probing. We make extensive use of the concept of a magnetic field in this chapter. Next we turn our attention to this issue of how to define the direction of the magnetic field.

29-4 Relating Magnetic Force and Field

In order to determine the direction of the magnetic field, we can fire a charged particle through a region of space where a magnetic field $\vec{B}$ is known to exist. If we shoot the charged particle in various directions, we find something surprising—the direction of $\vec{F}^{mag}$ is always perpendicular to the direction of $\vec{v}$ (Fig. 29-3). After many such trials we find that when the particle's velocity $\vec{v}$ is along a particular axis through the region of space, force $\vec{F}^{mag}$ is zero. Furthermore, we find that for all other directions of $\vec{v}$, the *magnitude* of $\vec{F}^{mag}$ depends on the direction of $\vec{v}$. In fact, it is proportional to $|\vec{v}|\sin\phi$ where ϕ is the angle between the zero-force axis and the direction of $\vec{v}$. Thinking back to our work on torque and angular momentum, these observations suggest that a cross product is involved. But a cross product of what two vectors?

Clearly, one of the two vectors involved in the cross product is the velocity vector. Our observation that the force is zero when the velocity is along a certain axis implies

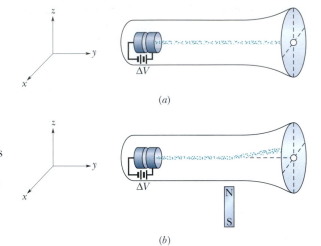

FIGURE 29-3 ■ (*a*) An electron beam is accelerated by a voltage source and travels through an evacuated glass tube to the center of a phosphorescent screen. (*b*) If a magnet is oriented vertically and placed just below the beam (along the $+z$ axis), the electrons are deflected horizontally along the $-x$ axis.

that the other vector must be aligned with this "zero magnetic force" axis. Referring back to our definition of the magnetic field magnitude, B, in Eq. 29-1, we note that the magnitude of the observed magnetic force is given by

$$F_{max}^{mag} = |q|vB,$$

where v is the particle speed and $|q|$ is the amount of charge the particle has. Suppose the direction of the magnetic field is taken to be along the "zero magnetic force" axis. We could then represent all of our observations with the following vector equation, known as the *magnetic force law* or **Lorentz force law:**

$$\vec{F}^{\,mag} = q\vec{v} \times \vec{B} \qquad \text{(magnetic force law).} \qquad (29\text{-}2)$$

That is, the force $\vec{F}^{\,mag}$ on the particle is equal to the charge q times the cross product of its velocity $\vec{v}$ and the magnetic field $\vec{B}$. If this expression is correct, the force on a negatively charged particle should be opposite in direction from the force on a positively charged particle. This does in fact turn out to be the case.

Furthermore, expressing the magnetic force on a charged particle moving through a magnetic field as $\vec{F}^{\,mag} = q\vec{v} \times \vec{B}$ requires that we adopt a standard convention for the *direction* of the magnetic field. That is,

> The direction of a magnetic field is defined to be related to the direction of the force on and the velocity of a positively charged particle by $\vec{F}^{\,mag} = q\vec{v} \times \vec{B}$.

Although this is not a very intuitive statement of how one goes about finding the direction of a magnetic field, we are forced to use it if we want to use $\vec{F}^{\,mag} = q\vec{v} \times \vec{B}$ to determine the magnitude and direction of the magnetic force on a moving charged particle.

Using the mathematical definition of a cross product to evaluate this expression, we see that we can write the magnitude of the magnetic force as

$$F^{mag} = |q\vec{v}||\vec{B}|\sin\phi = |q|vB\sin\phi, \qquad (29\text{-}3)$$

where ϕ is the smaller angle (the one whose value lies between $0°$ and $180°$) between the directions of velocity $\vec{v}$ and magnetic field $\vec{B}$.

We have seen that magnetic force and electric force are not the same. However, a magnetic force *is* exerted on a moving charged particle as well as on bar magnets. This suggests that there is a profound connection between electricity and magnetism—even though they are *not* the same thing. As it turns out, the theory of relativity, treated in Chapter 38, reveals a deep underlying connection between $\vec{E}$ and $\vec{B}$. Furthermore, much of the technology that makes our lives more comfortable today results from an understanding of this relationship. In Chapter 30, we show how moving electrical charges can create magnetic fields and in Chapter 31 we show an even deeper and more surprising link between electricity and magnetism (called Faraday's law). What we find is that a magnetic field can, if it changes in time, create an electric field without any electric charge present!

Finding the Magnetic Force on a Moving Charged Particle

Equation 29-3 reveals that the magnitude of the force $\vec{F}^{\text{mag}}$ acting on a particle in a magnetic field is proportional to the amount of charge $|q|$ and speed v of the particle. Thus, the force is equal to zero if the charge is zero or if the particle is stationary. Equation 29-3 also tells us that the magnitude of the force is zero if $\vec{v}$ and $\vec{B}$ are either parallel ($\phi = 0°$) or antiparallel ($\phi = 180°$), and the force is a maximum when $\vec{v}$ and $\vec{B}$ are perpendicular to each other.

Equation 29-2 tells us all this and the direction of $\vec{F}^{\text{mag}}$. From Section 12-4, we know that the cross product $\vec{v} \times \vec{B}$ in Eq. 29-2 is a vector that is perpendicular to the two vectors $\vec{v}$ and $\vec{B}$. The right-hand rule (Fig. 29-4a) specifies that the thumb of the right hand points in the direction of $\vec{v} \times \vec{B}$ when the fingers sweep $\vec{v}$ into $\vec{B}$. If q is positive, then (by Eq. 29-2) the force $\vec{F}^{\text{mag}}$ has the same sign as $\vec{v} \times \vec{B}$ and thus must be in the same direction. That is, for positive q, $\vec{F}^{\text{mag}}$ is directed along the thumb as in Fig. 29-4b. If q is negative, then the force $\vec{F}^{\text{mag}}$ and the cross product $\vec{v} \times \vec{B}$ have opposite signs and thus must be in opposite directions. For negative q, $\vec{F}^{\text{mag}}$ is directed opposite the thumb as in Fig. 29-4c.

Regardless of the sign of the charge, however,

> The force $\vec{F}^{\text{mag}}$ acting on a charged particle moving with velocity $\vec{v}$ through a magnetic field $\vec{B}$ is *always* perpendicular to $\vec{v}$ and $\vec{B}$.

Thus, $\vec{F}^{\text{mag}}$ *never* has a component parallel to $\vec{v}$. This means that $\vec{F}^{\text{mag}}$ cannot change the particle's speed $v = |\vec{v}|$ (and thus it cannot change the particle's kinetic energy). The force can change only the direction of $\vec{v}$ (and thus the direction of travel); only in this sense can $\vec{F}^{\text{mag}}$ accelerate the particle. If there are no other forces acting on the charged particle and the velocity of the particle is perpendicular to the direction of the magnetic field, this means that the particle will move in a circle. If the particle has a component perpendicular to the magnetic field *and* a component of velocity parallel

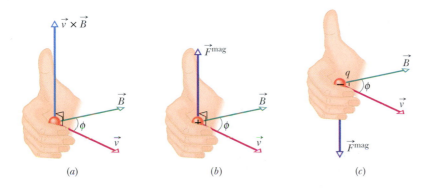

(a) (b) (c)

FIGURE 29-4 ■ (a) The right-hand rule (in which $\vec{v}$ is swept into $\vec{B}$ through the smaller angle ϕ between them) gives the direction of $\vec{v} \times \vec{B}$ as the direction of the thumb. (b) If q is positive, then the direction of $\vec{F}^{\text{mag}} = q\vec{v} \times \vec{B}$ is in the direction of $\vec{v} \times \vec{B}$. (c) If q is negative, then the direction of $\vec{F}^{\text{mag}}$ is opposite that of $\vec{v} \times \vec{B}$.

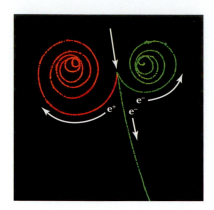

FIGURE 29-5 ■ Color enhanced tracks showing two electrons (e^-) and a positron (e^+) in a bubble chamber that is immersed in a uniform magnetic field that is directed out of the plane of the page.

to the magnetic field, the particle will move along a *helix* of constant radius. These paths are discussed in more detail in Section 29-5.

To develop a feeling for the relationship between the magnetic force on a moving charged particle and the magnetic field, $\vec{F}^{\,\text{mag}} = q\vec{v} \times \vec{B}$, consider Fig. 29-5. This figure shows some tracks left by charged particles moving rapidly through a *bubble chamber* at the Lawrence Berkeley Laboratory. The chamber, which is filled with liquid hydrogen, is immersed in a strong uniform magnetic field that is directed out of the plane of the figure. An incoming gamma ray particle—which leaves no track because it is uncharged—transforms into an electron (spiral track marked e^-) and a positron (track marked e^+) while it knocks an electron out of a hydrogen atom (long track marked e^-). At first these newly created charged particles are moving in the same direction as the gamma ray. As they move, they each experience a magnetic force of magnitude $F^{\text{mag}} = |q|vB$ and begin to move in a circular path given by $F^{\text{mag}} = mv^2/r$. Since $qvB = mv^2/r$, a particle has a path of radius $r = mv/|q|B$. You can use Eq. 29-2 and Fig. 29-4 to confirm that the three tracks made by these two negative particles and one positive particle curve in the proper directions. It is interesting to note that the electrons and positron do not move in a pure circle. Instead, they move in a shrinking spiral because they are slowed down through their interaction with the gas in the bubble chamber. This makes sense because $r = mv/|q|B$ and as each particle's speed, v, becomes smaller, so does its radius r. When this happens, the magnetic force, which is proportional to the particle's velocity, decreases and so the radius of the particle's path decreases.

What Produces a Magnetic Field?

We have discussed how a charged plastic rod produces a vector field—the electric field $\vec{E}$—at all points in the space around it. Similarly, a magnet produces a vector field—the **magnetic field** $\vec{B}$—at all points in the space around it. You get a hint of that magnetic field whenever you attach a note to a refrigerator door with a small magnet, or accidentally erase a computer disk by bringing it near a strong magnet. The magnet acts on the door or disk *by means of* its magnetic field.

In a common type of magnet, a wire coil is wound around an iron core and a current is sent through the coil; the strength of the magnetic field is determined by the size of the current. In industry, such **electromagnets** are used for sorting scrap metal (Fig. 29-1) among many other things. You are probably more familiar with **permanent magnets**—magnets, like the refrigerator-door type, that do not need current to have a magnetic field.

How then are magnetic fields set up? We know about two ways to create magnetic fields. (1) We observe that moving electrically charged particles, such as the current in a wire or charged beams of cosmic rays create magnetic fields. (2) We find that elementary particles such as protons, neutrons, and electrons have *intrinsic* magnetic moments that create magnetic fields. In Chapter 30 we discuss how moving charges create magnetic fields, and in Chapter 32 we consider the role of intrinsic magnetic moments in the creation of magnetic fields. In this chapter we stay focused on how to represent magnetic fields and how they influence charged particles that are moving.

The SI unit for $\vec{B}$ that follows from Eqs. 29-2 and 29-3 is the newton per coulomb-meter per second. For convenience, the SI unit for magnetic field is called the tesla (T):

$$1 \text{ tesla} = 1 \text{ T} = 1\frac{\text{newton}}{(\text{coulomb})(\text{meter/second})}.$$

Recalling that a coulomb per second is an ampere, we have

$$1 \text{ T} = 1\frac{\text{newton}}{(\text{coulomb/second})(\text{meter})} = 1\frac{\text{N}}{\text{A}\cdot\text{m}}. \tag{29-4}$$

TABLE 29-1
Some Approximate Magnetic Fields

At the surface of a neutron star	10^8 T
Near a big electromagnet	1.5 T
Near a small bar magnet	10^{-2} T
At Earth's surface	10^{-4} T
In interstellar space	10^{-10} T
Smallest value in a magnetically shielded room	10^{-14} T

An earlier (non-SI) unit for $\vec{B}$, that is still in common use is the *gauss* (G), and

$$1 \text{ tesla} = 10^4 \text{ gauss.} \qquad (29\text{-}5)$$

Table 29-1 lists the magnetic fields that occur in a few situations. Note that Earth's magnetic field near the planet's surface is about 10^{-4} T $(= 100\,\mu\text{T}$ or 1 gauss).

READING EXERCISE 29-1: The figure shows three situations in which a charged particle with velocity $\vec{v}$ travels through a uniform magnetic field $\vec{B}$. In each situation, what is the direction of the magnetic force $\vec{F}^{\text{mag}}$ on the particle?

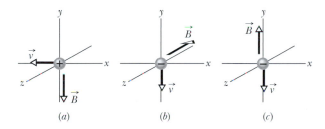

(a)   (b)   (c)   ■

Magnetic Field Lines

We can represent magnetic fields with field lines, as we did for electric fields. Similar rules apply; that is, (1) the direction of the tangent to a magnetic field line at any point gives the direction of $\vec{B}$ at that point, and (2) the spacing of the lines represents the magnitude of $\vec{B}$—the magnetic field is stronger where the lines are closer together, and conversely.

Figure 29-6a shows how the magnetic field near a *bar magnet* (a permanent magnet in the shape of a bar) can be represented by magnetic field lines. The lines all pass through the magnet, and they all form closed and continuous loops (even those that are not shown closed in the figure). They don't start or end anywhere. Since electric field lines begin and end on electric charges, this is consistent with our assumption that there are no magnetic charges (monopoles). As shown with field lines, the external magnetic effects of a bar magnet are strongest near its ends, where the field lines are most closely spaced. Thus, the magnetic field of the bar magnet in Fig. 29-6b collects the iron filings mainly near the two ends of the magnet. Overall, outside of the bar magnet the field lines look just like they would for an electric dipole, but inside the magnet they point in the opposite direction.

The (closed) field lines enter one end of a magnet and exit the other end. The end of a magnet from which the field lines emerge is called the *north pole* of the magnet; the other end, where field lines enter the magnet, is called the *south pole*. (Remember that the direction of the field line is related to the direction of the force on a moving positively charged particle.) Some of the magnets we use to fix notes on refrigerators

(a)   (b)

FIGURE 29-6 ■ (a) The magnetic field lines for a bar magnet. (b) A "cow magnet"—a bar magnet that is intended to be slipped down into the rumen (first stomach) of a cow to prevent accidentally ingested bits of scrap iron from reaching the cow's intestines. The iron filings at its ends reveal the directions of the magnetic field lines in the vicinity of the magnet.

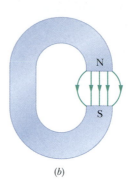

FIGURE 29-7 ▪ (a) A horseshoe magnet and (b) a C-shaped magnet. (Only a few of the possible of the external field lines are shown.)

(a)   (b)

are short bar magnets. Figure 29-7 shows two other common shapes for magnets: a *horseshoe magnet* and a magnet that has been bent around into the shape of a C so that the *pole faces* are facing each other. (The magnetic field between the pole faces can then be approximately uniform.) Regardless of the shape of the magnets, if we place two of them near each other we find:

> Opposite magnetic poles attract each other, and like magnetic poles repel each other.

Earth has a magnetic field that is produced in its core. We discuss current theories about the nature and origin of the Earth's magnetic field in Section 32-9. On Earth's surface, we can detect this magnetic field with a compass, which is essentially a slender bar magnet on a low-friction pivot. This bar magnet, or this needle, turns because its north pole end is attracted toward the Arctic region, or North Pole, of Earth. Thus, the *south* pole of Earth's magnetic field must be located toward the North Pole. Logically, we then should call the pole there a south pole. However, because we call that direction north, we are trapped into the statement that Earth has a *geomagnetic north pole* in that direction.

With more careful measurement we would find that in the northern hemisphere, the magnetic field lines of Earth generally point down into Earth and toward the Arctic. In the southern hemisphere, they generally point up out of Earth and away from the Antarctic—that is, away from Earth's *geomagnetic south pole*.

TOUCHSTONE EXAMPLE 29-1: Proton in a Magnetic Field

A uniform magnetic field $\vec{B}$, with magnitude 1.2 mT, is directed vertically upward throughout the volume of a laboratory chamber. A proton with kinetic energy 5.3 MeV enters the chamber, moving horizontally from south to north. What is the magnitude of the magnetic deflecting force acting on the proton as it enters the chamber? The proton mass is 1.67×10^{-27} kg. (Neglect Earth's magnetic field.)

SOLUTION ▪ Because the proton is charged and moving through a magnetic field, a magnetic force $\vec{F}^{\text{mag}}$ can act on it. The **Key Idea** here is that, because the initial direction of the proton's velocity is not along a magnetic field line, $\vec{F}^{\text{mag}}$ is not simply zero. To find the magnitude of $\vec{F}^{\text{mag}}$, we can use Eq. 29-3 provided we first find the proton's speed $|\vec{v}| = v$. We can find v from the given

kinetic energy, since $K = \frac{1}{2}mv^2$. Solving for $|\vec{v}|$, we find

$$v = \sqrt{\frac{2K}{m}} = \sqrt{\frac{(2)(5.3 \text{ MeV})(1.60 \times 10^{-13} \text{ J/MeV})}{1.67 \times 10^{-27} \text{ kg}}}$$

$$= 3.2 \times 10^7 \text{ m/s}.$$

Equation 29-3 then yields

$$F^{\text{mag}} = |q|vB_{\sin}\phi$$

$$= (1.60 \times 10^{-19} \text{ C})(3.2 \times 10^7 \text{ m/s})$$

$$\times (1.2 \times 10^{-3} \text{ T})(\sin 90°)$$

$$= 6.1 \times 10^{-15} \text{ N}. \qquad \text{(Answer)}$$

This may seem like a small force, but it acts on a particle of small mass, producing a large magnitude of acceleration; namely,

$$a = \frac{F^{mag}}{m} = \frac{6.1 \times 10^{-15}\,\text{N}}{1.67 \times 10^{-27}\,\text{kg}} = 3.7 \times 10^{12}\,\text{m/s}^2.$$

To find the direction of $\vec{F}^{mag}$, we use the **Key Idea** that $\vec{F}^{mag}$ has the direction of the cross product $q\vec{v} \times \vec{B}$. Because the charge q is positive, $\vec{F}^{mag}$ must have the same direction as $\vec{v} \times \vec{B}$, which can be determined with the right-hand rule for cross products (as in Fig. 29-4b). We know that $\vec{v}$ is directed horizontally from south to north and $\vec{B}$ is directed vertically up. The right-hand rule shows us that the deflecting force $\vec{F}^{mag}$ must be directed horizontally from west to east, as Fig. 29-8 shows. (The array of dots in the figure represents a magnetic field directed out of the plane of the figure. An array of Xs would have represented a magnetic field directed into that plane.)

If the charge of the particle were negative, the magnetic deflecting force would be directed in the opposite direction—that is, horizontally from east to west. This is predicted automatically by Eq. 29-2, if we substitute a negative value for q.

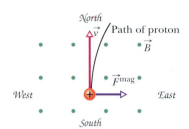

FIGURE 29-8 ■ An overhead view of a proton moving from south to north with velocity $\vec{v}$ in a chamber. A magnetic field is directed vertically upward in the chamber, as represented by the array of dots (which resemble the tips of arrows). The proton is deflected toward the east.

29-5 A Circulating Charged Particle

Remember that when we studied projectile motion we found that the (vertical) gravitational acceleration had no effect on the horizontal velocity of the projectile. Furthermore, when we studied uniform circular motion, we found that the (radial) centripetal acceleration only changed the direction of the object's velocity (keeping it moving in a circle), but did not speed it up or slow it down. This is a general relationship: The component of acceleration that is perpendicular to the direction of velocity only changes the direction of the velocity, not the magnitude.

We have a similar situation here. If we have a charged particle whose size is small enough to ignore, the magnetic force the particle feels is always perpendicular to its velocity and not its magnitude. As we established earlier, if the velocity and magnetic field are perpendicular (and there are no other forces on the particle), the particle will move in a circle.

If a particle moves in a circle at constant speed, we can be sure that the net force acting on the particle is constant in magnitude and is centripetal. That is, the force points toward the center of the circle, always perpendicular to the particle's velocity. Think of a stone tied to a string and whirled in a circle on a smooth horizontal surface, or of a satellite moving in a circular orbit around the Earth. In the first case, the tension in the string provides the necessary force and centripetal acceleration. In the second case, Earth's gravitational attraction provides the force and acceleration.

Figure 29-9 shows another example of a centripetal magnetic force: A beam of electrons is projected into a chamber by an *electron gun* G. The electrons enter in the plane of the page with speed v and move in a region of uniform magnetic field $\vec{B}$ directed out of the plane of the figure. As a result, a magnetic force $\vec{F}^{mag} = q\vec{v} \times \vec{B}$ continually deflects the electrons, and because the particle's velocity, $\vec{v}$, and the magnetic field it passes through, $\vec{B}$, are always perpendicular to each other, this deflection causes the electrons to follow a circular path. The path is visible in the photo because atoms of gas in the chamber emit light when some of the circulating electrons collide with them.

We would like to determine the parameters that characterize the circular motion of these electrons, or of any particle having an amount of charge $|q|$ and mass m moving perpendicular to a uniform magnetic field $\vec{B}$ at speed v. From Eq. 29-3, the force acting on the particle has a magnitude of $|q|vB$. From Newton's Second Law ($\vec{F} = m\vec{a}$) applied to uniform circular motion (Eq. 5-34),

$$F^{mag} = m\frac{v^2}{r}, \tag{29-6}$$

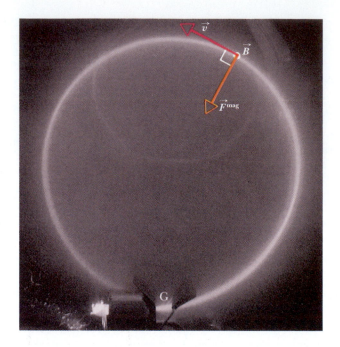

FIGURE 29-9 ■ Electrons circulating in a chamber containing gas at low pressure (their path is the glowing circle). A uniform magnetic field $\vec{B}$, pointing directly out of the plane of the page, fills the chamber. Note the radially directed magnetic force $\vec{F}^{\,\text{mag}}$; for circular motion to occur, $\vec{F}^{\,\text{mag}}$ must point toward the center of the circle. Use the right-hand rule for cross products to confirm that $\vec{F}^{\,\text{mag}} = q\vec{v} \times \vec{B}$ gives $\vec{F}^{\,\text{mag}}$ the proper direction. (Don't forget to incude the sign of q.)

we have

$$|q|vB = \frac{mv^2}{r}. \tag{29-7}$$

Solving for r, we find the radius of the circular path as

$$r = \frac{mv}{|q|B} \qquad \text{(radius of circular path)}. \tag{29-8}$$

The period T (the time for one full revolution) is equal to the circumference divided by the speed:

$$T = \frac{2\pi r}{v} = \frac{2\pi}{v}\frac{mv}{|q|B} = \frac{2\pi m}{|q|B} \qquad \text{(period)}. \tag{29-9}$$

The frequency f (the number of revolutions per unit time) is

$$f = \frac{1}{T} = \frac{|q|B}{2\pi m} \qquad \text{(frequency)}. \tag{29-10}$$

The angular frequency ω of the motion is then

$$\omega = 2\pi f = \frac{|q|B}{m} \qquad \text{(angular or cyclotron frequency)}. \tag{29-11}$$

The quantities T, f, and ω do not depend on the speed of the particle (provided that speed is much less than the speed of light). Fast particles move in large circles and slow ones in small circles, but all particles with the same charge-to-mass ratio q/m take the same time T (the period) to complete one round trip. A bigger velocity makes the particle travel in a larger circle. The increase in speed is exactly compensated by the increase in distance, so the time it takes to go around the circle is the same. We see later that this plays an important role in the construction of a charged particle accelerator known as a **cyclotron.** Using Eq. 29-2, you can show that if you are looking in the direction of $\vec{B}$, the direction of rotation for a positive particle is always counterclockwise; the direction for a negative particle is always clockwise.

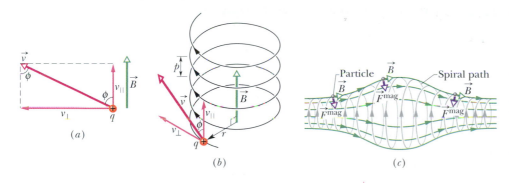

FIGURE 29-10 ■ (*a*) A charged particle moves in a uniform magnetic field $\vec{B}$, its velocity $\vec{v}$ making an angle ϕ with the field direction. (*b*) The particle follows a helical path, of radius r and pitch p. (*c*) A charged particle spiraling in a nonuniform magnetic field. (The particle can become trapped, spiraling back and forth between the strong field regions at either end.) Note that the magnetic force vectors at the left and right sides have a component pointing toward the center of the figure.

Helical Paths

As we discussed in regard to the electrons and positron in the bubble chamber of Figure 29-5, if the velocity of a charged particle moving through a magnetic field is changing, the particle will move in a shrinking spiral, rather than a circle. One way this can happen is for the particle to be slowed by frictional or other forces. Furthermore, if the velocity of a charged particle has a component parallel to the (uniform) magnetic field, the particle will move in a helical path about the direction of the field vector. Figure 29-10*a*, for example, shows the velocity vector $\vec{v}$ of such a particle resolved into two components, one parallel to $\vec{B}$ and one perpendicular to it:

$$v_{\parallel} = |\vec{v}|\cos\phi \quad \text{and} \quad v_{\perp} = |\vec{v}|\sin\phi. \tag{29-12}$$

The parallel component determines the *pitch p* of the helix—that is, the distance between adjacent turns (Fig. 29-10*b*). The perpendicular component determines the radius of the helix and is the quantity to be substituted for $|\vec{v}|$ in Eq. 29-8.

Figure 29-10*c* shows a charged particle spiraling in a nonuniform magnetic field. The more closely spaced field lines at the left and right sides indicate that the magnetic field is stronger there. When the field at an end is strong enough, the particle "reflects" from that end. If the particle reflects from both ends, it is said to be trapped in a *magnetic bottle*.

Confining Particles in a Tokomak Reactor

In the chapter opener we explained that in order to induce fusion reactions capable of releasing large amounts of energy, we must fuse light atoms together. To do this we need to confine ions having very high energy, and hence high temperature. Magnetic fields are ideal for containing the ions because both the ions and the electrons are charged and will spiral along magnetic field lines instead of hitting the walls of a containment vessel.

Scientists have not yet been able to confine charged particles at high enough temperatures to achieve controlled fusion. However, experiments reveal that one of the most effective configurations of magnetic field lines for containing the light atomic ions is shaped like a torus. A torus is basically a donut shape. The containment vessel of the Joint European Torus, commonly known as a tokomak, is shown at the beginning of this chapter. In a tokomak reactor, the magnetic field is produced by a series of magnetic coils that are evenly spaced around the torus-shaped containment vessel as shown in Fig. 29-11. The magnetic field lines form continuous loops inside the ring of the torus. In theory, when a tokomak is working properly, the high temperature ions and electrons should revolve in helical paths around the field lines. An ion can then travel in a continuous loop until it undergoes a fusion reaction with another ion.

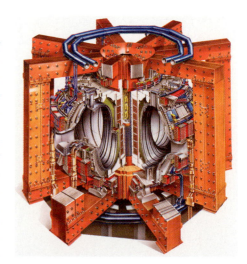

FIGURE 29-11 ■ A cutaway drawing of the JET tokomak showing the donut shaped containment vessel and surrounding magnetic coils.

Particles Trapped in the Earth's Magnetic Field

The terrestrial magnetic field acts as a magnetic bottle, trapping electrons and protons; the trapped particles form the *Van Allen radiation belts,* which loop well above the Earth's atmosphere between Earth's north and south geomagnetic poles. These particles bounce back and forth, from one end of this magnetic bottle to the other, within a few seconds.

When a large solar flare shoots additional energetic electrons and protons into the radiation belts, an electric field is produced in the region where electrons normally reflect. This field eliminates the reflection and instead drives electrons down into the atmosphere, where they collide with atoms and molecules of air, causing that air to emit light. This light forms the aurora—a curtain of light that hangs down to an altitude of about 100 km. Green light is emitted by oxygen atoms, and pink light is emitted by nitrogen molecules, but often the light is so dim that we perceive only white light.

READING EXERCISE 29-2: The figure shows the circular paths of two particles that travel at the same speed in a uniform magnetic field $\vec{B}$, which is directed into the page. One particle is a proton; the other is an electron (which is less massive). The relative sizes of the circles are not to scale. (a) Which particle follows the smaller circle, and (b) does that particle travel clockwise or counterclockwise? ■

TOUCHSTONE EXAMPLE 29-2: Mass Spectrometer

Figure 29-12 shows the essentials of a *mass spectrometer,* which can be used to measure the mass of an ion; an ion of mass m (to be measured) and charge q is produced in source S. The initially stationary ion is accelerated by the electric field due to a potential difference ΔV. The ion leaves S and enters a separator chamber in which a uniform magnetic field $\vec{B}$ is perpendicular to the path of the ion. The magnetic field causes the ion to move in a semicircle, striking (and thus altering) a photographic plate at distance x from the entry slit. Suppose that in a certain trial $B = 80.000$ mT and $\Delta V = 1000.0$ V, and ions of charge $q = +1.6022 \times 10^{-19}$ C strike

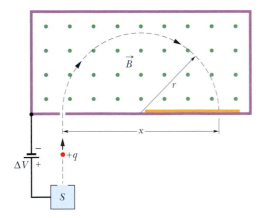

FIGURE 29-12 ■ Essentials of an early model of a mass spectrometer. A positive ion, after being accelerated from its source S by potential difference ΔV, enters a chamber of uniform magnetic field $\vec{B}$. There it travels through a semicircle of radius r and strikes a photographic plate at a distance x from where it entered the chamber.

the plate at $x = 1.6254$ m. What is the mass m of the individual ions, in unified atomic mass units (1 u = 1.6605×10^{-27} kg)?

SOLUTION ■ One **Key Idea** here is that, because the (uniform) magnetic field causes the (charged) ion to follow a circular path, we can relate the ion's mass m to the path's radius r with Eq. 29-8 ($r = m|\vec{v}|/|q\vec{B}|$). From Fig. 29-12 we see that $r = x/2$, and we are given the magnitude $|\vec{B}|$ of the magnetic field. However, we don't know the ion's speed v in the magnetic field, after it has been accelerated due to the potential difference ΔV.

To relate v and ΔV, we use the **Key Idea** that mechanical energy ($E^{\text{mec}} = K + U$) of the mass spectrometer system is conserved during the acceleration. When the ion emerges from the source, its kinetic energy is approximately zero. At the end of the acceleration, its kinetic energy is $\frac{1}{2}mv^2$. Also, during the acceleration, the positive ion moves through a change in potential of $-\Delta V$. Thus, because the ion has positive charge q, its potential energy changes by $-q\Delta V$. If we now write the conservation of the system's mechanical energy as

$$\Delta K + \Delta U = 0,$$

we get

$$\tfrac{1}{2}mv^2 - q\Delta V = 0$$

or

$$v = \sqrt{\frac{2|q\Delta V|}{m}}. \tag{29-13}$$

Substituting this into Eq. 29-8 gives us

$$r = \frac{mv}{|q\vec{B}|} = \frac{m}{|q|B}\sqrt{\frac{2|q\Delta V|}{m}} = \frac{1}{B}\sqrt{\frac{2m|\Delta V|}{|q|}}.$$

Thus,

$$x = 2r = \frac{2}{B}\sqrt{\frac{2m|\Delta V|}{|q|}}.$$

Solving this for m and substituting the given data yield

$$m = \frac{B^2|q|x^2}{8|\Delta V|}$$

$$= \frac{(0.080000\ \text{T})^2\,(1.6022\times10^{-19}\ \text{C})(1.6254\ \text{m})^2}{8(1000.0\ \text{V})}$$

$$= 3.3863\times10^{-25}\ \text{kg} = 203.93\ \text{u}. \qquad \text{(Answer)}$$

29-6 Crossed Fields: Discovery of the Electron

As we have seen, both an electric field $\vec{E}$ and a magnetic field $\vec{B}$ can produce a force on a charged particle. When the two fields are perpendicular to each other, they are said to be *crossed fields*. Here we shall examine what happens to charged particles— namely, electrons—as they move through crossed fields. We use as our example the experiment that led to the discovery of the electron in 1897 by J. J. Thomson at Cambridge University.

Figure 29-13 shows a modern, simplified version of Thomson's experimental apparatus—a *cathode ray tube* (which is like the picture tube in a standard television set). Charged particles (which we now know as electrons) are emitted by a hot filament at the rear of the evacuated tube and are accelerated by an applied potential difference ΔV. After the electrons pass through a slit in screen C, they form a narrow beam. They then pass through a region of crossed $\vec{E}$ and $\vec{B}$ fields, headed toward a fluorescent screen S, where they produce a spot of light (on a television screen the spot is part of the picture). The forces on the charged particles in the crossed-fields region can deflect them from the center of the screen. By controlling the magnitudes and directions of the fields, Thomson could thus control where the spot of light appeared on the screen. Recall that the force on a negatively charged particle due to an electric field is directed opposite the field. Thus, for the particular field arrangement of Fig. 29-13, electrons are forced up the page by the electric field $\vec{E}$ and down the page by the magnetic field $\vec{B}$; that is, the forces are *in opposition*. Thomson's procedure was equivalent to the following series of steps:

1. Set $\vec{E} = 0$ N/C and $\vec{B} = 0$ T and note the position of the spot on screen S due to the undeflected beam.

2. Turn on $\vec{E}$ and measure the resulting beam deflection.

3. Maintaining $\vec{E}$, now turn on $\vec{B}$ and adjust its value until the beam returns to the undeflected position. (With the forces in opposition, they can be made to cancel.)

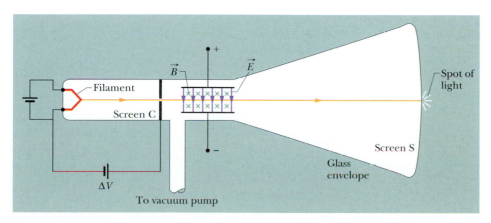

FIGURE 29-13 ■ A modern version of J. J. Thomson's apparatus for measuring the ratio of mass to the amount of charge of an electron. The electric field $\vec{E}$ is established by connecting a battery across the deflecting-plate terminals. The magnetic field $\vec{B}$ is set up by means of a current in a system of coils (not shown). The magnetic field shown is into the plane of the figure, as represented by the array of Xs (which resemble the feathered ends of arrows).

We discussed the deflection of a charged particle moving perpendicular to an electric field $\vec{E}$ between two plates (step 2 here) in Touchstone Example 23-4. We found that the magnitude of the deflection of the particle at the far end of the plates is

$$|\Delta y| = \frac{|q|EL^2}{2mv^2}, \tag{29-14}$$

where v is the particle's initial speed (which was v_x in Touchstone Example 23-4), m its mass, and q its charge, and L is the length of the plates. So long as the particle's deflection is small, we can apply this same equation to the beam of electrons in Fig. 29-13; if necessary, we can calculate the deflection by measuring the deflection of the beam on screen S and then working back to calculate the deflection y at the end of the plates. (Because the direction of the deflection is set by the sign of the particle's charge, Thomson was able to show that the particles lighting up his screen were negatively charged.)

When the two fields in Fig. 29-13 are adjusted so that the two deflecting forces cancel (step 3), we have from Eqs. 29-1 and 29-3,

$$|q|E = |q|vB\sin(90°) = |q|vB,$$

so the particle speed v is given by the ratio of the field magnitudes

$$v = \frac{E}{B}. \tag{29-15}$$

Thus, the crossed fields allow us to measure the speed of the charged particles passing through them. Substituting Eq. 29-15 for $|\vec{v}|$ in Eq. 29-14 and rearranging yield

$$\frac{m}{|q|} = \frac{B^2L^2}{2|\Delta y|E}, \tag{29-16}$$

in which all quantities on the right can be measured. Thus, the crossed fields allow us to measure the mass-charge amount ratio $m/|q|$ of the particles moving through Thomson's apparatus.

Thomson claimed that these particles are found in all matter. He also claimed that they are lighter than the lightest known atom (hydrogen) by a factor of more than 1000. (The exact ratio proved later to be 1836.15.) His $m/|q|$ measurement, coupled with the boldness of his two claims, is considered to be the moment of "discovery of the electron."

READING EXERCISE 29-3: The figure shows four directions for the velocity vector $\vec{v}$ of a positively charged particle moving through a uniform electric field $\vec{E}$ (directed out of the page and represented by an encircled dot) and a uniform magnetic field $\vec{B}$ (pointing to the left). (a) Rank directions 1, 2, 3, and 4 according to the magnitude of the net force on the particle, greatest first. (b) Of all four directions, which might result in a net force of zero?

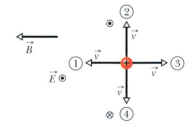

29-7 The Hall Effect

In Chapters 22 and 26 we claimed that currents in solid conductors are due to moving electrons, and that the positive nuclei are at rest. What evidence do we have for this claim? In the late 1870s, Edwin H. Hall, a 24-year-old graduate student at the

Johns Hopkins University, investigated the deflection of electric current passing through copper wire when the wire is placed in a magnetic field. The result of his work, which is called the **Hall effect** after him, allows us to answer important questions about the nature of charge carriers. For example, Hall's findings allowed him to determine whether charge carriers in a conductor are positive or negative. In addition, Hall's measurements enabled him to deduce the number of charge carriers per unit volume contained in a given conductor.

What happens to a current-carrying metal wire in a magnetic field if the charge carriers are positive and negative charges are at rest? Figure 29-14a shows a copper strip of width d, carrying a current i that is assumed to be made up of positive charge carriers (the convention at the time) moving from the top of the figure to the bottom. The charge carriers drift (with an average speed $|\langle\vec{v}\rangle|$) in the direction of the current, from top to bottom. At the instant shown in Fig. 29-14a, an external magnetic field $\vec{B}$, pointing into the plane of the figure, has just been turned on. From Eq. 29-2 we see that a deflecting magnetic force $\vec{F}^{\text{mag}}$ will act on each drifting positive charge, pushing it toward the right edge of the strip.

As time goes on, positive charges pile up on the right edge of the strip, leaving uncompensated negative charges in fixed positions at the left edge. The separation of positive and negative charges produces a constant electric field $\vec{E}$ within the strip, pointing from right to left. This field exerts an average electrostatic force $\langle\vec{F}^{\text{elec}}\rangle$ on a typical positive charge, tending to push it back toward the left.

An equilibrium quickly develops in which the electric force on each positive charge (pushing left) builds up until it just cancels the magnetic force (pushing right). When this happens, as Fig. 29-14b shows, the force due to $\vec{B}$ and the force due to $\vec{E}$ are in balance. The drifting positive charges then move along the strip toward the bottom of the page at an average velocity $\langle\vec{v}\rangle$, with no further collection of positive charge on the right edge of the strip and thus no further increase in the electric field $\vec{E}$.

A *Hall potential difference* ΔV is associated with the electric field across strip width d. Because the field is constant, we use Eq. 25-39 to get

$$|\Delta V| = Ed. \qquad (29\text{-}17)$$

By connecting a voltmeter across the width, we can measure the potential difference between the two edges of the strip. Moreover, the voltmeter can tell us which edge is at higher potential. This information, in turn, tells us whether our charge carriers are positive or negative.

So what do we find? For the situation of Fig. 29-14a, we find that the *left* edge is at *higher* potential, meaning we have a buildup of positive charge there. This result is inconsistent with our assumption that the charge carriers are positive.

Suppose we make the opposite assumption, that the charge carriers in current i are negative, as shown in Fig. 29-15. The negative charge carriers drift (with an average speed $|\langle\vec{v}\rangle|$) in the *opposite* direction of the conventional current, from bottom to top. You can use the magnetic force law (Eq. 29-2) to convince yourself that as these charge carriers move from bottom to top in the strip, they are pushed to the right edge by $\vec{F}^{\text{mag}}$ and thus that the *left* edge is at higher potential. Because that last statement is in fact what we actually observe with a voltmeter, we conclude that the charge carriers must be negative.

Now for the quantitative part. When the electric and magnetic forces are in balance (Fig. 29-14b), Eqs. 29-1 and 29-3 give us a relationship between the magnitudes of the electric and magnetic fields:

$$eE = e|\langle\vec{v}\rangle|B, \qquad (29\text{-}18)$$

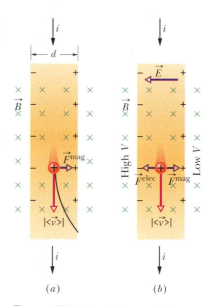

FIGURE 29-14 ■ What would happen if a positive current were to flow through a strip of copper immersed in a magnetic field $\vec{B}$? (a) As soon as the magnetic field is turned on, the positive charges follow a curved path as shown. (b) A short time later positive charges pile up on the right side of the strip. Thus, the right side of the strip has a higher potential than the left side. Since the higher potential is observed on the left not the right, we conclude that the *charge carriers are not positive.*

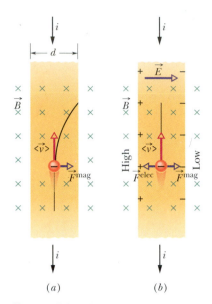

FIGURE 29-15 ■ What should happen when the conventional current, i, actually consists of a negative electron current flowing in the opposite direction? (*a*) As soon as the magnetic field is turned on, electrons follow the curved path shown. (*b*) A short time later negative charges pile up on the right side of the strip so that a higher potential develops on the left. Since this prediction matches experimental findings we must conclude that the *charge carriers are negative.*

where e is the amount of the charge on the electron. From Eq. 26-21, the average or drift speed $|\langle \vec{v} \rangle|$ is

$$|\langle \vec{v} \rangle| = \frac{|\vec{J}|}{ne} = \frac{|i|}{neA}, \tag{29-19}$$

in which $|\vec{J}|(=|i|/A)$ is the current density in the strip, A is the cross-sectional area of the strip, and n is the *number density* of charge carriers (their number per unit volume).

In Eq. 29-18, substituting $|\Delta V|/d$ for E (Eq. 29-17) and substituting for $|\langle \vec{v} \rangle|$ with the rightmost term in Eq. 29-19, we obtain

$$n = \frac{|i|B}{e\ell|\Delta V|}, \tag{29-20}$$

in which $\ell = A/d$ is the thickness of the strip. With this equation we can find n from measurable quantities.

It is also possible to use the Hall effect to measure directly the average or drift speed $|\langle \vec{v} \rangle|$ of the charge carriers, which you may recall is of the order of centimeters per hour. In this clever experiment, the metal strip is moved mechanically through the magnetic field in a direction opposite that of the drift velocity of the charge carriers. The speed of the moving strip is then adjusted until the Hall potential difference vanishes. At this condition, with no Hall effect, the velocity of the charge carriers *with respect to the laboratory frame* must be zero, so the velocity of the strip must be equal in magnitude but opposite in direction to the velocity of the negative charge carriers.

TOUCHSTONE EXAMPLE 29-3: Motional Potential Difference

Figure 29-16 shows a solid metal cube, of edge length $d = 1.5$ cm, moving in the positive y direction at a constant velocity $\vec{v}$ of magnitude 4.0 m/s. The cube moves through a uniform magnetic field $\vec{B}$ of magnitude 0.050 T directed toward positive z.

(a) Which cube face is at a lower electric potential and which is at a higher electric potential because of the motion through the field?

SOLUTION ■ One **Key Idea** here is that, because the cube is moving through a magnetic field $\vec{B}$, a magnetic force $\vec{F}^{\,\text{mag}}$ acts on its charged particles, including its conduction electrons. A second **Key Idea** is how $\vec{F}^{\,\text{mag}}$ causes an electric potential difference between certain faces of the cube. When the cube first begins to move

through the magnetic field, its electrons do also. Because each electron has charge $q = -e$ and is moving through a magnetic field with velocity $\vec{v}$, the magnetic force $\vec{F}^{\,\text{mag}}$ acting on it is given by Eq. 29-2. Because q is negative, the direction of $\vec{F}^{\,\text{mag}}$ is opposite the cross product $\vec{v} \times \vec{B}$, which is in the positive direction of the x axis in Fig. 29-16. Thus, $\vec{F}^{\,\text{mag}}$ acts in the negative direction of the x axis, toward the left face of the cube (which is hidden from view in Fig. 29-16).

Most of the electrons are fixed in place in the molecules of the cube. However, because the cube is a metal, it contains conduction electrons that are free to move. Some of those conduction electrons are deflected by $\vec{F}^{\,\text{mag}}$ to the left cube face, making that face negatively charged and leaving the right face positively charged. This charge separation produces an electric field $\vec{E}$ directed from the positively charged right face to the negatively charged left face. Thus, the left face is at a lower electric potential, and the right face is at a higher electric potential.

(b) What is the potential difference between the faces of higher and lower electric potential?

SOLUTION ■ The **Key Ideas** here are these:

1. The electric field $\vec{E}$ created by the charge separation produces an electric force $\vec{F}^{\,\text{elec}} = q\vec{E}$ on each electron. Because q is

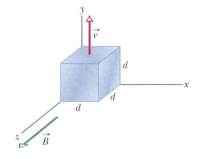

FIGURE 29-16 ■ A solid metal cube of edge length d moves at constant velocity $\vec{v}$ through a uniform magnetic field $\vec{B}$.

negative, this force is directed opposite the field $\vec{E}$—that is, toward the right. Thus on each electron, $\vec{F}^{\text{elec}}$ acts toward the right and $\vec{F}^{\text{mag}}$ acts toward the left.

2. When the cube had just begun to move through the magnetic field and the charge separation had just begun, the magnitude of $\vec{E}$ began to increase from zero. Thus, the magnitude of $\vec{F}^{\text{elec}}$ also began to increase from zero and was initially smaller than the magnitude $\vec{F}^{\text{mag}}$. During this early stage, the net force on any electron was dominated by $\vec{F}^{\text{mag}}$, which continuously moved additional electrons to the left cube face, increasing the charge separation.

3. However, as the charge separation increased, eventually magnitude $|\vec{F}^{\text{elec}}|$ became equal to magnitude $|\vec{F}^{\text{mag}}|$. The net force on any electron was then zero, and no additional electrons were moved to the left cube face. Thus, the magnitude of $\vec{F}^{\text{elec}}$ could not increase further, and the electrons were then in equilibrium.

We seek the potential difference ΔV between the left and right cube faces after equilibrium was reached (which occurred quickly). We can obtain the magnitude of ΔV with Eq. 29-17 ($|\Delta V| = Ed$)

provided we first find the magnitude $|\vec{E}| = E$ of the electric field at equilibrium. We can do so with the equation for the balance of force magnitudes ($|\vec{F}^{\text{elec}}| = |\vec{F}^{\text{mag}}|$).

For F^{elec}, we substitute $|q|E$. For F^{mag}, we substitute $|q|vB\sin\phi$ from Eq. 29-3. From Fig. 29-16, we see that the angle ϕ between v and B is 90°; so $\sin\phi = 1$. We can now write ($F^{\text{elec}} = F^{\text{mag}}$) as

$$|q|E = |q|vB\sin 90° = |q|vB.$$

This gives us $E = vB$, so Eq. 29-17 ($|\Delta V| = Ed$) becomes

$$|\Delta V| = |V_{\text{left}} - V_{\text{right}}| = vBd. \qquad (29\text{-}21)$$

Substituting known values gives us

$$|\Delta V| = (4.0 \text{ m/s})(0.050 \text{ T})(0.015 \text{ m})$$
$$= 0.0030 \text{ V} = 3.0 \text{ mV}.$$

Since the left face of the cube has excess negative charges, the right face is at a higher potential than the left face by 3.0 mV. (Answer)

29-8 Magnetic Force on a Current-Carrying Wire

We have just seen that a magnetic field exerts a sideways force on electrons moving in a wire. This force must then be transmitted to the wire itself, because the conduction electrons cannot escape sideways out of the wire.

In Fig. 29-17a, a vertical wire, carrying no current and fixed in place at both ends, extends through the gap between the vertical pole faces of a magnet represented by the shaded circle. The magnetic field between the faces is directed outward from the page. In Fig. 29-17b, a current is sent upward through the wire; the wire deflects to the right. In Fig. 29-17c, we reverse the direction of the current and the wire deflects to the left.

Figure 29-18 shows what happens inside the wire of Fig. 29-17. We see one of the conduction electrons, drifting downward with an assumed average (drift) speed $|\langle\vec{v}\rangle|$. Equation 29-3, in which we must put $\phi = 90°$, tells us that a force of magnitude $F^{\text{mag}} = e|\langle\vec{v}\rangle|B$ must act on a typical electron. From Eq. 29-2 we see that this force must be directed to the right. We expect then that the wire as a whole will experience a force to the right, in agreement with Fig. 29-17b.

If, in Fig. 29-18, we were to reverse *either* the direction of the magnetic field *or* the direction of the current, the force on the wire would reverse, being directed now to the left. Note too that it does not matter whether we consider negative charges drifting downward in the wire (the actual case) or positive charges drifting upward. The direction of the deflecting force on the wire is the same. We are safe then in dealing with a current of positive charge.

Consider a length L of the wire in Fig. 29-18. All the conduction electrons in this section of wire will drift past a plane that is parallel to xx' (shown in Fig. 29-18) in a time $\Delta t = L/|\langle\vec{v}\rangle|$. Thus, in that time the charge that will pass through the plane is given by

$$q = i\Delta t = i\frac{L}{|\langle\vec{v}\rangle|}.$$

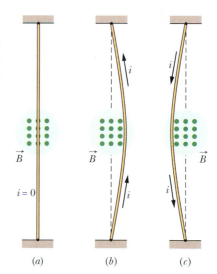

FIGURE 29-17 ■ A flexible wire passes between the pole faces of a magnet (only the farther pole face is shown). (a) Without current in the wire, the wire is straight. (b) With upward current, the wire is deflected rightward. (c) With downward current, the deflection is leftward. Connections for getting the current into one end of the wire and out of the other are not shown.

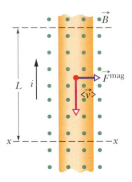

FIGURE 29-18 ■ A close-up view of a section of the wire of Fig. 29-17b. The current direction is upward, which means that electrons drift downward. A magnetic field that emerges from the plane of the page causes the electrons and the wire to be deflected to the right.

Substituting this into Eq. 29-3 yields the following expressions for the magnitude of the magnetic force

$$F^{mag} = |q| |\langle \vec{v} \rangle| B \sin\phi = \frac{|i| L |\langle \vec{v} \rangle| B}{|\langle \vec{v} \rangle|} \sin 90°$$

or

$$F^{mag} = |i| LB. \qquad (29\text{-}22)$$

This equation gives the magnetic force that acts on a length L of straight wire carrying a current i and immersed in a magnetic field $\vec{B}$ that is perpendicular to the wire.

If the magnetic field is *not* perpendicular to the wire, as in Fig. 29-19, the magnetic force is given by a generalization of Eq. 29-22:

$$\vec{F}^{mag} = |i| \vec{L} \times \vec{B} \qquad \text{(force on a current).} \qquad (29\text{-}23)$$

Here $\vec{L}$ is a *length vector* that has magnitude $|\vec{L}|$ and is directed along the wire segment in the direction of the (conventional) current. The magnitude of the magnetic field is

$$F^{mag} = |i| LB \sin\phi, \qquad (29\text{-}24)$$

where ϕ is the smaller angle between the directions of $\vec{L}$ and $\vec{B}$. The direction of $\vec{F}^{mag}$ is that of the cross product $\vec{L} \times \vec{B}$, because we take current i to be a positive quantity. Equation 29-23 tells us that $\vec{F}^{mag}$ is always perpendicular to the plane defined by $\vec{L}$ and $\vec{B}$, as indicated in Fig. 29-19.

Equation 29-23 is equivalent to Eq. 29-2 in that either can be taken as the defining equation for $\vec{B}$. In practice, we define $\vec{B}$ from Eq. 29-23. It is much easier to measure the magnetic force acting on a wire than that on a single moving charge.

If a wire is not straight or the field is not uniform, we can imagine it broken up into small straight segments and apply Eq. 29-23 to each short segment $d\vec{L}$. The force on the wire as a whole is then the vector sum of all the forces on the segments that make it up. In the differential limit, we can write

$$d\vec{F}^{mag} = i\, d\vec{L} \times \vec{B}, \qquad (29\text{-}25)$$

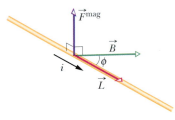

FIGURE 29-19 ■ A wire carrying current i makes an angle ϕ with magnetic field $\vec{B}$. The wire has length L in the field and length vector $\vec{L}$ (in the direction of the current). A magnetic force $\vec{F}^{mag} = i\vec{L} \times \vec{B}$ acts on the wire.

and we can find the resultant force on any given arrangement of currents by integrating Eq. 29-25 over that arrangement.

In using Eq. 29-25, bear in mind that there is no such thing as an isolated current-carrying wire segment of length $d\vec{L}$. There must always be a way to introduce the current into the segment at one end and take it out at the other end.

READING EXERCISE 29-4: The figure shows a current i through a wire in a uniform magnetic field $\vec{B}$, as well as the magnetic force $\vec{F}^{mag}$ acting on the wire. The field is oriented so that the magnitude force is a maximum. In what direction is the field?

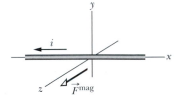

TOUCHSTONE EXAMPLE 29-4: Levitating a Wire

A straight, horizontal length of copper wire has a current $i = 28$ A through it. If this current is directed out of the page as shown in Fig. 29-20, what are the magnitude and direction of the minimum magnetic field $\vec{B}$ needed to suspend the wire—that is, to balance the gravitational force on it? The linear density (mass per unit length) is 46.6 g/m.

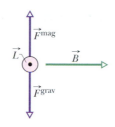

FIGURE 29-20 ■ A current-carrying wire (shown in cross section) can be made to "float" in a magnetic field. The current in the wire emerges from the plane of the page, and the magnetic field is directed to the right.

SOLUTION ■ One **Key Idea** is that, because the wire carries a current, a magnetic force $\vec{F}^{\text{mag}}$ can act on the wire if we place it in a magnetic field $\vec{B}$. To balance the downward gravitational force $\vec{F}^{\text{grav}}$ on the wire, we want $\vec{F}^{\text{mag}}$ to be directed upward (Fig. 29-20).

A second **Key Idea** is that the direction of $\vec{F}^{\text{mag}}$ is related to the directions of $\vec{B}$ and the wire's length vector $\vec{L}$ by Eq. 29-23. Because $\vec{L}$ is directed horizontally (and the current is taken to be positive), Eq. 29-23 and the right-hand rule for cross products tell us that $\vec{B}$ must be horizontal and rightward (in Fig. 29-20) to give the required upward $\vec{F}^{\text{mag}}$.

The magnitude of $\vec{F}^{\text{mag}}$ is given by Eq. 29-24 ($|\vec{F}^{\text{mag}}| = |i\vec{L}||\vec{B}|\sin\phi$). Because we want $\vec{F}^{\text{mag}}$ to balance $\vec{F}^{\text{grav}}$, we want

$$|i|LB\sin\phi = mg, \qquad (29\text{-}26)$$

where mg is the magnitude of $\vec{F}^{\text{grav}}$ and m is the mass of the wire. We also want the minimal field magnitude B for $\vec{F}^{\text{mag}}$ to balance $\vec{F}^{\text{grav}}$. Thus, we need to maximize $\sin\phi$ in Eq. 29-26. To do so, we set $\phi = 90°$, thereby arranging for $\vec{B}$ to be perpendicular to the wire. We then have $\sin\phi = 1$, so Eq. 29-26 yields a magnetic field magnitude of

$$|\vec{B}| = B = \frac{mg}{|i|L\sin\phi} = \frac{(m/L)g}{|i|}. \qquad (29\text{-}27)$$

We write the result this way because we know m/L, the linear density of the wire. Substituting known data then gives us a magnitude of

$$B = \frac{(46.6 \times 10^{-3}\ \text{kg/m})(9.8\ \text{m/s}^2)}{28\ \text{A}}$$
$$= 1.6 \times 10^{-2}\ \text{T}. \qquad \text{(Answer)}$$

This is about 160 times the strength of Earth's magnetic field. As stated in the second paragraph of this solution, the right-hand rule tells us that B must point to the right.

29-9   Torque on a Current Loop

Much of the world's work is done by electric motors. The forces that do this work are magnetic. In principle a direct current motor can be constructed from a single loop of current-carrying wire that is immersed in a magnetic field and is attached to a battery. If the current were to flow through the loop in the same direction all the time, the magnetic field would push on this loop in one direction at one instant of time, but would reverse the direction of the force when the loop was rotated halfway around. We would get a vibration that would quickly damp out. We can, however, get a continuous rotation if we use a connection, called a commutator, that reverses the current direction when the loop has gone halfway around (Fig. 29-21). Then, the force will continue to push the loop in the same direction and the motor will spin. Although many essential details have been omitted, the figure does suggest how the action of a magnetic field on a current loop produces rotary motion. To understand how the dc motor works in detail, we need to understand how a magnetic field can cause a current-carrying wire loop to rotate by exerting a torque on it.

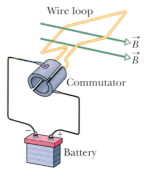

FIGURE 29-21 ■ The elements of an electric motor. A rectangular loop of wire, carrying a current and free to rotate about a fixed axis, is placed in a magnetic field. Magnetic forces on the wire produce a torque that rotates it. A commutator reverses the direction of the current every half-revolution so that the torque always acts in the same direction.

How a Current Loop Can Experience a Torque

Figure 29-22a shows a front view of a rectangular loop of sides a and b. The loop is carrying a current i and is immersed in a uniform magnetic field $\vec{B}$. We start our consideration of the torque on the loop with a special case in which the plane of the loop is parallel to the magnetic field as shown in Fig. 29-22a.

Let's use Eq. 29-24 to find the forces on each side of the loop for our special case. For sides 1 and 3 the vector $\vec{L}$ points in the direction of the current and has magnitude a. The angle between $\vec{L}$ and $\vec{B}$ for these is $\phi = 0°$. Thus, the magnitude of the forces acting on this side is

$$F_1^{\text{mag}} = F_3^{\text{mag}} = |i|aB\sin 0° = 0\ \text{N}.$$

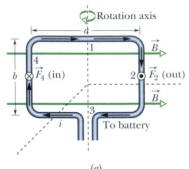

(a)
Front view (maximum torque)

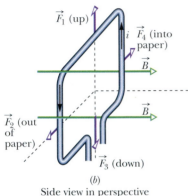

(b)
Side view in perspective
(no torque)

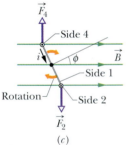

(c)
Top view (intermediate torque)

FIGURE 29-22 ■ A rectangular loop, of length a and width b and carrying a current i, is located in a uniform magnetic field. A torque $\vec{\tau}$ that is perpendicular to the magnetic field acts on the loop. The angle, ϕ, perpendicular or normal to the plane of the loop and the B-field varies. (a) The plane of the loop is aligned with the magnetic field so that $\phi = 90°$. (b) A perspective drawing of the loop after it has rotated to $\phi = 0°$ due to the torque exerted on it by the magnetic field. (c) A top view of the loop when it is part way between $\phi = 90°$ (part a) and $\phi = 0°$ (part b).

The situation is different for sides 2 and 4. For them, $\vec{L}$, which has magnitude b, is perpendicular to $\vec{B}$ so $\phi = 90°$. Thus, the forces $\vec{F}_2$ and $\vec{F}_4$ have the common magnitude given by

$$F_2^{\text{mag}} = F_4^{\text{mag}} = |i|\,bB\sin 90° = |i|\,bB. \qquad (29\text{-}28)$$

However, since the direction of the current is different on each of these sides, the right-hand rule tells us that these two forces point in opposite directions. The vector $\vec{F}_2$ points out of the page while the vector $\vec{F}_4$ points into the page. However, as Fig. 29-22a shows, these two forces do *not* share the same line of action so they *do* produce a net torque. The torque tends to rotate the loop toward an orientation for which the plane of the loop is perpendicular to the direction of the magnetic field $\vec{B}$. At $\phi = 90°$ that torque has a moment arm of magnitude $a/2$ about the central axis of the loop. The magnitude of the torque due to forces $\vec{F}_2$ and $\vec{F}_4$ is then (see Fig. 29-22a),

$$\tau_{90} = \frac{a}{2}F_2 + \frac{a}{2}F_4 = |i|\frac{a}{2}(bB) + |i|\frac{a}{2}(bB) = |i|\,abB. \qquad (29\text{-}29)$$

As the coil in Fig. 29-22a starts to rotate, the moment arm between sides 2 and 4 decreases, and it reaches zero when the loop is in the position shown in Fig. 29-22b. In general, the torque on the loop is given by

$$\tau' = |i|\,abB\sin\phi, \qquad (29\text{-}30)$$

where ϕ is the smaller angle normal to the area subtended by the loop and the external magnetic field (Fig. 22-22c).

Suppose we replace the single loop of current with a *coil* of N loops, or *turns*. Further, suppose that the turns are wound tightly enough that they can be approximated as all having the same dimensions and lying in a plane. Then the turns form a *flat coil* and a torque $\vec{\tau}'$ with the magnitude found in Eq. 29-29 acts on each of the turns. The total torque on the coil then has magnitude

$$\tau = N\tau' = N|i|\,AB\sin\phi, \qquad (29\text{-}31)$$

in which $A(= ab)$ is the area enclosed by the coil. Equation 29-31 holds for all flat coils, no matter what their shape, provided the magnetic field is uniform.

How a DC Motor Works

Consider the operation of a motor like that shown in Fig. 29-21. When the coil is at the point where the plane of the coil is perpendicular to the field direction so $\phi = 0°$, the polarity of the battery is suddenly reversed. Since the coil is accelerated by the initial torque on it, it sails past the point where $\phi = 0°$, and a new torque takes over and continues to rotate the coil in the same direction. This automatic reversal of the current occurs every half cycle and is accomplished with a commutator that electrically connects the rotating coil with the stationary contacts connected to the battery (or other power source).

29-10 The Magnetic Dipole Moment

We can describe the current-carrying coil of the preceding section with a single vector $\vec{\mu}$, its magnetic dipole moment. The direction of the magnetic dipole $\vec{\mu}$ is determined by another right hand rule similar to the one shown in Fig. 29-4. If you wrap your right hand around the coil in the direction of the positive current, your thumb points in the direction of the magnetic dipole $\vec{\mu}$.

We define the magnitude of $\mu = |\vec{\mu}|$ as

$$\mu = N|i|A \qquad \text{(magnetic moment magnitude)}, \qquad (29\text{-}32)$$

in which N is the number of turns in the coil, $|i|$ is the magnitude current through the coil, and A is the area enclosed by each turn of the coil. (Equation 29-32 tells us that the unit of $\vec{\mu}$ is the ampere-square meter.) Using $\vec{\mu}$, we can rewrite Eq. 29-31 for the magnitude of the torque on the coil due to a magnetic field as

$$\tau = |\vec{\mu}||\vec{B}|\sin\phi = \mu B \sin\phi, \qquad (29\text{-}33)$$

in which ϕ is the smallest angle between the vectors $\vec{\mu}$ and $\vec{B}$.

We can generalize this to the vector relation

$$\vec{\tau} = \vec{\mu} \times \vec{B}, \qquad (29\text{-}34)$$

which reminds us very much of the corresponding equation for the torque exerted by an *electric* field on an *electric* dipole—namely, Eq. 23-37:

$$\vec{\tau} = \vec{p} \times \vec{E}.$$

In each case the torque exerted by the external field—either magnetic or electric—is equal to the vector product of the corresponding dipole moment and the field vector.

A magnetic dipole in an external magnetic field has a **magnetic potential energy** that depends on the dipole's orientation in the field. For electric dipoles,

$$U(\theta) = -\vec{p} \cdot \vec{E}.$$

In strict analogy, we can write for the magnetic case

$$U(\theta) = -\vec{\mu} \cdot \vec{B}. \qquad (29\text{-}35)$$

A magnetic dipole has its lowest energy $(-|\vec{\mu}||\vec{B}|\cos\theta = -\mu B)$ when its dipole moment $\vec{\mu}$ is lined up with the magnetic field (Fig. 29-23). It has its highest energy $(-\mu B\cos 180° = +\mu B)$ when the vector $\vec{\mu}$ is directed opposite the field.

When a magnetic dipole rotates in the presence of a magnetic field from an initial orientation θ_1 to another orientation θ_2, the work $W_{B\to\mu}$ done on the dipole by the magnetic field is

$$W_{B\to\mu} = -\Delta U = -(U_2 - U_1), \qquad (29\text{-}36)$$

where U_2 and U_1 are calculated with Eq. 29-35. If an external torque acts on the dipole during the change in its orientation, then work $W_{\text{ext}\to\mu}$ is done on the dipole by the external torque. *If the dipole is stationary* before and after the change in its orientation, then work $W_{\text{ext}\to\mu}$ is the negative of the work done on the dipole by the field. Thus,

$$W_{\text{ext}\to\mu} = -W_{B\to\mu} = U_2 - U_1. \qquad (29\text{-}37)$$

So far, we have identified only a current-carrying coil as a magnetic dipole. However, a simple bar magnet is also a magnetic dipole, as is a rotating sphere of charge. Earth itself is (approximately) a magnetic dipole. And, most subatomic particles, including the electron, the proton, and the neutron, have magnetic dipole moments. As you will see in Chapter 32, all these quantities can be viewed as current loops. For comparison, some approximate magnetic dipole moments are shown in Table 29-2.

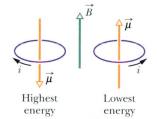

FIGURE 29-23 ■ The orientations of highest and lowest energy of a magnetic dipole in an external magnetic field $\vec{B}$. In each case, the direction of the current i determines the direction of the magnetic dipole moment $\vec{\mu}$ shown in Fig. 29-23 via the right-hand rule.

TABLE 29-2

Some Magnetic Dipole Moments

A small bar magnet	5 J/T
Earth	8.0×10^{22} J/T
A proton	1.4×10^{-26} J/T
An electron	9.3×10^{-24} J/T

READING EXERCISE 29-5: The figure shows four orientations, at angle θ, of a magnetic dipole moment $\vec{\mu}$ in a magnetic field. Rank the orientations according to (a) the magnitude of the torque on the dipole and (b) the potential energy of the dipole, greatest first.

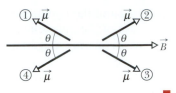

TOUCHSTONE EXAMPLE 29-5: Coil in an External Magnetic Field

Figure 29-24 shows a circular coil with 250 turns, an area A of 2.52×10^{-4} m^2, and a current of 100 μA. The coil is at rest in a uniform magnetic field of magnitude $|\vec{B}| = 0.85$ T, with its magnetic dipole moment $\vec{\mu}$ initially aligned with $\vec{B}$.

(a) In Fig. 29-24, what is the direction of the current in the coil?

SOLUTION ■ The **Key Idea** here is to apply the right-hand rule to the coil by curling your fingers around the current in the coil so your right thumb points in the $\vec{\mu}$ direction. Thus, in the wires on the near side of the coil—those we see in Fig. 29-24—the current is from top to bottom.

(b) How much work would the torque applied by an external agent have to do on the coil to rotate it 90° from its initial orien-

tation, so that $\vec{\mu}$ is perpendicular to $\vec{B}$ and the coil is again at rest?

SOLUTION ■ The **Key Idea** here is that the work $W_{ext \to \mu}$ done by the applied torque would be equal to the change in the coil's potential energy due to its change in orientation. From Eq. 29-37 ($W_{ext \to \mu} = U_2 - U_1$), we find

$$W_{ext \to \mu} = U(90°) - U(0°)$$

$$= -\mu B \cos 90° - (-\mu B \cos 0°) = 0 + \mu B$$

$$= \mu B.$$

Substituting for $\vec{\mu}$ from Eq. 29-32 ($\mu = N|i|A$), we find that

$$W_{ext \to \mu} = (N|i|AB)$$

$$= (250)(100 \times 10^{-6}\text{ A})(2.52 \times 10^{-4}\text{ m}^2)(0.85\text{ T})$$

$$= 5.356 \times 10^{-6}\text{ J} \approx 5.4\ \mu\text{J}. \qquad \text{(Answer)}$$

FIGURE 29-24 ■ A side view of a circular coil carrying a current and oriented so that its magnetic dipole moment $\vec{\mu}$ is aligned with magnetic field $\vec{B}$.

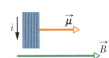

29-11 The Cyclotron

Physicists have been able to use their understanding of how charged particles behave in magnetic fields to develop devices that can accelerate protons to high speeds. These high-energy protons are extremely useful to scientists for several reasons. Collisions between energetic protons and matter allow them to learn about the nature of atomic and subatomic particles. High-energy protons and ions can also be used to create new radioactive elements. In addition, physicians can use high-energy protons to destroy tumors in cancer patients. In 1939, E. O. Lawrence was awarded a Nobel Prize in physics for the development of the cyclotron—the first of many magnetic accelerators capable of accelerating protons, ions, and electrons.

The principles that govern the operation of the cyclotron are quite simple. Figure 29-9 showed experimental evidence that a charged particle projected into an evacuated chamber perpendicular to a uniform magnetic field moves in a circular orbit. We used the magnetic force law (Eq. 29-2) to derive the frequency of revolution of the orbit. In Eq. 29-10 we found that $f = |q|B/2\pi m$. This is known as the cyclotron frequency, and its derivation had a rather surprising outcome. The frequency, f, with which a charged particle moves in its circular orbit depends only on its charge, its mass, and the magnetic field strength. So f is independent of speed. This is because a particle with low speed moves in a small circle whereas one with a higher speed moves in a larger circle. The particle speeds and orbital sizes are related in such a way that all charged particles take the same amount of time to make a revolution in a uniform magnetic field. (At least this is true for all speeds that are well below the speed of light.) Lawrence used the fact that the orbital frequency of a charged particle does not depend on its speed in the design of the cyclotron.

The original cyclotron was first used to accelerate protons. It consisted of two hollow semicircular disks shaped more or less like a capital D as shown in Fig. 29-25. In early cyclotrons, the dees, as they are called, were made of copper sheeting. The diameter of a dee was only about one meter. The dees were then placed in a vacuum chamber and oriented perpendicular to a large uniform magnetic field having a strength of a few teslas. There was a small gap between them. The dees were connected to an electrical oscillator that can alternate the potential difference across the gap between them at exactly the same frequency as an orbiting proton would have in the magnetic field. This arrangement is shown in Fig. 29-26.

To begin the operation of the original cyclotron, an oscillator was set at the cyclotron frequency. Then hydrogen gas was leaked into the vacuum chamber. Next a beam of high-energy electrons was injected into the center of the chamber so that other electrons were knocked out of hydrogen atoms. This ionization process produced protons. At a time when the oscillator caused the left dee to be at a lower potential than the right dee, the proton received a kick in the direction of the right dee. It moved into the right dee where the electric field was zero. However, the magnetic field penetrated the dee and caused the proton to start into a small, low-speed orbit. Only half a cycle later the proton reached the gap again. Since the oscillator was tuned to the cyclotron frequency, the potential of the right dee was now lower than that of the left dee and the proton got another kick as it crossed the gap. The proton then proceeded into another circular orbit that involved a larger speed and radius, given by Eq. 29-8,

$$r = \frac{mv}{|q|B}.$$

When the proton reached the gap again it completed one full cycle but so had the alternating voltage oscillator. Thus the proton got another kick. This process continued, with the circulating proton always being in step with the oscillations of the dee potential. When the proton finally spiraled out to the edge of the dee system, a deflector plate sent it out through a portal. The path of such a proton is shown in Fig. 29-27.

Recall that the key to the operation of the cyclotron is that the frequency f at which the proton circulates in the field (and that does not depend on its speed) must be equal to the fixed frequency f_{osc} of the electrical oscillator, or

$$f = f_{osc} \quad \text{(resonance condition)}. \tag{29-38}$$

This *resonance condition* says that, if the energy of the circulating proton is to increase, energy must be fed to it at a frequency f_{osc} that is equal to the natural frequency f at which the proton circulates in the magnetic field.

Combining Eqs. 29-10 and 29-38 allows us to write the resonance condition as

$$|q|B = 2\pi m f_{osc}. \tag{29-39}$$

For the proton, q and m are fixed. The oscillator (we assume) is designed to work at a single fixed frequency f_{osc}. We can then either "tune" the cyclotron by varying either $\vec{B}$ or f_{osc} until Eq. 29-39 is satisfied. Then many protons can circulate through the magnetic field and emerge as a beam.

If the cyclotron is powerful enough to accelerate protons, electrons, or ions to speeds close to that of light, relativistic effects come into play. In such cases the simple resonance condition between orbital and oscillator frequencies no longer hold. More sophisticated magnetic field-based high-energy accelerators called synchrotrons and betatrons have been designed. We introduce relativistic effects in Chapter 38 where we discuss special relativity.

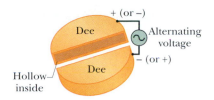

FIGURE 29-25 ■ Cyclotron dees are hollow semicircular metal containers that are open along their diameters.

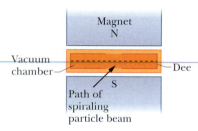

FIGURE 29-26 ■ Cutaway view of dees placed between the poles of a large electromagnet. The dotted line shows the plane in which the paths of the particles orbit.

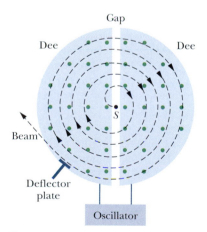

FIGURE 29-27 ■ Top view of dees showing the path of a charged particle beam in a cyclotron. Each time the particle passes through the gap three things happen: (1) the particle gets a kick and is accelerated to a higher speed, (2) the oscillator changes the sign of the gap's potential difference, and (3) the particle goes into a new semicircular orbit with a larger radius than before.

TOUCHSTONE EXAMPLE 29-6: Cyclotron

Suppose a cyclotron is operated at an oscillator frequency of 12 MHz and has a dee radius $R = 53$ cm.

(a) What is the magnitude of the magnetic field needed for deuterons to be accelerated in the cyclotron? A deuteron is the nucleus of deuterium, an isotope of hydrogen. It consists of a proton and a neutron and thus has the same charge as a proton. Its mass is $m = 3.34 \times 10^{-27}$ kg.

SOLUTION ■ The **Key Idea** here is that, for a given oscillator frequency f_{osc}, the magnetic field magnitude B required to accelerate any particle in a cyclotron depends on the ratio m/q of mass to charge for the particle, according to Eq. 29-39. For deuterons and the oscillator frequency $f_{osc} = 12$ MHz, we find

$$B = \frac{2\pi m f_{osc}}{q} = \frac{(2\pi)(3.34 \times 10^{-27} \text{ kg})(12 \times 10^6 \text{ s}^{-1})}{1.60 \times 10^{-19} \text{ C}}$$

$$= 1.57 \text{ T} \approx 1.6 \text{ T}. \qquad \text{(Answer)}$$

Note that, to accelerate protons, B would have to be reduced by a factor of 2, providing the oscillator frequency remained fixed at 12 MHz.

(b) What is the resulting kinetic energy of the deuterons?

SOLUTION ■ One **Key Idea** here is that the kinetic energy $\frac{1}{2}mv^2$ of a deuteron exiting the cyclotron is equal to the kinetic energy it had just before exiting, when it was traveling in a circular path with a radius approximately equal to the radius R of the cyclotron dees. A second **Key Idea** is that we can find the speed v of the deuteron in that circular path with Eq. 29-8 ($r = mv/|q|B$). Solving that equation for v, substituting R for r, and then substituting known data, we find

$$v = \frac{Rq B}{m} = \frac{(0.53 \text{ m})(1.60 \times 10^{-19} \text{ C})(1.57 \text{ T})}{3.34 \times 10^{-27} \text{ kg}}$$

$$= 3.99 \times 10^7 \text{ m/s}.$$

This speed corresponds to a kinetic energy of

$$K = \frac{1}{2}mv^2$$

$$= \frac{1}{2}(3.34 \times 10^{-27} \text{ kg})(3.99 \times 10^7 \text{ m/s})^2 \qquad \text{(Answer)}$$

$$= 2.7 \times 10^{-12} \text{ J},$$

or about 17 MeV.

Problems

SEC. 29-3 ■ DEFINING A MAGNETIC FIELD $\vec{B}$

1. Alpha Particle An alpha particle travels at a velocity $\vec{v}$ of magnitude 550 m/s through a uniform magnetic field $\vec{B}$ of magnitude 0.045 T. (An alpha particle has a charge of $+3.2 \times 10^{-19}$ C and a mass of 6.6×10^{-27} kg.) The angle between $\vec{v}$ and $\vec{B}$ is 52°. What are the magnitudes of (a) the force $\vec{F}^{\text{mag}}$ acting on the particle due to the field and (b) the acceleration of the particle due to $\vec{F}^{\text{mag}}$? (c) Does the speed of the particle increase, decrease, or remain equal to 550 m/s?

2. TV Camera An electron in a TV camera tube is moving at 7.20×10^6 m/s in a magnetic field of strength 83.0 mT. (a) Without knowing the direction of the field, what can you say about the greatest and least magnitudes of the force acting on the electron due to the field? (b) At one point the electron has an acceleration of magnitude 4.90×10^{14} m/s². What is the angle between the electron's velocity and the magnetic field?

3. Proton Traveling A proton traveling at 23.0° with respect to the direction of a magnetic field of strength 2.60 mT experiences a magnetic force of 6.50×10^{-17} N. Calculate (a) the proton's speed and (b) its kinetic energy in electron-volts.

4. Force on Charges An electron that has velocity

$$\vec{v} = (2.0 \times 10^6 \text{ m/s})\hat{i} + (3.0 \times 10^6 \text{ m/s})\hat{j}$$

moves through the magnetic field $\vec{B} = (0.030 \text{ T})\hat{i} - (0.15 \text{ T})\hat{j}$. (a) Find the force on the electron. (b) Repeat your calculation for a proton having the same velocity.

5. Television Tube Each of the electrons in the beam of a television tube has a kinetic energy of 12.0 keV. The tube is oriented so that the electrons move horizontally from geomagnetic south to geomagnetic north. The vertical component of Earth's magnetic field points down and has a magnitude of 55.0 μT. (a) In what direction will the beam deflect? (b) What is the magnitude of the acceleration of a single electron due to the magnetic field? (c) How far will the beam deflect in moving 20.0 cm through the television tube?

SEC. 29-5 ■ A CIRCULATING CHARGED PARTICLE

6. Accelerated from Rest An electron is accelerated from rest by a potential difference of 350 V. It then enters a uniform magnetic field of magnitude 200 mT with its velocity perpendicular to the field. Calculate (a) the speed of the electron and (b) the radius of its path in the magnetic field.

7. Field Perpendicular to Beam A uniform magnetic field is applied perpendicular to a beam of electrons moving at 1.3×10^6 m/s. What is the magnitude of the field if the electrons travel in a circular arc of radius 0.35 m?

8. Heavy Ions Physicist S. A. Goudsmit devised a method for measuring the masses of heavy ions by timing their periods of revolution in a known magnetic field. A singly charged ion of iodine makes 7.00 rev in a field of 45.0 mT in 1.29 ms. Calculate its mass, in atomic mass units. (Actually, the method allows mass measurements to be carried out to much greater accuracy than these approximate data suggest.)

9. Kinetic Energy An electron with kinetic energy 1.20 keV circles in a plane perpendicular to a uniform magnetic field. The orbit radius is 25.0 cm. Find (a) the speed of the electron, (b) the magnetic field, (c) the frequency, and (d) the period of the motion.

10. Circular Path An alpha particle ($q = +2e$, $m = 4.00$ u) travels in a circular path of radius 4.50 cm in a uniform magnetic field with magnitude $B = 1.20$ T. Calculate (a) its speed, (b) its period of revolution, (c) its kinetic energy in electron-volts, and (d) the potential difference through which it would have to be accelerated to achieve this energy.

11. Frequency of Revolution (a) Find the frequency of revolution of an electron with an energy of 100 eV in a uniform magnetic field of magnitude 35.0 μT. (b) Calculate the radius of the path of this electron if its velocity is perpendicular to the magnetic field.

12. Source of Electrons A source injects an electron of speed $v = 1.5 \times 10^7$ m/s into a uniform magnetic field of magnitude $B = 1.0 \times 10^{-3}$ T. The velocity of the electron makes an angle $\theta = 10°$ with the direction of the magnetic field. Find the distance d from the point of injection at which the electron next crosses the field line that passes through the injection point.

13. Beam of Electrons A beam of electrons whose kinetic energy is K emerges from a thin-foil "window" at the end of an accelerator tube. There is a metal plate a distance d from this window and perpendicular to the direction of the emerging beam (Fig. 29-28). Show that we can prevent the beam from hitting the plate if we apply a uniform magnetic field $\vec{B}$ such that its magnitude is

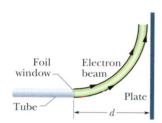

FIGURE 29-28 ■ Problem 13.

$$B \geq \sqrt{\frac{2mK}{e^2 d^2}},$$

in which m and e are the electron mass and charge. How should $\vec{B}$ be oriented?

14. Proton, Deuteron, Alpha A proton, a deuteron ($q = +e$, $m = 2.0$ u), and an alpha particle ($q = +2e$, $m = 4.0$ u) with the same kinetic energies enter a region of uniform magnetic field $\vec{B}$, moving perpendicular to $\vec{B}$. Compare the radii of their circular paths.

15. Nuclear Experiment In a nuclear experiment a proton with kinetic energy 1.0 MeV moves in a circular path in a uniform magnetic field. What energy must (a) an alpha particle ($q = +2e$, $m = 4.0$ u) and (b) a deuteron ($q = +e$, $m = 2.0$ u) have if they are to circulate in the same circular path?

16. Uniform Magnetic Field A proton of charge $+e$ and mass m enters a uniform magnetic field $\vec{B} = B\hat{i}$ with an initial velocity $\vec{v} = v_{1x}\hat{i} + v_{1y}\hat{j}$. Find an expression in unit-vector notation for its velocity $\vec{v}$ at any later time t.

17. Mass Spectrometer A certain commercial mass spectrometer (see Touchstone Example 29-2) is used to separate uranium ions of mass 3.92×10^{-25} kg and charge 3.20×10^{-19} C from related species. The ions are accelerated through a potential difference of 100 kV and then pass into a uniform magnetic field, where they are bent in a path of radius 1.00 m. After traveling through 180° and passing through a slit of width 1.00 mm and height 1.00 cm, they are collected in a cup. (a) What is the magnitude of the (perpendicular)

magnetic field in the separator? If the machine is used to separate out 100 mg of material per hour, calculate (b) the current of the desired ions in the machine and (c) the thermal energy produced in the cup in 1.00 h.

18. Half Circle In Fig 29-29, a charged particle moves into a region of uniform magnetic field B, goes through half a circle, and then exits that region. The particle is either a proton or an electron (you must decide which). It spends 130 ns within the region. (a) What is the magnitude $|\vec{B}|$? (b) If the particle is sent back through the magnetic field (along the same initial path) but with 2.00 times its previous kinetic energy, how much time does it spend within the field?

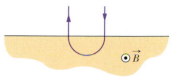

FIGURE 29-29 ■ Problem 18.

19. Positron A positron with kinetic energy 2.0 keV is projected into a uniform magnetic field $\vec{B}$ of magnitude 0.10 T, with its velocity vector making an angle of 89° with $\vec{B}$. Find (a) the period, (b) the pitch p, and (c) the radius r of its helical path.

20. Neutral Particle A neutral particle is at rest in a uniform magnetic field $\vec{B}$. At time $t = 0$ it decays into two charged particles, each of mass m. (a) If the charge of one of the particles is $+q$, what is the charge of the other? (b) The two particles move off in separate paths, both of which lie in the plane perpendicular to $\vec{B}$. At a later time the particles collide. Express the time from decay until collision in terms of m, $|\vec{B}|$, and $|q|$.

SEC. 29-6 ■ CROSSED FIELDS: DISCOVERY OF THE ELECTRON

21. Horizontal Motion An electron with kinetic energy 2.5 keV moves horizontally into a region of space in which there is a downward-directed uniform electric field of magnitude 10 kV/m. (a) What are the magnitude and direction of the (smallest) uniform magnetic field that will cause the electron to continue to move horizontally? Ignore the gravitational force, which is small. (b) Is it possible for a proton to pass through the combination of fields undeflected? If so, under what circumstances?

22. At One Instant A proton travels through uniform magnetic and electric fields. The magnetic field is $\vec{B} = (-2.5 \text{ mT})\hat{i}$. At one instant the velocity of the proton is $\vec{v} = (2000 \text{ m/s})\hat{j}$. At that instant, what is the magnitude of the net force acting on the proton if the electric field is (a) $(4.0 \text{ V/m})\hat{k}$ and (b) $(4.0 \text{ V/m})\hat{i}$?

23. Potential Difference An electron is accelerated through a potential difference of 1.0 kV and directed into a region between two parallel plates separated by 20 mm with a potential difference of 100 V between them. The electron is moving perpendicular to the electric field of the plates when it enters the region between the plates. What magnitude of uniform magnetic field, applied perpendicular to both the electron path and the electric field, will allow the electron to travel in a straight line?

24. Electric and Magnetic Field An electric field of magnitude 1.50 kV/m and a magnetic field of 0.400 T act on a moving electron to produce no net force. (a) Calculate the minimum speed $|\vec{v}|$ of the electron. (b) Draw a set of vectors $\vec{E}$, $\vec{B}$, and $\vec{v}$ that could yield the net force.

25. Ion Source An ion source is producing ions of ^{6}Li (mass = 6.0 u), each with a charge of $+e$. The ions are accelerated by a potential difference of 10 kV and pass horizontally into a region in which there is a uniform vertical magnetic field of magnitude $|\vec{B}| = 1.2$ T. Calculate the strength of the smallest electric field, to be set up over the same region, that will allow the ^{6}Li ions to pass through undeflected.

26. Initial Velocity An electron has an initial velocity of $(12.0 \text{ km/s})\hat{j} + (15.0 \text{ km/s})\hat{k}$ and a constant acceleration of $(2.00 \times 10^{12} \text{ m/s}^2)\hat{i}$ in a region in which uniform electric and magnetic fields are present. If $\vec{B} = (400 \ \mu\text{T})\hat{i}$, find the electric field $\vec{E}$.

SEC. 29-7 ■ THE HALL EFFECT

27. Field Ratio (a) In Fig 29-14, show that the ratio of the magnitudes of the Hall electric field $\vec{E}$ to the electric field $\vec{E}^{\text{curr}}$ responsible for moving charge (the current) along the length of the strip is

$$\frac{E}{E^{\text{curr}}} = \frac{B}{ne\rho}$$

where ρ is the resistivity of the material and n is the number density of the charge carriers and e is the amount of charge on the electron. (b) Compute this ratio numerically for Problem 28. (See Table 26-2.)

28. Strip of Copper A strip of copper 150 μm wide is placed in a uniform magnetic field $\vec{B}$ of magnitude 0.65 T, with $\vec{B}$ perpendicular to the strip. A current $i = 23$ A is then sent through the strip such that a Hall potential difference ΔV appears across the width of the strip. Calculate ΔV. (The number of charge carries per unit volume for copper is 8.47×10^{28} electrons/m^3.)

29. Metal Strip A metal strip 6.50 cm long, 0.850 cm wide, and 0.760 mm thick moves with constant velocity $\vec{v}$ through a uniform magnetic field of magnitude $|\vec{B}| = 1.20$ mT directed perpendicular to the strip, as shown in Fig. 29-30. A potential difference of 3.90 μV is measured between points x and y across the strip. Calculate the speed $|\vec{v}|$.

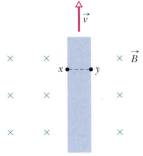

FIGURE 29-30 ■ Problem 29.

SEC. 29-8 ■ MAGNETIC FORCE ON A CURRENT-CARRYING WIRE

30. A Wire Carries a Current A wire 1.80 m long carries a current of 13.0 A and makes an angle of 35.0° with a uniform magnetic field of magnitude $B = 1.50$ T. Calculate the magnitude of the magnetic force on the wire.

31. Horizontal Conductor A horizontal conductor that is part of a power line carries a current of 5000 A from south to north. The magnitude of the Earth's magnetic field is 60.0 μT. The field is directed toward the north and is inclined downward at 70° to the horizontal. Find the magnitude and direction of the magnetic force on 100 m of the conductor due to Earth's field.

32. Along the x Axis A wire 50 cm long lying along the x axis carries a current of 0.50 A in the positive x direction. It passes through a magnetic field $\vec{B} = (0.0030 \text{ T})\hat{j} + (0.0100 \text{ T})\hat{k}$. Find the magnetic force on the wire.

33. A Wire of Length A wire of 62.0 cm length and 13.0 g mass is suspended by a pair of flexible leads in a uniform magnetic field of magnitude 0.440 T (Fig. 29-31). What are the magnitude and direction of the current required to remove the tension in the supporting leads?

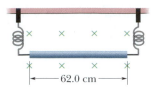

FIGURE 29-31 ■ Problem 33.

34. Electric Train Consider the possibility of a new design for an electric train. The engine is driven by the force on a conducting axle due to the vertical component of Earth's magnetic field. To produce the force, current is maintained down one rail, through a conducting wheel, through the axle, through another conducting wheel, and then back to the source via the other rail. (a) What amount of current is needed to provide a modest force of magnitude 10kN? Take the vertical component of Earth's field to be 10 μT and the length of the axle to be 3.0 m. (b) At what rate would electric energy be lost for each ohm of resistance in the rails? (c) Is such a train totally or just marginally unrealistic?

35. Copper Rod A 1.0 kg copper rod rests on two horizontal rails 1.0 m apart and carries a current of 50 A from one rail to the other. The coefficient of static friction between rod and rails is 0.60. What is the magnitude of the smallest magnetic field (not necessarily vertical) that would cause the rod to slide?

SEC. 29-9 ■ TORQUE ON A CURRENT LOOP

36. Current Loop A single-turn current loop, carrying a current of 4.00 A, is in the shape of a right triangle with sides 50.0, 120, and 130 cm. The loop is in a uniform magnetic field of magnitude 75.0 mT whose direction is parallel to the current in the 130 cm side of the loop. (a) Find the magnitude of the magnetic force on each of the three sides of the loop. (b) Show that the total magnetic force on the loop is zero.

37. Rectangular Coil Figure 29-32 shows a rectangular 20-turn coil of wire, of dimensions 10 cm by 5.0 cm. It carries a current of 0.10 A and is hinged along one long side. It is mounted in the xy plane, at 30° to the direction of a uniform magnetic field of magnitude 0.50 T. Find the magnitude and direction of the torque acting on the coil about the hinge line.

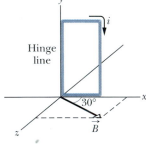

FIGURE 29-32 ■ Problem 37.

38. Arbitrarily Shaped Coil Prove that the relation $\tau = N|i|AB \sin \phi$ (Eq. 29-31) holds for closed loops of arbitary shape and not only for rectangular loops as in Fig. 29-22. (*Hint:* Replace the loop of arbitrary shape with an assembly of adjacent long, thin, approximately rectangular loops that are nearly equivalent to the loop of arbitrary shape as far as the distribution of current is concerned.)

39. Show That A length L of wire carries a current i. Show that if the wire is formed into a circular coil, then the magnitude of the maximum torque in a given magnetic field is developed when the coil has one turn only. Also show that maximum torque has the magnitude $\tau = L^2iB/4\pi$.

40. Zero Total Force A closed wire loop with current i is in a uniform magnetic field $\vec{B}$, with the plane of the loop at angle θ to the direction of $\vec{B}$. Show that the total magnetic force on the loop is zero. Does your proof also hold for a nonuniform magnetic field?

41. Wire Ring Figure 29-33 shows a wire ring of radius a that is perpendicular to the general direction of a radially symmetric, diverging magnetic field. The magnetic field at the ring is everywhere of the same magnitude $|\vec{B}|$, and its direction at the ring everywhere makes an angle θ with a normal to the plane of the ring. The twisted lead wires have no effect on the problem. Find the magnitude and direction of the force the field exerts on the ring if the ring carries a positive current i.

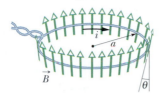

FIGURE 29-33 ■
Problem 41.

42. Maximum Torque A particle of charge q moves in a circular wire loop of radius a with speed $|\vec{v}|$. Find the maximum torque exerted on the loop by a uniform magnetic field of magnitude $|\vec{B}|$.

43. Wooden Cylinder Figure 29-34 shows a wooden cylinder with mass $m = 0.250$ kg and length $L = 0.100$ m, with $N = 10.0$ turns of wire wrapped around it longitudinally, so that the plane of the wire coil contains the axis of the cylinder. Also the plane of the coil is parallel to the inclined plane. There is a vertical, uniform magnetic field of magnitude 0.500 T. What is the least amount of current $|i|$ through the coil that will prevent the cylinder from rolling down a plane inclined at an angle θ to the horizontal?

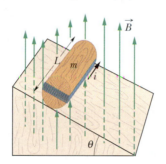

FIGURE 29-34 ■
Problem 43.

SEC. 29-10 ■ THE MAGNETIC DIPOLE MOMENT

44. Earth's Moment The magnitude of magnetic dipole moment of Earth is 8.00×10^{22} J/T. Assume that this is produced by charges flowing in Earth's molten outer core. If the radius of their circular path is 3500 km, calculate the amount of current associated with each moving charge.

45. Calculate the Current A circular coil of 160 turns has a radius of 1.90 cm. (a) Calculate the current that results in a magnetic dipole moment of 2.30 A · m^2. (b) Find the maximum magnitude of torque that the coil, carrying this current, can experience in a uniform 35.0 mT magnetic field.

46. Moment and Torque A circular wire loop whose radius is 15.0 cm carries an amount of current of 2.60 A. It is placed so that the normal to its plane makes an angle of 41.0° with a uniform magnetic field of magnitude 12.0 T. (a) Calculate the magnitude of the magnetic dipole moment of the loop. (b) What is the magnitude of torque that acts on the loop?

47. Right Triangle A current loop, carrying an amount of current of 5.0 A, is in the shape of a right triangle with sides 30, 40, and 50 cm. The loop is in a uniform magnetic field of magnitude 80 mT whose direction is parallel to the current in the 50 cm side of the loop. Find the magnitude of (a) the magnetic dipole moment of the loop and (b) the torque on the loop.

48. Wall Clock A stationary circular wall clock has a face with a radius of 15 cm. Six turns of wire are wound around its perimeter; the wire carries a current of 2.0 A in the clockwise direction. The clock is located where there is a constant, uniform external magnetic field of magnitude 70 mT (but the clock still keeps perfect time). At exactly 1:00 P.M., the hour hand of the clock points in the direction of the external magnetic field. (a) After how many minutes will the minute hand point in the direction of the torque on the winding due to the magnetic field? (b) Find the torque magnitude.

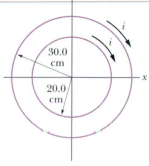

49. Concentric Loops Two concentric, circular wire loops, of radii 20.0 and 30.0 cm, are located in the xy plane; each carries a clockwise current of 7.00 A (Fig. 29-35). (a) Find the magnitude of the net magnetic dipole moment of this system. (b) Repeat for reversed current in the inner loop.

FIGURE 29-35 ■
Problem 49.

50. ABCDEFA Figure 29-36 shows a current loop $ABCDEFA$ carrying a current $i = 5.00$ A. The sides of the loop are parallel to the coordinate axes, with $AB = 20.0$ cm, $BC = 30.0$ cm, and $FA = 10.0$ cm. Calculate the magnitude and direction of the magnetic dipole moment of this loop. (*Hint:* Imagine equal and opposite currents i in the line segment AD; then treat the two rectangular loops $ABCDA$ and $ADEFA$.)

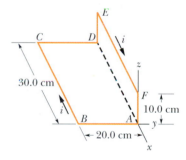

FIGURE 29-36 ■ Problem 50.

51. Circular Loop A circular loop of wire having a radius of 8.0 cm carries a current of 0.20 A. A vector of unit length and parallel to the dipole moment $\vec{\mu}$ of the loop is given by $0.60\hat{i} - 0.80\hat{j}$. If the loop is located in a uniform magnetic field given by $\vec{B} = (0.25\ \text{T})\hat{i} + (0.30\ \text{T})\hat{k}$, find (a) the torque on the loop (in unit-vector notation) and (b) the magnetic potential energy of the loop.

Additional Problems

52. Permanent Magnet You can observe that a permanent magnet can exert forces on moving charges or currents. (a) If a magnet exerts a force on a moving charge, would the magnet experience any forces? Explain. (b) In the case of the gravitational or electrostatic interaction between two objects, each object has a common property, such as mass in the case of gravitational interaction or excess charge in the case of the electrostatic interaction. A permanent magnet and a moving electron seem very different. Can you think of any way that they might have a common property? Explain.

53. U-Shaped Magnet An electron having a velocity of magnitude v enters a region between the poles of a U-shaped magnet. This region has a uniform magnetic field, $\vec{B}$, pointing out of the paper in the positive z direction as shown in Fig. 29-37.

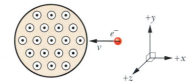

FIGURE 29-37 ▪ Problem 53.

(a) If the magnetic field points out of the paper, where is the north pole of the magnet—in front of or behind the image shown? (b) Use the right-hand rule to find the direction of force on the electron as it passes into the region where the magnetic field is uniform. (c) Sketch the path of the electron, assuming that the magnetic field is relatively weak. (d) If the speed of the electron is 4.79×10^6 m/s and the magnitude of the magnetic field is 0.234 T, what is the magnitude of the force on the electron?

54. A Velocity Selector A group of physicists at Argonne National Laboratory in Illinois wants to bombard metals with monoenergetic beams of alpha particles to study radiation damage. (Alpha particles are helium nuclei, which consist of two neutrons and two protons and thus have a net charge of $+2e$ where e is the amount of the charge on the electron.) They have managed to create a beam of alpha particles from the decay of radioactive elements, but some of the alpha particles lose energy as they collide with other atoms in the source. As a new physicist assigned to the group you have been asked to use a velocity selector to select only the alpha particles in the beam that are close to one velocity and get rid of the others. The velocity selector consists of: (1) a power supply capable of delivering large potential differences between capacitor plates and (2) a large permanent magnet that has a uniform magnetic field perpendicular to the beam. The setup for the velocity selector is shown in Fig. 29-38. The direction of the B-field is out of the paper. Your magnet has a field of 0.22 T and the capacitor plates have a spacing of 2.5 cm. You are asked to figure out how the velocity selector works and then tell your group what voltage to put across the capacitor plates to select a velocity of 4.2×10^6 m/s.

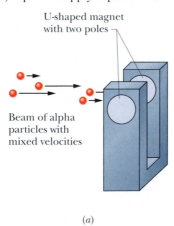

U-shaped magnet with two poles

Beam of alpha particles with mixed velocities

(a)

FIGURE 29-38a ▪ Problem 54.

This is your first job and you feel overwhelmed by the assignment, but you calm down and begin to analyze the situation one step at a time. You come up with the following:

(a) The magnet is oriented so its magnetic field is out of the paper in the diagram you are given, so you use the right-hand rule to determine the direction of the magnetic force on an alpha particle passing from left to right into the magnetic field. What direction did you come up with for the force?

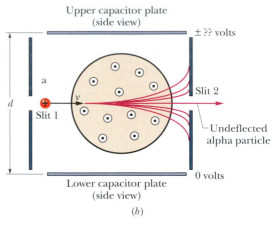

Upper capacitor plate
(side view)

± ?? volts

a

d

Slit 1

v

Slit 2

Undeflected alpha particle

0 volts

Lower capacitor plate
(side view)

(b)

FIGURE 29-38b ▪ Enlarged view of the region between the poles of the magnet showing a single alpha particle having a speed v entering the region of uniform magnetic field.

(b) You realize that by using $\vec{F}^{\text{mag}} = q\vec{v} \times \vec{B}$, you can calculate the magnitude of force on an alpha particle moving at speed v just as it enters the uniform magnetic field as a function of the charge on the alpha particle and the magnitude of the magnetic field B. What is the expression for the magnitude of the force in terms of e, v, and B?

(c) You realize that you might be able to put just the right voltage across the two capacitor plates so that the electrical force on a given alpha particle will be equal in magnitude and opposite in direction to the Lorentz magnetic force. Then any alpha particles with just the right velocity will pass straight through the poles of the magnet without being deflected. First you think about whether the voltage on the upper capacitor plate should be positive or negative to give a canceling force. What do you decide?

(d) Next you realize that if you know the electric field between the plates and the charge on the alpha particle then you can compute the electrical force on it. What is the relationship between the electrical force $\vec{F}^{\text{elec}}$, charge, q, and electric field $\vec{E}$?

(e) Finally, you use the fact that the magnitude of the electric field between capacitor plates is given by $E = |\Delta V|/d$ where d is the spacing between the plates. Show that the voltage needed to have the electrical force and the magnetic force be "equal and opposite" can be calculated using the equation $|\Delta V| = vBd$. Calculate the voltage needed.

55. Region A—Region B Figure 29-39 shows a charged particle that is moving in the positive x direction when it encounters region A with a uniform magnetic field. Its path is bent in a half-circle and then moves into region B also with a uniform magnetic field. The particle undergoes another half revolution. Finally it passes between two charged capacitor plates and is deflected downward in the negative y direction.

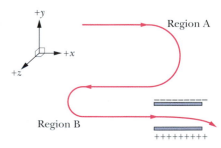

Region A

Region B

FIGURE 29-39 ▪ Problem 55.

(a) Is the charge positive or negative? Explain.
(b) What is the direction of the magnetic field in region A? Explain.
(c) What is the direction of the magnetic field in region B? Explain.
(d) Which region has the larger magnetic field, A or B? Explain.

56. A Mass Spectrometer It is possible to accelerate ions to a known kinetic energy in an electric field. Sometimes chemists and physicists do this as part of a method to identify the chemical elements present in a beam of ions by determining the mass of each ion. This can be done by bending the ion beam in a uniform magnetic field and measuring the radius of the semicircular path each ion takes. A device that does this is called a mass spectrometer. A schematic of a mass spectrometer is shown in Fig. 29-40.

Boron is the fifth element in the periodic table so it always has 5 protons. However, different isotopes of boron have 3, 5, 6, 7, or 8 neutrons in addition to the 5 protons to make up boron-8, boron-10, boron-11 and so on. As a research chemist for the Borax Company you have been asked

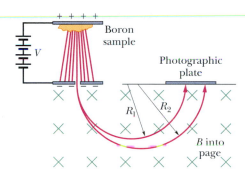

FIGURE 29-40 ■ Problem 56.

to use a mass spectrometer to determine the relative abundance of different isotopes of boron in a sample of boron obtained from a mine near Death Valley in California. You decide to accelerate a beam of singly charged boron ions (i.e., those that have lost one of their orbital electrons). You use an accelerating potential difference of -2.68×10^3 volts. The boron beam then enters a uniform magnetic field you set up to have a magnetic flux density of 0.182 T in a direction perpendicular to the direction of the boron beam. You observe two bright spots on your photographic plate with the spot corresponding to a radius of 13.0 cm having four times the intensity of the one corresponding to a radius of 13.6 cm. There are very faint spots at 11.6 cm, 14.2 cm, and 14.8 cm. Which isotope of boron has approximately 80% abundance? Which one has about 20% abundance? Which ones are present in only trace amounts? Please show all your reasoning and calculations. *Hints:* (1) An atomic mass unit is given by 1.66×10^{-27} kg, which is close to the mass of the proton and neutron. (2) Find the velocity of each isotope of boron in meters per second just after it has been accelerated by the potential difference of -2.68×10^3 volts. (3) It is helpful to do the calculations for each of the five isotopes on a spreadsheet.

57. Bubble Chamber Tracks Energetic gamma rays like those coming from outer space can disappear near a heavy nucleus producing a rapidly moving pair of particles consisting of an electron and a positron. (A positron is a small positively charged particle that has the same mass and amount of charge as an electron). This process is called pair production. A device called a bubble chamber allows one to observe the path taken by electron–positron pairs produced by gamma rays. The study of bubble chamber tracks in the presence of magnetic fields has revealed a great deal about high-energy gamma rays, the processes of pair production, and the loss of

energy by electrons and positrons. A sample bubble chamber track is shown in Fig. 29-41.

(a) If the magnetic field is uniform pointing into the paper, which trajectory (the upper one or the lower one) shows the motion of the positron? Explain your reasoning. (b) In which part of the spiral does the positron have the greatest energy—the large radius part or the small radius part? Explain the reasons for your answer. (c) Is the electron moving faster, slower, or at the same speed as the positron at the point in time when the two particles are created? Cite the evidence for your answer. (d) Suppose the bubble chamber photograph in Fig. 29-41 is an enlargement of the actual event so that the length L is actually only 2.4×10^{-3} m. Show that the radius of curvature of the electron path just after the electron is created is approximately 0.8×10^{-3} m. *Hint:* Measure L in picture units to find a scale factor and then measure the appropriate feature of the electron path in picture units and use the scale factor to find R in meters. (e) Use the Lorentz force law and the expression for centripetal force to find the equation relating the speed of the electron to B, R, e, and m. (f) Suppose the magnitude of the magnetic field in the bubble chamber is $B = 0.54$ T. Calculate the approximate speed of the electron when it is first created in the bubble chamber.

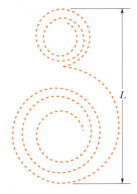

FIGURE 29-41 ■
Problem 57.

58. Three Force Fields We have studied three long-range forces: gravity, electricity, and magnetism. Compare and contrast these three forces giving at least one feature that all three forces have in common, and at least one feature that distinguishes each force.

59. Comparing $\vec{E}$ and $\vec{B}$ Fields We have studied two fields: electric and magnetic. Explain why we introduce the idea of field, and compare and contrast the electric and magnetic fields. In your comparison, be certain to discuss at least one similarity and one difference.

60. Anti-matter Ion Cosmic Rays An international consortium is presently building a device to look for anti-matter nuclei in cosmic rays to help us decide whether there are galaxies made of anti-matter. Anti-matter is just like ordinary matter except the basic particles (anti-protons and anti-electrons) have opposite charge from ordinary matter counterparts. Anti-protons are negative, and anti-electrons (positrons) are positive.

A schematic of the device is shown in Fig. 29-42. A cosmic ray—say, a carbon nucleus or an anti-carbon nucleus—enters the device at the left where its position and velocity are measured. It then passes through a (reasonably uniform) magnetic field. Its path is bent in one direction if its charge is positive and in the opposite direction it its charge is negative. Its deflection is measured as it goes out of the device.

(a) In Fig. 29-42, what is the direction of the magnetic field? How do you know?
(b) Which path is followed by each particle in the device? How do you know?
(c) If you were given the magnetic field, B, the size of the device, D, the amount of charge on the incoming particle, q, and the mass of the incoming particle, M, would this be enough to calculate the displacement of the charge, d? If so, describe briefly how you would

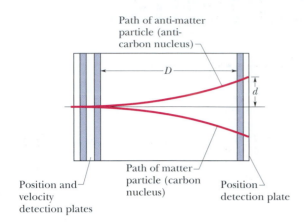

Path of anti-matter particle (anti-carbon nucleus)

Path of matter particle (carbon nucleus)

Position and velocity detection plates

Position detection plate

FIGURE 29-42 ■ Problem 60.

do it (but don't do it). If not, explain what additional information you would need (but don't estimate it).

61. Magnets and Charge A bar magnet is hung from a string through its center as shown in Fig. 29-43. A charged rod is brought up slowly into the position shown. In what direction will the magnet tend to rotate? Suppose the charged rod is replaced by a bar magnet with the north pole on top. In what direction will the magnet tend to rotate? Is there a difference between what happens to the hanging magnet in the two situations? Explain why you either do or do not think so.

FIGURE 29-43 ■ Problem 61.

* C14 is a radioactive isotope of carbon that behaves chemically almost identically to its more common but slightly lighter sibling, C12. The amount of C14 in the atmosphere stays about constant since it is being produced continually by cosmic rays. Once carbon from the air is bound into an organic substance, the C14 will decay with half of them vanishing every 5730 years. The ratio of C14 to C12 in an organic substance therefore tells how long ago it died.

62. Buying a Mass Spectrometer You are assigned the task of working with a desktop-sized magnetic spectrometer for the purpose of measuring the ratio of C^{12} to C^{14} atoms in a sample in order to determine the sample's age. For this problem, let's concentrate on the magnet that will perform the separation of masses. Suppose you have burned and vaporized the sample so that the carbon atoms are in a gas. You now pass this gas through an "ionizer" that on the average strips one electron from each atom. You then accelerate the ions by putting them through an electrostatic accelerator—two capacitor plates with small holes that permit the ions to enter and leave. (From the University of Washington Physics Education Group)

The two plates are charged so that they are at a voltage difference of ΔV volts. The electric field produced by charges on the capacitor plates accelerates the ions to an energy of $q\Delta V$. These are then introduced into a nearly constant, vertical magnetic field. If we ignore gravity, the magnetic field will cause the charged particles to follow a circular path in a horizontal plane. The radius of the circle will depend on the atom's mass. (Assume the whole device will be placed inside a vacuum chamber.)

Answer these three questions about how the device works.

(a) We want to keep the voltage at a moderate level. If ΔV is 1000 volts, how big of a magnetic field would we require to have a plausible tabletop-sized instrument? Is this a reasonable magnetic field to have with a tabletop-sized magnet?

(b) Do the C^{12} and C^{14} atoms hit the collection plate far enough apart? (If they are not separated by at least a few millimeters at the end of their path, we will have trouble collecting the atoms in separate bins.)

(c) Can we get away with ignoring gravity? (*Hint:* Calculate the time it would take the atom to travel its semicircle and calculate how far it would fall in that time.)

30 | Magnetic Fields Due to Currents

This is the way we presently launch materials into space. However, when we begin mining the Moon and the asteroids, where we will not have a source of fuel for such conventional rockets, we shall need a more effective way. Electromagnetic launchers may be the answer. A small prototype, the *electromagnetic rail gun*, can accelerate a projectile from rest to a speed of 10 km/s (36 000 km/h) within 1 ms.

How can such rapid acceleration possibly be accomplished?

The answer is in this chapter.

30-1 Introduction

When people first began to study magnetism scientifically (say, starting from Gilbert's *Treatise de Magnete* in 1600), they focused on the properties of magnets. For example, they studied lodestones (pieces of iron naturally magnetized by the Earth's magnetic field) and found that magnets interact with other magnets through an "action-at-a-distance" force that we now call magnetism. Magnetism was found to be a third distinct noncontact force to add to the list of the two already known: gravity and electricity.

As we learned in the previous chapter, *stationary* electric charges and magnets do not interact (except for the polarization effects that stationary charges can induce in all objects). However, *moving* electric charges do experience a force in the presence of a magnet. Since magnets can exert forces on other magnets, could it be that moving charges behave like magnets?

We have postulated the existence of an entity called the magnetic field in order to introduce a magnetic force law that provides a mathematical description of the force that a permanent magnet can exert on moving electrical charges. Newton's Third Law states that whenever one object exerts a force on another object, the latter object exerts an equal and opposite force on the former. So, if a magnet exerts a force on a current-carrying wire, shouldn't the wire exert an equal and opposite force on the magnet? The symmetry demanded by Newton's Third Law leads us to predict that if moving charges feel forces as they pass through magnetic fields, then they should be capable of exerting forces on the sources of these magnetic fields. In the early 19th century the Danish physicist Hans Christian Oersted demonstrated that an electric current does indeed exert forces on a magnet in its vicinity.

In this chapter we describe how to determine the magnetic fields associated with current-carrying wires and the forces they exert on other wires and magnets. We begin with a summary of Oersted's observations of magnetic phenomena associated with current-carrying wires. We also discuss the work of Biot and Savart, two French scientists. Biot and Savart made a series of careful observations to formulate a mathematical expression describing the magnetic field from a short segment of current-carrying wire, doing for magnetism what Coulomb did for electricity. Next we show how the Biot–Savart law and an alternative law known as Ampère's law (much as Gauss' law was an alternative to Coulomb's) can be used to calculate the magnetic fields and forces associated with various configurations of current-carrying wires. The ability to make such calculations has had a tremendous impact on the design of devices ranging from electric toothbrushes to gigantic particle accelerators.

30-2 Magnetic Effects of Currents—Oersted's Observations

The Earth has a relatively weak magnetic field that interacts with magnets. This phenomenon was exploited for navigational purposes through the development of the compass—a small bar magnet suspended so it pivots freely. Hence a compass is a sensitive magnetic field detector. Oersted and other scientists used the orientation of a compass to detect magnetic fields and determine their directions. By convention, the north-seeking pole of a magnet points in the direction of the magnetic field at its location.

In 1820, H. C. Oersted reported on a famous experiment connecting magnetism with electric currents. He placed a conducting wire along the north–south line of the Earth's magnetic field and laid a compass on top of the wire. The needle pointed along the wire (and the Earth's north–south line). When Oersted connected the ends of the wire to the terminals of a battery, the compass needle swung *perpendicular* to the wire as shown in Fig. 30-1, demonstrating that moving charges in a wire affect a compass in the same way a magnet does. Oersted also noticed that when the direction of the current is reversed, the compass needle flips so it points in the opposite direction.

Oersted found that moving charged particles, such as a current in a wire, create magnetic fields. Oersted's observation was especially surprising because this was the first known instance in which the force on an object (in this case the compass) was not observed to act along a line connecting it with the source of the force (in this case the wire). Within a week of the time that Oersted announced his observations, a French physicist, André Marie Ampère, began to refine them. Ampère noted that the magnetic field lines lay in concentric circles around the wire. His careful observations revealed that a long current-carrying wire sets up a magnetic field that orients small compass magnets so they are tangent to a circle centered on the wire that lies in a plane perpendicular to the wire. The alignment of iron filings, which act like small compasses, is shown in Fig. 30-2. Drawing the direction of the compass needle alignments at many different points that completely surround the wire results in an image of concentric circles like those shown in Fig. 30-3.

Ampère also developed a graphic way of relating the direction of conventional current (that is, traveling from the positive to the negative terminal of a battery) and the orientation of the magnetic field, which is indicated by the direction of the north pole of a compass needle. Ampère stated his **right-hand rule** as follows:

> Encircle the wire with the fingers of the right hand, thumb extended in the direction of positive current. The fingers then point in the direction of deflection of the north pole.

This right-hand rule is shown graphically in Fig. 30-4.

You can easily replicate the following observations made by Oersted, Ampère, and many others in the early 19th century using a battery, wire, a piece of cardboard, and one or more small compasses:

- The compass needles are more strongly deflected when they are close to the wire than when they are far from the wire.

- For a given current, the amount of needle deflection depends only on the needle's radial distance from the wire.

- At a given radial distance from the wire, increasing the current in the wire increases the needle deflection.

- The direction of the needle deflection flips (change by 180°) if you reverse the direction of the current flow.

- Drawing the directions of the needle orientations at many different points that completely surround the wire results in an image of concentric circles like those shown in Fig. 30-3.

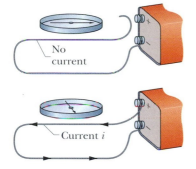

FIGURE 30-1 ■ Oersted's experiment showing how a compass needle becomes aligned in a direction that is perpendicular to the direction of the current in a length of wire.

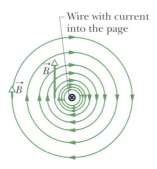

FIGURE 30-2 ■ Iron filing slivers that have been sprinkled onto cardboard collect in concentric circles when a strong current is sent through the central wire. The filings are magnetized and align themselves like tiny compasses in the direction of the magnetic field produced by the current.

FIGURE 30-3 ■ The magnetic field lines produced by a current in a long straight wire form concentric circles around the wire. Here the current is into the page, as indicated by the ×. The field lines are farther apart as the distance from the wire increases, signifying a decrease in the magnitude of the field with distance.

READING EXERCISE 30-1: In each of the following situations, assume that the magnetic field associated with a current-carrying wire can point up, down, left, right, into the page, or out of the page. (a) If the direction of the conventional current in the wire is out of the page, what is the direction of the magnetic field it generates at point 1? (b) At point 2? (c) If the direction of the conventional current in the wire is into the page, what is the direction of the magnetic field it generates at point 1? (d) At point 2?

•1
(a)
(c)
•1
⊙ •2
Wire (b)
⊗ •2
Wire (d)

■

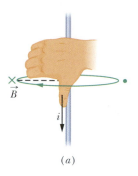

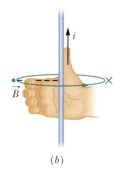

FIGURE 30-4 ■ Ampère's right-hand rule gives the direction of the magnetic field relative to the conventional current in a wire. (a) The situation of Fig. 30-3, seen from the side. The magnetic field $\vec{B}$ at any point to the left of the wire is perpendicular to the dashed radial line and directed into the page, in the direction of the fingertips, as indicated by the ×. (b) If the current is reversed, $\vec{B}$ at any point to the left is still perpendicular to the dashed radial line but now is directed out of the page, as indicated by the dot.

30-3 Calculating the Magnetic Field Due to a Current

It is very useful to be able to compute the net magnetic field created by a current-carrying wire. We would also like to be able to do this either for long straight wires or for any wire no matter how it bends around.

Two French physicists named Biot and Savart (rhymes with "Leo and bazaar") were able to develop a mathematical description of the magnetic field in the vicinity of a short segment of current-carrying wire. To do this, these investigators made a set of very clever experimental measurements:

• First, the two investigators positioned magnets around their experimental setup in order to cancel out the local magnetic field of the Earth.

• Next they placed sharp bends in a current-carrying wire so they could observe the approximate effect that an "isolated" short element of wire would have.

• Then they ran a known current through the wire and measured the direction of the magnetic field produced by the small wire segment at various locations using the final orientation of the suspended compass needles.

• Finally, they measured the relative magnitude of the torque on the suspended compass needles before they reached their final orientation and thus the relative force applied to the needles. In doing this, they were actually making measurements of the strength of the field at various locations.

Given what we know of the observations summarized in the previous section, it is not surprising that Biot and Savart found that the magnitude of the magnetic field contribution $|d\vec{B}| = dB$ is directly proportional to the amount of the current $|i|$ and the length of the small segment of wire. They also found that the magnitude of the magnetic field at a point P in space decreases as the inverse square of the distance between the segment of wire and point P. The two investigators proposed that the magnitude of the field contribution $d\vec{B}$ produced at a point P by a segment of wire $d\vec{s}$ carrying a current i is

$$dB = \frac{\mu_0}{4\pi} \frac{|i\,ds|\sin\phi}{r^2}, \tag{30-1}$$

where $d\vec{s}$ is a vector of magnitude ds equal to the length of the piece of wire and direction given by the direction of the current. ϕ is the angle between the directions of $d\vec{s}$ and $\vec{r}$, where $\vec{r}$ is the vector that extends from $d\vec{s}$ to point P. (See Figure 30-5b.) The symbol μ_0 is called the *magnetic constant* (or *permeability*). By definition its value in SI units is exactly

$$\mu_0 \equiv 4\pi \times 10^{-7}\,\text{T} \cdot \text{m/A} \approx 1.26 \times 10^{-6}\,\text{T} \cdot \text{m/A} \quad \text{(magnetic constant).} \tag{30-2}$$

Equation 30-1 is similar in many ways to that found for the differential electric field from a small segment of wire holding static charge described by Eq. 23-21. However, the perpendicular relationship between the direction of a segment of wire and the magnetic field it produces is a new phenomenon. Fortunately, it turns out that a vector crossproduct can be used to find the direction of the magnetic field contribution. The direction of $d\vec{B}$, shown as being into the page in Fig. 30-5b, is the same as that given by the cross product $d\vec{s} \times \vec{r}$. We can therefore recast Eq. 30-1 in vector form as

$$d\vec{B} = \frac{\mu_0}{4\pi} \frac{i\,d\vec{s} \times \vec{r}}{r^3} \quad \text{(Biot–Savart law).} \tag{30-3}$$

This vector equation is known as the **Biot–Savart law.** The law, which was experimentally deduced, is an inverse-square law (the exponent in the denominator of Eq. 30-3 is 3 only because of the factor $\vec{r}$ in the numerator). How can we use this law to calculate the net magnetic field $\vec{B}$ produced at a point by various distributions of current?

If our goal is to calculate the magnetic field that is produced by a given current *distribution* based on the field produced by *segments* of the distribution, perhaps we should use the same basic procedure we used in Chapter 23 to calculate the electric field produced by a given distribution of charged particles. Let us quickly review that basic procedure. We first mentally divide the charge distribution into charge elements dq, as is done for a charge distribution of arbitrary shape in Fig. 30-5a. We then calculate the field $d\vec{E}$ produced at some point P by a single charge element. Because the electric fields contributed by different elements can be superimposed, we calculate the net field $\vec{E}$ at P by summing, via integration, the contributions $d\vec{E}$ from all the elements.

Recall that we express the magnitude of $d\vec{E}$ as

$$|dE| = k\,\frac{|dq|}{r^2},\qquad (30\text{-}4)$$

in which r is the distance between the charge element dq and point P. For a positively charged element, the direction of $d\vec{E}$ is that of $\vec{r}$, where $\vec{r}$ is the vector that extends from the charge element dq to the point P. Using $\vec{r}$, we can rewrite Eq. 30-4 in vector form as

$$d\vec{E} = k\,\frac{dq}{r^3}\,\vec{r},\qquad (30\text{-}5)$$

which indicates that the direction of the vector $d\vec{E}$ produced by a positively charged element is the direction of the vector $\vec{r}$. Note that just as is the case for the Biot-Savart law this is an inverse-square law ($d\vec{E}$ depends on inverse r^2) in spite of the r^3 term in the denominator. This is because the $\vec{r}$ term in the numerator cancels one of the r's in the denominator.

We can use the same basic procedure to calculate the magnetic field due to a current. Figure 30-5b shows a wire of arbitrary shape carrying a current i. We want to find the magnetic field $\vec{B}$ at a nearby point P. We first mentally divide the wire into differential elements ds and then define for each element a length vector $d\vec{s}$ that has length ds and whose direction is the direction of the current in ds. We can then define a differential *current-length element* to be $i\,d\vec{s}$; we wish to calculate the field $d\vec{B}$ produced at P by a single current-length element. From experiment we find that magnetic fields, like electric fields, can be superimposed to find a net field. Thus, we can calculate the net field $\vec{B}$ at P by summing contributions for discrete sources or by integrating the contributions $d\vec{B}$ from all the current-length elements in a continuous source. However, this summation (or integration) is more challenging than the process associated with electric fields because of a complexity. The charge element dq that produces an electric field is a scalar, but a current-length element $i\,d\vec{s}$ that produces a magnetic field is the product of a scalar and a vector.

Magnetic Field Due to a Current in a Long Straight Wire

Shortly we shall use the law of Biot and Savart to prove that the magnitude of the magnetic field at a perpendicular distance R from a long (infinite) straight wire carrying a current i is given by

$$B = \frac{\mu_0|i|}{2\pi R}\qquad\text{(long straight wire).}\qquad (30\text{-}6)$$

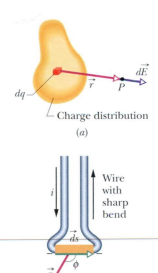

(a)

(b)

FIGURE 30-5 ■ (a) A charge element dq produces a differential electric field $d\vec{E}$ at point P. (b) A current-length element $i\,d\vec{s}$, isolated by sharp bends in the wire, produces a differential magnetic field $d\vec{B}$ at point P. The $\times$ (the tail of an arrow) at the dot for point P indicates that $d\vec{B}$ is directed into the page there for the special case where i and $d\vec{s}$ are parallel. If a small magnetic compass needle is used to detect the magnetic field, then its north pole points into the page.

The field magnitude $B = |\vec{B}|$ in Eq. 30-6 depends only on the amount of current and the perpendicular distance R of the point from the wire. We shall show in our derivation that the field lines of $\vec{B}$ form concentric circles around the wire, as Fig. 30-3 shows and as the iron filings in Fig. 30-2 suggest. The increase in the spacing of the lines in Fig. 30-3 with increasing distance from the wire represents the $1/R$ decrease in the magnitude of $\vec{B}$ predicted by $B = \mu_0|i|/2\pi R$ (Eq. 30-6). The lengths of the two vectors $\vec{B}$ in Fig. 30-3 also show the $1/R$ decrease when we use Ampère's right-hand rule for finding the direction of the magnetic field set up by a current-length element, such as a section of a long wire. What we are really doing is describing the orientation of concentric circles centered on the wire. A careful review of Fig. 30-3 yields two additional points that are often quite useful in solving magnetic field problems. Namely, the magnetic field $\vec{B}$ due to a current-carrying wire at any point is *tangent to a magnetic field line* and it is *perpendicular to a dashed radial line connecting the point and the current.*

Proof of Equation 30-6

Figure 30-6, which is just like Fig. 30-5b except that now the wire is straight and of infinite length, illustrates the task at hand; we seek the field $\vec{B}$ at point P, a perpendicular distance R from the wire. The magnitude of the differential magnetic field produced at P by the current-length element $|i\,d\vec{s}\,|$ located a distance r from P is given by Eq. 30-1:

$$|d\vec{B}| = \frac{\mu_0}{4\pi} \frac{|i\,d\vec{s}\,|\sin\phi}{r^2}.$$

Since the direction of $d\vec{s}$ is always in the direction of the current, we find that the direction of $d\vec{B}$ in Fig. 30-6 (given by $d\vec{s} \times \vec{r}$) is into the page.

Note that $d\vec{B}$ at point P has this same direction (into the page) for all the current-length elements into which the wire can be divided. Thus, we can find the magnitude of the magnetic field produced at P by the current-length elements in the upper half of the infinitely long wire by integrating dB in Eq. 30-1 from 0 to ∞.

Now consider a current-length element in the lower half of the wire, one that is as far below P as $d\vec{s}$ is above P. By Eq. 30-6, the magnetic field produced at P by this current-length element has the same magnitude and direction as that from $i\,d\vec{s}$ in Fig. 30-6. Further, the magnetic field produced by the lower half of the wire is exactly the same as that produced by the upper half. To find the magnitude of the total magnetic field $\vec{B}$ at P, we need only multiply the result of our integration by 2. We get

$$B = 2\int_0^\infty dB = \frac{\mu_0|i|}{2\pi} \int_0^\infty \frac{|\sin\phi\,ds|}{r^2}. \tag{30-7}$$

The variables ϕ, s, and r in this equation are not independent but (see Fig. 30-6) are related by

$$r = \sqrt{s^2 + R^2}$$

and

$$\sin\phi = \sin(\pi - \phi) = \frac{R}{\sqrt{s^2 + R^2}}.$$

Using these substitutions along with the solution to integral 19 in Appendix E, Eq. 30-7 describing the magnitude of the magnetic field becomes

$$B = \frac{\mu_0|i|}{2\pi} \int_0^\infty \frac{R\,ds}{(s^2 + R^2)^{3/2}} = \frac{\mu_0|i|}{2\pi R}\left[\frac{s}{(s^2 + R^2)^{1/2}}\right]_0^\infty.$$

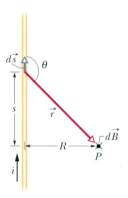

FIGURE 30-6 ▪ Calculating the magnetic field produced by a current i in a long straight wire. Using either Ampère's right-hand rule or $d\vec{s} \times \vec{r}$, we find $d\vec{B}$ at P is directed into the page as shown.

Substituting the limits in the expression above gives a *B*-field magnitude of

$$B = \frac{\mu_0 |i|}{2\pi R}, \qquad \text{(infinite straight wire),} \qquad (30\text{-}8)$$

which is the relation we set out to prove. Note that the magnitude of the magnetic field at *P* due to either the lower half or the upper half of the infinite wire in Fig. 30-6 is half this value; that is,

$$B = \frac{\mu_0 |i|}{4\pi R} \qquad \text{(semi-infinite straight wire).} \qquad (30\text{-}9)$$

Magnetic Field Due to a Current in a Circular Arc of Wire

To find the magnetic field produced at a point by a current in a curved wire, we would again use Eq. 30-1 to write the magnitude of the field produced by a single current-length element, and we would again integrate to find the net field produced by all the current-length elements. That integration can be difficult, depending on the shape of the wire; it is fairly straightforward, however, when the wire is a circular arc and the point is the center of curvature.

Figure 30-7*a* shows such an arc-shaped wire with central angle ϕ_C, radius *R*, and center *C*, carrying current *i*. At *C*, each current-length element $i\,d\vec{s}$ of the wire produces a magnetic field element of magnitude *dB* given by Eq. 30-1. Moreover, as Fig. 30-7*b* shows, no matter where the element is located on the wire, the angle ϕ between the vectors $d\vec{s}$ and $\vec{r}$ is 90°; also, $r = R$. Thus, by substituting *R* for *r* and 90° for ϕ, we obtain from Eq. 30-1,

$$dB = \frac{\mu_0}{4\pi} \frac{|i|\,ds \sin 90°}{R^2} = \frac{\mu_0}{4\pi} \frac{|i|\,ds}{R^2}. \qquad (30\text{-}10)$$

The field at *C* due to each current-length element in the circular arc has this same magnitude.

An application of the right-hand rule anywhere along the wire (as in Fig. 30-7*c*) will show that all the differential fields $d\vec{B}$ have the same direction at *C*—directly out of the page. Thus, the total field at *C* is simply the sum (via integration) of all the fields $d\vec{B}$. We use the identity $ds = R\,d\phi$ to change the variable of integration from *ds* to $d\phi$ and obtain, from Eq. 30-10, a magnitude of

$$B = \int dB = \int_0^{\phi_C} \frac{\mu_0}{4\pi} \frac{|i|\,R\,d\phi}{R^2} = \frac{\mu_0 |i|}{4\pi R} \int_0^{\phi_C} d\phi.$$

Integrating, we find that

$$B = \frac{\mu_0 |i|\phi_C}{4\pi R} \qquad \text{(at center of circular arc).} \qquad (30\text{-}11)$$

Note that this equation gives us the magnitude of the magnetic field *only* at the center of curvature of a circular arc of current. When you insert data into the equation, you must be careful to express ϕ_C in radians rather than degrees. For example, to find the magnitude of the magnetic field at the center of a full circle of current,

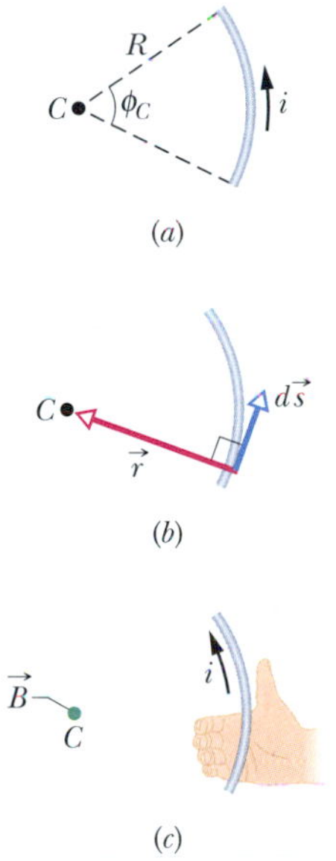

(a)

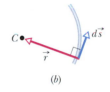

(b)

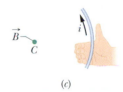

(c)

FIGURE 30-7 ▪ (*a*) A wire in the shape of a circular arc with center *C* carries current *i*. (*b*) For any element of wire along the arc, the angle between the directions of $d\vec{s}$ and $\vec{r}$ is 90°. (*c*) Determining the direction of the magnetic field at the center *C* due to the current in the wire; the field is out of the page, in the direction of the fingertips, as indicated by the colored dot at *C*.

you would substitute 2π for ϕ_C in Eq. 30-11, finding

$$B = \frac{\mu_0 |i|(2\pi)}{4\pi R} = \frac{\mu_0 |i|}{2R} \qquad \text{(at center of full circle).} \qquad (30\text{-}12)$$

READING EXERCISE 30-2: A uniform magnetic field is directed toward the right in the plane of the paper as shown in the diagram that follows. A wire oriented perpendicular to the plane of the paper carries a current i. Suppose that the resultant magnetic field at point 1 due to a superposition of the uniform magnetic field of magnitude $|\vec{B}|$ and the magnetic field of the wire at point 1 is zero. (a) Is the direction of the current in the wire into or out of the paper? Explain how you arrived at your conclusion. (b) Assume that point 2 lies at the same distance from the center of the wire as point 1 and that the length of the vector assigned to represent the magnitude of the uniform external magnetic field is that shown to the left. Construct a vector arrow showing the length and direction of the resultant magnetic field vector at point 2. Explain how you deduced what the vector should be. (Adapted from A. Arons, *Homework and Test Questions for Introductory Physics Teaching,* John Wiley and Sons, 1947.)

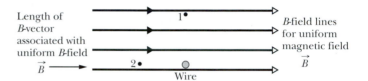

Length of
B-vector
associated with
uniform B-field

$\vec{B}$

1 •

2 •

Wire

B-field lines
for uniform
magnetic field

$\vec{B}$

TOUCHSTONE EXAMPLE 30-1: An Arc and Two Straight Lines

The wire in Fig. 30-8a carries a current i and consists of a circular arc of radius R and central angle $\pi/2$ rad, and two straight sections whose extensions intersect the center C of the arc. What magnetic field $\vec{B}$ does the current produce at C?

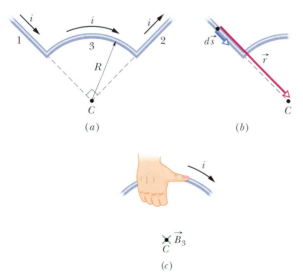

FIGURE 30-8 ■ (a) A wire consists of two straight sections (1 and 2) and a circular arc (3), and carries current i. (b) For a current-length element in section 1, the angle between $\vec{ds}$ and $\vec{r}$ is zero. (c) Determining the direction of magnetic field $\vec{B_3}$ at C due to the current in the circular arc; the field is into the page there.

SOLUTION ■ One **Key Idea** here is that we can find the magnetic field $\vec{B}$ at point C by applying the Biot–Savart law of Eq. 30-3 to the wire. A second **Key Idea** is that the application of Eq. 30-3 can be simplified by evaluating $\vec{B}$ separately for the three distinguishable sections of the wire—namely, (1) the straight section at the left, (2) the straight section at the right, and (3) the circular arc.

Straight sections. For any current-length element in section 1, the angle ϕ between $\vec{ds}$ and $\vec{r}$ is zero (Fig. 30-8b), so Eq. 30-1 gives us

$$|d\vec{B_1}| = \frac{\mu_0}{4\pi}\frac{|i\,\vec{ds}|\sin\phi}{r^2} = \frac{\mu_0}{4\pi}\frac{|i\,\vec{ds}|\sin 0}{r^2} = 0 \text{ T}.$$

Thus, the current along the entire length of wire in straight section 1 contributes no magnetic field at C:

$$\vec{B_1} = 0 \text{ T}.$$

The same situation prevails in straight section 2, where the angle ϕ between $\vec{ds}$ and $\vec{r}$ for any current-length element is 180°. Thus,

$$\vec{B_2} = 0 \text{ T}.$$

Circular arc. The **Key Idea** here is that application of the Biot–Savart law to evaluate the magnetic field at the center of a circular arc leads to Eq. 30-11 ($|B| = \mu_0 |i|\phi/4\pi R$). Here the central angle ϕ of the arc is $\pi/2$ rad. Thus from Eq. 30-11, the magnitude of the magnetic field $\vec{B_3}$ at the arc's center C is

$$|\vec{B_3}| = \frac{\mu_0 |i|(\pi/2)}{4\pi R} = \frac{\mu_0 |i|}{8R}.$$

To find the direction of $\vec{B}_3$, we apply the right-hand rule displayed in Fig. 30-4. Mentally grasp the circular arc with your right hand as suggested in Fig. 30-8c, with your thumb in the direction of the current. The direction in which your fingers curl around the wire indicates the direction of the magnetic field lines around the wire. In the region of point C (inside the circular arc), your fingertips point *into the plane* of the page. Thus, $\vec{B}_3$ is directed into that plane.

Net field. Generally, when we must combine two or more magnetic fields to find the net magnetic field, we must combine the

fields as vectors and not simply add their magnitudes. Here, however, only the circular arc produces a magnetic field at point C. Thus, we can write the magnitude of the net field $\vec{B}$ as

$$|\vec{B}| = |\vec{B}_1 + \vec{B}_2 + \vec{B}_3| = 0 + 0 + \left|\frac{\mu_0 i}{8R}\right| = \left|\frac{\mu_0 i}{8R}\right|. \quad \text{(Answer)}$$

The direction of $\vec{B}$ is the direction of $\vec{B}_3$—namely, into the plane of Fig. 30-8.

TOUCHSTONE EXAMPLE 30-2: Two Long Parallel Wires

Figure 30-9a shows two long parallel wires carrying currents i_1 and i_2 in opposite directions. What are the magnitude and direction of the net magnetic field at point P? Assume the following values: $i_1 = 15$ A, $i_2 = 32$ A, and $d = 5.3$ cm.

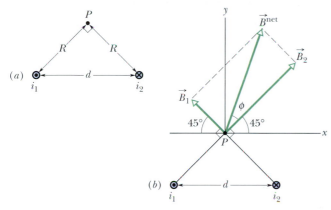

FIGURE 30-9 ■ (a) Two wires carry currents i_1 and i_2 in opposite directions (out of and into the page). Note the right angle at P. (b) The separate fields $\vec{B}_1$ and $\vec{B}_2$ are combined vectorially to yield the net field $\vec{B}^{\text{net}}$.

SOLUTION ■ One **Key Idea** here is that the net magnetic field $\vec{B}$ at point P is the vector sum of the magnetic fields due to the currents in the two wires. A second **Key Idea** is that we can find the magnetic field due to any current by applying the Biot–Savart law to the current. For points near the current in a long straight wire, that law leads to Eq. 30-6.

In Fig. 30-9a, point P is distance R from both currents i_1 and i_2. Thus, Eq. 30-6 tells us that at point P those currents produce magnetic fields $\vec{B}_1$ and $\vec{B}_2$ with magnitudes

$$B_1 = \frac{\mu_0 |i_1|}{2\pi R} \quad \text{and} \quad B_2 = \frac{\mu_0 |i_2|}{2\pi R}.$$

In the right triangle of Fig. 30-9a, note that the base angles (between sides R and d) are both 45°. Thus, we may write $\cos 45° = R/d$ and replace R with $d \cos 45°$. Then the field magnitudes B_1 and B_2 become

$$B_1 = \frac{\mu_0 |i_1|}{2\pi d \cos 45°} \quad \text{and} \quad B_2 = \frac{\mu_0 |i_2|}{2\pi d \cos 45°}.$$

We want to combine $\vec{B}_1$ and $\vec{B}_2$ to find their vector sum, which is the net field $\vec{B}^{\text{net}}$ at P. To find the directions of $\vec{B}_1$ and $\vec{B}_2$, we apply the right-hand rule of Fig. 30-4 to each current in Fig. 30-9a. For wire 1, with current out of the page, we mentally grasp the wire with the right hand, with the thumb pointing out of the page. Then the curled fingers indicate that the field lines run counterclockwise. In particular, in the region of point P, they are directed upward to the left. Recall that the magnetic field at a point near a long, straight current-carrying wire must be directed perpendicular to a radial line between the point and the current. Thus, $\vec{B}_1$ must be directed upward to the left as drawn in Fig. 30-9b. (Note carefully the perpendicular symbol between vector $\vec{B}_1$ and the line connecting point P and wire 1.)

Repeating this analysis for the current in wire 2, we find that $\vec{B}_2$ is directed upward to the right as drawn in Fig. 30-9b. (Note the perpendicular symbol between vector $\vec{B}_2$ and the line connecting point P and wire 2.)

We can now vectorially add $\vec{B}_1$ and $\vec{B}_2$ to find the net magnetic field $\vec{B}^{\text{net}}$ at point P, either by using a vector-capable calculator or by resolving the vectors into components and then combining the components of $\vec{B}^{\text{net}}$. However, in Fig. 30-9b, there is a third method: Because $\vec{B}_1$ and $\vec{B}_2$ are perpendicular to each other, they form the legs of a right triangle, with $\vec{B}^{\text{net}}$ as the hypotenuse. The Pythagorean theorem then gives us

$$B^{\text{net}} = \sqrt{B_1^2 + B_2^2} = \frac{\mu_0}{2\pi d(\cos 45°)}\sqrt{i_1^2 + i_2^2}$$

$$= \frac{(4\pi \times 10^{-7}\,\text{T} \cdot \text{m/A})\sqrt{(15\,\text{A})^2 + (32\,\text{A})^2}}{(2\pi)(5.3 \times 10^{-2}\,\text{m})(\cos 45°)}$$

$$= 1.89 \times 10^{-4}\,\text{T} \approx 190\,\mu\text{T}. \quad \text{(Answer)}$$

The angle ϕ between the directions of $\vec{B}^{\text{net}}$ and $\vec{B}_2$ in Fig. 30-9b follows from

$$\phi = \tan^{-1}\frac{B_1}{B_2},$$

which, with B_1 and B_2 as given above, yields

$$\phi = \tan^{-1}\frac{i_1}{i_2} = \tan^{-1}\frac{15\,\text{A}}{32\,\text{A}} = 25°.$$

The angle between the direction of $\vec{B}^{\text{net}}$ and the x axis shown in Fig. 30-9b is then

$$\phi + 45° = 25° + 45° = 70°. \quad \text{(Answer)}$$

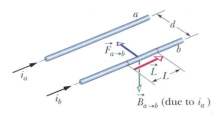

FIGURE 30-10 ■ Two parallel wires carrying currents in the same direction attract each other. $\vec{B}_{a \to b}$ is the magnetic field at wire b produced by the current in wire a. $\vec{F}_{a \to b}$ is the resulting force acting on wire b because it carries current in field $\vec{B}_{a \to b}$.

30-4 Force Between Parallel Currents

Back in 1820, when Ampère was first replicating Oersted's observations, he predicted that two current-carrying wires in parallel would exert forces on each other. This is a logical consequence of the Biot–Savart law, which quantifies the magnetic field surrounding a current-carrying wire, and the magnetic force law, which describes the force on a current in the presence of a magnetic field. Indeed, Ampère observed that there is a mutual interaction between the two wires. In other words, each wire exerts a force on the other. As shown in Fig. 30-10, the application of the right-hand rules that accompany the Biot–Savart law (Eq. 30-3) and the expression for the magnetic force on a current (Eq. 29-22) lead us to predict that wires that carry currents in the same direction will attract, whereas wires that carry currents in opposite directions will repel. It is interesting that the attractions and repulsions are opposite to the electrostatic and magnetic relationships, where unlike charges or poles attract and like charges or poles repel.

We can use the two equations just mentioned to derive a third equation that describes the forces between two parallel current-carrying wires. Why do we want to determine these interaction forces? Three reasons come to mind. First, we can compare the measurement of these forces to the forces predicted by our third equation to verify the Biot–Savart law. Second, these mutual interaction forces enable us to define the ampere as the SI unit of current. Finally, by understanding the nature of these forces we can design an electromagnetic launcher (like that mentioned in the "puzzler" on the first page of this chapter).

Figure 30-10 shows two parallel wires, separated by a distance d and carrying currents i_a and i_b. The first step in analyzing the forces between these wires is to find an expression for the force on wire b due to the current in wire a. The current in wire a produces a magnetic field $\vec{B}_{a \to b}$ at the location of wire b, and it is this magnetic field produced by wire a that actually causes wire b to experience a force denoted as $\vec{F}_{a \to b}$. According to Eq. 30-6, the magnitude of $B_{a \to b}$ at every point along wire b is

$$B_{a \to b} = \frac{\mu_0 |i_a|}{2 \pi d}. \tag{30-13}$$

The right-hand rule tells us that the direction of $\vec{B}_{a \to b}$ at wire b is down, as shown in Fig. 30-10.

Now that we have determined the magnetic field vector, we can find the force that wire a produces on wire b. The expression for the force on a length of current-carrying wire (Eq. 29-22) tells us that the force on wire b is

$$\vec{F}_{a \to b} = i_b \vec{L} \times \vec{B}_{a \to b}, \tag{30-14}$$

where $\vec{L}$ is the length vector (direction given by the direction of current i) of the wire. In Fig. 30-10 the vectors $\vec{L}$ and $\vec{B}_{a \to b}$ are perpendicular, so using Eqs. 30-13 and 30-14, we can express the magnitude of the force on wire b due to the current in wire a as

$$F_{a \to b} = |i_b| L B_{a \to b} \sin 90° = \frac{\mu_0 L |i_a i_b|}{2 \pi d}. \tag{30-15}$$

The direction of $\vec{F}_{a \to b}$ is the direction of the cross product $\vec{L} \times \vec{B}_{a \to b}$. Applying the right-hand rule for cross products to $\vec{L}$ and $\vec{B}_{a \to b}$ in Fig. 30-10, we find that $\vec{F}_{a \to b}$ points directly toward wire a, as shown.

The general procedure for finding the force on a current-carrying wire is this:

> To find the force on a current-carrying wire due to a second current-carrying wire, first find the field due to the second wire at the site of the first wire. Then find the force on the first wire due to that field.

We could now use this procedure to compute the force on wire *a* due to the current in wire *b*. We would find that the force has the same magnitude but is in the opposite direction. This is true regardless of whether the currents are the same or in opposite directions. Once again, Newton's Third Law holds:

> Parallel currents attract, and antiparallel currents repel.

The forces acting between currents in parallel wires provide us with the basis for defining the ampere, which is one of the seven SI base units. It is appropriately named after André Marie Ampère, who was the first to demonstrate the forces acting between parallel currents. The official SI definition, adopted in 1946, is:

> The **ampere** is that constant current which, if maintained in two straight, parallel conductors of infinite length, of negligible circular cross section, and placed 1 m apart in a vacuum, would produce between each of these conductors a force equal to 2×10^{-7} newton per meter of length.

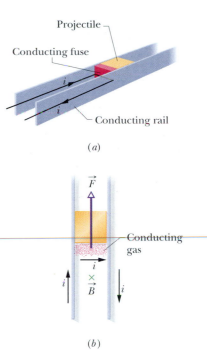

FIGURE 30-11 ■ (a) A rail gun, as a current *i* is set up in it. The current rapidly causes the conducting fuse to vaporize. (b) The current produces a magnetic field $\vec{B}$ between the rails, and the field causes a force $\vec{F}$ to act on the conducting gas, which is part of the current path. The gas propels the projectile along the rails, launching it.

Rail Gun

A rail gun is a device in which a magnetic force can accelerate a projectile to a high speed in a short time. The basics of a rail gun are shown in Fig. 30-11a. A large current flows in a circuit consisting of two conducting rails joined by a conducting "fuse" (such as a narrow piece of copper) between the rails, and then back to the current source along the second rail. The projectile to be fired lies on the far side of the fuse and fits loosely between the rails. Immediately after the current is established, the fuse element melts and vaporizes, creating a conducting gas between the rails where the fuse had been.

The right-hand rule of Fig. 30-4 shows that the current in the rails of Fig. 30-11a produces a magnetic field that is directed downward between the rails. The net magnetic field $\vec{B}$ exerts a force $\vec{F}$ on the gas due to the current *i* through the gas (Fig. 30-11b). Using Eq. 30-14 and the right-hand rule for cross products, we find that $\vec{F}$ points outward along the rails. As the gas is forced outward along the rails, it pushes the projectile, accelerating it by as much as $5 \times 10^7 \, \text{m/s}^2$ or $(5 \times 10^6 \, \text{g})$, and then launches it with a speed of 10 km/s, all within less than one millisecond.

READING EXERCISE 30-3: The figure shows three long, straight, parallel, equally spaced wires with identical amounts of current either into or out of the page. Rank the wires according to the magnitude of the force on each due to the currents in the other two wires, greatest first. ■

30-5 Ampère's Law

We can find the net electric field due to *any* distribution of charges with the inverse-square law for the differential field $d\vec{E}$ (Eq. 30-5), but if the distribution is complicated, we may have to use a computer. Recall, however, that if the distribution has planar, cylindrical, or spherical symmetry, we can apply Gauss' law to find the net electric field with considerably less effort.

Similarly, we can find the net magnetic field due to any distribution of currents with the inverse-square law for the differential field $d\vec{B}$ (Eq. 30-3), but again we may have to use a computer for a complicated distribution. However, if the distribution has enough symmetry, we can apply *Ampère's law* to find the magnetic field with considerably less effort. This law, which can be derived from the Biot–Savart law, has traditionally been credited to André Marie Ampère (1775–1836), for whom the SI unit of current is named. However, the law actually was advanced by English physicist James Clerk Maxwell.

Ampère's law is

$$\oint \vec{B} \cdot d\vec{s} = \mu_0 i^{\text{enc}} \qquad \text{(Ampère's law).} \qquad (30\text{-}16)$$

The circle on the integral sign means that the scalar (or dot) product $\vec{B} \cdot d\vec{s}$ is to be integrated around an imaginary *closed* loop, called an *Ampèrian loop*. The current i^{enc} on the right is the *net* current encircled by that loop.

In Gauss' law we choose a closed surface on which to evaluate the integral. The integral flux is proportional to the net charge enclosed by the surface. In Ampère's law, we choose a closed loop on which to evaluate the integral. The integral is proportional to the net current passing through the loop.

To see the meaning of the scalar product $\vec{B} \cdot d\vec{s}$ and its integral, let us first apply Ampère's law to the general situation shown in Fig. 30-12. This figure depicts the cross sections of three long straight wires that carry currents i_1, i_2, and i_3 either directly into or directly out of the page. An arbitrary Ampèrian loop lying in the plane of the page encircles two of the currents but not the third. The counterclockwise direction marked on the loop indicates the arbitrarily chosen direction of integration for Eq. 30-16.

To apply Ampère's law, we mentally divide our imaginary loop into short, nearly straight, directed pieces, $d\vec{s}$. The direction of each of these pieces is tangent to the loop along the direction of integration. Assume that at the location of the element $d\vec{s}$ shown in Fig. 30-12, the net magnetic field due to the three currents is $\vec{B}$. Because the wires are perpendicular to the page, we know that the magnetic field at $d\vec{s}$ due to each current is in the plane of Fig. 30-12; thus, the net magnetic field $\vec{B}$ at $d\vec{s}$ must also be in that plane. However, we do not know the orientation of $\vec{B}$ within the plane. In Fig. 30-12, $\vec{B}$ is arbitrarily drawn at an angle ϕ to the direction of $d\vec{s}$.

The scalar product $\vec{B} \cdot d\vec{s}$ on the left side of Eq. 30-16 is then equal to $(B \cos\phi)\,ds$. Thus, Ampère's law can be written as

$$\oint \vec{B} \cdot d\vec{s} = \oint (B \cos \phi)\, ds = \mu_0 i^{\text{enc}}. \qquad (30\text{-}17)$$

We can now interpret the scalar product $\vec{B} \cdot d\vec{s}$ as being the product of a length ds of the Ampèrian loop and the field component $B \cos \phi$ that is tangent to the loop. Then we can interpret the integration as being the summation of all such products around the entire loop.

When we can actually perform this integration, we do not need to know the direction of $\vec{B}$ before integrating. Instead, we arbitrarily assume $\vec{B}$ to be generally in the direction of integration (as in Fig. 30-12). Then we use the following curled fingers-straight thumb right-hand rule to assign a plus sign or a minus sign to each of the currents that make up the net encircled current i^{enc}:

> **CURLED-STRAIGHT RIGHT-HAND RULE FOR AMPÈRE'S LAW:** Curl your right hand around the Ampèrian loop, with the fingers pointing in the direction of integration. A current through the loop in the general direction of your outstretched thumb is assigned a plus sign, and a current generally in the opposite direction is assigned a minus sign.

Finally, we solve Eq. 30-17 for the magnitude of $\vec{B}$. Once we have chosen a coordinate system to describe the system, we can use Ampère's law right-hand rule to decide whether $\vec{B}$ is positive or negative.

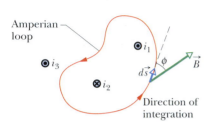

FIGURE 30-12 ■ Ampère's law applied to an arbitrary Ampèrian loop that encircles two long straight wires but excludes a third wire. Note the directions of the currents.

In Fig. 30-13 we apply the curled-straight rule for Ampère's law to the situation of Fig. 30-12. With the indicated counterclockwise direction of integration, the net current encircled by the loop is

$$i^{\text{enc}} = i_1 - i_2.$$

(Current i_3 is not encircled by the loop.) We can then rewrite Eq. 30-17 as

$$\left|\oint (B \cos \phi) ds\right| = \mu_0 \left|(i_2 - i_1)\right|. \qquad (30\text{-}18)$$

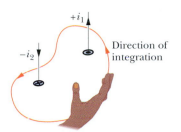

FIGURE 30-13 ■ A right-hand rule for Ampère's law, to determine the signs for currents encircled by an Ampèrian loop. The situation is that of Fig. 30-12.

You might wonder why, since current i_3 contributes to the magnetic-field magnitude B on the left side of Eq. 30-18, it is not needed on the right side. The answer is that the contributions of current i_3 to the magnetic field cancel out because the integration in Eq. 30-18 is made around the full loop. In contrast, the contributions of an encircled current to the magnetic field do not cancel out.

We cannot solve Eq. 30-18 for the magnitude B of the magnetic field, because for the situation of Fig. 30-12 we do not have enough information to solve the integral. However, we do know the magnitude of the integral; it must be equal to the value of $\mu_0 \left|(i_1 - i_2)\right|$, which is set by the net current passing through the loop. Next we apply Ampère's law to two situations in which symmetry does allow us to solve the integrals and determine the magnetic fields.

The Magnetic Field Outside a Long Straight Wire with Current

Figure 30-14 shows a long straight wire that carries current i (assumed to be uniformly distributed) that points directly out of the page. The equation for the magnetic field magnitude, B, produced by a long straight wire (Eq. 30-6) tells us that B depends only on the radial distance from the wire. That is, the field $\vec{B}$ has cylindrical symmetry about the wire. We can take advantage of that symmetry to simplify the integral in Ampère's law (Eqs. 30-16 and 30-17) if we encircle the wire with a concentric circular Ampèrian loop of radius r, as in Fig. 30-14. The magnetic field $\vec{B}$ then has the same magnitude B at every point on the loop. We shall integrate counterclockwise, so that $\vec{ds}$ has the direction shown in Fig. 30-14.

We can further simplify the quantity $B \cos \phi$ in Eq. 30-17 by noting that $\vec{B}$ is tangent to the loop at every point along the loop, as is $\vec{ds}$. Thus, $\vec{B}$ and $\vec{ds}$ are parallel at each point on the loop. Then at every point the angle ϕ between $\vec{ds}$ and $\vec{B}$ is 0° (so $\cos \phi = +1$). The magnitude of the integral in Eq. 30-17 then becomes

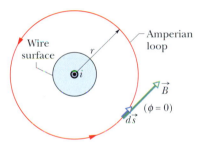

FIGURE 30-14 ■ Using Ampère's law to find the magnetic field produced by a current i in a long straight wire. The Ampèrian loop is a concentric circle that lies outside the wire.

$$\left|\oint \vec{B} \cdot \vec{ds}\right| = \oint (B \cos \phi) ds = B \left|\oint ds\right| = B(2\pi r).$$

Note that $\oint ds$ above is the summation of all the line segment lengths ds around the circular loop; that is, it simply gives the circumference $2\pi r$ of the loop.

The right side of Ampère's law becomes $+\mu_0 \left|i\right|$ and we then have

$$B(2\pi r) = \mu_0 i$$

or

$$B = \frac{\mu_0 \left|i\right|}{2\pi r}. \qquad (30\text{-}19)$$

With a slight change in notation, this is Eq. 30-6, which we derived earlier—with considerably more effort—using the Biot-Savart law. We know that the correct direction of $\vec{B}$ must be the counterclockwise one shown in Fig. 30-14 when i is positive. When i is negative, the correct direction for $\vec{B}$ is clockwise.

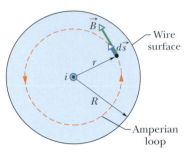

FIGURE 30-15 ■ Using Ampère's law to find the magnetic field that a current i produces inside a long straight wire of circular cross section. The current is uniformly distributed over the cross section of the wire and emerges from the page. An Ampèrian loop is drawn inside the wire.

The Magnetic Field Inside a Long Straight Wire with Current

Figure 30-15 shows the cross section of a long straight wire of radius R that carries a uniformly distributed current i either directly out of the page or directly into the page. Because the current is uniformly distributed over a cross section of the wire, the magnetic field $\vec{B}$ that it produces must have cylindrical symmetry. Thus, to find the magnetic field at points inside the wire, we can again use an Amperian loop of radius r, as shown in Fig. 30-15, where now $r < R$. Symmetry again requires that $\vec{B}$ is tangent to the loop, as shown, so the left side of Ampère's law again yields

$$\oint \vec{B} \cdot d\vec{s} = B \left| \oint d\vec{s} \right| = B(2\pi r). \tag{30-20}$$

To find the right side of Ampère's law, we note that because the current is uniformly distributed, the current i^{enc} encircled by the loop is proportional to the area encircled by the loop; that is,

$$i^{\text{enc}} = i \frac{\pi r^2}{\pi R^2} = i \frac{r^2}{R^2}. \tag{30-21}$$

Then Ampère's law gives us

$$B(2\pi r)\mu_0 |i^{\text{enc}}| = \mu_0 |i| \frac{r^2}{R^2}$$

or

$$B = \left(\frac{\mu_0 |i|}{2\pi R^2} \right) r. \tag{30-22}$$

Thus, inside the wire, the magnitude B of the magnetic field is proportional to r; that magnitude is zero at the center and a maximum at the surface, where $r = R$. Note that Eqs. 30-19 and 30-22 give the same value for B at $r = R$; that is, the expressions for the magnetic field outside the wire and inside the wire yield the same result at the surface of the wire.

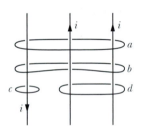

READING EXERCISE 30-4: The figure shows three equal currents i (two parallel and one antiparallel) and four Amperian loops. Rank the loops according to the magnitude of $\oint \vec{B} \cdot d\vec{s}$ along each, greatest first. ■

TOUCHSTONE EXAMPLE 30-3: Hollow Conducting Cylinder

Figure 30-16a shows the cross section of a long hollow conducting cylinder with inner radius $a = 2.0$ cm and outer radius $b = 4.0$ cm. The cylinder carries a current out of the page, and the current density in the cross section is given by $|\vec{J}| = cr^2$, with $c = 3.0 \times 10^6$ A/m^4 and r in meters. What is the magnitude of the magnetic field $\vec{B}$ at a point that is 3.0 cm from the central axis of the cylinder?

SOLUTION ■ The point at which we want to evaluate $\vec{B}$ is inside the material of the conducting cylinder, between its inner and outer radii. We note that the current distribution has cylindrical symmetry (it is the same all around the cross section for any given radius). Thus, the **Key Idea** here is that the symmetry allows us to use Ampère's law to find $\vec{B}$ at the point. We first draw the Ampèrian loop shown in Fig. 30-16b. The loop is concentric with the cylinder and has radius $R = 3.0$ cm, because we want to evaluate $\vec{B}$ at that distance from the cylinder's central axis.

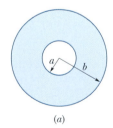

(a)

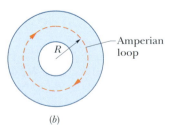

(b)

FIGURE 30-16 ■ (a) Cross section of a conducting cylinder of inner radius a and outer radius b. (b) An Ampèrian loop of radius R is added to compute the magnetic field at points that are a distance R from the central axis.

Next, we must compute the current i^{enc} that is encircled by the Ampèrian loop. However, a second **Key Idea** is that we *cannot* set up a proportionality as in Eq. 30-21, because here the current is not uniformly distributed. Instead, we must integrate the current density from the cylinder's inner radius a to the loop radius r. Since $\vec{J}$ and $d\vec{A}$ are parallel, $\vec{J} \cdot d\vec{A} = J\,dA$, so

$$|i^{enc}| = \left| \int \vec{J} \cdot d\vec{A} \right| = \left| \int_a^R cr^2(2\pi r\,dr) \right|$$

$$= \left| 2\pi c \int_a^R r^3\,dr \right| = \left| 2\pi c \left[\frac{r^4}{4} \right]_a^R \right|$$

since $\quad |i^{enc}| = \dfrac{\pi c(R^4 - a^4)}{2}, \quad$ since $R > a.$

The direction of integration indicated in Fig. 30-16b is (arbitrarily) clockwise. Applying the right-hand rule for Ampère's law to that loop, we find that we should take i^{enc} as negative because the current is directed out of the page but our thumb is directed into the page.

We next evaluate the left side of Ampère's law exactly as we did in Fig. 30-15, and we again obtain Eq. 30-20. Then Ampère's law,

$$\oint \vec{B} \cdot d\vec{s} = \mu_0 i^{enc},$$

gives us

$$(B \cos \phi)(2\pi R) = -\frac{\mu_0 \pi c}{2}(R^4 - a^4),$$

where $\cos \phi = \cos 0° = +1$ if $\vec{B}$ is parallel to $d\vec{s}$ and $\cos \phi = \cos 180° = -1$ if $\vec{B}$ is antiparallel to $d\vec{s}$. Solving for $(B \cos\phi)$ for $\phi = 180°$ and substituting known data yield

$$(B \cos \phi) = -\frac{\mu_0 c}{4R}(R^4 - a^4)$$

$$-B = -\frac{(4\pi \times 10^{-7}\,\text{T} \cdot \text{m/A})(3.0 \times 10^6\,\text{A/m}^4)}{4(0.030\,\text{m})}$$

$$\times\,[(0.030\,\text{m})^4 - (0.020\,\text{m})^4]$$

$$-B = -2.0 \times 10^{-5}\,\text{T}.$$

Thus, the magnetic field $\vec{B}$ at a point 3.0 cm from the central axis is

$$B = 2.0 \times 10^{-5}\,\text{T} \qquad\qquad \text{(Answer)}$$

and forms magnetic field lines that are directed opposite our direction of integration, hence counterclockwise in Fig. 30-16b.

30-6 Solenoids and Toroids

Magnetic Field of a Solenoid

We now turn our attention to another situation in which Ampère's law proves useful. It concerns the magnetic field produced by the current in a long, tightly wound helical coil of wire. Such a coil is called a **solenoid** (Fig. 30-17). Solenoids are very common electrical devices that are important in many technological applications.

To make the calculation simpler here, we will assume that the length of the solenoid is much greater than the diameter. Figure 30-18 shows a section through a portion of a "stretched-out" solenoid. The solenoid's magnetic field is the vector sum of the fields produced by the individual turns (loops) that make up the solenoid. For points very close to each turn, the wire behaves magnetically almost like a long straight wire, and the lines of $\vec{B}$ there are almost concentric circles. Figure 30-18 suggests that the field tends to cancel between adjacent turns. It also suggests that, at points inside the solenoid and reasonably far from the wire, $\vec{B}$ is approximately

FIGURE 30-17 ■ A solenoid carrying current i.

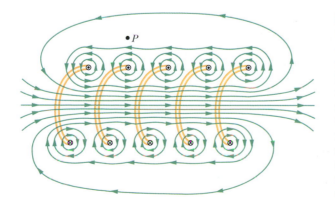

FIGURE 30-18 ■ A vertical cross section through the central axis of a "stretched-out" solenoid. The back portions of five turns are shown, as are the magnetic field lines due to a current through the solenoid. Each turn produces circular magnetic field lines near it. Near the solenoid's axis, the field lines combine into a net magnetic field that is directed along the axis. The closely spaced field lines there indicate a strong magnetic field. Outside the solenoid the field lines are widely spaced; the field there is very weak.

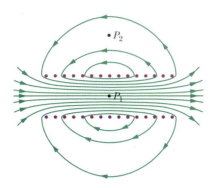

FIGURE 30-19 ■ Magnetic field lines for a real solenoid of finite length. The field is strong and uniform at interior points such as P_1 but relatively weak at external points such as P_2.

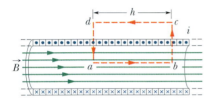

FIGURE 30-20 ■ Application of Ampère's law to a section of a long ideal solenoid carrying a current i. The Ampèrian loop is the rectangle $abcd$.

parallel to the (central) solenoid axis. In the limiting case of an *ideal solenoid,* which is infinitely long and consists of tightly packed (*close-packed*) turns of square wire, the field inside the coil is uniform and parallel to the solenoid axis.

At points above the solenoid, such as P in Fig. 30-18, the field set up by the upper parts of the solenoid turns (marked $\odot$) is directed to the left (as drawn near P) and tends to cancel the field set up by the lower parts of the turns (marked $\otimes$), which is directed to the right (not drawn). In the limiting case of an ideal solenoid, the magnetic field outside the solenoid is zero. Taking the external field to be zero is an excellent assumption for a real solenoid if its length is much greater than its diameter and if we consider external points such as point P that are not near either end of the solenoid. The direction of the magnetic field along the solenoid axis is given by a curled-straight right-hand rule: Grasp the solenoid with your right hand so that your fingers follow the direction of the current in the windings; your extended right thumb then points in the direction of the axial magnetic field.

Figure 30-19 shows the lines of $\vec{B}$ for a real solenoid. The spacing of the lines of $\vec{B}$ in the central region shows that the field inside the coil is fairly strong and uniform over the cross section of the coil. The external field, however, is relatively weak.

Let us now apply Ampère's law,

$$\oint \vec{B} \cdot d\vec{s} = \mu_0 i^{\,\text{enc}}, \tag{30-23}$$

to the ideal solenoid of Fig. 30-20, where $\vec{B}$ is uniform within the solenoid and zero outside it, using the rectangular Amperian loop $abcda$. We write $\oint \vec{B} \cdot d\vec{s}$ as the sum of four integrals, one for each loop segment:

$$\oint \vec{B} \cdot d\vec{s} = \int_a^b \vec{B} \cdot d\vec{s} + \int_b^c \vec{B} \cdot d\vec{s} + \int_c^d \vec{B} \cdot d\vec{s} + \int_d^a \vec{B} \cdot d\vec{s}. \tag{30-24}$$

The first integral on the right of Eq. 30-24 is Bh, where B is the magnitude of the uniform field $\vec{B}$ inside the solenoid and h is the (arbitrary) length of the segment from a to b. The second and fourth integrals are zero because for every element ds of these segments, $\vec{B}$ either is perpendicular to ds or is zero, and thus $\vec{B} \cdot d\vec{s}$ is zero. The third integral, which is taken along a segment that lies outside the solenoid, is zero because $\vec{B} = 0$ at all external points. Thus, $\oint \vec{B} \cdot d\vec{s}$ for the entire rectangular loop has the value Bh.

The net current $i^{\,\text{enc}}$ encircled by the rectangular Ampèrian loop in Fig. 30-20 is not the same as the current i in the solenoid windings because the windings pass more than once through this loop. Let n be the number of turns per unit length of the solenoid; then the loop encloses nh turns, so

$$i^{\,\text{enc}} = i(nh).$$

Ampère's law then gives us

$$Bh = \mu_0 |i| nh,$$

or
$$B = n\mu_0 |i| \qquad \text{(inside ideal solenoid).} \tag{30-25}$$

Although we derived Eq. 30-25 for an infinitely long ideal solenoid, it holds quite well for actual solenoids if we apply it only at interior points, well away from the solenoid ends. Equation 30-25 is consistent with the experimental fact that the magnetic field magnitude $|\vec{B}| = B$ within a solenoid does not depend on the diameter or the length of the solenoid and that B is uniform over the solenoidal cross section. A solenoid thus

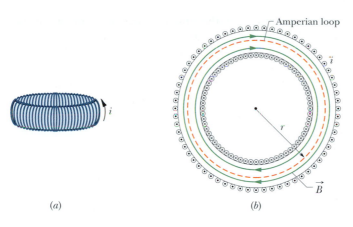

Amperian loop

i

r

$\vec{B}$

(a)

(b)

FIGURE 30-21 ■ (*a*) A toroid carrying a current *i*. (*b*) A horizontal cross section of the toroid. The interior magnetic field (inside the doughnut-shaped tube) can be found by applying Ampère's law with the Ampèrian loop shown.

provides a practical way to set up a known uniform magnetic field for experimentation, just as a parallel-plate capacitor provides a practical way to set up a known uniform electric field.

Magnetic Field of a Toroid

Figure 30-21*a* shows a **toroid,** which may be described as a solenoid bent into the shape of a hollow doughnut. What magnetic field $\vec{B}$ is set up at its interior points (within the hollow of the doughnut)? We can find out from Ampère's law and the symmetry of the toroid.

From the symmetry, we see that the lines of $\vec{B}$ form concentric circles inside the toroid, directed as shown in Fig. 30-21*b*. Let us choose a concentric circle of radius *r* as an Ampèrian loop and traverse it in the clockwise direction. Ampère's law (Eq. 30-16) yields

$$B(2\pi r) = N\mu_0 |i|,$$

where *i* is the current in the toroid windings (and is positive for those windings enclosed by the Ampèrian loop) and *N* is the total number of turns. This gives

$$B = \frac{N\mu_0 |i|}{2\pi} \frac{1}{r} \qquad \text{(toroid).} \qquad (30\text{-}26)$$

In contrast to the situation for a solenoid, $\vec{B}$ is not constant over the cross section of a toroid. With Ampère's law, it is easy to show that $\vec{B} = 0$ for points outside an ideal toroid (as if the toroid were made from an ideal solenoid).

The direction of the magnetic field within a toroid follows from our curled-straight right-hand rule: Grasp the toroid with the fingers of your right hand curled in the direction of the current in the windings; your extended right thumb points in the direction of the magnetic field.

30-7 A Current-Carrying Coil as a Magnetic Dipole

So far we have examined the magnetic fields produced by current in a long straight wire, a solenoid, and a toroid. We turn our attention here to the field produced by a coil carrying a current. You saw in Section 29-10 that such a coil behaves as a magnetic dipole in that, if we place it in an external magnetic field $\vec{B}$, a torque $\vec{\tau}$ given by

$$\vec{\tau} = \vec{\mu} \times \vec{B} \qquad (30\text{-}27)$$

acts on it. Here $\vec{\mu}$ is the magnetic dipole moment of the coil and has the magnitude NiA, where N is the number of turns (or loops), i is the current in each turn, and A is the area enclosed by each turn.

Recall that the direction of $\vec{\mu}$ is given by a curled-straight right-hand rule: Grasp the coil so that the fingers of your right hand curl around it in the direction of the current; your extended thumb then points in the direction of the dipole moment $\vec{\mu}$.

Magnetic Field of a Coil

We turn now to the other aspect of a current-carrying coil as a magnetic dipole. What magnetic field does *it* produce at a point in the surrounding space? The problem does not have enough symmetry to make Ampère's law useful, so we must turn to the Biot–Savart law. For simplicity, we first consider only a coil with a single circular loop and only points on its central axis, which we take to be a z axis. We shall show that the magnetic field at such points only has a z-component, B_z which is given by

$$\vec{B} = B_z\hat{k} = \frac{\mu_0 iR^2}{2(R^2 + z^2)^{3/2}}\hat{k}, \tag{30-28}$$

where R is the radius of the circular loop and z is the distance of the point in question from the center of the loop. Furthermore, the direction of the magnetic field $\vec{B}$ is the same as the direction of the magnetic dipole moment $\vec{\mu}$ of the loop.

For axial points far from the loop, we have $z \gg R$ in Eq. 30-28. With that approximation, the equation for the z-component of $\vec{B}$, which is a function of z only, reduces to

$$B_z \approx \frac{\mu_0 iR^2}{2z^3}.$$

Recalling that πR^2 is the area A of the loop and extending our result to include a coil of N turns, we can write this equation as

$$B_z = \frac{\mu_0}{2\pi}\frac{NiA}{z^3}.$$

Further, since $\vec{B}$ and $\vec{\mu}$ have the same direction, we can write the equation in vector form, substituting from $\mu = NiA$ (Eq. 29-32):

$$\vec{B} = B_z\hat{k} = \frac{\mu_0}{2\pi}\frac{\vec{\mu}}{z^3} \qquad \text{(current-carrying coil).} \tag{30-29}$$

Note that the magnetic constant μ_0 and the magnetic moment vector $\vec{\mu}$ are completely different quantities with different units. The choice of the symbol μ to represent both quantities is unfortunate.

In summary, we have two ways in which we can regard a current-carrying coil as a magnetic dipole: (1) it experiences a torque when we place it in an external magnetic field; (2) it generates its own intrinsic magnetic field, given by Eq. 30-29 for distant points along its axis. Figure 30-22 shows some magnetic field lines for a current loop; one side of the loop acts as a north pole (in the direction of $\vec{\mu}$) and the other side as a south pole, as suggested by the lightly drawn magnet in the figure.

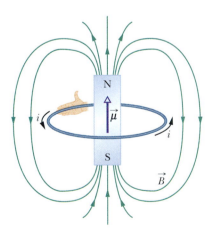

FIGURE 30-22 ■ A current loop produces a magnetic field like that of a bar magnet and thus has associated north and south poles. The magnetic dipole moment $\vec{\mu}$ of the loop, given by a curled-straight right-hand rule, points from the south pole to the north pole, in the direction of the field $\vec{B}$ within the loop.

Proof of Equation 30-28

Figure 30-23 shows the back half of a circular loop of radius R carrying a current i. Consider a point P on the axis of the loop, a distance z from its plane. Let us apply the Biot–Savart law to a differential element $d\vec{s}$ of the loop, located at the left side of the loop. The length vector $d\vec{s}$ for this element points perpendicularly out of the page. The angle θ between $d\vec{s}$ and $\vec{r}$ in Fig. 30-23 is $90°$; the plane formed by these two vectors is perpendicular to the plane of the figure and contains both $\vec{r}$ and $d\vec{s}$. Using the Biot–Savart law and the right-hand rule, we see that the differential field $d\vec{B}$ produced at point P by the current in this element is perpendicular to this plane. Thus $d\vec{B}$ lies in the plane of the figure, perpendicular to $\vec{r}$ (as indicated in Fig. 30-23).

Let us resolve $d\vec{B}$ into two components: $d\vec{B}_\parallel$ along the axis of the loop and $d\vec{B}_\perp$ perpendicular to this axis. From the symmetry, the vector sum of all the perpendicular components $d\vec{B}_\perp$ due to all the loop elements ds is zero. This leaves only the axial components $d\vec{B}_\parallel$ and we have the magnitude of the axial component given by

$$B_\parallel = \int dB_\parallel.$$

For the element $d\vec{s}$ in Fig. 30-23, the Biot–Savart law (Eq. 30-1) tells us that the magnitude of the axial magnetic field component at distance r is

$$dB_\parallel = \frac{\mu_0}{4\pi} \frac{i\, ds \sin 90°}{r^2}.$$

We also have

$$dB_\parallel = dB \cos \alpha.$$

Combining these two relations, we obtain

$$dB_\parallel = \frac{\mu_0 i \cos \alpha\, ds}{4\pi r^2}. \tag{30-30}$$

Figure 30-23 shows that r and α are not independent but are related to each other. Let us express each in terms of the variable z, the distance between point P and the center of the loop. The relations are

$$r = \sqrt{R^2 + z^2} \tag{30-31}$$

and

$$\cos \alpha = \frac{R}{r} = \frac{R}{\sqrt{R^2 + z^2}}. \tag{30-32}$$

Substituting Eqs. 30-31 and 30-32 into Eq. 30-30, we find

$$dB_\parallel = \frac{\mu_0 i R}{4\pi (R^2 + z^2)^{3/2}}\, ds.$$

Note that i, R, and z have the same values for all elements $d\vec{s}$ around the loop, so when we integrate this equation, we find that the magnitude of the axial field component is given as

$$B_\parallel = \oint dB_\parallel$$

$$= \frac{\mu_0 i R}{4\pi (R^2 + z^2)^{3/2}} \oint ds,$$

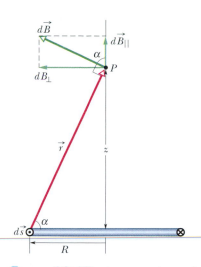

FIGURE 30-23 ■ A current loop of radius R. The plane of the loop is perpendicular to the page and only the back half of the loop is shown. We use the law of Biot and Savart to find the magnetic field at point P on the central axis of the loop.

or, since $\oint ds$ is simply the circumference $2\pi R$ of the loop, the axial or z-component of the magnetic field is

$$\vec{B} = B_z\hat{k} = \frac{\mu_0 iR^2}{2(R^2 + z^2)^{3/2}}\hat{k},$$

which is Eq. 30-28, the relation we sought to prove.

READING EXERCISE 30-5: The figure here shows four arrangements of circular loops of radius r or $2r$, centered on vertical axes (perpendicular to the loops) and carrying identical currents in the directions indicated. Assume the sizes of the loops are exaggerated and that $z \gg R$. Rank the arrangements according to the magnitude of the net magnetic field at the dot, midway between the loops on the central axis, greatest first.

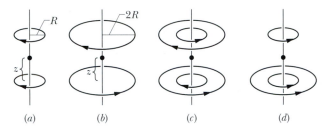

(a)   (b)   (c)   (d)

Problems

SEC. 30-3 ■ CALCULATING THE MAGNETIC FIELD DUE TO A CURRENT

1. Surveyor A surveyor is using a magnetic compass 6.1 m below a power line in which there is a steady current of 100 A. (a) What is the magnitude of the magnetic field at the site of the compass due to the power line? (b) Will this interfere seriously with the compass reading? The horizontal component of the Earth's magnetic field at the site is 20 μT.

2. Electron Gun The electron gun in a traditional television tube fires electrons of kinetic energy 25 keV at the screen in a circular beam 0.22 mm in diameter; 5.6×10^{14} electrons arrive each second. Calculate the magnitude of the magnetic field produced by the beam at a point 1.5 mm from the beam axis.

3. Philippines At a certain position in the Philippines, the magnitude of the Earth's magnetic field of 39 μT is horizontal and directed due north. Suppose the net field is zero exactly 8.0 cm above a long, straight, horizontal wire that carries a constant current. What are (a) the size and (b) the direction of the current?

4. Locate Points A long wire carrying a current of 100 A is placed in a uniform external magnetic field of 5.0 mT. The wire is perpendicular to this magnetic field. Locate the points at which the net magnetic field is zero.

5. Particle with Positive Charge A particle with positive charge q is a distance d from a long straight wire that carries a current i; the particle is traveling with speed $|\vec{v}|$ perpendicular to the wire. What are the direction and magnitude of the force on the particle if it is moving (a) toward and (b) away from the wire?

6. Semicircular Arcs A straight conductor carrying a current i splits into identical semicircular arcs as shown in Fig. 30-24. What is

the magnitude of the magnetic field at the center C of the resulting circular loop?

7. Two Semi-Infinite A wire carrying current i has the configuration shown in Fig. 30-25. Two semi-infinite straight sections, both tangent to the same circle, are connected by a circular arc, of central angle ϕ, along the circumference of the circle, with all sections lying in the same plane. What must ϕ be in order for $|\vec{B}|$ to be zero at the center of the circle?

8. Use Biot–Savart Use the Biot–Savart law to calculate the magnitude and direction of the magnetic field $\vec{B}$ at C, the common center of the semicircular arcs AD and HJ in Fig. 30-26a. The two arcs, of radii R_2 and R_1, respectively, form part of the circuit $ADJHA$ carrying current i.

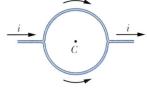

FIGURE 30-24 ■ Problem 6.

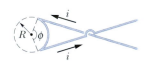

FIGURE 30-25 ■ Problem 7.

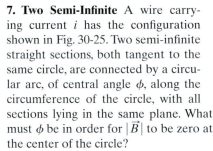

(a)

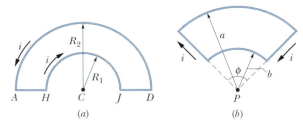

(b)

FIGURE 30-26 ■ Problems 8 and 9.

9. Curved Segments In the circuit of Fig. 30-26b, the curved segments are arcs of circles of radii a and b with common center P. The straight segments are along radii. Find the magnitude and direction of the magnetic field $\vec{B}$ at point P, assuming a current i in the circuit.

10. Magnitude and Directions The wire shown in Fig. 30-27 carries current i. What are the magnitude and direction of the magnetic field $\vec{B}$ produced at the center C of the semicircle by (a) each straight segment of length L, (b) the semicircular segment of radius R, and (c) the entire wire?

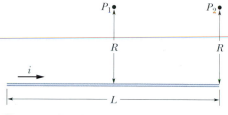

FIGURE 30-27 ■
Problem 10.

11. Straight Wire In Fig. 30-28, a straight wire of length L carries current i. Show that the magnitude of the magnetic field $\vec{B}$ produced by this segment at P_1, a distance R from the segment along a perpendicular bisector, is

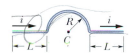

FIGURE 30-28 ■ Problems 11 and 13.

$$B = \frac{\mu_0 |i|}{2\pi R} \frac{L}{(L^2 + 4R^2)^{1/2}}.$$

Show that this expression for $|\vec{B}|$ reduces to an expected result as $L \to \infty$.

12. Square Loop A square loop of wire of edge length a carries current i. Using the results of Problem 11, show that, at the center of the loop, the magnitude of the magnetic field produced by the current is

$$B = \frac{2\sqrt{2}\,\mu_0 |i|}{\pi a}.$$

13. Length L In Fig. 30-28, a straight wire of length L carries current i. Show that

$$B = \frac{\mu_0 |i|}{4\pi R} \frac{L}{(L^2 + R^2)^{1/2}}$$

gives the magnitude of the magnetic field $\vec{B}$ produced by the wire at P_2, a perpendicular distance R from one end of the wire.

14. Rectangular Loop Using the results of Problem 11, show that the magnitude of the magnetic field produced at the center of a rectangular loop of wire of length L and width W, carrying a current i, is

$$B = \frac{2\mu_0 |i|}{\pi} \frac{(L^2 + W^2)^{1/2}}{LW}.$$

15. Square Loop Two A square loop of wire of edge length a carries current i. Using the results of Problem 11, show that the magnitude of the magnetic field produced at a point on the axis of the loop and a distance x from its center is

$$B(x) = |\vec{B}(x)| = \frac{4\mu_0 |i| a^2}{\pi(4x^2 + a^2)(4x^2 + 2a^2)^{1/2}}.$$

Prove that this result is consistent with the result of Problem 12.

16. Length a In Fig. 30-29, a straight wire of length a carries a current i. Show that the magnitude of the magnetic field produced by the current at point P is $B = \sqrt{2}\mu_0 |i|/8\pi a$.

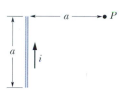

FIGURE 30-29 ■
Problem 16.

17. Two Wires Two wires, both of length L, are formed into a circle and a square, and each carries current i. Show that the square produces a greater magnetic field at its center than the circle produces at its center. (See Problem 12.)

18. Magnetic Field Find the magnitude and direction of the magnetic field $\vec{B}$ at point P in Fig. 30-30. (See Problem 16.)

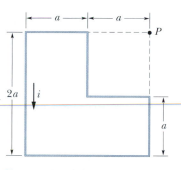

FIGURE 30-30 ■ Problem 18.

19. Long Thin Ribbon Figure 30-31 shows a cross section of a long thin ribbon of width w that is carrying a uniformly distributed total current i into the page. Calculate the magnitude and direction of the magnetic field $\vec{B}$ at a point P in the plane of the ribbon at a distance d from its edge. (*Hint:* Imagine the ribbon to be constructed from many long, thin, parallel wires.)

FIGURE 30-31 ■
Problem 19.

20. Find Magnitude and Direction Find the magnitude and direction of the magnetic field $\vec{B}$ at point P in Fig. 30-32, for $|i| = 10$ A and $a = 8.0$ cm. (See Problems 13 and 16.)

21. Perpendicular Bisector Figure 30-33 shows two very long straight wires (in cross section) that each carry currents of 4.00 A directly out of the page. Distance $d_1 = 6.00$ m and distance $d_2 = 4.00$ m. What is the magnitude of the net magnetic field at point P, which lies on a perpendicular bisector to the wires?

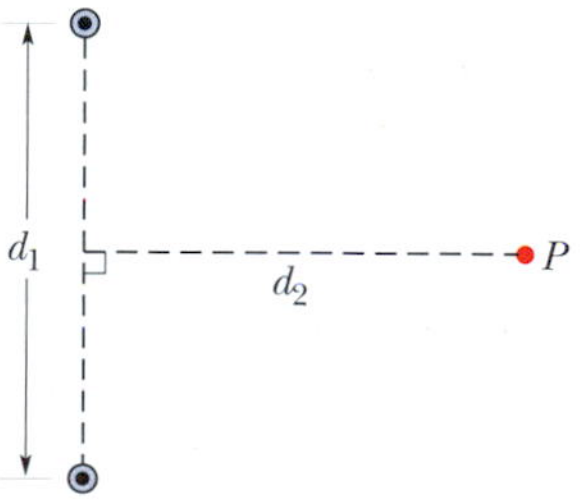

22. Greatest and 10% In Fig. 30-34, point P is at perpendicular distance $R = 2.00$ cm from a very long straight wire carrying a current. The magnetic field $\vec{B}$ set up at point P is due to contributions from all the identical current-length elements $i\,d\vec{s}$ along the wire. What is the distance s to the current-length element that makes (a) the greatest contribution to field $\vec{B}$ and (b) 10% of the greatest contribution?

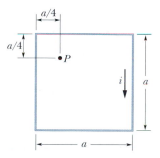

FIGURE 30-32 ■
Problem 20.

FIGURE 30-33 ■
Problem 21.

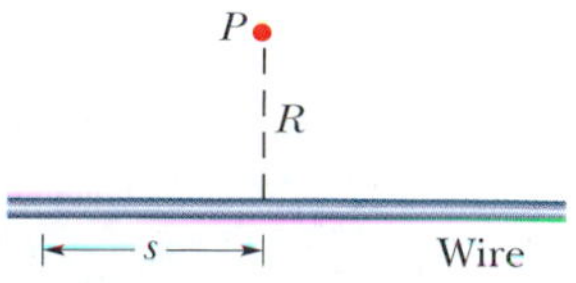

FIGURE 30-34 ■
Problem 22.

SEC. 30-4 ■ FORCE BETWEEN TWO PARALLEL CURRENTS

23. Two Parallel Wires Two long parallel wires are 8.0 cm apart. What equal currents must be in the wires if the magnetic field halfway between them is to have a magnitude of 300 μT? Answer for both (a) parallel and (b) antiparallel currents.

24. *i* and *3i* Two long parallel wires a distance *d* apart carry currents of *i* and *3i* in the same direction. Locate the point or points at which their magnetic fields cancel.

25. Two Parallel Wires Two Two long, straight, parallel wires, separated by 0.75 cm, are perpendicular to the plane of the page as shown in Fig. 30-35. Wire 1 carries a current of 6.5 A into the page. What must be the current (magnitude and direction) in wire 2 for the resultant magnetic field at point *P* to be zero?

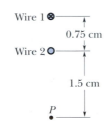

Wire 1

0.75 cm

Wire 2

1.5 cm

P

FIGURE 30-35 ■
Problem 25.

26. Five Parallel Wires Figure 30-36 shows five long parallel wires in the *xy* plane. Each wire carries a current *i* = 3.00 A in the positive *x* direction. The separation between adjacent wires is *d* = 8.00 cm. In unit-vector notation, what are the magnitude and direction of the magnetic force per meter exerted on each of these five wires by the other wires?

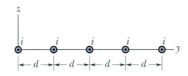

FIGURE 30-36 ■ Problem 26.

27. Four Long Wires Four long copper wires are parallel to each other, their cross sections forming the corners of a square with sides *a* = 20 cm. A 20 A current exists in each wire in the direction shown in Fig. 30-37. What are the magnitude and direction of $\vec{B}$ at the center of the square?

28. Four Currents Form a Square Four identical parallel currents *i* are arranged to form a square of edge length *a* as in Fig. 30-37, *except* that they are *all* out of the page. What is the force per unit length (magnitude and direction) on any one wire?

FIGURE 30-37 ■
Problems 27, 28, and 29.

29. Force per Unit Length In Fig. 30-37, what is the force per unit length acting on the lower left wire, in magnitude and direction, with the current directions as shown? The currents are *i*.

30. Idealized Schematic Figure 30-38 is an idealized schematic drawing of a rail gun. Projectile *P* sits between two wide rails of circular cross section; a source of current sends current through the rails and through the (conducting) projectile itself (a fuse is not used). (a) Let *w* be the distance between the rails, *R* the radius of

the rails, and *i* the current. Show that the magnitude of the force on the projectile is directed to the right along the rails and is given approximately by

$$F = |\vec{F}| = \frac{i^2 \mu_0}{\pi} \ln \frac{w + R}{R}.$$

(b) If the projectile starts from the left end of the rails at rest, find the speed *v* at which it is expelled at the right. Assume that $|i|$ = 450 kA, *w* = 12 mm, *R* = 6.7 cm, *L* = 4.0 m, and the mass of the projectile is *m* = 10 g.

31. Rectangular Loop Two In Fig. 30-39, the long straight wire carries a current of 30 A and the rectangular loop carries a current of 20 A. Calculate the resultant force acting on the loop. Assume that *a* = 1.0 cm, *b* = 8.0 cm, and *L* = 30 cm.

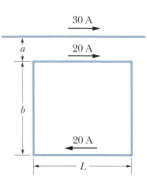

30 A

20 A

a

b

20 A

L

FIGURE 30-39 ■
Problem 31.

SEC. 30-5 ■ AMPÈRE'S LAW

32. Eight Wires Eight wires cut the page perpendicularly at the points shown in Fig. 30-40. A wire labeled with the integer *k* (*k* = 1, 2, . . . , 8) carries the current *ki*. For those with odd *k*, the current is out of the page; for those with even *k*, it is into the page. Evaluate $\oint \vec{B} \cdot d\vec{s}$ along the closed path in the direction shown.

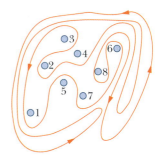

FIGURE 30-40 ■
Problem 32.

33. Eight Conductors Each of the eight conductors in Fig. 30-41 carries 2.0 A of current into or out of the page. Two paths are indicated for the line integral $\oint \vec{B} \cdot d\vec{s}$. What is the value of the integral for the path (a) at the left and (b) at the right?

 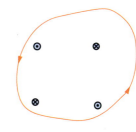

FIGURE 30-41 ■ Problem 33.

34. Cross Section of a Cylindrical Conductor Figure 30-42 shows a cross section of a long cylindrical conductor of radius *a*, carrying a uniformly distributed current *i*. Assume that *a* = 2.0 cm and *i* = 100 A, and plot the magnitude of the magnetic field $|\vec{B}(r)| = B(r)$ over the range $0 < r < 6.0$ cm.

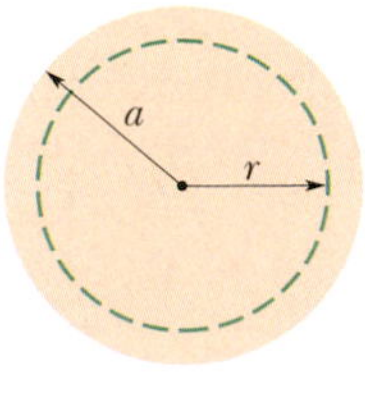

35. Cannot Drop to Zero Show that a uniform magnetic field $\vec{B}$ cannot drop abruptly to zero (as is suggested by the lack of field lines

a

r

FIGURE 30-42 ■
Problem 34.

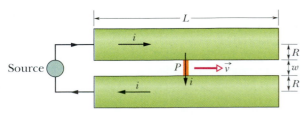

L

i

Source

P $\vec{v}$

i

R

w

R

FIGURE 30-38 ■ Problem 30.

to the right of point a in Fig. 30-43) as one moves perpendicular to $\vec{B}$, say along the horizontal arrow in the figure. (*Hint:* Apply Ampère's law to the rectangular path shown by the dashed lines.) In actual magnets "fringing" of the magnetic field lines always occurs, which means that $\vec{B}$ approaches zero in a gradual manner. Modify the field lines in the figure to indicate a more realistic situation.

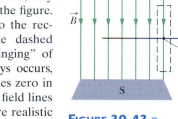

FIGURE 30-43
Problem 35.

36. Two Square Conducting Loops Two square conducting loops carry currents of 5.0 and 3.0 A as shown in Fig. 30-44. What is the value of $\oint \vec{B} \cdot d\vec{s}$ for each of the two closed paths shown?

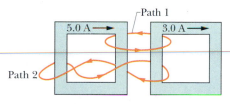

FIGURE 30-44 ▪ Problem 36.

37. Current Density The current density inside a long, solid, cylindrical wire of radius a is in the direction of the central axis and varies linearly with radial distance r from the axis according to $|\vec{J}| = |\vec{J}_0| r/a$. Find the magnitude and direction of the magnetic field inside the wire.

38. Uniformly Distributed Current A long straight wire (radius = 3.0 mm) carries a constant current distributed uniformly over a cross section perpendicular to the axis of the wire. If the magnitude of the current density is 100 A/m², what are the magnitudes of the magnetic fields (a) 2.0 mm from the axis of the wire and (b) 4.0 mm from the axis of the wire?

39. Cylindrical Hole Figure 30-45 shows a cross section of a long cylindrical conductor of radius a containing a long cylindrical hole of radius b. The axes of the cylinder and hole are parallel and are a distance d apart; a current i is uniformly distributed over the tinted area. (a) Use superposition to show that the magnitude of the magnetic field at the center of the hole is

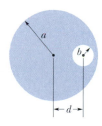

FIGURE 30-45 ▪
Problem 39.

$$B = \frac{\mu_0 |i| d}{2\pi(a^2 - b^2)}.$$

(b) Discuss the two special cases $b = 0$ and $d = 0$. (*Hint:* Regard the cylindrical hole as resulting from the superposition of a complete cylinder (no hole) carrying a current in one direction and a cylinder of radius b carrying a current in the opposite direction, both cylinders having the same current density.)

40. Circular Pipe A long circular pipe with outside radius R carries a (uniformly distributed) current i into the page as shown in Fig. 30-46. A wire runs parallel to the pipe at a distance of $3R$ from

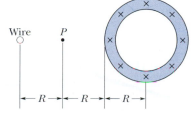

FIGURE 30-46 ▪ Problem 40.

center to center. Find the amount and direction of the current in the wire such that the net magnetic field at point P has the same magnitude as the net magnetic field at the center of the pipe but is in the opposite direction.

41. Conducting Sheet Figure 30-47 shows a cross section of an infinite conducting sheet lying in the x-y plane, carrying a current per unit x-length of λ; the current emerges perpendicularly out of the page. (a) Use the Biot–Savart law and symmetry to show that for all points P above the sheet, and all points P' below it, the magnetic field $\vec{B}$ is parallel to the sheet and directed as shown. (b) Use Ampère's law to prove that $B = \frac{1}{2}\mu_0 |\lambda|$ at all points P and P'.

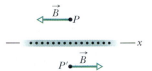

FIGURE 30-47 ▪ Problems 41 and 48.

42. Field at P is Zero Figure 30-48 shows, in cross section, two long straight wires; the 3.0 A current in the right-hand wire is out of the page. What are the size and direction of the current in the left-hand wire if the net magnetic field at point P is to be zero?

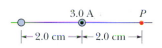

FIGURE 30-48 ▪
Problem 42.

SEC. 30-6 ▪ SOLENOIDS AND TOROIDS

43. Field Inside Solenoid A 200-turn solenoid having a length of 25 cm and a diameter of 10 cm carries a current of 0.30 A. Calculate the magnitude of the magnetic field $\vec{B}$ inside the solenoid.

44. Field Inside Solenoid Two A solenoid that is 95.0 cm long has a radius of 2.00 cm and a winding of 1200 turns; it carries a current of 3.60 A. Calculate the magnitude of the magnetic field inside the solenoid.

45. Toroid A toroid having a square cross section, 5.00 cm on a side, and an inner radius of 15.0 cm has 500 turns and carries a current of magnitude 0.800 A. (It is made up of a square solenoid—instead of a round one as in Fig. 30-21—bent into a doughnut shape.) What is the magnitude of the magnetic field inside the toroid at (a) the inner radius and (b) the outer radius of the toroid?

46. Length of Wire A solenoid 1.30 m long and 2.60 cm in diameter carries a current of 18.0 A. The magnitude of the magnetic field inside the solenoid is 23.0 mT. Find the length of the wire forming the solenoid.

47. Field Inside Toroid In Section 30-6, we showed that the magnitude of the magnetic field at any radius r *inside* a toroid is given by

$$B = \frac{\mu_0 |i| N}{2\pi r}.$$

Show that as you move from any point just inside a toroid to a point just outside, the magnitude of the *change* in $\vec{B}$ that you encounter is just $\mu_0 |\lambda|$. Here $|\lambda|$ is the amount of current per unit length along a circumference of radius r within the toroid. Compare this with the similar result found in Problem 48. Isn't the equality surprising?

48. Solenoid as Cylindrical Conductor Treat an ideal solenoid as a thin cylindrical conductor whose current per unit length, measured parallel to the cylinder axis, is λ. (a) By doing so, show that the magnitude of the magnetic field inside an ideal solenoid can be written as $B = \mu_0|\lambda|$. This is the value of the *change* in $\vec{B}$ that you encounter as you move from inside the solenoid to outside, through the solenoid wall. (b) Show that the same change occurs as you move through an infinite flat current sheet such as that of Fig. 30-47 (see Problem 41). Does this equality surprise you?

49. Direction of Field A long solenoid with 10.0 turns/cm and a radius of 7.00 cm carries a current of 20.0 mA. A current of 6.00 A exists in a straight conductor located along the central axis of the solenoid. (a) At what radial distance from the axis will the direction of the resulting magnetic field be at 45.0° to the axial direction? (b) What is the magnitude of the magnetic field there?

50. Find Current in Solenoid A long solenoid has 100 turns/cm and carries current i. An electron moves within the solenoid in a circle of radius 2.30 cm perpendicular to the solenoid axis. The speed of the electron is $0.0460c$ (c = speed of light). Find the amount of current $|i|$ in the solenoid.

SEC. 30-7 ■ A CURRENT-CARRYING COIL AS A MAGNETIC DIPOLE

51. Magnetic Dipole What is the magnetic dipole moment $\vec{\mu}$ of the solenoid described in Problem 43?

52. One Turn Coil Figure 30-49a shows a length of wire carrying a current i and bent into a circular coil of one turn. In Fig. 30-49b the same length of wire has been bent more sharply, to give a coil of two turns, each of half the original radius. (a) If B_a and B_b are the magnitudes of the magnetic fields at the centers of the two coils, what is the ratio B_b/B_a? (b) What is the ratio of the magnitude of the dipole moments, μ_b/μ_a of the coils?

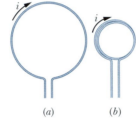

(a) (b)

FIGURE 30-49 ■ Problem 52.

53. Student's Electromagnet A student makes a short electromagnet by winding 300 turns of wire around a wooden cylinder of diameter $d = 5.0$ cm. The coil is connected to a battery producing a current of 4.0 A in the wire. (a) What is the magnetic moment of this device? (b) At what axial distance $z \gg d$ will the magnetic field of this dipole have the magnitude 5.0 μT (approximately one-tenth that of the Earth's magnetic field)?

54. Helmholtz Figure 30-50 shows an arrangement known as a Helmholtz coil. It consists of two circular coaxial coils, each of N turns and radius R, separated by a distance R. The two coils carry equal currents i in the same direction. Find the magnitude of the net magnetic field at P, midway between the coils.

55. Field as a Function of Distance Two 300-turn coils of radius R each

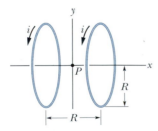

FIGURE 30-50 ■ Problems 54, 55, and 57.

carry a current i. They are arranged a distance R apart, as in Fig. 30-50. For $R = 5.0$ cm and $i = 50$ A, plot the magnitude $|B(x)| = B(x)$ of the net magnetic field as a function of distance x along the common x axis over the range $x = -5$ cm to $x = +5$ cm, taking $x = 0$ at the midpoint P. (Such coils provide an especially uniform field $\vec{B}$ near point P.) (*Hint:* See Eq. 30-28.)

56. Square Current Loop The magnitude $B(x)$ of the magnetic field at points on the axis of a square current loop of side a is given in Problem 15. (a) Show that the axial magnetic field of this loop, for $x \gg a$, is that of a magnetic dipole (see Eq. 30-29). (b) What is the magnitude of the magnetic dipole moment of this loop?

57. Let the Separation Be In Problem 54 (Fig. 30-50), let the separation of the coils be a variable s (not necessarily equal to the coil radius R). (a) Show that the first derivative of the magnitude of the net magnetic field of the coils (dB/dx) vanishes at the midpoint P regardless of the value of s. Why would you expect this to be true from symmetry? (b) Show that the second derivative (d^2B/dx^2) also vanishes at P, provided $s = R$. This accounts for the uniformity of B near P for this particular coil separation.

58. abcdefgha A conductor carries a current of 6.0 A along the closed path $abcdefgha$ involving 8 of the 12 edges of a cube of side 10 cm as shown in Fig. 30-51. (a) Why can one regard this as the superposition of three square loops: $bcfgb$, $abgha$, and $cdefc$? (*Hint:* Draw currents around those square loops.) (b) Use this superposition to find the magnetic dipole moment $\vec{\mu}$ (magnitude and direction) of the closed path. (c) Calculate the magnitude and direction of the magnetic field $\vec{B}$ at the points $(x, y, z) = (0.0$ m, 5.0 m, 0.0 m) and $(5.0$ m, 0.0 m, 0.0 m).

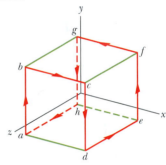

FIGURE 30-51 ■ Problem 58.

59. What Torque A circular loop of radius 12 cm carries a current of 15 A. A flat coil of radius 0.82 cm, having 50 turns and a current of 1.3 A, is concentric with the loop. (a) What magnetic field $\vec{B}$ (magnitude and direction) does the loop produce at its center? (b) What torque acts on the coil? Assume that the planes of the loop and coil are perpendicular and that the magnetic field due to the loop is essentially uniform throughout the volume occupied by the coil.

60. Two Different Arcs A length of wire is formed into a closed circuit with radii a and b, as shown in Fig. 30-52 and carries a current i. (a) What are the magnitude and direction of $\vec{B}$ at point P? (b) Find the magnetic dipole moment of the circuit.

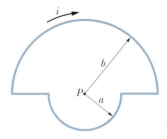

FIGURE 30-52 ■ Problem 60.

Additional Problems

61. Cross Section of a Wire Figure 30-53 shows the cross section of a wire that is perpendicular to the plane of the paper. Suppose a compass is placed at location A, which is a distance r from the wire. The compass points in the direction shown in the diagram. (a) Resketch the diagram and draw arrows to show what direction you expect the compass to point if it were moved to locations B and C. *Note:* Use the symbol $\odot$ if the flow is out of the page and the symbol $\otimes$ if the flow is into the page. (b) Indi-

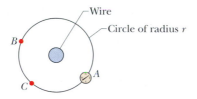

FIGURE 30-53 ■ Problem 61.

cate in what direction *positive current* is flowing through the wire and describe the rule you are using to deduce the direction of current in the wire. (c) What is the direction of the flow of *electrons* through the wire?

62. Wires in a B-Field A uniform magnetic field is directed toward the right in the plane of the paper as shown in Fig. 30-54. A wire lying perpendicular to the plane of the paper at location A carries

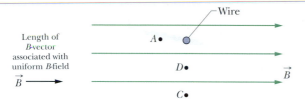

FIGURE 30-54 ■ Problem 62.

a current i. Suppose that the resultant magnetic field at point D due to a superposition of the uniform magnetic field of magnitude B and the magnetic field of the wire of magnitude B_w is zero. (a) Is the direction of the current in the wire into or out of the paper? Explain how you arrived at your conclusion. (b) Assume that point A lies at the same distance from the center of the wire as point D and that the length of the vector assigned to represent the magnitude of the uniform external magnetic field is that shown on the right. Construct a vector diagram showing the net magnetic field vector B_A^{net} at point A. (c) Assume that point C is twice the distance from the center of the wire as point D. Construct a vector diagram showing the net magnetic field vector, B_C^{net}, at point C. (Adapted from A. Arons, Homework and Test Questions for Introductory Physics Teaching, John Wiley and Sons, 1994.)

63. Earth's Field The magnitude of the Earth's magnetic field, B, at either geomagnetic pole, is about 7×10^{-5} T. Using a model in which you assume that this field is produced by a single current loop at the equator, determine the current that would generate such a field ($R_e = 6.37 \times 10^6$ m). *Hint:* The magnitudes of the magnetic field due to a single current loop of radius R at a distance R from its center and perpendicular to the plane of the loop is given by the equation

$$B = \frac{\mu_0 |i|}{2\sqrt{8}R}$$

and see Fig. 30-55.

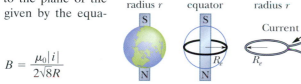

FIGURE 30-55 ■ Problem 63.

64. Comparing Electric and Magnetic Forces One In this problem we consider situations corresponding to three different long thin lines of matter containing charges: 1. A copper wire carrying an electric current from left to right, 2. A long amber rod that has been rubbed with fur and has a uniform excess of negative charge, and 3.

Additional Problems **885**

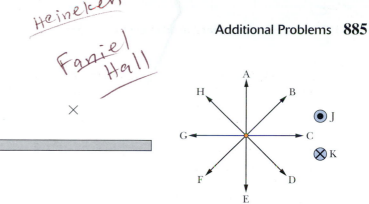

FIGURE 30-56 ■ Problem 64.

A beam of electrons passing from left to right through a vacuum inside a cathode ray tube. The direction of the electric current and of the electron flow are from left to right. Figure 30–56 shows a location marked x and a set of directions with labels on the right.

(a) For each of the three lines of matter, indicate in what direction the electric and magnetic fields at the location x would point. To indicate the direction, use one of the letters associated with a directional arrow on the "compass" in Fig. 30-56. If any of the fields are zero, write 0.

(b) Now consider placing a positive charge at the location x. In one case it is stationary, while in a second case it is moving in the direction C (to the right). Indicate the direction nearest to the total force the charge would feel. (Ignore gravity and air resistance.) Do this for all three lines of matter and for both cases.

65. Comparing Electric and Magnetic Forces Two Figure 30-57 shows a long wire carrying a current i to the right and a long amber rod with a charge density (charge/unit length) of λ. Assume that i and λ are both positive.

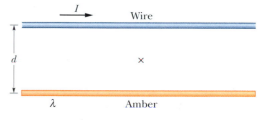

FIGURE 30-57 ■ Problem 65.

(a) The two are separated by a distance d. The point marked x is halfway between them. Copy this figure onto your paper and draw arrows to represent the following. (Be sure to label your arrows clearly to show which one is which.)

  i. the direction of the magnetic field at the point marked x
  ii. the direction of the electric field at the point marked x
  iii. the direction of the electric force that a positive charge q placed at x would feel
  iv. the direction of the magnetic force that a positive charge q placed at x would feel if it were moving to the right.

(b) The current, i, is $+10$ A, the charge density, λ, is -1 nC/m ($= 10^{-9}$ C/m) (note that it is negative), and the distance between the wires is 40 cm. At the instant shown, a proton with charge $q = 1.6 \times 10^{-19}$ C is moving into the page with a speed $v = 10^6$ m/s. Ignoring gravity, what is the magnitude and direction of the net force the proton feels at that time?

66. Direction of Magnetic Forces Figure 30-58 shows a cross section of four long parallel wires (labeled A through D) taken in a plane perpendicular to the wires. One or more of the wires may be carrying a current. If a wire carries a current, i_0, it is in the direction indicated and has strength $|i_0|$.

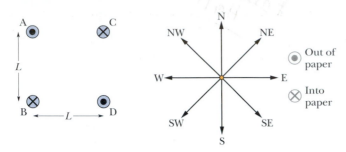

FIGURE 30-58 ■ Problem 66.

For each of the four vector quantities listed in (i) through (iv) below give the direction of the quantity. To indicate the direction, use one of the directions on the "compass" in Fig. 30-58. If the magnitude of the quantity is zero, write "0." If it is nonzero but in none of the indicated directions, write "Other."

i. Only wires B and D are carrying current. The direction of the force on wire D is_____.
ii. Only wires B and D are carrying current. The direction of the force felt by an electron traveling in the E direction (on the compass) is_____.
iii. Only wires B and D are carrying current. The direction of the force felt by an electron traveling in the N direction (on the compass) is_____.
iv. All four wires are carrying current. The direction of the net force felt by wire A is_____.

67. Magnetic Forces and Fields Figure 30-59 shows a cross section of three long parallel wires (labeled A through C) taken in a plane perpendicular to the wires. One or more of the wires may be carrying a current. If a wire carries a current, i_0, it is in the direction indicated and has strength $|i_0|$. For each of the five vector quantities (1) through (5) shown, indicate the direction of the quantity on the compass in Fig. 30-59. If the magnitude of the quantity is zero, write "0." If the result is not zero but points in a direction other than one of those indicated, write "other."

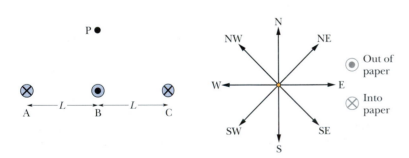

FIGURE 30-59 ■ Problem 67.

1. The magnetic field at point P if only wire A is carrying a current
2. The magnetic field at wire C if only wire A is carrying a current
3. The magnetic force on wire C if only wires A and C carry currents
4. The magnetic force on wire C if only wire A is carrying a current
5. The magnetic force on a proton at P traveling to the right (i.e., in direction E) if only wire B is carrying a current.

68. Right-Hand Rules During our discussions of magnetism and rotation we have encountered a number of different right-hand

rules for obtaining the direction or sign of various quantities. Describe three right-hand rules. In your discussion of each one, include a statement of the equation or law in which the rule is applied, and whether the rule is "fundamental" or derived from a more basic principle.

69. Magnetic Forces Figure 30-60 shows parts of two long, current-carrying wires labeled 1 and 2. The wires lie in the same plane and cross at right angles at the point indicated. When carrying a current, each wire carries the same amount of current in the direction shown. At the right is shown

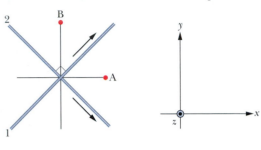

FIGURE 30-60 ■ Problem 69.

a set of coordinate directions for describing the direction of vectors.

For each of the vectors discussed, indicate the direction of the vector using the coordinate system shown. For example, you might specify "the $+x$ direction" or "the $-z$ direction" or "in the x-y plane at 45° between the $+x$ and $+y$ directions." If the magnitude of the vector requested is zero, write "0."

(a) The direction of the force on a positively charged ion at the point B moving in the $+y$ direction if only wire 1 carries current
(b) The direction of the force on a positively charged ion at the point B moving in the $-z$ direction if both wires carry current
(c) The direction of the force on a positively charged ion at the point A moving in the $+x$ direction if only wire 2 carries current

For the next two parts of the problem, select which answer is correct if both wires carry current.
(d) The magnetic force on wire 1 will

i. push it in the $-z$ direction
ii. push it in the $+z$ direction
iii. tend to rotate it clockwise about the joining point
iv. tend to rotate it counterclockwise about the joining point
v. none of the above

(e) The magnetic force on wire 2 will

i. push it in the $-z$ direction
ii. push it in the $+z$ direction
iii. tend to rotate it clockwise about the joining point
iv. tend to rotate it counterclockwise about the joining point
v. none of the above

70. Constrained to a Circle Figure 30-61 shows, in cross section, two long straight wires held against a plastic cylinder of radius 20.0 cm. Wire 1 carries current $i_1 = 60.0$ mA out of the page and is fixed in place at the left side of the cylinder. Wire 2 carries current $i_2 = 40.0$ mA out of the page and can be moved around the cylinder. At what angle θ_2 should wire 2 be positioned such that the net magnetic field at the origin from the two currents has a magnitude of 80.0 nT?

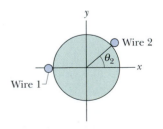

FIGURE 30-61 ■ Problem 70.

71. Element Length Figure 30-62a shows an element of length $ds = 1.00 \ \mu m$ in a very long straight wire carrying current. The current in that element sets up a differential magnetic field $d\vec{B}$ at points in the surrounding space. Figure 30-62b gives the magnitude dB of the field in pico-Teslas (10^{-12} T) for points 2.5 cm from the element, as a function of angle θ between the wire and a straight line to the point. What is the magnitude of the magnetic field set up by the entire wire at perpendicular distance 2.5 cm from the wire?

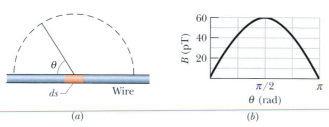

(a)   (b)

FIGURE 30-62 ■ Problem 71.

72. Where Is Wire 2 Two long straight thin wires with current lie against an equally long plastic cylinder, at radius $R = 20.0$ cm from the cylinder's central axis. Figure 30-63a shows, in cross section, the cylinder and wire 1 but not wire 2. With wire 2 fixed in place, wire 1 is moved around the cylinder, from angle $\theta_1 = 0°$ to angle $\theta_1 = 180°$, through the first and second quadrants of the xy coordinate system. The net magnetic field $\vec{B}$ at the center of the cylinder is measured as a function of θ_1. Figure 30-63b gives the x-component B_x of that field in micro-Teslas (10^{-6} T) and Fig. 30-63c gives the y-component B_y, both as functions of θ_1. (a) At what angle θ_2 is wire 2 located? What are the size and direction of the currents in (b) wire 1 and (c) wire 2?

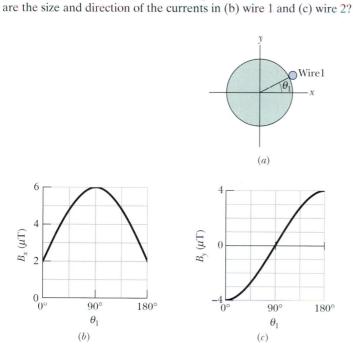

(a)

(b)   (c)

FIGURE 30-63 ■ Problem 72.

73. The Ratio of Currents Figure 30-64a shows, in cross section, two long, parallel wires carrying current and separated by distance L. The ratio $|i_1/i_2|$ of their current amounts is 4.00; the directions of the currents are not indicated. Figure 30-64b shows the y-compo-

nent B_y in nano-Teslas (10^{-9} T) of their net magnetic field along the x axis to the right of wire 2. (a) At what value of $x > 0$ is B_y maximum? (b) If $|i_2| = 3$ mA, what is the value of the maximum? What are the directions of (c) current i_1 and (d) current i_2?

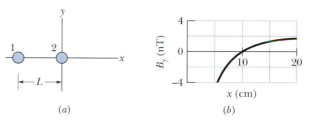

(a)   (b)

FIGURE 30-64 ■ Problem 73.

74. Same Radius Different Current In Fig. 30-65a two circular loops, with different currents but the same radius of 4.0 cm, are centered on a y axis. They are initially separated by distance $L = 3.0$ cm, with loop 2 positioned at the origin of the axis. The currents in the two loops produce a net magnetic field at the origin, with y-component B_y. That component is to be measured as loop 2 is gradually moved in the positive direction of the y axis. Figure 30-65b gives B_y in micro-Teslas (10^{-6} T) as a function of the position y of loop 2. The curve approaches an asymptote of $B_y = 7.20 \ \mu T$ as $y \rightarrow \infty$. What are (a) current i_1 in loop 1 and (b) current i_2 in loop 2?

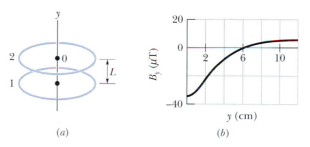

(a)   (b)

FIGURE 30-65 ■ Problem 74.

75. How Many Revolutions An electron is shot into one end of a solenoid, as it enters the uniform magnetic field within the solenoid, its speed is 800 m/s and its velocity vector makes an angle of 30° with the central axis of the solenoid. The solenoid carries 4.0 A and has 8000 turns along its length. How many revolutions does the electron make along its helical path within the solenoid by the time it emerges from the solenoid's opposite end? (In a real solenoid, where the field is not uniform at the two ends, the number of revolutions would be slightly less than the answer here.)

76. Force per Unit Length Two Figure 30-66 shows wire 1 in cross section; the wire is long and straight, carries a current of 4.00 mA out of the page, and is at distance $d_1 = 2.4$ cm from a surface. Wire 2, which is parallel to wire 1 and also long, is at horizontal distance $d_2 = 5.0$ cm from wire 1 and carries a current of 6.80 mA into the page. What is the x component of the magnetic force per unit length on wire 2 due to the current in wire 1?

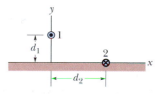

FIGURE 30-66 ■ Problem 76.

31 | Induction and Maxwell's Equations

The General Motors EV1 electric car was marketed in the southwest with two generations of vehicles in 1997 and 1999. Production was completed in 1999 and all leases have been assigned. The EV1 had no engine, tailpipe, valves, pistons, timing belts, or crankshaft. The EV1 came with an inductive charging system in which there was no metal-to-metal connection. The charger, which plugged into a 220-volt outlet, had a paddle which, when inserted into the charge port at the front of the car, provided the electricity to re-charge the batteries.

How can electric car batteries be charged without making electrical contact with the power source?

The answer is in this chapter.

31-1 Introduction

In the previous chapter we discovered that the moving charges that make up electric currents create magnetic fields. We also learned that both permanent magnets and moving charges can exert forces on each other. These discoveries have powerful practical consequences. They allow us to build electromagnets to create large magnetic fields. More significantly, they enable us to harness the forces these large magnetic fields can exert on moving charges to create electric motors capable of moving massive objects.

In 1820, when Oersted observed that electric currents create magnetic fields, a number of prominent scientists began to look for ways to use magnetic fields to create currents. For more than a decade, scientists searched for current induced by static magnetic fields and failed to find it. By 1831, both Michael Faraday (Fig 31-1) and an American physicist, Joseph Henry, had discovered that a *changing* magnetic field is required to induce electric current. This phenomenon is called **electromagnetic induction.**

The discovery of electromagnetic induction, usually credited to Faraday, was of tremendous technological importance. Induction made it possible to create electric power from motion. Indeed, by the end of the 19th century, systems had been developed for the generation and transmission of electric power. Applications of Faraday's and Henry's discoveries are found in the design of thousands of electrical devices including transformers, high-speed trains, inductive battery chargers, and electric guitar pickups.

Although the practical benefits of the discovery of induction are tremendous, so is its impact on science. Many scientists view Faraday's law of induction as one of the most profound laws in all of classical physics because it "closed the loop" between magnetism and electricity. By combining Faraday's law with Ampère's law, we can understand how electricity and magnetism can be treated as complimentary aspects of the same phenomenon. By the middle of the 19th century, James Clerk Maxwell incorporated the ideas of Faraday and others into a famous set of four equations describing electromagnetic phenomena. In this chapter you will learn about the characteristics of electromagnetic induction and about Maxwell's synthesis of electromagnetic interactions.

READING EXERCISE 31-1: Why did it take so long for scientists working in the early 19th century to actually observe magnetic induction? ■

FIGURE 31-1 ■ Michael Faraday, a famous English scientist, is credited with the discovery of electromagnetic induction.

31-2 Induction by Motion in a Magnetic Field

Let us start our treatment of electromagnetic induction by considering what happens if we move a coil of conducting wire at a constant velocity through a uniform magnetic field and then out of the field as shown in Fig. 31-2. This is not the observation made by Faraday. We will describe that later. Notice that the diagram shows the plane of the coil is always perpendicular to the direction of the magnetic field. Under what conditions can a current be induced? We will consider this situation from both an experimental and a theoretical perspective.

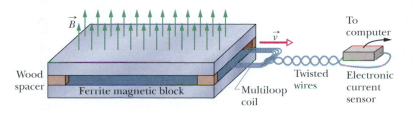

FIGURE 31-2 ■ It is not difficult to measure the current induced in a coil of wire while it is being pulled out of the gap between a pair of ferrite blocks separated by wooden spacers. The magnetic field in the central area between the magnetic blocks is essentially uniform. The ends of a multi-loop coil are connected to an electronic current sensor.

FIGURE 31-3 ■ A computer data acquisition system is used to measure the induced current 200 times a second as the coil shown in Fig. 31-2 is pulled steadily out of the uniform magnetic field in the central part of the gap between the two magnetic ferrite blocks. From 0.0 s to 0.3 s the entire coil is in the uniform magnetic field. After 0.9 s the coil is entirely outside the magnetic field. Between 0.3 s and 0.9 s part of the coil is in the *B*-field and part is outside of it.

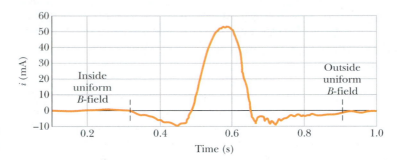

A Conductor Moving Through a Magnetic Field — Observation

If we connect the ends of the coil to an ammeter, we see the needle jump back and forth a bit erratically during the time that the coil is passing out of the gap between the magnets. When the whole area of the coil is still in the central part of the gap between the magnets, the ammeter needle points to zero. When the coil has completely emerged from the region of space influenced by the magnets, the ammeter needle points to zero once again. This current jump can be seen in more detail using an electronic current sensor as shown in Fig. 31-3.

Both casual observation with a sensitive ammeter and the data gathered using an electronic current sensor show that for this situation:

• When the coil is not moving, there is no induced current no matter what the steady magnetic field is like at its location.

• When the coil is moving through a region where the magnetic field is entirely uniform or zero, there is no induced current.

• When the coil is moving through a region where the steady magnetic field is not uniform, a current is induced.

We can draw the following conclusion from these observations:

> **OBSERVATION:** When a conducting loop moves perpendicular to a magnetic field, a current will be induced whenever the coil experiences a *changing* magnetic field.

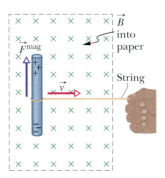

FIGURE 31-4 ■ A piece of wire is pulled through a uniform magnetic field at a constant velocity and becomes polarized.

A Conductor Moving Through a Magnetic Field — Theory

Although it's not obvious without reflection, you are capable of predicting that a *changing* magnetic field is required to induce an electric current in a moving coil. This induction is a natural consequence of the magnetic force laws described in Eqs. 29-2 and 29-23.

Straight Conductor Moving in a Uniform $\vec{B}$-Field: Let's start by using the force law to predict what happens to a straight piece of conducting wire if we pull it at a constant velocity in a direction perpendicular to a uniform magnetic field (Fig. 31-4). Each of the charges in the conductor experiences a force given by the magnetic force law, $\vec{F}^{\text{mag}} = q\vec{v} \times \vec{B}$ (Eq. 29-2). The direction of the force is given by the right-hand rule for cross products as shown in Fig. 31-5. Since there are mobile electrons in metals, these electrons will move toward the bottom of the wire, exposing fixed excess positive charge at the top of the wire. Thus, a current will flow in the wire for an instant until the electric field created by the charge separation opposes any further electron flow.

FIGURE 31-5 ■ The right-hand rule for the magnetic force law provides an "upward" force on positive changes and a "downward" force on mobile negative charges present in the wire shown in Fig. 31-4.

Loop Moving in a Uniform $\vec{B}$-Field: Perhaps we can induce a current by forming a closed loop instead of a single length of wire. This doesn't help because both the left and right segments (*a* and *c* in Fig. 31-6) are perpendicular to the motion, so they become polarized in the same manner. This merely results in excess charge piling up on the top and bottom segments (*b* and *d* as shown in Fig. 31-6).

Loop Moving from a Uniform $\vec{B}$-Field to No Field: The easiest way to create a current using the polarization caused by the magnetic force law is to pull our loop in such a way that segment *a* is inside the magnetic field and segment *c* is not. Now what happens? We can see from Fig. 31-7 that the electrons in segment *a*, which is the trailing loop segment, continue to have magnetic forces exerted on them. But there are no forces on electrons in segment *c* of the loop because these electrons are not in the magnetic field. The only force that contributes to the current flow in the loop is the force on the left segment of wire, so the electrons in this segment of wire are pushed downward. The result is a net flow of electrons in a counterclockwise direction. Since "conventional current" as defined in Chapter 26 represents the flow of positive charge carriers, conventional current flow would be clockwise as shown in Fig. 31-7.

Nonuniform $\vec{B}$-Field: Theoretically we still expect to be able to induce a current in our loop in any nonuniform magnetic field. For example, suppose the magnetic field in Fig. 31-7 is weaker (but not necessarily zero) on the right side of the loop (near segment *c*) than it is at the left (near segment *a*). In this case, the magnetic forces on electrons in the left and right segments of the loop will no longer be equal and the forces on the charges in one of the segments will overpower those on the charges in the other segments. This will cause a net current to be induced.

Our theoretical considerations enable us to conclude that by applying the magnetic force law, we can predict the results of the observations presented in the first part of this section: When a conducting loop moves perpendicular to a magnetic field, then a current will be induced whenever the coil experiences a *changing* magnetic field through it.

READING EXERCISE 31-2: In the discussion above, we determined that the forces on the electrons in the top and bottom segments of the wire loop shown in Fig. 31-6 did not contribute to the current flow. Why is this the case? ∎

READING EXERCISE 31-3: Suppose the magnetic field shown in Fig. 31-6 varies continuously in such a way that it is always stronger on the right than it is on the left. What will be the direction of the resulting (conventional) current in the loop? Explain. ∎

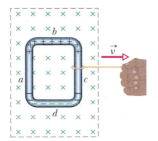

FIGURE 31-6 ∎ A wire loop is pulled by a string through a uniform magnetic field at a constant velocity. Although excess charge accumulates on the top and bottom segments (*b* and *d*), no current is induced in the loop.

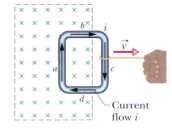

FIGURE 31-7 ∎ A wire loop is pulled by a string through a region where the magnetic field is uniform on one side and zero on the other. Electrons from segment *a* are allowed to flow counterclockwise around the loop. The conventional current flow is clockwise.

31-3 Induction by a Changing Magnetic Field

Michael Faraday made a significant contribution to physics when he asked: What happens if instead of moving the wire loop in a magnetic field we keep the wire loop *stationary* and move a magnet toward or away from the loop to create a "moving" or changing magnetic field? One might argue that since the electrons in the wire are not moving in this case, the velocity of the loop segments used in the magnetic force law expression is zero and so there should be no force on the electrons and therefore no current. On the other hand, in many ways these two situations are the same. In order to answer his question, Faraday made observations similar to those discussed below.

Observation 1, with a magnet: Figure 31-8 shows a conducting loop connected to a sensitive ammeter. Since there is no battery or other source of emf included, there is no current in the circuit. What happens if we move a bar magnet toward the loop? We observe that a current suddenly appears in the circuit! But the current disappears as soon as the magnet stops moving. If we then move the magnet away, a current again

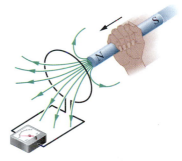

FIGURE 31-8 ∎ An idealized setup showing a current meter registering nonzero currents in a stationary wire loop when a magnet is moving near the loop. (Typically a multiturn loop is needed to generate a detectable current.)

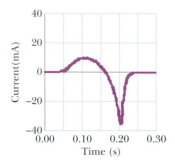

FIGURE 31-9 ■ The current induced as a magnet is dropped through a stationary multiturn coil (like that shown in Fig. 31-8). A computer data acquisition system is used to record current data at 2000 points/second.

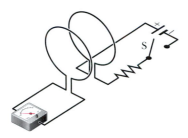

FIGURE 31-10 ■ An idealized setup showing an ammeter registering a current in the left-hand wire loop while switch S is being closed or opened (to turn the current in the right-hand loop on and off). No motion of the coils is involved. Faraday made essentially the same observation using multi-loop coils.

suddenly appears, but now in the opposite direction. If we experimented for a while, we would observe the following:

1. A current appears only if there is relative motion between the loop and the magnet (one must move relative to the other, but it doesn't matter which one); the current disappears when the relative motion between them ceases. See Fig. 31-9 for a graph of the current induced by a magnet dropped through a stationary coil.

2. Faster motion produces a greater current.

3. If moving the magnet's north pole toward the loop causes, say, clockwise current, then moving the north pole away causes counterclockwise current. Moving the south pole toward or away from the loop also causes current, but in the reversed direction.

We call the current produced in the loop an **induced current;** the work done per unit charge to produce that current (to move the conduction electrons that constitute the current) is called an **induced emf,** and the process of producing the current and emf is called induction. Currents that are caused by batteries in a circuit and those caused by induction in a wire loop are the same—mobile electrons are flowing through wires.

Observation 2, replacing the magnet with a current-carrying coil: Let us now perform a second observation. For this observation we use the apparatus of Fig. 31-10, with the two conducting loops close to each other but not touching. If we close switch S to turn on a current in the right-hand loop, the meter suddenly and briefly registers a current—an induced current—in the left-hand loop. If we then open the switch, another sudden and brief induced current appears in the left-hand loop, but in the opposite direction. We get an induced current (and thus an induced emf) only when the current in the right-hand loop is changing (either when turning on or off) and not when it is constant (even if the current is large). The outcome of this second observation is not surprising. We know from Ampère's law (Chapter 30) that the magnitude of the magnetic field surrounding a current-carrying wire increases as the current increases and its direction changes when the direction of current changes.

Faraday also noticed that the actual amount of magnetic field present at the area enclosed by the loop does not matter. Instead, the values of the induced emf and induced current are determined by the *rate* at which the amount changes.

When we pull all of these observations together, the way Faraday did, we conclude that

> Induced emf and current are present whenever the magnetic field present in the area subtended by the conducting loop *changes* for any reason.
>
> The amount of induced emf and current increases as a function of the rate of change of the magnetic field present at the area subtended by the loop.

Charging an Electric Car by Induction

Suppose we replace the switch in Fig. 31-10 with a source of current in the right-hand loop that varies over time sinusoidally. Then we create a magnetic field at the location of the left-hand loop that is also changing sinusoidally in time. This time-varying magnetic field then induces current in the left-hand loop that varies with time as well. By using some circuitry to filter out the negative current, we can use this induced current to charge a battery even though there is *no electrical contact* between the right- and left-hand loops. This type of noncontact charging is also used for charging familiar devices such as electric toothbrushes. Although an actual charger for an electric toothbrush or car like that discussed in the chapter-opening puzzler has more loops of wire and electrical circuits in it, it works on the same principle of electromagnetic induction discovered by Henry and Faraday in the early 19th century.

One drawback of inductive charging is that it is slower than direct charging. This is not a problem for electric toothbrush charging, but it is for electric car charging. This is probably one reason why the inductively charged General Motors EV1 cars like the one shown in the chapter puzzler have been taken off the market. You will learn more about the practical applications of induction in Chapter 32.

READING EXERCISE 31-4: Can the magnetic force law be used to explain why a current appears in a stationary loop when a bar magnet is brought close to it? If so, use your understanding of this force law to explain how this happens. If not, justify why not. ■

READING EXERCISE 31-5: Consider the induced current data shown in Fig. 31-9. The magnet is accelerating as it falls through the stationary coil. The magnet is dropped in free fall. The extrema of currents are about +8mA and −35mA. Why is the negative extremum larger? ■

31-4 Faraday's Law

We can enhance the predictive power of Faraday's qualitative observations by developing a mathematical formulation of electromagnetic induction. The mathematical expression that describes electromagnetic induction is commonly known as **Faraday's law.** Although we derive Faraday's law for a simplified situation using concepts and laws that we have already introduced, it can be applied to virtually any situation.

Magnetic Flux

To begin we use the concept of magnetic flux to quantify the amount of magnetic field at the area enclosed by a loop. In Chapter 24, in a similar situation, we needed to calculate the amount of an electric field present on a surface. There we determined electric flux for a small element of essentially flat area in Eq. 24-2 as $\Phi^{elec} = \vec{E} \cdot \Delta\vec{A}$ (the dot product of the normal vector representing a small area and the electric field vector at the location of the area). By analogy, the *magnetic flux* at the surface of a small area element $\Delta\vec{A}$ that is located in a magnetic field $\vec{B}$ is defined as

$$\Phi^{mag} = \vec{B} \cdot \Delta\vec{A} \qquad \text{(magnetic flux at an area } \Delta\vec{A}\text{)}. \qquad (31\text{-}1)$$

Simply put, the flux of magnetic field Φ^{mag} at an area element A is the product of the area element and the component of the field *perpendicular* to it for a uniform magnetic field. The validity of this basic definition depends on the assumption that the magnetic field $\vec{B}$ is uniform over the surface element $\Delta\vec{A}$. If the field varies over the area, we must break the area up into little pieces in such a way that the field will be about constant for each piece. We then calculate the flux in each little piece and perform an integration to add up all the little contributions in analogy to the more general definition of electric flux.

From Eq. 31-1, we see that the SI unit for magnetic flux is the tesla-square meter, which is called the weber (abbreviated Wb):

$$1 \text{ weber } = 1 \text{ Wb} = 1 \text{ T} \cdot \text{m}^2. \qquad (31\text{-}2)$$

Using our simplified formulation of magnetic flux, we are now ready to derive Faraday's law.

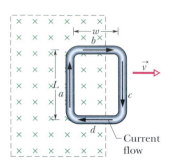

FIGURE 31-11 ■ A wire loop is moving at a constant velocity through a region where the magnetic field is uniform on one side and zero on the other. While this is happening, the magnetic flux at the area subtended by the coil is decreasing at a constant rate.

A Simplified Derivation of Faraday's Law

Consider the simple situation depicted in Fig. 31-7 in which a wire loop is being pulled out of a uniform magnetic field at a constant velocity. Next we derive the relationship between the emf induced in the loop and the rate of change of the magnetic flux enclosed by the loop. To help us with the derivation we have redrawn the situation and introduced symbols for the dimensions of the loop and the axis along which it moves in Fig. 31-11.

According to the magnetic force law, each charge in the left part of the loop (segment a) will experience a force of magnitude $F^{mag} = qvB$. As the positive and negative charges separate, an electric field of magnitude

$$E = \frac{F^{mag}}{|q|} = vB \tag{31-3}$$

will be generated. If segment a has a length L, then the potential difference of induced emf across it is given by

$$\mathscr{E} = EL = vBL. \tag{31-4}$$

Next we need to relate the right side of Eq. 31-4 to the rate at which the magnetic flux at the area subtended by the loop is decreasing as it moves out of the uniform B-field. If we designate the loop as being pulled in the x direction, then its velocity component can be expressed as $v_x = dx/dt$. Note that the area of the moving loop is decreasing at a rate given by $dA/dt = -L\,dx/dt = -Lv_x$. Since the magnetic field that subtends the left part of the area enclosed by the loop is constant, the rate of change of the magnetic flux at the loop can be expressed as

$$\frac{d\Phi^{mag}}{dt} = \frac{d(BA)}{dt} = B\frac{dA}{dt} = -v_x BL. \tag{31-5}$$

Combining Eqs. 31-4 with 31-5, we get an expression for Faraday's law for a single loop or coil,

$$\mathscr{E} = -\frac{d\Phi^{mag}}{dt} \quad \text{(Faraday's law for a single-turn coil)}. \tag{31-6}$$

As you will see in the next section, the induced emf $\mathscr{E}$ tends to oppose the flux change, and the minus sign indicates that opposition. Faraday's law can also be expressed in words:

> The amount of the emf $\mathscr{E}$ induced in a conducting loop is equal to the rate at which the magnetic flux Φ^{mag} at the area enclosed by the loop changes with time.

If we change the magnetic flux at a coil of N turns, an induced emf appears in every turn and the total emf induced in the coil is the sum of these individual induced emfs. If the coil is tightly wound (closely packed), so that the same magnetic flux Φ^{mag} is present in each turn, the total emf induced in the coil is

$$\mathscr{E} = -N\frac{d\Phi^{mag}}{dt} \quad \text{(Faraday's law for an N-turn coil)}. \tag{31-7}$$

Although we have used simple geometry to derive Faraday's law (Eq. 31-7), experiments (such as the one shown in Fig. 31-12) have verified that the mathematical expression we have derived is true for any situation where the flux enclosed by a set

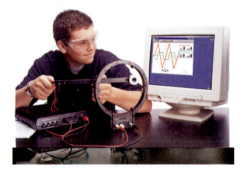

FIGURE 31-12 ■ It is quite easy to verify Faraday's law with modern apparatus and computer data acquisition systems. Here a student holds a small multiturn pickup coil inside a larger field coil that is generating a "sawtooth" magnetic field that increases and then decreases continuously. The B-field is shown on the jagged dark red trace on the computer screen. The induced current in the pickup coil is shown by the squarish lighter green trace. (Photo courtesy of PASCO scientific.)

of conducting loops or coils is changing. In fact, there are many ways to change the magnetic flux at a coil and thus induce emfs and currents:

1. Change the magnitude B of the magnetic field within the coil.

2. Change the area of the coil, or the portion of that area that happens to lie within the magnetic field (for example, by expanding the coil or sliding it out of the field).

3. Change the angle between the direction of the magnetic field $\vec{B}$ and the area of the coil (for example, by rotating the coil so that $\vec{B}$ is first perpendicular to the plane of the coil and then is along that plane).

Later on in the chapter we will derive a more general form of Faraday's law that relates flux change to electric field induction even when no charges or conducting loops are present.

READING EXERCISE 31-6: The graph gives the magnitude $B(t)$ of a magnetic field that exists throughout the area subtended by a conducting loop, perpendicular to the plane of the loop. Although it changes with time, at any particular instant the magnetic field is uniform over the area of the loop. (a) Rank the five time intervals (a, b, c, d, and e) shown on the graph according to the amount of the emf $|\mathcal{E}|$ induced in the loop, greatest first. (b) Explain your reasoning. ■

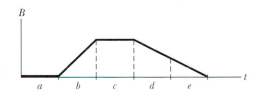

TOUCHSTONE EXAMPLE 31-1: Coil in a Long Solenoid

The long solenoid S shown (in cross section) in Fig. 31-13 has 220 turns/cm and carries a current $i = 1.5$ A; its diameter D is 3.2 cm. At its center we place a 130-turn, closely packed coil C of diameter $d = 2.1$ cm. The current in the solenoid is reduced to zero at a steady rate in 25 ms. What is the size of the emf $|\mathcal{E}|$ that is induced in coil C while the current in the solenoid is changing?

SOLUTION ■ The **Key Ideas** here are these:

1. Because coil C is located in the interior of the solenoid, it lies within the magnetic field produced by current i in the solenoid; thus, there is a magnetic flux Φ^{mag} present in coil C.

2. Because current i decreases, flux Φ^{mag} also decreases.

3. As Φ^{mag} decreases, emf $\mathcal{E}$ is induced in coil C, according to Faraday's law.

Because coil C consists of more than one turn, we apply Faraday's law in the form of Eq. 31-7 ($\mathcal{E} = -N\,d\Phi^{\mathrm{mag}}/dt$), where the number

of turns N is 130 and $d\Phi^{\mathrm{mag}}/dt$ is the rate at which the flux in each turn changes.

Because the current in the solenoid decreases at a steady rate, flux Φ^{mag} also decreases at a steady rate and we can write $d\Phi^{\mathrm{mag}}/dt$ as $\Delta\Phi^{\mathrm{mag}}/\Delta t$. Then, to evaluate $\Delta\Phi^{\mathrm{mag}}$, we need the final and initial flux. The final flux Φ_f^{mag} is zero because the final current in the solenoid is zero. To find the initial flux Φ_i^{mag}, we need two more **Key Ideas:**

4. The flux at the area enclosed by each turn of coil C depends on the area A and orientation of that turn in the solenoid's magnetic field $\vec{B}$. Because $\vec{B}$ is uniform and directed perpendicular to area A, the flux is given by Eq. 31-1 ($\Phi^{\mathrm{mag}} = BA$).

5. The magnitude B of the magnetic field in the interior of a solenoid depends on the solenoid's current i and its number n of turns per unit length, according to Eq. 30-25 ($B = n\mu_0|i|$).

For the situation of Fig. 31-13, A is $\frac{1}{4}\pi d^2 (= 3.46 \times 10^{-4}$ m^2) and n is 220 turns/cm, or 22 000 turns/m. Substituting Eq. 30-25 into Eq. 31-1 then leads to

$$\Phi_i^{\mathrm{mag}} = BA = (n\mu_0|i|)A$$

$$= (22\,000 \text{ turns/m})(4\pi \times 10^{-7} \text{ T·m/A})(1.5 \text{ A})(3.46 \times 10^{-4} \text{ m}^2)$$

$$= 1.44 \times 10^{-5}\,\text{Wb}.$$

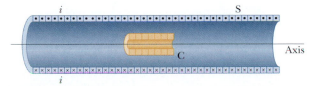

FIGURE 31-13 ■ A coil C is located inside solenoid S, which carries current i.

Now we can write

$$\frac{d\Phi^{\text{mag}}}{dt} = \frac{\Delta\Phi^{\text{mag}}}{\Delta t} = \frac{\Phi_f^{\text{mag}} - \Phi_i^{\text{mag}}}{\Delta t}$$

$$= \frac{(0 - 1.44 \times 10^{-5}\ \text{Wb})}{25 \times 10^{-3}\ \text{s}}$$

$$= -5.76 \times 10^{-4}\ \text{Wb/s} = -5.76 \times 10^{-4}\ \text{V}.$$

We are interested only in the size of the emf, so we ignore the minus signs here and in Eq. 31-7, writing

$$|\mathcal{E}| = \left|N\frac{d\Phi^{\text{mag}}}{dt}\right| = (130\ \text{turns})(5.76 \times 10^{-4}\ \text{V})$$

$$= 7.5 \times 10^{-2}\ \text{V} = 75\ \text{mV}. \qquad \text{(Answer)}$$

31-5 Lenz's Law

Soon after Faraday proposed his law of induction, Heinrich Friedrich Lenz devised a rule—now known as Lenz's law—for determining the direction of an induced current in a loop:

> An induced current has a direction such that the magnetic field due to the current opposes the change in the magnetic flux that has induced the current.

It is important to notice that it is the *change* in the flux that determines the direction of the induced current rather than the direction of the magnetic field or motion. Furthermore, the direction of an induced emf is that of the induced current. To get a feel for Lenz's law, let us apply it in two different but equivalent ways to Fig. 31-14, where the north pole of a magnet is being moved toward a conducting loop.

1.   Opposition to Flux Change. In Fig. 31-14, with the magnet initially distant, there is no magnetic flux at the area encircled by the loop. As the north pole of the magnet then nears the loop with its magnetic field $\vec{B}$ directed *toward the left*, the flux at the loop increases. To oppose this increase in flux, the induced current i must set up its own field $\vec{B}_i$ *directed toward the right* inside the loop, as shown in Fig. 31-15a; then the rightward flux of field $\vec{B}_i$ opposes the increasing leftward flux of field $\vec{B}$. The right-hand rule of Fig. 30-19 then tells us that i must be counterclockwise in Fig. 31-15a.

2.   Opposition to Pole Movement. The approach of the magnet's north pole in Fig. 31-14 increases the magnetic flux in the loop and thereby induces a current in the loop. From Fig. 30-22, we know that the loop then acts as a magnetic dipole with a south pole and a north pole, and that its magnetic dipole moment $\vec{\mu}$ is directed from south to north. To oppose the magnetic flux increase being caused by the approaching magnet, the loop's north pole (and thus $\vec{\mu}$) must face *toward* the approaching north pole so as

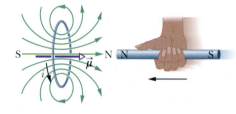

FIGURE 31-14 ■ Lenz's law at work. As the magnet is moved toward the loop, a current is induced in the loop. The current produces its own magnetic field, with magnetic dipole moment $\vec{\mu}$ oriented so as to oppose the motion of the magnet. Thus, the induced current must be counterclockwise as shown.

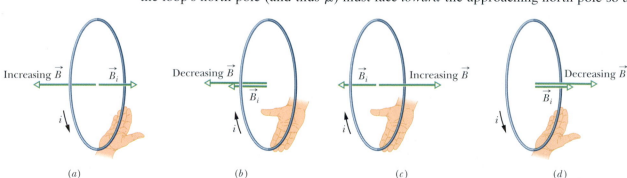

(a)   (b)   (c)   (d)

FIGURE 31-15 ■ The current i induced in a loop has the direction such that the current's magnetic field $\vec{B}_i$ opposes the change in the magnetic field $\vec{B}$ inducing i. The field $\vec{B}_i$ is always directed opposite an increasing field $\vec{B}$ shown in (a) and (c) and in the same direction as a decreasing field $\vec{B}$ shown in (b) and (d). The curled-straight right-hand rule gives the direction of the induced current based on the direction of the induced field.

to repel it (Fig. 31-14). Then the curled-straight right-hand rule for $\vec{\mu}$ (Fig. 30-22) tells us that the current induced in the loop must be counterclockwise in Fig. 31-14.

If we next pull the magnet away from the loop, a current will again be induced. Now, however, the loop will have a south pole facing the retreating north pole of the magnet, so as to oppose the retreat. Thus, the induced current will be clockwise.

As we noted above, be careful to remember that the flux of $\vec{B_i}$ always opposes the *change* in the flux of $\vec{B}$, but that does not always mean that $\vec{B_i}$ points opposite $\vec{B}$. For example, if we pull the magnet away from the loop in Fig. 31-14, the flux Φ^{mag} from the magnet is still directed to the left at the area subtended by the loop, but it is now decreasing. The flux of $\vec{B_i}$ must now be to the left inside the loop, to oppose the *decrease* in Φ^{mag}, as shown in Fig. 31-15b. Thus, $\vec{B_i}$ and $\vec{B}$ are now in the same direction.

Figures 31-15c and d show the situations in which the south pole of the magnet approaches and retreats from the loop, respectively. Figure 31-16 is a photo of a demonstration of Lenz's law in action.

Electric Guitars

Soon after rock began in the mid-1950s, guitarists switched from acoustic guitars to electric guitars—but it was Jimi Hendrix who first used the electric guitar as an electronic instrument. He was able to create new sounds that continue to influence rock music today. What is it about an electric guitar that enabled Hendrix to make different sounds?

Whereas an acoustic guitar depends for its sound on the acoustic resonance produced in the hollow body of the instrument by the oscillations of the strings, an electric guitar like that being played by Hendrix in Fig. 31-17 is a solid instrument, so there is no body resonance. Instead, the oscillations of the metal strings are sensed by electric "pickups" that send signals to an amplifier and a set of speakers.

The basic construction of a pickup is shown in Fig. 31-18. Wire connecting the instrument to the amplifier is coiled around a small magnet. The magnetic field of the magnet produces a north and south pole in the section of the metal string just above the magnet. That section of string then has its own magnetic field. When the string is plucked and thus made to oscillate, its motion relative to the coil changes the flux of its magnetic field at the area encircled by the coil, inducing a current in the coil. As the string oscillates toward and away from the coil, the induced current changes direction at the same frequency as the string's oscillations, thus relaying the frequency of oscillation to the amplifier and speaker.

On a Stratocaster©, there are three groups of pickups, placed near the bridge at the end of the wide part of the guitar body. The group closest to the bridge better detects the high-frequency oscillations of the strings; the group farthest from the near end better detects the low-frequency oscillations. By throwing a toggle switch on the guitar, the musician can select which group or which pair of groups will send signals to the amplifier and speakers.

To gain further control over his music, Hendrix sometimes rewrapped the wire in the pickup coils of his guitar to change the number of turns. In this way, he altered the amount of emf induced in the coils and thus their relative sensitivity to string oscillations. Even without this additional measure, you can see that the electric guitar offers far more control over the sound that is produced than can be obtained with an acoustic guitar.

READING EXERCISE 31-7: Lenz's law states: "An induced current has a direction such that the magnetic field due to the current opposes the change in the magnetic flux that induces the current." (a) Suppose there is a magnetic field directed into the plane of this page and that the strength of the field is decreasing. Would a magnetic field that opposes this change in magnetic flux be directed into the page, out of the page, or in some other direction? Explain your reasoning. (b) Suppose that there is a magnetic field directed into the plane of this page that is increasing in strength. Would a magnetic field that opposes this change in magnetic flux be directed into the page, out of the page, or in some other direction? Explain your reasoning. ■

FIGURE 31-16 ■ This demonstration of Lenz's law occurs when an electromagnet is switched on suddenly. The current induced in a metal ring opposes the electromagnet's current. The repulsive forces between the magnet and the ring cause the ring to jump more than a meter. (Photo courtesy of PASCO scientific.)

FIGURE 31-17 ■ Jimi Hendrix playing his Fender Stratocaster©. This guitar has three groups of six electric pickups each (within the wide part of the body). A toggle switch (at the bottom of the guitar) allows the musician to determine which group of pickups sends signals to an amplifier and thus to a speaker system.

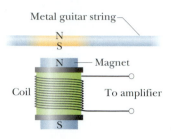

Metal guitar string

N
S
N — Magnet
Coil   To amplifier
S

FIGURE 31-18 ■ A side view of an electric guitar pickup. When the metal string (which acts like a magnet) oscillates, it causes a variation in magnetic flux that induces a current in the coil.

READING EXERCISE 31-8: The figure shows three situations in which identical circular conducting loops are in uniform magnetic fields that are either increasing (Inc) or decreasing (Dec) in magnitude at identical rates. In each, the dashed line coincides with a diameter. (a) Rank the situations according to the amount of the current induced in the loops, greatest first. (b) Explain your reasoning.

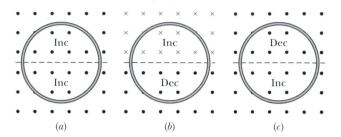

(a)   (b)   (c)

TOUCHSTONE EXAMPLE 31-2: Induced Emf

Figure 31-19 shows a conducting loop consisting of a half-circle of radius $r = 0.20$ m and three straight sections. The half-circle lies in a uniform magnetic field $\vec{B}$ that is directed out of the page; the field magnitude is given by $B = (4.0 \text{ T/s}^2)t^2 + (2.0 \text{ T/s})t + 3.0 \text{ T}$. An ideal battery with $\mathscr{E}_{\text{bat}} = 2.0$ V is connected to the loop. The resistance of the loop is 2.0 Ω.

(a) What are the amount and direction of the emf $\mathscr{E}^{\text{ind}}$ induced around the loop by field $\vec{B}$ at $t = 10$ s?

SOLUTION ■ One **Key Idea** here is that, according to Faraday's law, $\mathscr{E}^{\text{ind}}$ is equal to the negative rate $d\Phi^{\text{mag}}/dt$ at which the magnetic flux at the area encircled by the loop changes. A second **Key Idea** is that the flux at the loop depends on the loop's area A and its orientation in the magnetic field $\vec{B}$. Because $\vec{B}$ is uniform and is perpendicular to the plane of the loop, the flux is given by Eq. 31-1 ($\Phi^{\text{mag}} = BA$). Using this equation and realizing that only the field magnitude B changes in time (not the area A), we rewrite Faraday's law, Eq. 31-6, as

$$\left|\mathscr{E}^{\text{ind}}\right| = \left|\frac{d\Phi^{\text{mag}}}{dt}\right| = \left|\frac{d(BA)}{dt}\right| = A\left|\frac{dB}{dt}\right|.$$

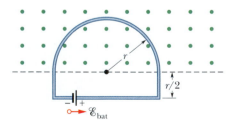

FIGURE 31-19 ■ A battery is connected to a conducting loop consisting of a half-circle of radius r that lies in a uniform magnetic field. The field is directed out of the page; its magnitude is changing.

A third **Key Idea** is that, because the flux penetrates the loop only within the half-circle, the area A in this equation is $\frac{1}{2}\pi r^2$. Substituting this and the given expression for B yields

$$\left|\mathscr{E}^{\text{ind}}\right| = A\left|\frac{dB}{dt}\right| = \frac{\pi r^2}{2}\frac{d}{dt}[(4.0 \text{ T/s}^2)t^2 + (2.0 \text{ T/s})t + 3.0 \text{ T}]$$

$$= \frac{\pi r^2}{2}[(8.0 \text{ T/s}^2)t + (2.0 \text{ T/s})].$$

At $t = 10$ s, then,

$$\left|\mathscr{E}^{\text{ind}}\right| = \frac{\pi(0.20 \text{ m})^2}{2}[(8.0 \text{ T/s})(10 \text{ s}) + (2.0 \text{ T/s})]$$

$$= 5.152 \text{ V} \approx 5.2 \text{ V}. \quad \text{(Answer)}$$

To find the direction of $\mathscr{E}^{\text{ind}}$, we first note that in Fig. 31-19 the flux at the loop is out of the page and increasing. Then the **Key Idea** here is that the induced field B^{ind} (due to the induced current) must oppose that increase, and thus be into the page. Using the curled-straight right-hand rule (Fig. 30-8c), we find that the induced current *contribution* must be clockwise around the loop. The induced emf $\mathscr{E}^{\text{ind}}$ must then also be clockwise.

(b) What is the current in the loop at $t = 10$ s?

SOLUTION ■ The **Key Idea** here is that two emfs tend to move charges around the loop. The induced $\mathscr{E}^{\text{ind}}$ tends to drive a current clockwise around the loop; the battery's $\mathscr{E}_{\text{bat}}$ tends to drive a current counterclockwise. Because $\mathscr{E}^{\text{ind}}$ is greater than $\mathscr{E}_{\text{bat}}$, the net emf $\mathscr{E}^{\text{net}}$ is clockwise, and thus so is the current. To find the current at $t = 10$ s, we use $i = \mathscr{E}/R$:

$$i = \frac{\mathscr{E}^{\text{net}}}{R} = \frac{\mathscr{E}^{\text{ind}} - \mathscr{E}_{\text{bat}}}{R}$$

$$= \frac{5.152 \text{ V} - 2.0 \text{ V}}{2.0 \Omega} = 1.58 \text{ A} \approx 1.6 \text{ A}. \quad \text{(Answer)}$$

31-6 Induction and Energy Transfers

Let us return to the simple situation we considered in Fig. 31-7. What are the consequences of the fact that a clockwise current is induced when the loop is pulled to the right and a counterclockwise current is induced when the loop is pushed to the left? If one pushes the loop back and forth (right and left), the result is an alternating current in the loop. This is current just like the current in our household electric system. It is a current that could run a motor, light a bulb, or provide heating through the resistive dissipation. If it took no effort on our part to push the loop back and forth, we could solve the energy crisis. Of course, it does take effort (work) on our part to push and pull the loop back and forth.

If you want to drag a metal loop out of a magnetic field at a constant velocity, you have to exert a force on the loop to balance the magnetic force associated with the charges moving in the magnetic field. This requires you to do work on the loop, but doing work adds energy to a system. We certainly cannot violate the principle of conservation of energy. So, where does this energy go? One place the energy could go is into an increase in the internal energy of the loop's wires. Since we observe a temperature rise in the wires, we conclude that the work done has been transformed into thermal energy—one form of internal energy. This makes sense. There is a current i in the loop that has some resistance R, and we learned in Section 26-7 that the electric power dissipation (or rate of thermal energy increase in the wires) is given by

$$P = i^2R \qquad \text{(resistive dissipation)}. \qquad \text{(Eq. 26-11)}$$

How does this rate of energy loss compare to the rate we are doing work? Perhaps they are the same. In that case, we might conclude that the work we do in moving the loop is transformed into thermal energy in the loop. Let's work out the details.

Figure 31-11 shows a situation involving induced current. A rectangular loop of wire of width L has one end in a uniform external magnetic field that is directed perpendicularly into the plane of the loop. This field may be produced, for example, by a large electromagnet. The dashed lines in the figure show the assumed limits of the magnetic field; the fringing of the field at its edges is neglected. You are asked to pull this loop to the right at a constant velocity $\vec{v}$.

In the situation of Fig. 31-11, the flux of the field at the loop is changing with time. Let us now calculate the rate at which you do mechanical work as you pull steadily on the loop. The amount of work done by a force $\vec{F}$ in moving a loop a small distance $d\vec{x}$ in a time dt is

$$dW = \vec{F} \cdot d\vec{x}.$$

For simplicity, let us consider a force $\vec{F}$, which is completely in the direction of the displacement $d\vec{x}$. Then

$$dW = \vec{F} \cdot d\vec{x} = F\,dx.$$

The rate of doing work (which is called the *power P*) is

$$P = \frac{dW}{dt} = F\frac{dx}{dt}.$$

So
$$P = Fv, \qquad (31\text{-}8)$$

where v is the speed at which we move the loop.

Suppose that we wish to find an expression for the power, P, in terms of the magnitude B of the magnetic field and the characteristics of the loop—namely, its resistance

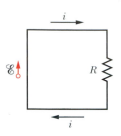

FIGURE 31-20 ■ A circuit diagram for the loop of Fig. 31-7 while it is moving.

R to current and its dimension L. As you move the loop to the right in Fig. 31-11, the portion of its area within the magnetic field decreases. Thus, the flux at the loop also decreases and, according to Lenz's law, a current is produced in the loop. It is the presence of this current that causes the force that opposes your pull.

To find the amount of the current, we first apply Faraday's law for a single loop in conjunction with Eq. 31-4. We can write the amount of this emf as

$$|\mathcal{E}| = \left|\frac{d\Phi^{\text{mag}}}{dt}\right| = BvL. \tag{31-9}$$

Figure 31-20 shows the loop depicted in Fig. 31-7, as a circuit. The induced emf, $\mathcal{E}$, is represented on the left, and the collective resistance R of the loop is represented on the right. The direction of the induced current i is shown as in Fig. 31-7, and we have already established that $\mathcal{E}$ must have the same direction as the conventional current, i.

To find the amount of the induced current, we cannot apply the loop rule for potential differences in a circuit because, as you will see in Section 31-7, we cannot define a potential difference for an induced emf. However, we can apply the equation $i = \mathcal{E}/R$. With Eq. 31-9, the current amount becomes

$$|i| = \frac{BvL}{R}. \tag{31-10}$$

Because three segments of the loop in Fig. 31-7 carry this current through the magnetic field, sideways deflecting forces act on those segments. From Chapter 29, we know that the magnitude of such a deflecting force is given in general notation by

$$F_d = |i\vec{L} \times \vec{B}|. \tag{31-11}$$

The deflecting forces acting on segments a, b, and d of the loop shown in Fig. 31-7 can be denoted as $\vec{F}_a$, $\vec{F}_b$, and $\vec{F}_d$. Application of the right-hand rule to each of these segments shows that the forces are perpendicular to each segment and point outward from the loop. Note, however, that from the symmetry, $\vec{F}_b$ and $\vec{F}_d$ are oppositely directed and equal in magnitude, so they cancel. This leaves only $\vec{F}_a$, which is directed opposite the force $\vec{F}$ you apply to the loop. Therefore, $\vec{F} = -\vec{F}_a$.

Using Eq. 31-11 to obtain the magnitude of $\vec{F}_a$ and noting that the angle between $\vec{B}$ and the length vector $\vec{L}$ for the left segment is 90°, we can write

$$F = F_a = |i|BL \sin 90° = |i|BL. \tag{31-12}$$

Substituting Eq. 31-10 for i in Eq. 31-12 then gives us

$$F = \frac{B^2vL^2}{R}. \tag{31-13}$$

Since B, L, and R are constants, the speed v at which you move the loop is constant if the magnitude F of the force you apply to the loop is also constant.

By substituting Eq. 31-13 into Eq. 31-8, we find the rate at which you do work on the loop as you pull it out of the magnetic field:

$$P = Fv = \frac{B^2v^2L^2}{R} \qquad \text{(rate of doing work).} \tag{31-14}$$

To complete our analysis, let us find the rate at which internal energy appears in the loop as you pull it along at constant speed. We calculate it from Eq. 26-11,

$$P = i^2R. \tag{31-15}$$

Substituting for *i* from Eq. 31-10, we find

$$P = \left(\frac{BvL}{R}\right)^2 R = \frac{B^2 v^2 L^2}{R} \qquad \text{(rate of internal energy gain),} \qquad (31\text{-}16)$$

which is exactly equal to the rate at which you are doing work on the loop (Eq. 31-14). Thus, the work that you do in pulling the loop through the magnetic field is transferred to thermal energy in the loop, manifesting itself as a small increase in the loop's temperature.

Eddy Currents

Suppose we replace the conducting loop of Fig. 31-7 with a solid conducting plate as shown in Fig. 31-21*a*. If we then move the plate out of the magnetic field, the relative motion of the field and the conductor again induces a current in the conductor. Thus, we again encounter an opposing force and must do work because of the induced current. With the plate, however, the conduction electrons making up the induced current do not follow one path as they do with the loop. Instead, the electrons swirl about within the plate as if they were caught in an eddy (or whirlpool) of water. Such a current is called an *eddy current* and can be represented as in Fig. 31-21*a as if* it followed a single path.

Eddy currents are used to cook food on an induction stove. To do this an oscillating current is sent through a conducting coil that lies just below the cooking surface. The magnetic field produced by that current oscillates and induces an oscillating current in the conducting cooking pan. Because the pan has some resistance to that current, the electrical energy of the current is continuously transformed to the pan's energy, resulting in a temperature increase of the pan and the food in it. What's amazing is that the stove itself might not get hot at all—only the pan.

As with the conducting loop of Fig. 31-7, the current induced in the plate results in mechanical energy being dissipated as it increases the pan's thermal energy. The dissipation is more apparent in the arrangement of Fig. 31-21*b*; a conducting plate, free to rotate about a pivot, is allowed to swing down through a magnetic field like a pendulum. Each time the plate enters and leaves the field, a portion of its mechanical energy is transferred to its thermal energy. After several swings, no mechanical energy remains and the warmed-up plate just hangs from its pivot.

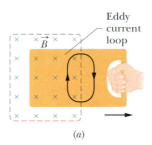

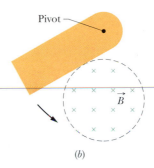

FIGURE 31-21 ■ (*a*) As you pull a solid conducting plate out of a magnetic field, *eddy currents* are induced in the plate. A typical loop of eddy current is shown; it has the same clockwise sense of circulation as the current in the conducting loop of Fig. 31-7. (*b*) A conducting plate is allowed to swing like a pendulum about a pivot and into a region of magnetic field. As it enters and leaves the field, eddy currents are induced in the plate.

READING EXERCISE 31-9: The figure shows four wire loops, with edge lengths of either *L* or 2*L*. All four loops will move through a region of uniform magnetic field $\vec{B}$ (directed out of the page) at the same constant velocity. (a) Rank the four loops according to the maximum amount of the emf induced as they move through the field, greatest first. (b) Explain your reasoning.

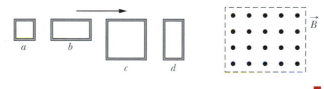

31-7 Induced Electric Fields

Let us place a copper ring of radius *r* in a uniform external magnetic field, as in Fig. 31-22*a*. The field—neglecting fringing—fills a cylindrical volume of radius *R*. Suppose that we increase the strength of this field at a steady rate, perhaps by increasing—in an appropriate way—the current in the windings of the electromagnet that produces the field. The magnetic flux at the ring will then change at a steady rate and—by Faraday's law—an induced emf and thus an induced current will appear in the ring. From Lenz's law we can deduce that the direction of the induced current is counterclockwise in Fig. 31-22*a*.

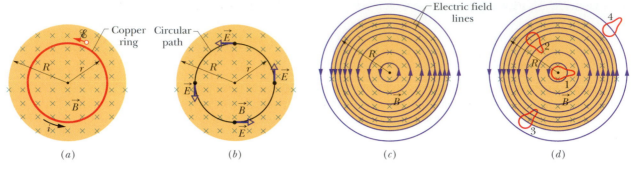

FIGURE 31-22 ■ (*a*) If the magnetic field increases at a steady rate, a constant induced current appears, as shown, in the copper ring of radius *r*. (*b*) An induced electric field exists even when the ring is removed; the electric field is shown at four points. (*c*) The complete picture of the induced electric field, displayed as field lines. (*d*) Four similar closed paths that enclose identical areas. Equal emfs are induced around paths 1 and 2, which lie entirely within the region of the changing magnetic field. A smaller emf is induced around path 3, which only partially lies in that region. No emf is induced around path 4, which lies entirely outside the magnetic field.

If there is a current in the copper ring, an electric field must be present along the ring; an electric field is needed to do the work of moving the conduction electrons. Moreover, the electric field must have been produced by the changing magnetic flux. This **induced electric field** $\vec{E}$ is just as real as an electric field produced by static charges; either field will exert a force $q\vec{E}$ on a particle of charge q.

By this line of reasoning, we are led to a more general and informative restatement of Faraday's law of induction:

> A changing magnetic field produces an electric field.

The striking feature of this statement is that the electric field is induced even if there is no copper ring.

To fix these ideas, consider Fig. 31-22*b*, which is just like Fig. 31-22*a* except the copper ring has been replaced by a hypothetical circular path of radius *r*. We assume, as previously, that the magnetic field $\vec{B}$ is increasing in magnitude at a constant rate dB/dt. The electric field induced at various points around the circular path must—from the symmetry—be tangent to the circle, as Fig. 31-22*b* shows.* Hence, the circular path is an electric field line. There is nothing special about the circle of radius *r*, so the electric field lines produced by the changing magnetic field must be a set of concentric circles, as in Fig. 31-22*c*.

As long as the magnetic field is increasing with time, the electric field represented by the circular field lines in Fig. 31-22*c* will be present. If the magnetic field remains constant with time, there will be no induced electric field and thus no electric field lines. If the magnetic field is decreasing with time (at a constant rate), the electric field lines will still be concentric circles as in Fig. 31-22*c*, but they will now have the opposite direction. All this is what we have in mind when we say: A changing magnetic field produces an electric field.

A Reformulation of Faraday's Law

Consider a particle of charge q moving around the circular path of Fig. 31-22*b*. The work W done on it in one revolution by the induced electric field is $q\mathscr{E}$, where $\mathscr{E}$ is the

---

* Arguments of symmetry would also permit the lines of $\vec{E}$ around the circular path to be radial, rather than tangential. However, such radial lines would imply that there are free charges, distributed symmetrically about the axis of symmetry, on which the electric field lines could begin or end; there are no such charges.

induced emf—that is, the work done per unit charge in moving the test charge around the path. From another point of view, the work is

$$\int \vec{F} \cdot d\vec{s} = qE(2\pi r),$$ (31-17)

where $|q|E$ is the magnitude of the force acting on the test charge and $2\pi r$ is the distance over which that force acts. Setting these two expressions for W equal to each other and canceling q, we find that

$$|\mathscr{E}| = 2\pi r E.$$ (31-18)

More generally, we can rewrite Eq. 31-17 to give the work done on a particle of charge q moving along any closed path:

$$W = \oint \vec{F} \cdot d\vec{s} = q \oint \vec{E} \cdot d\vec{s}.$$ (31-19)

(The circle indicates that the integral is to be taken around the closed path.) Substituting $q\mathscr{E}$ for W, we find that

$$\mathscr{E} = \oint \vec{E} \cdot d\vec{s}.$$ (31-20)

This integral reduces at once to Eq. 31-18 if we evaluate it for the special case of Fig. 31-22b.

With Eq. 31-20, we can expand the meaning of induced emf. Previously, induced emf meant the work per unit charge done in maintaining current due to a changing magnetic flux, or it meant the work done per unit charge on a charged particle that moves around a closed path in a changing magnetic flux. However, we can see in Fig. 31-22b and Eq. 31-20 that an induced emf can exist without the need of a current or particle: An induced emf is the sum—via integration—of quantities $\vec{E} \cdot d\vec{s}$ around a closed path, where $\vec{E}$ is the electric field induced by a changing magnetic flux and $d\vec{s}$ is a differential length vector along the closed path.

If we combine Eq. 31-20 with Faraday's law in Eq. 31-6 ($\mathscr{E} = -d\Phi^{mag}/dt$), we can rewrite Faraday's law as

$$\oint \vec{E} \cdot d\vec{s} = -\frac{d\Phi^{mag}}{dt} \qquad \text{(Faraday's law, general formula)}.$$ (31-21)

This equation says simply that a changing magnetic field induces an electric field. The changing magnetic field appears on the right side of this equation, the electric field on the left.

Faraday's law in the form of Eq. 31-21 can be applied to *any* closed path that can be drawn in a changing magnetic field. But $\oint \vec{E} \cdot d\vec{s}$ can only be evaluated for symmetrical situations. Figure 31-22d, for example, shows four such paths, all having the same shape and area but located in different positions in the changing field. For paths 1 and 2, the induced emfs $\mathscr{E}(=\oint \vec{E} \cdot d\vec{s})$ are equal because these paths lie entirely in the magnetic field and thus have the same value of $d\Phi^{mag}/dt$. This is true even though the electric field vectors at points along these paths are different, as indicated by the patterns of electric field lines in the figure. For path 3 the induced emf is smaller because the enclosed flux Φ^{mag} (hence, $d\Phi^{mag}/dt$) is smaller, and for path 4 the induced emf is zero, even though the electric field is not zero at any point on the path.

A New Look at Electric Potential

Induced electric fields are produced not by static charges but by a changing magnetic flux. Although electric fields produced in either way exert forces on charged particles, there is an important difference between them. The difference is not in the way they affect charges at a given point (the electric force on a charge q in this field is still qE), but in their global properties. Their field lines behave differently and there is a problem defining the electric potential associated with induced electric fields. The simplest evidence of this difference is that the field lines of induced electric fields form closed loops, as in Fig. 31-22c. Field lines produced by static charges never do so but rather must start on positive charges and end on negative charges. Since the induced fields are not caused by charges, there is no place for the field lines to start or end. Instead, they form closed loops, similar to those of magnetic fields. (But these are still electric fields! They act on stationary charges whereas magnetic fields don't.)

So, a varying *magnetic field* is accompanied by circular *electric field lines*. An electric current is known to be accompanied by circular magnetic field lines. But is an electric *current* the only source of circular magnetic field lines? Might it be possible that a varying *electric field* is accompanied by a circulating *magnetic field*? This is a question we will consider in the next chapter.

What we are immediately concerned with is that the electric field lines make closed loops, which has a powerful implication for trying to define an electrostatic potential. Since the potential difference equals the work per unit charge, if we carry a charge around a loop of electric field line, the $\vec{E}$ field always acts in the direction of motion, so every small step we make makes a positive contribution to the work. But since the field follows a loop, we can come back to our starting point after having only done positive work! The implication is:

> Electric potential has meaning only for electric fields that are produced by static charges; it has no meaning for electric fields that are produced by induction.

You can understand this statement quantitatively by considering what happens to a charged particle that makes a single journey around the circular path in Fig. 31-22b. It starts at a certain point and, on its return to that same point, has experienced an emf $\mathcal{E}$ of, let us say, 5 V; that is, work of 5 J/C has been done on the particle, and thus the particle should then be at a point that is 5 V greater in potential. However, that is impossible because the particle is back at the same point, which cannot have two different values of potential. We must conclude that potential has no meaning for electric fields that are set up by changing magnetic fields.

We can take a more formal look by recalling Eq. 25-16, which defines the potential difference between two points 1 and 2 in an electric field $\vec{E}$:

$$V_2 - V_1 = -\int_1^2 \vec{E} \cdot d\vec{s}. \tag{31-22}$$

In Chapter 25 we had not yet encountered Faraday's law of induction, so the electric fields involved in the derivation of Eq. 25-16 were those due to static charges. If 1 and 2 in Eq. 31-22 are the same point, the path connecting them is a closed loop, V_1 and V_2 are identical, and Eq. 31-22 reduces to

$$\oint \vec{E} \cdot d\vec{s} = 0 \text{ V}. \tag{31-23}$$

However, when a changing magnetic flux is present, this integral is *not* zero but is $-d\Phi^{\text{mag}}/dt$, as Eq. 31-21 asserts. Thus, assigning electric potential to an induced electric field leads us to a contradiction. We must conclude that electric potential difference is path dependent for the electric fields associated with induction.

READING EXERCISE 31-10: The figure shows five lettered regions in which a uniform magnetic field extends either directly out of the page (as in region a) or into the page. The field is increasing in magnitude at the same steady rate in all five regions; the regions are identical in area. Also shown are four numbered paths along which $\oint \vec{E} \cdot d\vec{s}$ has the magnitudes given below in terms of a unit "mag." Determine whether the magnetic fields in regions b through e are directed into or out of the page.

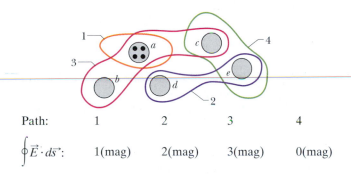

Path:	1	2	3	4
$\oint \vec{E} \cdot d\vec{s}$:	1(mag)	2(mag)	3(mag)	0(mag)

TOUCHSTONE EXAMPLE 31-3: Inducing an Electric Field

In Fig. 31-22b, take $R = 8.5$ cm and $dB/dt = 0.13$ T/s.

(a) Find an expression for the magnitude E of the induced electric field at points within the magnetic field, at radius r from the center of the magnetic field. Evaluate the expression for $r = 5.2$ cm.

SOLUTION ■ The **Key Idea** here is that an electric field is induced by the changing magnetic field, according to Faraday's law. To calculate the field magnitude E, we apply Faraday's law in the form of Eq. 31-21. We use a circular path of integration with radius $r \leq R$ because we want E for points within the magnetic field. We assume from the symmetry that $\vec{E}$ in Fig. 31-22b is tangent to the circular path at all points. The path vector $d\vec{s}$ is also always tangent to the circular path, so the dot product $\vec{E} \cdot d\vec{s}$ in Eq. 31-21 must have the magnitude $E \, ds$ at all points on the path. We can also assume from the symmetry that E has the same value at all points along the circular path. Then the left side of Eq. 31-21 becomes

$$\oint \vec{E} \cdot d\vec{s} = \oint E \, ds = E \oint ds = E(2\pi r). \qquad (31\text{-}24)$$

(The integral $\oint ds$ is the circumference $2\pi r$ of the circular path.)

Next, we need to evaluate the right side of Eq. 31-21. Because $\vec{B}$ is uniform over the area A encircled by the path of integration and is directed perpendicular to that area, the magnetic flux is given by Eq. 31-1:

$$\Phi^{\text{mag}} = BA = B(\pi r^2). \qquad (31\text{-}25)$$

Substituting this and Eq. 31-24 into Eq. 31-21 and dropping the minus sign, we find that the magnitude of the electric field is

$$E(2\pi r) = (\pi r^2)\frac{dB}{dt}$$

or

$$E = \frac{r}{2}\frac{dB}{dt}. \qquad \text{(Answer) } (31\text{-}26)$$

Equation 31-26 gives the magnitude of the electric field at any point for which $r \leq R$ (that is, within the magnetic field). Substituting given values yields, for the magnitude of $\vec{E}$ at $r = 5.2$ cm,

$$E = \frac{(5.2 \times 10^{-2}\ \text{m})}{2}(0.13\ \text{T/s})$$

$$= 0.0034\ \text{V/m} = 3.4\ \text{mV/m}. \qquad \text{(Answer)}$$

(b) Find an expression for the magnitude E of the induced electric field at points that are outside the magnetic field, at radius r. Evaluate the expression for $r = 12.5$ cm.

SOLUTION ■ The **Key Idea** of part (a) applies here also, except that we use a circular path of integration with radius $r = R$, because we want to evaluate E for points outside the magnetic field. Proceeding as in (a), we again obtain Eq. 31-24. However, we do not then obtain Eq. 31-25, because the new path of integration is now outside the magnetic field, and we need this **Key Idea**: The magnetic flux encircled by the new path is only that in the area πR^2 of the magnetic field region. Therefore,

$$\Phi^{\text{mag}} = BA = B(\pi R^2). \qquad (31\text{-}27)$$

Substituting this and Eq. 31-24 into Eq. 31-21 (without the minus sign) and solving for the magnitude of $\vec{E}$ yield

$$E = \frac{R^2}{2r}\frac{dB}{dt}. \qquad \text{(Answer) } (31\text{-}28)$$

Since E is not zero here, we know that an electric field is induced even at points that are outside the changing magnetic field, an

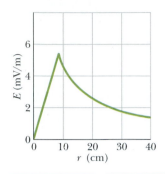

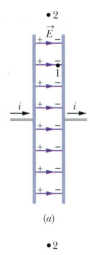

(a)

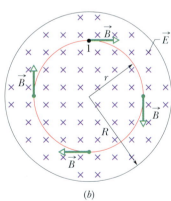

(b)

FIGURE 31-24 ■ (a) A circular parallel-plate capacitor, shown in side view, is being charged by a constant current i. (b) A view from within the capacitor, toward the plate at the right. The electric field $\vec{E}$ is uniform, is directed into the page (toward the plate), and grows in magnitude as the charge on the capacitor increases. The magnetic field $\vec{B}$ induced by this changing electric field is shown at four points on a circle with a radius r less than the plate radius R.

FIGURE 31-23 ■ A plot of the induced electric field $E(r)$ for the conditions given in Touchstone Example 31-3.

important result that (as you shall see in Section 32-5) makes transformers possible. With the given data, Eq. 31-28 yields the magnitude of $\vec{E}$ at $r = 12.5$ cm:

$$E = \frac{(8.5 \times 10^{-2} \text{ m})^2}{(2)(12.5 \times 10^{-2} \text{ m})}(0.13 \text{ T/s})$$

$$= 3.8 \times 10^{-3} \text{ V/m} = 3.8 \text{ mV/m}. \qquad \text{(Answer)}$$

Equations 31-26 and 31-28 give the same result, as they must, for $r = R$. Figure 31-23 shows a plot of $E(r)$ based on these two equations.

31-8 Induced Magnetic Fields

Let's consider a region in space where no electric currents are present. As we have seen, a changing magnetic flux induces an electric field, and we end up with Faraday's law of induction in the form

$$\oint \vec{E} \cdot d\vec{s} = -\frac{d\Phi^{\text{mag}}}{dt} \qquad \text{(Faraday's law of induction).} \qquad (31\text{-}29)$$

Here $\vec{E}$ is the electric field induced along a closed loop by the changing magnetic flux Φ^{mag} encircled by that loop. Because symmetry is often so powerful in physics, we should be tempted to ask whether induction can occur in the opposite sense; that is, can a changing electric flux induce a magnetic field?

The answer is that it can; furthermore, the equation governing the induction of a magnetic field is almost symmetric with Eq. 31-21. We often call it Maxwell's law of induction after James Clerk Maxwell, and we write it as

$$\oint \vec{B} \cdot d\vec{s} = \mu_0 \varepsilon_0 \frac{d\Phi^{\text{elec}}}{dt} \qquad \text{(Maxwell's law of induction—no currents).} \qquad (31\text{-}30)$$

Here $\vec{B}$ is the magnetic field induced along a closed loop by the changing electric flux Φ^{elec} in the region encircled by that loop.

As an example of this sort of induction, we consider the charging of a parallel-plate capacitor with circular plates, as shown in Fig. 31-24a. (Although we shall focus on this particular arrangement, a changing electric flux will always induce a magnetic field whenever it occurs.) We assume that the charge on the capacitor is being increased at a steady rate by a constant current i in the connecting wires. Then the amount of the electric field between the plates must also be increasing at a steady rate.

Figure 31-24b is a view of the right-hand plate of Fig. 31-24a from between the plates. The electric field is directed into the page. Let us consider a circular loop through point 1 in Figs. 31-24a and b, concentric with the capacitor plates and with a radius smaller than that of the plates. Because the electric field at the area subtended by the loop is changing, the electric flux at the loop must also be changing. According to Eq. 31-22, this changing electric flux induces a magnetic field around the loop.

Experiment proves that a magnetic field $\vec{B}$ *is* indeed induced around such a loop, directed as shown. This magnetic field has the same magnitude at every point around the loop and thus has circular symmetry about the central axis of the capacitor plates.

If we now consider a larger loop—say, through point 2 outside the plates in Figs. 31-24a and b—we find that a magnetic field is induced around that loop as well. Thus, while the electric field is changing, magnetic fields are induced between the plates, both inside and outside the gap. When the electric field stops changing, these induced magnetic fields disappear.

Although Eq. 31-30 is similar to Eq. 31-29, the equations differ in two ways. First, Eq. 31-30 has the two extra symbols, μ_0 and ε_0, but they appear only because we employ SI units. Second, Eq. 31-30 lacks the minus sign of Eq. 31-29. That difference in sign means that the induced electric field $\vec{E}$ and the induced magnetic field $\vec{B}$ have opposite directions when they are produced in otherwise similar situations.

To see this opposition of directions, examine Fig. 31-25, in which an increasing magnetic field $\vec{B}$, directed into the page, induces an electric field $\vec{E}$. The induced field $\vec{E}$ is counterclockwise, whereas the induced magnetic field $\vec{B}$ in Fig. 31-24b is clockwise.

Ampère – Maxwell Law

Now recall that the left side of Eq. 31-30, the integral of the dot product $\vec{B} \cdot d\vec{s}$ around a closed loop, appears in another equation—namely, Ampère's law:

$$\oint \vec{B} \cdot d\vec{s} = \mu_0 i^{\text{enc}} \qquad \text{(Ampère's law)}, \qquad (31\text{-}31)$$

where i^{enc} is the current encircled by the closed loop. Thus, our two equations that specify the magnetic field $\vec{B}$ produced by means other than a magnetic material (that is, by a current and by a changing electric field) give the field in exactly the same form. We can combine the two equations into the single equation

$$\oint \vec{B} \cdot d\vec{s} = \mu_0 \varepsilon_0 \frac{d\Phi^{\text{elec}}}{dt} + \mu_0 i^{\text{enc}} \qquad \text{(Ampère–Maxwell law)}. \qquad (31\text{-}32)$$

When there is a current but no change in electric flux (such as with a wire carrying a constant current), the first term on the right side of Eq. 31-32 is zero, and Eq. 31-32 reduces to Eq. 31-31, Ampère's law. When there is a change in electric flux but no current (such as inside or outside the gap of a charging capacitor), the second term on the right side of Eq. 31-32 is zero, and Eq. 31-32 reduces to Eq. 31-30, Maxwell's law of induction.

READING EXERCISE 31-11: Referring back to Chapter 30, where we first studied Ampère's law, describe how we found the direction of the magnetic field produced by a current. What did the magnetic field lines look like for a long, straight, current-carrying wire? Discuss any connections or similarities between the case of the current-carrying wire and the case shown in Fig. 31-24. ■

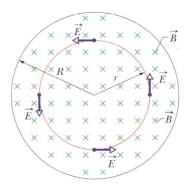

FIGURE 31-25 ■ A uniform magnetic field $\vec{B}$ in a circular region. The field, directed into the page, is increasing in magnitude. The electric field $\vec{E}$ induced by the changing magnetic field is shown at four points on a circle concentric with the circular region. Compare this situation with that of Fig. 31-24b.

TOUCHSTONE EXAMPLE 31-4: Inducing a Magnetic Field

A parallel-plate capacitor with circular plates of radius R is being charged as in Fig. 31-24a.

(a) Derive an expression for the magnitude of the magnetic field at radii r for the case $r \leq R$.

SOLUTION ■ The **Key Idea** here is that a magnetic field can be set up by a current and by induction due to a changing electric flux; both effects are included in Eq. 31-32. There is no current between the capacitor plates of Fig. 31-24, but the electric flux there is

changing. Thus, Eq. 31-32 reduces to

$$\oint \vec{B} \cdot d\vec{s} = \mu_0 \varepsilon_0 \frac{d\Phi^{\text{elec}}}{dt}. \qquad (31\text{-}33)$$

We shall separately evaluate the left and right sides of this equation.

Left side of Eq. 31-33: We choose a circular Ampèrian loop with a radius $r \leq R$ as shown in Fig. 31-24, because we want to evaluate the magnetic field for $r \leq R$—that is, inside the capacitor. The magnetic field $\vec{B}$ at all points along the loop is tangent to the loop, as is the path element $d\vec{s}$. Thus, $\vec{B}$ and $d\vec{s}$ are either parallel or antiparallel at each point of the loop. For simplicity, assume they are parallel (the choice does not alter our outcome here). Then

$$\oint \vec{B} \cdot d\vec{s} = \oint B \cdot ds \cos 0° = \oint B\,ds.$$

Due to the circular symmetry of the plates, we can also assume that $\vec{B}$ has the same magnitude at every point around the loop. Thus, B can be taken outside the integral on the right side of the above equation. The integral that remains is $\oint ds$, which simply gives the circumference $2\pi r$ of the loop. The left side of Eq. 31-33 is then $(B)(2\pi r)$.

Right side of Eq. 31-33: We assume that the electric field $\vec{E}$ is uniform between the capacitor plates and directed perpendicular to the plates. Then the electric flux Φ^{elec} encircled by the Ampèrian loop is EA, where A is the area encircled by the loop within the electric field. Thus, the right side of Eq. 31-33 is $\mu_0\varepsilon_0\,d(EA)/dt$.

Substituting our results for the left and right sides into Eq. 31-33, we get

$$B(2\pi r) = \mu_0\varepsilon_0\frac{d(EA)}{dt}.$$

Because A is a constant, we write $d(EA)$ as $A\,dE$, so we have

$$B(2\pi r) = \mu_0\varepsilon_0 A\frac{dE}{dt}. \qquad (31\text{-}34)$$

We next use this **Key Idea:** The area A that is encircled by the Ampèrian loop within the electric field is the full area πr^2 of the loop, because the loop's radius r is less than (or equal to) the plate radius R. Substituting πr^2 for A in Eq. 31-34 and solving the result for B give us, for $r \le R$,

$$B = \frac{\mu_0\varepsilon_0 r}{2}\frac{dE}{dt}. \qquad \text{(Answer) (31-35)}$$

This equation tells us that, inside the capacitor, B increases linearly with increased radial distance r, from zero at the center of the plates to a maximum value at the plate edges (where $r = R$).

(b) Evaluate the field magnitude B for $r = R/5 = 11.0$ mm and $dE/dt = 1.50 \times 10^{12}$ V/m·s.

SOLUTION ■ From the answer to (a), we have

$$B = \tfrac{1}{2}\mu_0\varepsilon_0 r\frac{dE}{dt}$$

$$= \tfrac{1}{2}(4\pi \times 10^{-7}\text{ T·m/A})(8.85 \times 10^{-12}\text{ C}^2/\text{N·m}^2)$$

$$\times (11.0 \times 10^{-3}\text{ m})(1.50 \times 10^{12}\text{ V/m·s})$$

$$= 9.18 \times 10^{-8}\text{ T}. \qquad \text{(Answer)}$$

(c) Derive an expression for the induced magnetic field for the case $r \ge R$.

SOLUTION ■ Our procedure is the same as in (a) except we now use an Ampèrian loop with a radius r that is greater than the plate radius R, to evaluate B outside the capacitor. Evaluating the left and right sides of Eq. 31-33 again leads to Eq. 31-34. However, we then need this subtle **Key Idea:** The electric field exists only between the plates, *not* outside the plates. Thus, the area A that is encircled by the Ampèrian loop in the electric field is *not* the full area πr^2 of the loop. Rather, A is only the plate area πR^2.

Substituting πR^2 for A in Eq. 31-34 and solving the result for B give us, for $r \ge R$,

$$B = \frac{\mu_0\varepsilon_0 R^2}{2r}\frac{dE}{dt}. \qquad \text{(Answer) (31-36)}$$

This equation tells us that, outside the capacitor, B decreases with increased radial distance r, from a maximum value at the plate edges (where $r = R$). By substituting $r = R$ into Eqs. 31-35 and 31-36, you can show that these equations are consistent; that is, they give the same maximum value of B at the plate radius.

The magnitude of the induced magnetic field calculated in (b) is so small that it can scarcely be measured with simple apparatus. This is in sharp contrast to the magnitudes of induced electric fields (Faraday's law), which can be measured easily. This experimental difference exists partly because induced emfs can easily be multiplied by using a coil of many turns. No technique of comparable simplicity exists for multiplying induced magnetic fields. In any case, the experiment suggested by this sample problem has been done, and the presence of the induced magnetic fields has been verified quantitatively.

31-9 Displacement Current

If you compare the two terms on the right side of Eq. 31-32, you will see that the product $\varepsilon_0(d\Phi^{\text{elec}}/dt)$ in the first term must have the units associated with a current. Since no charge actually flows, historically, that product has been treated as being a fictitious current called the **displacement current** i^{dis}:

$$i^{\text{dis}} = \varepsilon_0\frac{d\Phi^{\text{elec}}}{dt} \qquad \text{(displacement current)}. \qquad (31\text{-}37)$$

"Displacement" is a poorly chosen term in that nothing is being displaced, but we are stuck with the word. Nevertheless, we can now rewrite Eq. 31-32 as

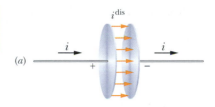

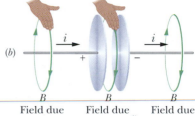

Field due   Field due   Field due
to current i to current i^{dis} to current i

$$\oint \vec{B} \cdot d\vec{s} = \mu_0 i_{dis}^{enc} + \mu_0 i^{enc} \quad \text{(Ampère–Maxwell law)}, \tag{31-38}$$

in which i_{dis}^{enc} is the displacement current that is encircled by the integration loop.

Let us again focus on a charging capacitor with circular plates, as in Fig. 31-26a. The real current i that is charging the plates changes the electric field $\vec{E}$ between the plates. The fictitious displacement current i^{dis} between the plates is associated with that changing field $\vec{E}$. Let us relate these two currents.

The amount of excess charge $|q|$ on each of the plates at any time is related to the magnitude $|\vec{E}| = E$ of the field between the plates at that time by Eq. 28-4:

$$|q| = \varepsilon_0 AE, \tag{31-39}$$

in which A is the plate area. To get the real current i, we differentiate Eq. 31-39 with respect to time, finding

$$\frac{d|q|}{dt} = |i| = \varepsilon_0 A \frac{dE}{dt}. \tag{31-40}$$

To get the displacement current i^{dis}, we can use Eq. 31-37. Assuming that the electric field $\vec{E}$ between the two plates is uniform (we neglect any fringing), we can replace the electric flux Φ^{elec} in that equation with EA. Then Eq. 31-37 becomes

$$|i^{dis}| = \varepsilon_0 \left| \frac{d\Phi^{elec}}{dt} \right| = \varepsilon_0 \left| \frac{d(EA)}{dt} \right| = \varepsilon_0 A \left| \frac{dE}{dt} \right|. \tag{31-41}$$

Comparing Eqs. 31-40 and 31-41, we see that the real current i charging the capacitor and the fictitious displacement current i^{dis} between the plates have the same value:

$$i^{dis} = i \quad \text{(displacement current in a capacitor).} \tag{31-42}$$

Thus, we can consider the fictitious displacement current i^{dis} to be simply a continuation of the real current i from one plate, across the capacitor gap, to the other plate. Because the electric field is uniformly spread over the plates, the same is true of this fictitious displacement current i^{dis}, as suggested by the spread of current arrows in Fig. 31-26a. Although no charge actually moves across the gap between the plates, the idea of the fictitious current i^{dis} can help us to quickly find the direction and magnitude of an induced magnetic field, as follows.

Finding the Induced Magnetic Field

In Chapter 30 we found the direction of the magnetic field produced by a real current i by using the right-hand rule of Fig. 30-4. We can apply the same rule to find the direction of an induced magnetic field produced by a fictitious displacement current i^{dis}, as shown in the center of Fig. 31-26b for a capacitor.

We can also use i^{dis} to find the magnitude of the magnetic field induced by a charging capacitor with parallel circular plates of radius R. We simply consider the space between the plates to be an imaginary circular wire of radius R carrying the imaginary current i^{dis}. Then, from Eq. 30-22, the magnitude of the magnetic field at a point inside the capacitor at radius r from the center is

$$B = \left(\frac{\mu_0 |i^{dis}|}{2\pi R^2} \right) r \quad \text{(inside a circular capacitor).} \tag{31-43}$$

FIGURE 31-26 ■ (a) The displacement current i^{dis} between the plates of a capacitor that is being charged by a current i. (b) The right-hand rule for finding the direction of the magnetic field around a wire with a real current (as at the left) also gives the magnetic field direction around a displacement current (as in the center).

Similarly, from Eq. 30-19, the magnitude of the magnetic field at a point outside the capacitor at radius r is

$$B = \frac{\mu_0 |i^{\text{dis}}|}{2\pi r} \qquad \text{(outside a circular capacitor)}. \qquad (31\text{-}44)$$

READING EXERCISE 31-12: Discuss the ways in which it is useful for us to think of the quantity $\varepsilon_0 \, d\Phi^{\text{elec}}/dt$ as a current. ∎

TOUCHSTONE EXAMPLE 31-5: Displacement Current

The circular parallel-plate capacitor in Touchstone Example 31-4 is being charged with a current i.

(a) Between the plates, what is the magnitude of $\oint \vec{B} \cdot d\vec{s}$, in terms of μ_0 and i, at a radius $r = R/5$ from their center?

SOLUTION ∎ The first **Key Idea** of Touchstone Example 31-4a holds here too. However, now we can replace the product $\varepsilon_0 \, d\Phi^{\text{elec}}/dt$ in Eq. 31-32 with a fictitious displacement current i^{dis}. Then integral $\oint \vec{B} \cdot d\vec{s}$ is given by Eq. 31-38, but because there is no real current i between the capacitor plates, the equation reduces to

$$\oint \vec{B} \cdot d\vec{s} = \mu_0 \, i^{\text{enc}}_{\text{dis}}. \qquad (31\text{-}45)$$

Because we want to evaluate $\oint \vec{B} \cdot d\vec{s}$ at radius $r = R/5$ (within the capacitor), the integration loop encircles only a portion $i^{\text{enc}}_{\text{dis}}$ of the total displacement current i^{dis}. A second **Key Idea** is to assume that i^{dis} is uniformly spread over the full plate area. Then the portion of the displacement current encircled by the loop is proportional to the area encircled by the loop:

$$\frac{(\text{encircled displacement current } i^{\text{enc}}_{\text{dis}})}{(\text{total displacement current } i^{\text{dis}})} = \frac{\text{encircled area } \pi r^2}{\text{full plate area } \pi R^2}.$$

This gives us a current magnitude of

$$i^{\text{enc}}_{\text{dis}} = i^{\text{dis}} \frac{\pi r^2}{\pi R^2}.$$

Substituting this into Eq. 31-45, we obtain

$$\oint \vec{B} \cdot d\vec{s} = \mu_0 i^{\text{dis}} \frac{\pi r^2}{\pi R^2}. \qquad (31\text{-}46)$$

Now substituting $i^{\text{dis}} = i$ (from Eq. 31-42) and $r = R/5$ into Eq. 31-46 leads to

$$\oint \vec{B} \cdot d\vec{s} = \mu_0 i \frac{(R/5)^2}{R^2} = \frac{\mu_0 i}{25}. \qquad \text{(Answer)}$$

(b) In terms of the maximum induced magnetic field, what is the magnitude of the magnetic field induced at $r = R/5$, inside the capacitor?

SOLUTION ∎ The **Key Idea** here is that, because the capacitor has parallel circular plates, we can treat the space between the plates as an imaginary wire of radius R carrying the imaginary current i^{dis}. Then we can use Eq. 31-43 to find the induced magnetic field magnitude B at any point inside the capacitor. At $r = R/5$, that equation yields

$$B = \left(\frac{\mu_0 |i^{\text{dis}}|}{2\pi R^2} \right) r = \frac{\mu_0 |i^{\text{dis}}|(R/5)}{2\pi R^2} = \frac{\mu_0 |i^{\text{dis}}|}{10\pi R}. \qquad (31\text{-}47)$$

The maximum field magnitude B^{max} within the capacitor occurs at $r = R$. It is

$$B^{\text{max}} = \left(\frac{\mu_0 |i^{\text{dis}}|}{2\pi R^2} \right) R = \frac{\mu_0 |i^{\text{dis}}|}{2\pi R}. \qquad (31\text{-}48)$$

Dividing Eq. 31-47 by Eq. 31-48 and rearranging the result, we find

$$B = \frac{B^{\text{max}}}{5}, \qquad \text{(Answer)}$$

We should be able to obtain this result with a little reasoning and less work. Equation 31-43 tells us that inside the capacitor, B increases linearly with r. Therefore, a point $\frac{1}{5}$ the distance out to the full radius R of the plates, where B^{max} occurs, should have a field B that is $\frac{1}{5} B^{\text{max}}$.

31-10 Gauss' Law for Magnetic Fields

In this chapter and the two that precede it, we have investigated several fundamental aspects of electricity and magnetism. Furthermore, we have seen many ways in which magnetism and electricity are connected. When combined as a set of laws, these ideas provide a framework from which we can understand all of the electromagnetic phenomena that fill our world, much like Newton's laws do in regard to forces and motion.

However, there remains one last idea that we must discuss before our view of electromagnetism is complete. This idea is contained in an idea known as *Gauss' law for magnetic fields*. Gauss' law for magnetic fields is a formal way of saying that magnetic monopoles do not exist. The law asserts that the net magnetic flux Φ^{mag} at any closed Gaussian surface is zero:

$$\Phi^{mag} = \oint \vec{B} \cdot d\vec{A} = 0 \qquad \text{(Gauss' law for magnetic fields).} \qquad (31\text{-}49)$$

Contrast this with Gauss' law for electric fields,

$$\Phi^{elec} = \oint \vec{E} \cdot d\vec{A} = \frac{q^{enc}}{\varepsilon_0} \qquad \text{(Gauss' law for electric fields).}$$

In both equations, the integral is taken over a *closed* Gaussian surface. Gauss' law for electric fields says that this integral (the net electric flux at the surface) is proportional to the net electric charge q^{enc} enclosed by the surface. Gauss' law for magnetic fields says that there can be no net magnetic flux at the surface because there can be no net "magnetic charge" (individual magnetic poles) enclosed by the surface. The simplest magnetic structure that can exist and thus be enclosed by a Gaussian surface is a dipole, which consists of both a source and a sink for the field lines. Thus, there must always be as much magnetic flux into the surface as out of it, and the net magnetic flux must always be zero.

Gauss' law for magnetic fields holds for more complicated structures than a magnetic dipole, and it holds even if the Gaussian surface does not enclose the entire structure. Gaussian surface II near the bar magnet of Fig. 31-27 encloses no poles, and we can easily conclude that the net magnetic flux at it is zero. Gaussian surface I is more difficult to understand. It may seem to enclose only the north pole of the magnet because it encloses the label N and not the label S. However, a south pole must be associated with the lower boundary of the surface, because magnetic field lines enter the surface there. (The enclosed section is like one piece of the broken cylindrical magnet in Fig. 31-28.) Thus, Gaussian surface I encloses a magnetic dipole and the net flux at the surface is zero.

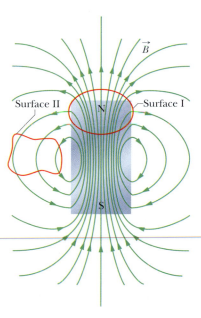

FIGURE 31-27 ▪ The field lines for the magnetic field $\vec{B}$ of a short bar magnet. The red curves represent cross sections of closed, three-dimensional Gaussian surfaces.

FIGURE 31-28 ▪ If you break a magnet, each fragment becomes a separate magnet, with its own north and south poles.

READING EXERCISE 31-13: The figure below shows four closed surfaces with flat top and bottom faces and curved sides. The table gives the areas A of the faces and the magnitudes B of the uniform and perpendicular magnetic fields at those faces; the units of A and B are arbitrary but consistent. (a) Rank the surfaces according to the magnitudes of the magnetic flux at their curved sides, greatest first. (b) Explain your reasoning.

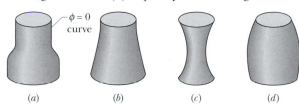

Surface	A_{top}	B_{top}, direction	A_{bot}	B_{bot}, direction
a	2	6, outward	4	3, inward
b	2	1, inward	4	2, inward
c	2	6, inward	2	8, outward
d	2	3, outward	3	2, outward

31-11 Maxwell's Equations in a Vacuum

Many 18th and 19th century scientists contributed to our understanding of electricity and magnetism including Franklin, Coulomb, Gauss, Oersted, Biot, Savart, Lorentz, Ampère, Henry, Faraday, and Maxwell. But it was James Clerk Maxwell who reformulated many of the basic equations describing electric and magnetic effects we have already presented. A special case of Maxwell's equations are shown in Table 31-1 for situations in which no dielectric or magnetic materials are present.

It is amazing that these four rather compact equations can be used to *derive a complete description of all electromagnetic interactions that were understood by the end of the 19th century*. Taken together they describe a diverse range of phenomena, from how a compass needle points north to how a car starts when you turn the ignition key. They have been used to design electric motors, cyclotrons, television transmitters and receivers, telephones, fax machines, radar, and microwave ovens.

In addition, many of the equations you have seen since Chapter 22 can be derived from Maxwell's equations. Perhaps the most exciting intellectual outcome of Maxwell's equations is their prediction of electromagnetic waves and our eventual understanding of the self-propagating nature of these waves that will be introduced in Chapter 34. Maxwell's picture of electromagnetic wave propagation was not fully appreciated until scientists abandoned the idea that all waves had to propagate through an elastic medium and accepted Einstein's theory of special relativity formulated in the early part of the 20th century.

Because we now know that visible light is a form of electromagnetic radiation, these equations provide the basis for many of the equations you will see in Chapters 34 through 37, which introduce you to optics and optical devices such as telescopes and eyeglasses.

The significance of Maxwell's equations should not be underestimated. Richard Feynman, a leading famous 20th-century physicist, recognized this when he stated:

> Now we realize that the phenomena of chemical interaction and ultimately of life itself are to be understood in terms of electromagnetism . . . . The electrical forces, enormous as they are, can also be very tiny, and we can control them and use them in many ways . . . From a long view of the history of mankind—seen from, say, ten thousand years from now—there can be little doubt that the most significant event of the nineteenth century will be judged as Maxwell's discovery of the laws of electrodynamics.

TABLE 31-1
Maxwell's Equations for Vacuum[a]

Name	Equation	
Gauss' law for electricity (Eq. 24-7)	$\oint \vec{E} \cdot d\vec{A} = q/\varepsilon_0$	Relates net electric flux to net enclosed electric charge
Gauss' law for magnetism (Eq. 31-49)	$\oint \vec{B} \cdot d\vec{A} = 0$	Relates net magnetic flux to net enclosed magnetic charge
Faraday's law (Eq. 31-7)	$\oint \vec{E} \cdot d\vec{s} = -\dfrac{d\Phi^{\text{mag}}}{dt}$	Relates induced electric field to changing magnetic flux
Ampère–Maxwell law (Eq. 31-32)	$\oint \vec{B} \cdot d\vec{s} = \mu_0\varepsilon_0\dfrac{d\Phi^{\text{elec}}}{dt} + \mu_0 i$	Relates induced magnetic field to changing electric flux and to current

[a]Written on the assumption that no dielectric or magnetic materials are present.

READING EXERCISE 31-14: Discuss several ways in which Gauss' law for electricity and Gauss' law for magnetism are similar. Discuss several ways in which they are different.

READING EXERCISE 31-15: Discuss several ways in which Faraday's law and the Ampère–Maxwell law are similar. Discuss several ways in which they are different.

Problems

SEC. 31-4 ■ FARADAY'S LAW

1. UHF Antenna A UHF television loop antenna has a diameter of 11 cm. The magnetic field of a TV signal is normal to the plane of the loop and, at one instant of time, its magnitude is changing at the rate 0.16 T/s. The magnetic field is uniform. What emf is induced in the antenna?

2. Small Loop A small loop of area A is inside of, and has its axis in the same direction as, a long solenoid of n turns per unit length and current i. If $i = I^{\max} \sin \omega t$, find the magnitude of the emf induced in the loop.

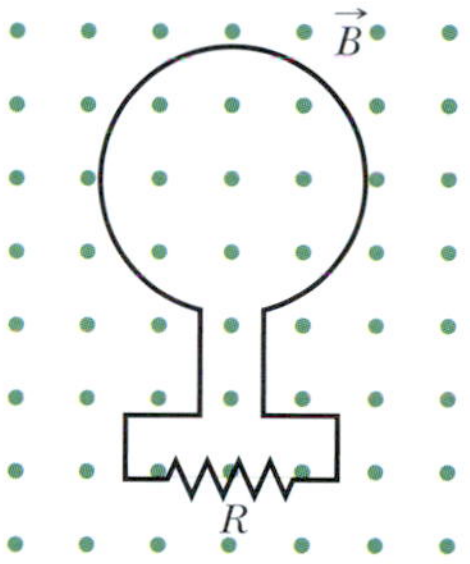

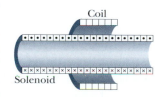

FIGURE 31-29 ■ Problems 3 and 13.

3. Magnetic Flux The magnetic flux encircled by the loop shown in Fig. 31-29 increases according to the relation $\Phi^{\text{mag}} = (6.0 \text{ mWb/s}^2)t^2 + (3.7 \text{ mWb/s})t$. (a) What is the magnitude of the emf induced in the loop when $t = 2.0$ s? (b) What is the direction of the current through R?

4. Calculate emf The magnitude of the magnetic field encircled by a single loop of wire, 12 cm in radius and of 8.5 Ω resistance, changes with time as shown in Fig. 31-30. Calculate the magnitude of the emf in the loop as a function of time. Consider the time intervals (a) $t_1 = 0.0$ s to $t_2 = 2.0$ s, (b) $t_2 = 2.0$ s to $t_3 = 4.0$ s, (c) $t_3 = 4.0$ s to $t_4 = 6.0$ s. The (uniform) magnetic field is perpendicular to the plane of the loop.

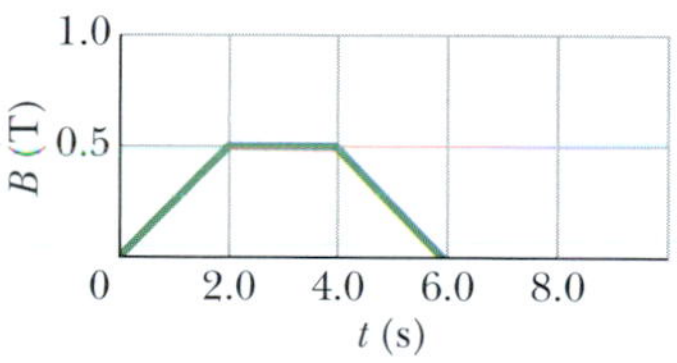

FIGURE 31-30 ■ Problem 4.

5. Uniform Magnetic Field A uniform magnetic field is normal to the plane of a circular loop 10 cm in diameter and made of copper wire (of diameter 2.5 mm). (a) Calculate the resistance of the wire. (See Table 26-2.) (b) At what rate must the magnetic field change with time if an induced current of 10 A is to appear in the loop?

6. Current in Solenoid The current in the solenoid of Touchstone Example 31-1 changes, not as stated there, but according to $i = (3.0 \text{ A/s})t + (1.0 \text{A/s}^2)t^2$. (a) Plot the induced emf in the coil from $t_1 = 0.0$ s to $t_2 = 4.0$ s. (b) The resistance of the coil is 0.15 Ω. What is the current in the coil at $t = 2.0$ s?

FIGURE 31-31 ■ Problem 7.

7. Coil Outside Solenoid In Fig. 31-31 a 120-turn coil of radius 1.8 cm and resistance 5.3 Ω is placed *outside* a solenoid like that of Touchstone Example 31-1. If the current in the solenoid is changed as in that sample problem, what current appears in the coil while the solenoid current is being changed?

8. Elastic Conducting Material An elastic conducting material is stretched into a circular loop of 12.0 cm radius. It is placed with its plane perpendicular to a uniform 0.800 T magnetic field. When released, the radius of the loop starts to shrink at an instantaneous rate of 75.0 cm/s. What magnitude of emf is induced in the loop at that instant?

9. Square Loop A square loop of wire is held in a uniform, magnetic field 0.24 T directed perpendicularly to the plane of the loop. The length of each side of the square is decreasing at a constant rate of 5.0 cm/s. What emf is induced in the loop when the length is 12 cm?

10. Rectangular Loop A rectangular loop (area = 0.15 m²) turns in a uniform magnetic field, $B = 0.20$ T. When the angle between the field and the normal to the plane of the loop is $\pi/2$ rad and increasing at 0.60 rad/s, what emf is induced in the loop?

SEC. 31-5 ■ LENZ'S LAW

11. Two Parallel Loops Though not to scale, Fig. 31-32 shows two parallel loops of wire with a common axis. The smaller loop (radius r) is above the larger loop (radius R) by a distance $x \gg R$. Consequently, the magnetic field due to the current i in the larger loop is nearly constant throughout the smaller loop. Suppose that x is increasing at the constant rate of $dx/dt = v$. (a) Determine the magnetic flux at the area bounded by the smaller loop as a function of x. (*Hint:* See Eq. 30-29.) In the smaller loop, find (b) the induced emf and (c) the direction of the induced current.

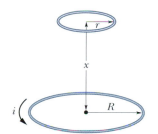

FIGURE 31-32 ■ Problem 11.

12. Circular Loop In Fig. 31-33, a circular loop of wire 10 cm in diameter (seen edge-on) is placed with its normal at an angle $\theta = 30°$ with the direction of a uniform magnetic field $\vec{B}$ of magnitude 0.50 T. The loop is then rotated such that the normal rotates in a cone about the field direction at the constant rate of 100 rev/min; the angle θ remains unchanged during the process. What is the emf induced in the loop?

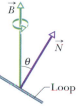

FIGURE 31-33 ■ Problem 12.

13. Flux At Loop   In Fig. 31-29 let the flux encircled by the loop be $\Phi^{\text{mag}}(0)$ at time $t_1 = 0$. Then let the magnetic field $\vec{B}$ vary in a continuous but unspecified way, in both magnitude and direction, so that at time t_2 the flux is represented by $\Phi^{\text{mag}}(t_2)$. (a) Show that the net charge $q(t_2)$ that has passed through resistor R in time t_2 is

$$q(t_2) = \frac{1}{R}[\Phi^{\text{mag}}(0) - \Phi^{\text{mag}}(t_2)]$$

and is independent of the way $\vec{B}$ has changed. (b) If $\Phi^{\text{mag}}(t_2) = \Phi^{\text{mag}}(0)$ in a particular case, we have $q(t_2) = 0$. Is the induced current necessarily zero throughout the interval from 0 to t_2?

14. Big Loop, Little Loop   A small circular loop of area 2.00 cm² is placed in the plane of, and concentric with, a large circular loop of radius 1.00 m. The current in the large loop is changed uniformly from 200 A to −200 A (a change in direction) in a time of 1.00 s, beginning at $t_1 = 0$. (a) What is the magnitude of the magnetic field at the center of the small circular loop due to the current in the large loop at $t_1 = 0$ s, $t_2 = 0.500$ s, and $t_3 = 1.00$ s? (b) What is the magnitude of the emf induced in the small loop at $t_2 = 0.500$ s? (Since the inner loop is small, assume the field $\vec{B}$ due to the outer loop is uniform over the area of the smaller loop.)

15. Copper Wire on Wooden Core   One hundred turns of insulated copper wire are wrapped around a wooden cylindrical core of cross-sectional area $1.20 \times 10^{-3}\,\text{m}^2$. The two ends of the wire are connected to a resistor. The total resistance in the circuit is 13.0 Ω. If an externally applied uniform longitudinal magnetic field in the core changes from 1.60 T in one direction to 1.60 T in the opposite direction, how much charge flows through the circuit? (*Hint:* See Problem 13.)

16. Earth's Field   At a certain place, Earth's magnetic field has magnitude $|\vec{B}| = 0.590$ gauss and is inclined downward at an angle of 70.0° to the horizontal. A flat horizontal circular coil of wire with a radius of 10.0 cm has 1000 turns and a total resistance of 85.0 Ω. It is connected to a meter with 140 Ω resistance. The coil is flipped through a half-revolution about a diameter, so that it is again horizontal. How much charge flows through the meter during the flip? (*Hint:* See Problem 13.)

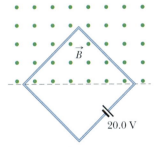

17. Square Loop   A square wire loop with 2.00 m sides is perpendicular to a uniform magnetic field, with half the area of the loop in the field as shown in Fig. 31-34. The loop contains a 20.0 V battery with negligible internal resistance. If the magnitude of the field varies with time according to $B = (0.0420\ \text{T}) - (0.870\ \text{T/s})t$, what are (a) the magnitude of the net emf in the circuit and (b) the direction of the current through the battery?

20.0 V

FIGURE 31-34 ■
Problem 17.

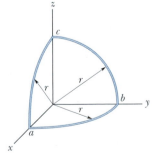

18. Three Circular Segments   A wire is bent into three circular segments, each of radius $r = 10$ cm, as shown in Fig. 31-35. Each segment is

FIGURE 31-35 ■
Problem 18.

a quadrant of a circle, *ab* lying in the *xy* plane, *bc* lying in the *yz* plane, and *ca* lying in the *zx* plane. (a) If a uniform magnetic field $\vec{B}$ points in the positive *x* direction, what is the magnitude of the emf developed in the wire when $\vec{B}$ increases at the rate of 3.0 mT/s in the *x* direction? (b) What is the direction of the current in segment *bc*?

19. Rectangular Coil   A rectangular coil of N turns and of length a and width b is rotated at frequency f in a uniform magnetic field $\vec{B}$, as indicated in Fig. 31-36. The coil is connected to co-rotating cylinders, against which metal brushes slide to make contact. If we arbitrarily define emf as being positive during the first quarter-turn, (a) show that the emf induced in the coil is given (as a function of time t) by

$$\mathcal{E} = 2\pi fNabB\sin(2\pi ft) = \mathcal{E}_0\sin(2\pi ft).$$

This is the principle of the commercial alternating-current generator. (b) Design a loop that will produce an emf with $\mathcal{E}_0 = 150$ V when rotated at 60.0 rev/s in a uniform magnetic field of 0.500 T.

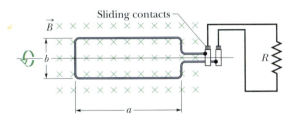

FIGURE 31-36 ■ Problem 19.

20. Semicircle   A stiff wire bent into a semicircle of radius a is rotated with frequency f in a uniform magnetic field, as suggested in Fig. 31-37. What are (a) the frequency and (b) the amplitude of the varying emf induced in the loop?

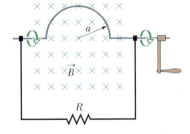

21. Electric Generator   An electric generator consists of 100 turns of wire formed into a rectangular loop 50.0 cm by 30.0 cm, placed entirely in a uniform magnetic field with magnitude $B = 3.50$ T. What is the maximum value of the emf produced when the loop is spun at 1000 rev/min about an axis perpendicular to $\vec{B}$?

FIGURE 31-37 ■ Problem 20.

22. Closed Circular Loop   In Fig. 31-38, a wire forms a closed circular loop, with radius $R = 2.0$ m and resistance 4.0 Ω. The circle is centered on a long straight wire; at time $t = 0$, the current in the long straight wire is 5.0 A rightward. Thereafter, the current changes according to $i = 5.0\ \text{A} - (2.0\ \text{A/s}^2)t^2$. (The straight wire is insulated, so there is no electrical contact between it and the wire of the loop.) What are the magnitude and direction of the current induced in the loop at times $t > 0$?

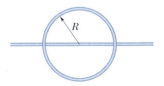

FIGURE 31-38 ■
Problem 22.

23. Square Loop Two In Fig. 31-39, the square loop of wire has sides of length 2.0 cm. A magnetic field is directed out of the page; its magnitude is given by $B = (4.0 \text{ T/m} \cdot \text{s}^2) \, t^2 y$, where B is in teslas, t is in seconds, and y is in meters. Determine the emf around the square at $t = 2.5$ s and indicate whether its direction is clockwise or counterclockwise.

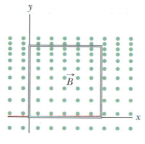

FIGURE 31-39 ▪ Problem 23.

24. Square Loop Three For the situation shown in Fig. 31-40, $a = 12.0$ cm and $b = 16.0$ cm. The current in the long straight wire is given by $i = (4.50 \text{ A/s}^2)t^2 - (10.0 \text{ A/s})t$, where i is in amperes and t is in seconds. (a) Find the magnitude of the emf in the square loop at $t = 3.00$ s. (b) Indicate whether the direction of the induced current in the loop is clockwise or counterclockwise at $t = 3.00$ s.

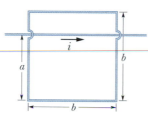

FIGURE 31-40 ▪ Problem 24.

25. Parallel Copper Wires Two long, parallel copper wires of diameter 2.5 mm carry currents of 10 A in opposite directions. (a) Assuming that their central axes are 20 mm apart, calculate the magnetic flux per meter of wire that exists in the space between those axes. (b) What fraction of this flux lies inside the wires? (c) Repeat part (a) for parallel currents.

26. Rectangular Wire Loop A rectangular loop of wire with length a, width b, and resistance R is placed near an infinitely long wire carrying current i, as shown in Fig. 31-41. The distance from the long wire to the center of the loop is r. Find (a) the magnitude of the magnetic flux encircled by the loop and (b) the amount of induced current in the loop $|i^{\text{ind}}|$ as it moves away from the long wire with velocity $\vec{v}$. (c) Indicate whether the induced current is clockwise or counterclockwise.

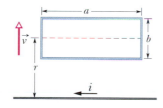

FIGURE 31-41 ▪ Problem 26.

SEC 31-6 ▪ INDUCTION AND ENERGY TRANSFERS

27. Internal Energy If 50.0 cm of copper wire (diameter = 1.00 mm) is formed into a circular loop and placed perpendicular to a uniform magnetic field that is increasing at the constant rate of 10.0 mT/s, at what rate does internal energy increase in the loop?

28. Loop Antenna A loop antenna of area A and resistance R is perpendicular to a uniform magnetic field $\vec{B}$. The field drops linearly to zero in a time interval Δt. Find an expression for the total internal energy added to the loop.

29. Rod on Rails A metal rod is forced to move with constant velocity $\vec{v}$ along two parallel metal rails, connected with a strip of metal at one end, as shown in Fig. 31-42. A magnetic field of magnitude $|\vec{B}| = 0.350$ T points out of the page. (a) If

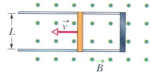

FIGURE 31-42 ▪ Problem 29 and Problem 31.

the rails are separated by 25.0 cm and the speed of the rod is 55.0 cm/s, what emf is generated? (b) If the rod has a resistance of 18.0 Ω and the rails and connector have negligible resistance, what is the current in the rod? (c) At what rate is mechanical energy being transformed to thermal energy?

30. Find Terminal Speed In Fig. 31-43, a long rectangular conducting loop, of width L, resistance R, and mass m, is hung in a horizontal, uniform magnetic field $\vec{B}$ that is directed into the page and that exists only above line aa. The loop is then dropped; during its fall, it accelerates until it reaches a certain terminal speed v_t. Ignoring air drag, find that terminal speed.

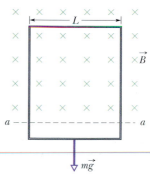

FIGURE 31-43 ▪ Problem 30.

31. Rod on Rails Two The conducting rod shown in Fig. 31-42 has length L and is being pulled along horizontal, frictionless conducting rails at a constant velocity $\vec{v}$. The rails are connected at one end with a metal strip. A uniform magnetic field $\vec{B}$, directed out of the page, fills the region in which the rod moves. Assume that $L = 10$ cm, $v = 5.0$ m/s, and $B = 1.2$ T. (a) What is the magnitude of the emf induced in the rod? (b) What is the magnitude and direction (clockwise or counterclockwise) of the current in the conducting loop? Assume that the resistance of the rod is 0.40 Ω and that the resistance of the rails and metal strip is negligibly small. (c) At what rate is thermal energy added to the rod? (d) What magnitude of force must be applied to the rod by an external agent to maintain its motion? (e) At what rate does this external agent do work on the rod? Compare this answer with the answer to (c).

32. Rods Bent into V Two straight conducting rails form a right angle where their ends are joined. A conducting bar in contact with the rails starts at the vertex at time $t = 0$ and moves with a constant velocity of magnitude 5.20 m/s along them, as shown in Fig. 31-44. A magnetic field of magnitude $B = 0.350$ T is directed out of the page. Calculate (a) the flux through the triangle formed by the rails and bar at $t = 3.00$ s and (b) the magnitude of emf around the triangle at that time. (c) If we write the emf as $\mathcal{E} = at^n$, where a and n are constants, what is the value of n?

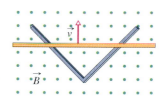

FIGURE 31-44 ▪ Problem 32.

33. Rod on Conducting Rails Two Figure 31-45 shows a rod of length L caused to move at constant speed v along horizontal conducting rails. The magnetic field in which the rod moves is *not uniform* but is provided by a current i in a long wire parallel to the rails. Assume that $v = 5.00$ m/s, $a = 10.0$ mm, $L = 10.0$ cm, and $i = 100$ A. (a) Calculate the magnitude of the emf induced in the rod. (b) What is the magnitude of the current in the conducting loop?

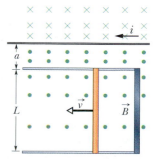

FIGURE 31-45 ▪ Problem 33.

Assume that the resistance of the rod is 0.400 Ω and that the resistance of the rails and the strip that connects them at the right is negligible. (c) At what rate is internal energy added to the rod? (d) What magnitude of force must be applied to the rod by an external agent to maintain its motion? (e) At what rate does this external agent do work on the rod? Compare this answer to that for (c).

SEC. 31-7 ■ INDUCED ELECTRIC FIELDS

34. Two Circular Regions Figure 31-46 shows two circular regions R_1 and R_2 with radii $r_1 = 20.0$ cm and $r_2 = 30.0$ cm. In R_1 there is a uniform magnetic field of magnitude $B_1 = 50.0$ mT into the page, and in R_2 there is a uniform magnetic field of magnitude $B_2 = 75.0$ mT out of the page (ignore any fringing of these fields). Both fields are decreasing at the rate of 8.50 mT/s. Calculate the integral $\oint \vec{E} \cdot d\vec{s}$ for each of the three dashed paths.

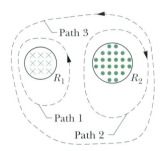

FIGURE 31-46 ■
Problem 34.

35. Long Solenoid A long solenoid has a diameter of 12.0 cm. When a current i exists in its windings, a uniform magnetic field of magnitude $B = 30.0$ mT is produced in its interior. By decreasing i, the field is caused to decrease at the rate of 6.50 mT/s. Calculate the magnitude of the induced electric field (a) 2.20 cm and (b) 8.20 cm from the axis of the solenoid.

36. Magnet Lab Early in 1981 the Francis Bitter National Magnet Laboratory at M.I.T. commenced operation of a 3.3-cm-diameter cylindrical magnet that produces a 30 T field, then the world's largest steady-state field. The field magnitude can be varied sinusoidally between the limits of 29.6 and 30.9 T at a frequency of 15 Hz. When this is done, what is the maximum value of the magnitude of the induced electric field at a radial distance of 1.6 cm from the axis? (*Hint:* See Touchstone Example 31-3.)

37. Drop to Zero Prove that the electric field $\vec{E}$ in a charged parallel-plate capacitor cannot drop abruptly to zero (as is suggested at point a in Fig. 31-47), as one moves perpendicular to the field, say, along the horizontal arrow in the figure. Fringing of the field lines always occurs in actual capacitors, which means that $\vec{E}$ approaches zero in a continuous and gradual way (see Problem 35 in Chapter 30). (*Hint:* Apply Faraday's law to the rectangular path shown by the dashed lines).

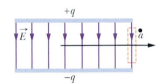

FIGURE 31-47 ■ Problem 37.

SEC 31-8 ■ INDUCED MAGNETIC FIELDS

38. Charging Capacitor Touchstone Example 31-4 describes the charging of a parallel-plate capacitor with circular plates of radius 55.0 mm. At what two radii r from the central axis of the capacitor is the magnitude of the induced magnetic field equal to 50% of its maximum value?

39. Induced Magnetic Field The induced magnetic field 6.0 mm from the central axis of a circular parallel-plate capacitor and

between the plates has magnitude of 2.0×10^{-7} T. The plates have radius 3.0 mm. At what rate $|d\vec{E}/dt|$ is the electric field magnitude between the plates changing?

40. Parallel-Plate Capacitor Suppose that a parallel-plate capacitor has circular plates with radius $R = 30$ mm and a plate separation of 5.0 mm. Suppose also that a sinusoidal potential difference with a maximum value of 150 V and a frequency of 60 Hz is applied across the plates. That is,

$$\Delta V = (150 \text{ V}) \sin[2\pi(60 \text{ Hz})t].$$

(a) Find $B^{\max}(R)$, the maximum value of the magnitude of the induced magnetic field that occurs at $r = R$. (b) Plot $B^{\max}(r)$ for $0 < r < 10$ cm.

41. Uniform Electric Flux Figure 31-48 shows a circular region of radius $R = 3.00$ cm in which a uniform electric flux is directed out of the page. The total electric flux enclosed by the region is given by $\Phi^{\text{elec}} = (3.00 \text{ mV} \cdot \text{m/s})t$, where t is time. What is the magnitude of the magnetic field that is induced at radial distances (a) 2.00 cm and (b) 5.00 cm?

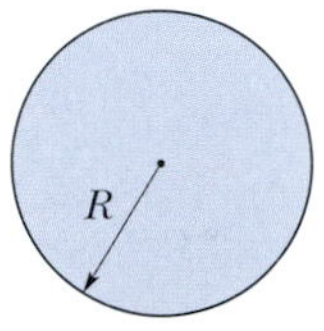

FIGURE 31-48 ■
Problems 41 through 44, and 57, 59, and 60.

42. Nonuniform Electric Flux Figure 31-48 shows a circular region of radius $R = 3.00$ cm in which an electric flux is directed out of the page. The flux encircled by a concentric circle of radius r is given by $\Phi^{\text{elec}} = (0.600 \text{ V} \cdot \text{m/s})(r/R)t$, where $r \leq R$ and t is time. What is the magnitude of the induced magnetic field at radial distances (a) 2.00 cm and (b) 5.00 cm?

43. Uniform Electric Field In Fig. 31-48, a uniform electric field is directed out of the page within a circular region of radius $R = 3.00$ cm. The magnitude of the electric field is given by $E = (4.5 \times 10^{-3} \text{ V/m} \cdot \text{s})t$, where t is time. What is the magnitude of the induced magnetic field at radial distances (a) 2.00 cm and (b) 5.00 cm?

44. Nonuniform Electric Field In Fig. 31-48, an electric field is directed out of the page within a circular region of radius $R = 3.00$ cm. The magnitude of the electric field is given by $E = (0.500 \text{ V/m} \cdot \text{s})(1 - r/R)t$, where t is the time and r is the radial distance $(r \leq R)$. What is the magnitude of the induced magnetic field at radial distances (a) 2.00 cm and (b) 5.00 cm?

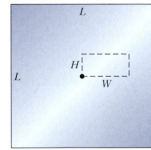

45. Discharging Capacitor A capacitor with square plates of edge length L is being discharged by a current of 0.75 A. Figure 31-49 is a head-on view of one of the plates from inside the capacitor. A dashed rectangular path is shown. If $L = 12$ cm, $W = 4.0$ cm, and $H = 2.0$ cm, what is the value of $\oint \vec{B} \cdot d\vec{s}$ around the dashed path?

FIGURE 31-49 ■
Problem 45.

46. Charging Capacitor The circuit in Fig. 31-50 consists of switch S, a 12.0 V ideal battery, a 20.0 MΩ resistor, and an air-filled capacitor. The capacitor has parallel circular plates of radius 5.00 cm,

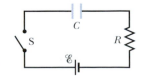

FIGURE 31-50 ■
Problem 46.

separated by 3.00 mm. At time $t = 0$ s, switch S is closed to begin charging the capacitor. The electric field between the plates is uniform. At $t = 250$ μs, what is the magnitude of the magnetic field within the capacitor, at radial distance 3.00 cm?

SEC. 31-9 ■ DISPLACEMENT CURRENT

47. Prove That Displacement Prove that the displacement current in a parallel-plate capacitor of capacitance C can be written as $i^{\text{dis}} = C(d\Delta V/dt)$, where ΔV is the potential difference between the plates.

48. At What Rate At what rate must the potential difference between the plates of a parallel-plate capacitor with a 2.0 μF capacitance be changed to produce a displacement current of 1.5 A?

49. Current Density For the situation of Touchstone Example 31-4, show that the magnitude of the current density of the displacement current is $J^{\text{dis}} = \varepsilon_0 \, (dE/dt)$ for $r \leq R$.

50. Being Discharged A parallel-plate capacitor with circular plates of radius 0.10 m is being discharged. A circular loop of radius 0.20 m is concentric with the capacitor and halfway between the plates. The displacement current through the loop is 2.0 A. At what rate is the magnitude of the electric field between the plates changing?

51. Displacement Current As a parallel-plate capacitor with circular plates 20 cm in diameter is being charged, the current density of the displacement current in the region between the plates is uniform and has a magnitude of 20 A/m². (a) Calculate the magnitude B of the magnetic field at a distance $r = 50$ mm from the axis of symmetry of this region. (b) Calculate dE/dt in this region.

52. Electric Field The magnitude of the electric field between the two circular parallel plates in Fig. 31-51 is $E = (4.0 \times 10^5 \text{ V} \cdot \text{m}) - (6.0 \times 10^4 \text{ V} \cdot \text{m/s})t$, with E in volts per meter and t in seconds. At $t = 0$ s, the field is upward as shown. The plate area is 4.0×10^{-2} m². For $t \geq 0$ s, (a) what are the magnitude and direction of the displacement current between the plates and (b) is the direction of the induced magnetic field clockwise or counterclockwise around the plates?

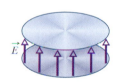

FIGURE 31-51 ■ Problem 52.

53. Magnitude of Electric Field The magnitude of a uniform electric field collapses to zero from an initial strength of 6.0×10^5 N/C in a time of 15 μs in the manner shown in Fig. 31-52. Calculate the amount of displacement current, $|i|$, through a 1.6 m² area perpendicular to the field, during each of the time intervals, a, b, and c shown on the graph. (Ignore the behavior at the ends of the intervals.)

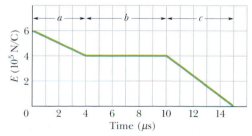

FIGURE 31-52 ■ Problem 53.

54. Displacement Current Two A parallel-plate capacitor with circular plates is being charged. Consider a circular loop centered on the central axis between the plates. The loop radius is 0.20 m, the plate radius is 0.10 m, and the displacement current through the loop is 2.0 A. What is the rate at which the magnitude of the electric field between the plates is changing?

55. Square Plates A parallel-plate capacitor has square plates 1.0 m on a side as shown in Fig. 31-53. A current of 2.0 A charges the capacitor, producing a uniform electric field $\vec{E}$ between the plates, with $\vec{E}$ perpendicular to the plates. (a) What is the displacement current i^{dis} through the region between the plates? (b) What is dE/dt in this region? (c) What is the displacement current through the square dashed path between the plates? (d) What is $\oint \vec{B} \cdot d\vec{s}$ around this square dashed path?

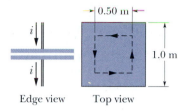

FIGURE 31-53 ■ Problem 55.

56. Consider a Loop A capacitor with parallel circular plates of radius R is discharging via a current of 12.0 A. Consider a loop of radius $R/3$ that is centered on the central axis between the plates. (a) How much displacement current is encircled by the loop? The maximum induced magnetic field has a magnitude of 12.0 mT. (b) At what radial distance from the central axis of the plate is the magnitude of the induced magnetic field 3.00 mT?

57. Uniform Displacement-Current Density. Figure 31-48 shows a circular region of radius $R = 3.00$ cm in which a displacement current is directed out of the page. The magnitude of the displacement current has a uniform density $J^{\text{dis}} = 6.00$ A/m². What is the magnitude of the magnetic field due to the displacement current at radial distances (a) 2.00 cm and (b) 5.00 cm?

58. Actual and Displacement Figure 31-54a shows current i that is produced in a wire of resistivity $1.62 \times 10^{-8} \Omega \cdot$ m in the direction indicated. The magnitude of the current versus time t is shown in Fig. 31-54b. Point P is at radius 9.00 mm from the wire's center. Determine the magnitude of the magnetic field at point P due to the real current i in the wire at (a) $t_1 = 20$ ms, (b) $t_2 = 40$ ms, (c) $t_3 = 60$ ms, and (d) $t_4 = 70$ ms. Next, assume that the electric field driving the current is confined to the wire. Then determine the magnitude of the magnetic field at point P due to the displacement current i^{dis} in the wire at (e) $t_1 = 20$ ms, (f) $t_2 = 40$ ms, (g) $t_3 = 60$ ms, and (h) $t_4 = 70$ ms. (i) When both magnetic fields are present at point P, what are their directions in Fig. 31-54a?

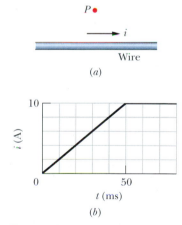

FIGURE 31-54 ■ Problem 58.

59. Nonuniform Displacement-Current Density. Figure 31-48 shows a circular region of radius $R = 3.00$ cm in which a displacement current is directed out of the page. The displacement current has a density of magnitude $J^{\text{dis}} = (4.00 \text{ A/m}^2)(1 - r/R)$, where r is the radial distance $r \leq R$. What is the magnitude of the magnetic field due to the displacement current at radial distances (a) 2.00 cm and (b) 5.00 cm?

60. Uniform Displacement Current. Figure 35-48 shows a circular region of radius $R = 3.00$ cm in which a uniform displacement current $i^{dis} = 0.500$ A is directed out of the page. What is the magnitude of the magnetic field due to the displacement current at radial distances (a) 2.00 cm and (b) 5.00 cm?

SEC. 31-10 ■ GAUSS' LAW FOR MAGNETIC FIELDS

61. Rolling a Sheet of Paper Imagine rolling a sheet of paper into a cylinder and placing a bar magnet near its end as shown in Fig. 31-55. (a) Sketch the magnetic field lines that pass

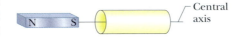

FIGURE 31-55 ■ Problem 61.

through the surface of the cylinder. (b) What can you say about the sign of $\vec{B} \cdot d\vec{A}$ for every area $d\vec{A}$ on the surface? (c) Does this result contradict Gauss' law for magnetism? Explain.

62. Die Suppose the magnetic flux at each of five faces of a die (singular of "dice") is given by $\Phi^{mag} = \pm N$ Wb, where $N(= 1$ to $5)$ is the number of spots on the face. The flux is positive (outward) for N even and negative (inward) for N odd. What is the flux at the sixth face of the die? Is it directed in or out?

63. Right Circular Cylinder A Gaussian surface in the shape of a right circular cylinder with end caps has a radius of 12.0 cm and a length of 80.0 cm. One end encircles an inward magnetic flux of 25.0 μWb. At the other end there is a uniform magnetic field of 1.60 mT, normal to the surface and directed outward. What is the net magnetic flux at the curved surface?

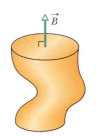

64. Weird Shape Figure 31-56 shows a closed surface. Along the flat top face, which has a radius of 2.0 cm, a magnetic field $\vec{B}$ of magnitude 0.30 T is directed outward. Along the flat bottom face, a magnetic flux of 0.70 mWb is directed outward. What are (a) the magnitude and (b) the net magnetic flux at the curved part of the surface?

FIGURE 31-56 ■ Problem 64.

Additional Problems

65. Power from a Tether A few years ago, the space shuttle *Columbia* tried an experiment with a tethered satellite. The satellite was released from the shuttle and slowly reeled out on a long conducting cable as shown in Fig. 31-57 (not to scale). For this problem we will make the following approximations:

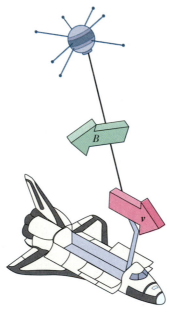

FIGURE 31-57 ■ Problem 65.

The shuttle is moving at a constant velocity.

The Earth's magnetic field is constant and uniform.

The line of the tether, the velocity of the system, and the magnetic field are all perpendicular to each other.

The Earth's field produces an emf from one end of the cable to the other. The idea is to use a system like this to generate electric power in space more efficiently than with solar panels.

(a) Explain why a voltage difference is produced.
(b) If the Earth's magnetic field is given by a magnitude $\vec{B}$, the shuttle–satellite system is moving with a velocity $\vec{v}$, and the tether has a length L, calculate the magnitude of the emf $\mathscr{E}$ from one end of the tether to the other.
(c) At the shuttle's altitude, the Earth's field is about 0.3 gauss and the shuttle's speed is about 7.5 km/s. The tether is 20 km long (!). What is the expected potential difference in volts?
(d) At the altitude of the shuttle, the thin atmosphere is lightly ionized, allowing a current of about 0.5 amps to flow from the satellite

back to the shuttle through the thin air. What is the resistance of the 20 km of ionized air?

66. Building a Generator The apparatus shown in Fig. 31-58 can be used to build a motor. This device can also be used to build a generator that will produce a voltage. (a) Explain the setup that one would use to make a motor and explain how it works. Do the same for the generator. (b) Estimate the maximum voltage that would be produced if you cranked the generator by hand. (*Hint:* As a comparison for estimating the strength of the bar magnet, the Earth's magnetic field at our location is about 0.4 gauss.)

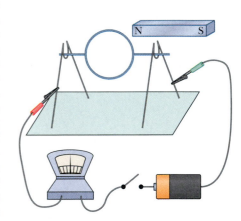

FIGURE 31-58 ■ Problem 66.

67. Faraday's Law Faraday's law describes the emf produced by magnetic fields in a variety of circumstances. State and discuss Faraday's law, being careful to include a discussion of different physical situations that may be described by the statement of the law.

68. Magnetic Field, Force, and Torque Figure 31-59 shows two long, current-carrying wires and a bar magnet. At the right is shown a compass specifying set of direction labels. For each of the vectors (a)–(e) below, select the direction label that best gives the direction of the item. If the magnitude of the item is zero, write 0. If none of the directions are correct, write N.

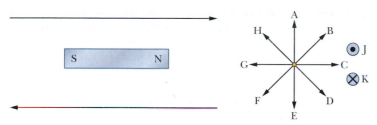

FIGURE 31-59 ■ Problem 68.

(a) The magnetic field due to the lower wire at the center of the upper wire
(b) The force on the lower wire due to the magnetic field from the upper wire
(c) The net torque acting on the upper wire
(d) The magnetic field due to the currents at the center of the magnet
(e) The net force acting on the lower wire due to the bar magnet

69. B Increases in Time In Fig. 31-60a, a uniform magnetic field $\vec{B}$ increases in magnitude with time t as given by Fig. 31-60b. A circular conducting loop of area 8.0×10^{-4} m^2 lies in the field, in the plane of the page. The amount of charge q that has passed point A on the loop is given in Fig. 31-60c as a function of t. What is the loop's resistance?

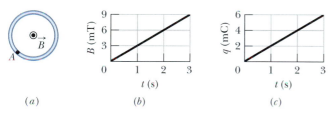

FIGURE 31-60 ■ Problem 69.

70. Circular Loop Around a Solenoid In Fig. 31-61a, a circular loop of wire is concentric with a solenoid and lies in a plane that is perpendicular to the solenoid's central axis. The loop has radius 6.00 cm. The solenoid has radius 2.00 cm, consists of 8000 turns per meter, and has a current i_{sol} that varies with time t as given in Fig. 31-61b. Figure 31-61c shows, as a function of time, the energy $E^{thermal}$ that is transformed to thermal energy in the loop. What is the loop's resistance?

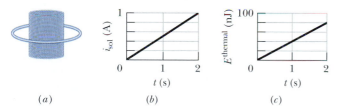

FIGURE 31-61 ■ Problem 70.

71. Magnitudes and Direction Figure 31-62a shows a wire that forms a rectangle and has a resistance of 5.0 mΩ. Its interior is split into three equal areas with different magnetic fields $\vec{B}_1$, $\vec{B}_2$, and $\vec{B}_3$ that are either directly out of or into the page, as indicated. The fields are uniform within each region. Figure 31-62b gives the change in the z components B_z of the three fields with time t. What are the magnitude and direction of the current induced in the wire?

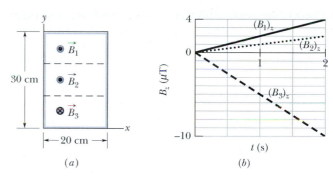

FIGURE 31-62 ■ Problem 71.

72. Two Concentric Regions Figure 31-63a shows two concentric circular regions in which uniform magnetic fields can change. Region 1, with radius $r_1 = 1.0$ cm, has an outward magnetic field $\vec{B}_1$ that is increasing in magnitude. Region 2, with radius $r_2 = 2.0$ cm, has an outward magnetic field $\vec{B}_2$ that may also be changing. Imagine that a conducting ring of radius R is centered on the two regions and then the emf $\mathscr{E}$ around the ring is determined. Figure 31-63b gives emf $\mathscr{E}$ as a function of the square of the ring's radius, R^2, to the outer edge of region 2. What are the rates of B-field magnitude change (a) dB_1/dt and (b) dB_2/dt? (c) Is the magnitude of $\vec{B}_2$ increasing, decreasing, or remaining constant?

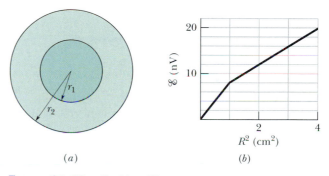

FIGURE 31-63 ■ Problem 72.

73. Pulled at Constant Speed Figure 31-64a shows a rectangular conducting loop of resistance $R = 0.020$ Ω, height $H = 1.5$ cm, and length $D = 2.5$ cm being pulled at constant speed $v = 40$ cm/s through two regions of uniform magnetic field. Figure 31-64b gives the current i induced in the loop as a function of the position x of the right side of the loop. For example, a current of 3.0 μA is induced clockwise as the loop enters region 1. What are the magnitudes and directions of the magnetic field in (a) region 1 and (b) region 2?

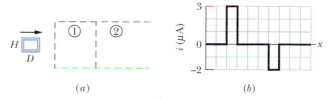

FIGURE 31-64 ■ Problem 73.

74. Plane Loop A plane loop of wire consisting of a single turn of area 8.0 cm² is perpendicular to a magnetic field that increases uniformly in magnitude from 0.50 T to 2.5 T in a time of 1.0 s. What is the resulting induced current if the coil has a total resistance of 2.0 Ω?

75. At What Rate Must *B* Change The plane of a rectangular coil of dimensions 5.0 cm by 8.0 cm is perpendicular to the direction of magnetic field *B*. If the coil has 75 turns and a total resistance of 8.0 Ω, at what rate must the magnitude of *B* change in order to induce a current of 0.10 A in the windings of the coil?

76. Rod on Rails 3 In the arrangement shown in Fig. 31-65, a conducting rod rolls to the right along parallel conducting rails connected on one end by a 6.0 Ω resistor. A 2.5 T magnetic field is directed *into* the paper. Let *L* = 1.2 m. Neglect the mass of the bar and friction. (a) Calculate the applied force required to move the bar to the right at a *constant* speed of 2.0 m/s. (b) At what rate is energy dissipated in the resistor?

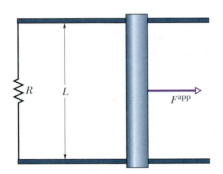

FIGURE 31-65 ■ Problem 76.

77. An Engineer An engineer has designed a setup with a small pickup coil placed in the center of a large field coil as shown in Fig. 31-66. Both coils have many turns of conducting wire. The field coil produces a magnetic field that is proportional in magnitude to the amount of current flowing through its wires. The pickup coil is smaller and its many turns can sense or "pick up" the changing magnetic field in the field coil. The pickup coil produces an emf that is proportional in magnitude to the rate of change of the magnetic field and the angle ϕ. Here ϕ is the angle between the normal to the field coil and the normal to the pickup coil. You have been hired as a consultant to check on the reliability of the engineer's work. You figure out how to use Faraday's law along with proportional reason-

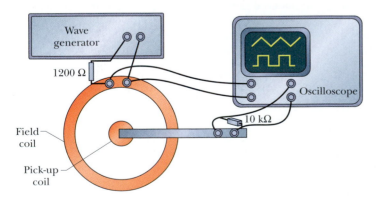

FIGURE 31-66a ■ Problem 77.

ing to check on the validity of the results that have been reported without doing any formal calculations or measurements. Sketches from the engineer's notebook are shown in Fig. 31-66b. (a) Look at the graph pair in Fig. 31-66b. Sketch the measured emf induced in the pickup coil if the engineer has adjusted the scope so the maximum emf is the first positive grid line and the minimum emf is on the first negative grid line. Assume that the normal to each of the coils is pointing in the same direction. (b) According to the engineer's notebook, she fed exactly the same pattern of current to the field coil but she turned the pickup coil so its normal makes an angle of +45° with respect to the normal to the plane of the field coil. Carefully sketch the pattern of emf observed in the pickup coil. What is the maximum and minimum amplitude of the emf in "grid" units? (c) What happens when she flips the pickup coil over around so its normal is 180° from the normal to the field coil? Sketch the emf and use the correct signs for the values of the induced emf for this situation. Explain the reasons for the shape and magnitude of your sketch in each case.

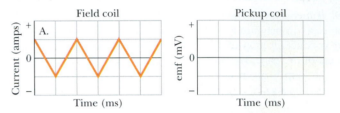

FIGURE 31-66b ■ Problem 77.

78. Engineer Task 2 You are still double-checking the work of the engineer from Problem 77. Consider the graph shown in Fig. 31-67. (a) What should our honest and competent engineer have reported for the pattern of emf values as a function of time? Assume that

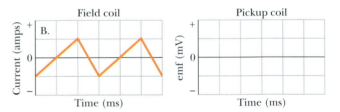

FIGURE 31-67 ■ Problem 78.

once again the normal to the pickup coil is in the same direction as the normal to the field coil. Please take care to sketch not only the shape of the emf graph but also its proper magnitude using the same gain setting on the oscilloscope as you did in Problem 77. Use a solid line for your sketch. (b) Suppose the engineer reduced the number of turns in the pickup coil by a factor of 2 and redid the measurements. Sketch a new graph showing the shape and proper magnitudes for the expected pickup coil emf using a dashed line. Explain the reasons for the shape and magnitude of your sketch in each case.

79. Engineer Task 3 You are still double-checking the work of the engineer from Problem 31-77. Assume that the number of turns in both the field and pickup coils is the same as in that problem, as is the oscilloscope setting. Consider the graph shown in Fig. 31-68. (a) What should our honest and competent engineer have reported for the pattern of current fed into the field coil as a function of time? Assume that once again the normal to the pickup coil is in the same

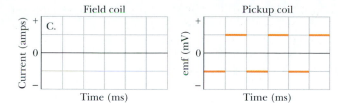

FIGURE 31-68 ■ Problem 79.

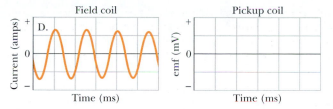

FIGURE 31-69 ■ Problem 80.

direction as the normal to the field coil. Please take care to sketch not only the shape of the emf graph but also its proper magnitude using the same gain setting on the oscilloscope as you did in Problem 31-77. Use a solid line for your sketch. (b) Suppose the engineer reduced the number of turns in the pickup coil by a factor of 2 and redid the measurements. Sketch a new graph showing the shape and proper magnitudes for emf in the field coil using a dashed line. Explain the reasons for the shape and magnitude of your sketch in each case.

80. Engineer Task 4 You are still double-checking the work of the engineer from Problem 31-77. Assume that the number of turns in both the field and pickup coils is the same as in that problem. Consider the graph shown in Fig. 31-69. What should our honest and competent engineer have reported for the pattern of emf induced in the pickup coil if the oscilloscope gain is adjusted to give a maximum value of emf of +2 oscilloscope grid units and a minimum value of −2 oscilloscope units? (*Hint:* What is the derivative of the sine function?) Explain the reason for the shape and magnitude of your sketch in each case.

81. Ring of Copper Figure 31-70 shows a ring of copper with its plane perpendicular to the axis of the nearby rod-shaped magnet. In which of the following situations will a current be induced in the ring? Choose all correct answers.

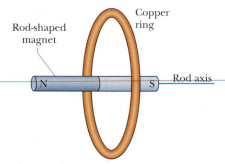

FIGURE 31-70 ■ Problem 81.

(a) The magnet is moved horizontally toward the left.

(b) The ring is moved away from the magnet.
(c) The ring is rotated around any of its diameters.
(d) The magnet is moved up or down.
(e) The ring is rotated around its center in the plane in which it lies.

32 | Inductors and Magnetic Materials

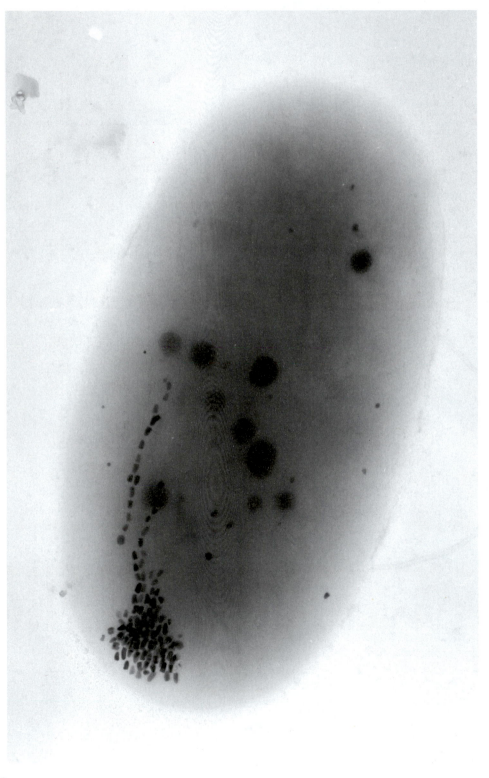

This is a microscopic view of a bacterium found in Australia that will swim to the muddy bottom of a pond to escape oxygen in its environment and find the nutrients it needs to survive. But if this bacterium were transported to a pond in the United States, it would swim to the top of the pond and die.

How does this bacterium know how to swim down in Australia but not in the U.S.?

The answer is in this chapter.

32-1 Introduction

In the previous chapter we described how an electric car or toothbrush could be charged without electrical contacts. Likewise the guitar pickup described in Section 31-5 amplifies sound. These devices make practical use of inductance. In this chapter we consider some additional practical uses of inductance phenomena in common electric circuit elements known as inductors and transformers. You will consider the basic behaviors of these elements in circuits where the voltage changes in time. Then you will move on to what appears to be an unrelated topic—the behavior of magnetic materials.

The simplest magnetic structure contained in magnetic materials is a magnetic dipole. We will trace the origin of magnetic dipoles, and the associated magnetic properties of materials back to atoms and electrons. You will then reconsider inductors and transformers and learn how magnetic materials can be used to enhance their performance. Some of the first inductors are pictured in Fig. 32-1.

Finally, we will discuss recent theories that enable us explain why the Earth behaves like a huge magnetic dipole, and we will consider the possible role induction plays in explaining the characteristics and changing nature of the Earth's magnetic field.

FIGURE 32-1 ■ The crude inductors with which Michael Faraday discovered the law of induction. In those days amenities such as insulated wire were not commercially available. It is said that Faraday insulated his wires by wrapping them with strips cut from one of his wife's petticoats.

32-2 Self-Inductance

Let's explore how the phenomenon of inductance introduced in the previous chapter can be useful in the design of electric circuits with changing currents. In Section 31-3 we saw that when two coils are near each other, a changing current in one of the coils can induce an emf in the other according to Faraday's law ($\mathcal{E} = -Nd\,\Phi^{\mathrm{mag}}/dt$). But if the second coil is part of an electric circuit, the current induced in it can also induce an emf in the first coil. This phenomenon, known as **mutual induction,** is used in the design of *inductive chargers*—noncontact charging systems like those used for electric toothbrushes and other devices. In multiple-loop coils the emfs produced by mutual induction are proportional to the number of loops in the coil. For this reason mutual induction is also used in the design of **transformers**—devices that can transform time-varying voltages to larger or smaller time-varying voltages.

In addition, when current in a single coil with one or more loops changes, this induces an emf in the *same* coil. This emf is produced as the result to the changing flux the coil produces in the area it encloses. This process is known as **self induction.** In general,

> A self-induced emf $\mathcal{E}_L$ appears in any coil whenever its current is changing.

According to Lenz's law the self-induced emf acts to oppose the change of current in the coil. For this reason, coils of wire called **inductors** (sometimes called "chokes") are useful in circuits whenever it is desirable to stabilize currents. In a circuit diagram, an inductor is denoted by a symbol that looks like helical loops of wire ( ). See Figs. 32-2 and 32-3. In addition, inductors can be combined with resistors and capacitors to modify the characteristics in circuits driven by oscillating voltage sources. In this chapter we consider the role of inductors in stabilizing currents. In the next chapter we will study the behavior of inductors in circuits with oscillating voltages.

A typical **inductor** consists of a wire that is coiled into a very large number of loops wrapped around a piece of hollow cardboard or perhaps a magnetic rod. Inductors come in many shapes. A common shape is a **solenoid,** which consists of a tightly

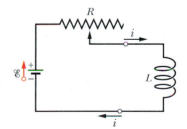

FIGURE 32-2 ■ If the current in a coil is changed by varying the contact position on a variable resistor, a self-induced emf $\mathcal{E}_L$ will appear in the coil while the current is changing.

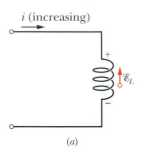

i (increasing)

(a)

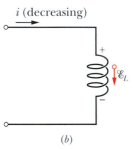

i (decreasing)

(b)

FIGURE 32-3 ■ The arrow and the + and − signs on either side of the inductor indicate the direction the emf $\mathscr{E}_L$ acts in relative to the direction of the current in the circuit alongside the coil. (*a*) The current *i* is increasing and the self-induced emf $\mathscr{E}_L$ appears along the coil in a direction such that it opposes the increase. (*b*) The current *i* is decreasing and the self-induced emf appears in a direction such that it opposes the decrease.

wound helical coil of wire—like the one Faraday wound shown in the lower part of Fig. 32-1. Because the magnetic field inside a solenoid is very uniform, it is not difficult to calculate the emfs created by current changes in solenoids. For this reason, we shall consider a solenoid as our basic type of inductor. Also, at first we assume that all inductors are air-core inductors that have no magnetic materials such as iron in their vicinity to distort their magnetic fields.

The Mathematics of Self-Inductance

We start our mathematical treatment of self-inductance with a solenoid-shaped inductor of length *l* and total number of loops *N*. When a charge flows through an inductor, the coil produces a magnetic field inside its coils whose strength is *directly proportional* to the current. For an ideal solenoid, the magnitude of the magnetic field is given by Eq. 30-25,

$$B = \mu_0 n |i| \qquad \text{(inside an ideal solenoid)}, \qquad \text{(Eq. 30-25)}$$

where μ_0 is the magnetic constant and *n* the number of turns per unit length.

This magnetic field yields an amount of flux over the area *A* enclosed by the coil of $|\Phi^{\text{mag}}| = BA = n(\mu_0 A |i|)$. Now, if we try to change the current by changing the resistance in the circuit shown in Fig. 32-2, then the magnetic field and hence the flux at the center of the coil changes. According to Faraday's law this change in flux will produce an emf in the coil given by Eq. 31-7 ($\mathscr{E} = -N d\Phi^{\text{mag}}/dt$). According to Lenz's law this emf will act to oppose the change in the current. Thus, if you close a switch that connects a voltage source to an inductor, the induced "back" emf will retard the rise in current through the circuit. An emf that acts to oppose a change in current is known as a **back emf.** Alternately, if a current already exists in a circuit then opening a switch will slow the rate of reduction of the current (Fig. 32-3). Applying Faraday's law and noting that the total number of turns *N* is the product of the turns per unit length, *n*, and the length, *l*, of the solenoid gives us

$$\mathscr{E}_L = -N\frac{d\Phi^{\text{mag}}}{dt} = -nl\frac{d\Phi^{\text{mag}}}{dt} = -\mu_0 A n^2 l \frac{di}{dt} \qquad \text{(solenoidal air-core inductor)}, \qquad (32\text{-}1)$$

where $\mathscr{E}_L$ is the self-induced emf in the solenoid.

If the solenoid is very much longer than its radius, then Eq. 32-1 expresses its inductance to a good approximation. However, we have neglected the spreading of the magnetic field lines near the ends of the solenoid, just as the parallel-plate capacitor formula $C = \varepsilon_0 A/d$ neglects the fringing of the electric field lines near the edges of the capacitor plates.

Equation 32-1 tells us that the amount of self-induced back emf is directly proportional to the rate of change of the current through the coil. The minus sign tells us that $\mathscr{E}_L$ is a back emf. It is customary to combine the product of constants (which for a solenoid is $\mu_0 A n^2 l$) and write this proportionality between the self-induced emf and the rate of current change as

$$\mathscr{E}_L = -L\frac{di}{dt}, \qquad (32\text{-}2)$$

where *L* is known as the self-inductance of the coil. As we learned in Chapter 31, the minus sign in the equation indicates that the emf acts to oppose the change in current. From Eq. 32-2 we see that when the inductance *L* is large, a large emf will be produced for a given rate of current change.

This combination of terms (such as area, length, and so on) that makes up the constant of proportionality, *L*, is only valid for a long solenoid. The terms will be different if the coil has a flat shape or if the inductor wire is wrapped around an iron core. In addition, since any electric circuit is basically a loop of some sort, all circuits have a certain amount of self-inductance even when no inductor is present. Self-inductance is usually negligible, but it can be significant when high-voltage circuits are switched on or off or when the circuit current oscillates at high frequencies. If we have a complicated geometry and cannot calculate inductance simply, the inductance L can be determined experimentally by measuring both the emf and the rate of change of current and taking the ratio of these quantities. Thus, for any geometry the **self-inductance** of an inductor or a circuit can be defined as the ratio of the induced emf to the rate of current change or

$$L \equiv -\frac{\mathcal{E}_L}{di/dt} \qquad \text{(self-inductance defined).} \qquad (32\text{-}3)$$

For any inductor having a self-inductance L, Eqs. 32-1 and 32-2 tell us that $N(d\phi^{mag}/dt) = L\, di/dt$. Thus we conclude that $\mathcal{E}_L = -Nd\, \Phi^{mag}/dt = -L\, di/dt$, so $Li = N\Phi^{mag}$, where N is the number of turns in the coil producing flux and i is the current in the coil producing the flux. The windings of the inductor are said to be *linked* by the shared flux, and the product $N\Phi^{mag}$ is called the *magnetic flux linkage*. This leads us to an alternate definition of inductance (which is equivalent to that given in Eq. 32-3):

$$L \equiv \frac{N\Phi^{mag}}{i} \qquad \text{(alternative definition of self-inductance).} \qquad (32\text{-}4)$$

The inductance L is thus a measure of the flux linkage produced by the inductor per unit of current.

Because the SI unit of magnetic flux is the tesla-square meter, the SI unit of inductance is the tesla square-meter per ampere ($T \cdot m^2/A$). We call this the **henry** (H), after American physicist Joseph Henry, the co-discoverer, with Faraday, of the law of induction. Thus,

$$1 \text{ henry} = 1 \text{ H} = 1 T \cdot m^2/A. \qquad (32\text{-}5)$$

> In any inductor (such as a flat coil, a solenoid, or a toroid) a self-induced emf appears whenever the current changes with time. The amount of the current has no influence on the amount of induced emf. Only the rate of change of the current matters.

You can find the *direction* of a self-induced emf from Lenz's law. The minus signs in Eqs. 32-2 and 32-3 indicate that—as the law states and Fig. 32-2 shows—the self-induced emf $\mathcal{E}_L$ has an orientation such that it opposes the change in current i.

Ideal Inductors

In Section 31-7 we saw that we cannot define an electric potential for an emf that is induced by a changing magnetic flux. This means that when a self-induced emf is produced, we cannot define an electric potential within the inductor itself. However, electric potentials can still be defined at points in a circuit that are not within the inductor—points where the electric fields are due to charge distributions.

Moreover, we can define a self-induced potential difference ΔV_L across an inductor (between its terminals, which we assume to be outside the region of changing flux). If the inductor is ideal so that its wire has negligible resistance, the amount of the measured voltage change ΔV_L is equal to the amount of the self-induced emf $\mathcal{E}_L$.

If, instead, the wire in the inductor has resistance R_L, we mentally separate the inductor into a resistance R_L (which we take to be outside the region of changing flux) and an ideal inductor of self-induced emf $\mathcal{E}_L$. As with a real battery of emf $\mathcal{E}$ and internal resistance R, the potential difference across the terminals of a real inductor then differs from the emf. Unless otherwise indicated, we assume here that inductors are ideal.

READING EXERCISE 32-1: (a) What happens to the inductance of a solenoid if: (a) the number of turns per unit length doubles, (b) the cross-sectional area enclosed by the windings doubles? ■

READING EXERCISE 32-2: The figure shows an emf $\mathscr{E}_L$ induced in a coil. Which of the following can describe the current through the coil: (a) constant and rightward, (b) constant and leftward, (c) increasing and rightward, (d) decreasing and rightward, (e) increasing and leftward, (f) decreasing and leftward? ■

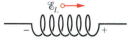

32-3 Mutual Induction

In this section we return to the case of two interacting coils, which we started discussing in the previous section. We saw earlier that if two coils are close together as in Fig. 32-4 (or Fig. 31-10), a steady current i in one coil will set up a magnetic flux Φ^{mag} at the other coil (*linking* the other coil). If we change the current, i, in the first coil with time, an emf $\mathscr{E}$ given by Faraday's law ($\mathscr{E} = -N d\,\Phi^{mag}/dt$) will be induced in the second coil. We called this process **mutual induction**, to suggest the mutual interaction of the two coils and to distinguish it from *self-induction*, in which only one coil is involved.

Let us look at mutual induction quantitatively. For any inductor having a self-inductance L, Eq. 32-3 tells us that

$$L \equiv -\frac{\mathscr{E}_L}{di/dt}$$

where i is the current in the coil producing the flux. Figure 32-4a shows two circular coils near each other that share a common central axis. Assume there is a steady current i_1 in coil 1, produced by the battery in the external circuit. This current creates a magnetic field represented by the lines of $\vec{B}_1$ in the figure. Coil 2 is connected to a

FIGURE 32-4 ■ Mutual induction. (*a*) If the current in coil 1 changes, an emf will be induced in coil 2. (*b*) If the current in coil 2 changes, an emf will be induced in coil 1.

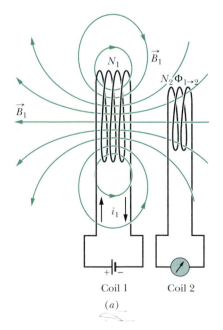

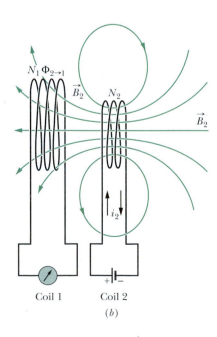

sensitive meter but contains no battery. A magnetic flux $\Phi_{1\to2}$ (the flux associated with the current in coil 1 that passes through coil 2) links the N_2 turns of coil 2.

Suppose that by external means we cause i_1 to vary with time. Then by analogy to the definition of self-inductance, we can write a mutual induction equation that is analogous to Eq. 32-2,

$$\mathscr{E}_2 = -M_{1\to2}\frac{di_1}{dt}.$$

This leads us to define the mutual inductance $M_{1\to2}$ of coil 2 due to coil 1 as

$$M_{1\to2} \equiv -\frac{\mathscr{E}_2}{di_1/dt} \qquad \text{(mutual inductance defined).} \qquad (32\text{-}6)$$

Once again we can formulate an alternate definition of mutual induction using the relationship between flux linkage in coil 2 and the current in coil 1, which is $M_{1\to2}i_1 = N_2\Phi_{1\to2}$. The factor N_2 is the number of turns in coil 2 and the factor $\Phi_{1\to2}$ is the magnetic flux present inside coil 2 due to coil 1. This allows us to define mutual inductance as

$$M_{1\to2} \equiv \frac{N_2\Phi_{1\to2}}{i_1} \qquad \text{(alternate definition of mutual inductance).} \qquad (32\text{-}7)$$

If we take the time derivative of all terms in the expression $M_{1\to2}i_1 = N_2\Phi_{1\to2}$ we can write

$$\mathscr{E}_2 = -M_{1\to2}\frac{di_1}{dt} = -N_2\frac{d\Phi_{1\to2}}{dt}. \qquad (32\text{-}8)$$

According to Faraday's law, the right side of this equation is just the amount of the emf $\mathscr{E}_2$ appearing in coil 2 due to the changing current in coil 1. As usual, the minus sign reminds us that induced emf acts to oppose the change in current.

Let us now interchange the roles of coils 1 and 2, as in Fig. 32-4b; that is, we set up a current i_2 in coil 2 by means of a battery, and this produces a magnetic flux $\Phi_{2\to1}$ that links coil 1. If we change i_2 with time, we have, by the arguments given above,

$$\mathscr{E}_1 = -M_{2\to1}\frac{di_2}{dt} = -N_1\frac{d\Phi_{2\to1}}{dt}. \qquad (32\text{-}9)$$

Thus, we see that the emf induced in either coil is proportional to the rate of change of current in the other coil. The proportionality constants $M_{1\to2}$ and $M_{2\to1}$ seem to be different. We assert, without proof, that they are in fact the same so that no subscripts are needed. (This conclusion is true but is not obvious.) Thus, we have

$$M_{1\to2} = M_{2\to1} = M, \qquad (32\text{-}10)$$

and we can rewrite Eqs. 32-9 and 32-10 as

$$\mathscr{E}_2 = -M\frac{di_1}{dt} \qquad (32\text{-}11)$$

and

$$\mathscr{E}_1 = -M\frac{di_2}{dt}. \qquad (32\text{-}12)$$

The induction is indeed mutual. The SI unit for M (as for L) is the henry.

TOUCHSTONE EXAMPLE 32-1: Two Coupled Coils

Figure 32-5 shows two circular close-packed coils, the smaller (radius R_2, with N_2 turns) being coaxial with the larger (radius R_1, with N_1 turns) and in the same plane.

(a) Derive an expression for the mutual inductance M for this arrangement of these two coils, assuming that $R_1 \gg R_2$.

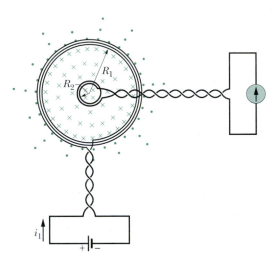

FIGURE 32-5 ■ A small coil is located at the center of a large coil. The mutual inductance of the coils can be determined by sending current i_1 through the large coil.

SOLUTION ■ The **Key Idea** here is that the mutual inductance M for these coils is the ratio of the flux linkage ($N\Phi$) through one coil to the current i in the other coil, which produces that flux linkage. Thus, we need to assume that currents exist in the coils; then we need to calculate the flux linkage in one of the coils.

The magnetic field through the larger coil due to the smaller coil is nonuniform in both magnitude and direction, so the flux in the larger coil due to the smaller coil is nonuniform and difficult to calculate. However, the smaller coil is small enough for us to assume that the magnetic field through it due to the larger coil is approximately uniform. Thus, the flux in it due to the larger coil is also approximately uniform. Hence, to find M we shall assume a current i_1 in the larger coil and calculate the flux linkage $N_2\Phi_{1\to2}$ in the smaller coil:

$$M_{1\to2} = \frac{N_2\Phi_{1\to2}}{i_1}. \qquad (32\text{-}13)$$

A second **Key Idea** is that the flux $\Phi_{1\to2}$ through each turn of the smaller coil is, from Eq. 31-1,

$$\Phi_{1\to2} = B_1A_2,$$

where B_1 is the magnitude of the magnetic field at points within the small coil due to the larger coil, and $A_2(=\pi R_2^2)$ is the area enclosed by the coil. Thus, the flux linkage in the smaller coil (with its N_2 turns) is

$$N_2\Phi_{1\to2} = N_2B_1A_2. \qquad (32\text{-}14)$$

A third **Key Idea** is that to find B_1 at points within the smaller coil, we can use Eq. 30-28, with z set to 0 because the smaller coil is in the plane of the larger coil. That equation tells us that each turn of the larger coil produces a magnetic field of magnitude $\mu_0 i_1/2R_1$ at points within the smaller coil. Thus, the larger coil (with its N_1 turns) produces a total magnetic field of magnitude

$$B_1 = N_1\frac{\mu_0 i_1}{2R_1} \qquad (32\text{-}15)$$

at points within the smaller coil.

Substituting Eq. 32-15 for B_1 and πR_2^2 for A_2 in Eq. 32-14 yields

$$N_2\Phi_{1\to2} = \frac{\pi\mu_0 N_1 N_2 R_2^2 i_1}{2R_1}.$$

Substituting this result into Eq. 32-7, and using Eq. 32-10, we find

$$M = M_{1\to2} = \frac{N_2\Phi_{1\to2}}{i_1} = \frac{\pi\mu_0 N_1 N_2 R_2^2}{2R_1}. \quad \text{(Answer) (32-16)}$$

Just as capacitance does not depend on the amount of charge on capacitor plates, mutual inductance, M, does not depend on the current in the coils.

(b) What is the value of M for $N_1 = N_2 = 1200$ turns, $R_2 = 1.1$ cm, and $R_1 = 15$ cm?

SOLUTION ■ Equation 32-16 yields

$$M = \frac{(\pi)(4\pi \times 10^{-7}\text{ H/m})(1200)(1200)(0.011\text{ m})^2}{(2)(0.15\text{ m})}$$

$$= 2.29 \times 10^{-3}\text{ H} \approx 2.3\text{ mH}. \qquad \text{(Answer)}$$

Consider the situation if we reverse the roles of the two coils—that is, if we produce a current i_2 in the smaller coil and try to calculate M from Eq. 32-7 in the form

$$M_{2\to1} = \frac{N_2\Phi_{2\to1}}{i_2}.$$

The calculation of $\Phi_{2\to1}$ (the nonuniform flux of the smaller coil's magnetic field encompassed by the larger coil) is not simple. If we were to do the calculation numerically using a computer, we would find M to be 2.3 mH, as above! This emphasizes that Eq. 32-10 ($M_{1\to2} = M_{2\to1} = M$) is not obvious.

32-4 *RL* Circuits (With Ideal Inductors)

In Section 28-9 we saw that if we suddenly switch an emf $\mathscr{E}$ on in a series circuit containing a resistor R and a capacitor C, the charge on the capacitor q does not build up immediately to its final equilibrium value $C\mathscr{E}$ but approaches it in an exponential fashion:

$$q = C\mathscr{E}(1 - e^{t/\tau_C}). \qquad (32\text{-}17)$$

The rate at which the charge builds up is determined by the capacitive time constant τ_C, defined in Eq. 28-42 as

$$\tau_C = RC. \qquad (32\text{-}18)$$

If we suddenly remove the emf from this same circuit, the charge does not immediately fall to zero but approaches zero in an exponential fashion:

$$q = q_0 e^{-t/\tau_C}. \qquad (32\text{-}19)$$

The time constant τ_C describes the fall of the charge as well as its rise and q_0 is the initial charge on the capacitor.

An analogous slowing of the rise (or fall) of the current occurs if we introduce an emf $\mathscr{E}$ into (or remove it from) a single-loop circuit containing a resistor R and an inductor L. We assume the inductor is ideal and has a resistance R_L that is much less than R. When the switch S in Fig. 32-6 is closed on a, for example, the current in the resistor starts to rise. If the inductor were not present, the current would rise rapidly to a steady value $\mathscr{E}/R$. Because of the inductor, however, a self-induced emf $\mathscr{E}_L$ appears in the circuit. As predicted from Lenz's law, this emf opposes the rise of the current. This means that it opposes the battery emf $\mathscr{E}$ in polarity. Thus the current in the resistor responds to the *difference* between two emfs, a constant one $\mathscr{E}$ due to the battery, and a variable one $\mathscr{E}_L(= -L\,di/dt)$ due to self-induction. As long as $\mathscr{E}_L$ is present, the current in the resistor will be less than $\mathscr{E}/R$. As time goes on, the rate at which the current increases becomes less rapid and the amount of the self-induced emf, which is proportional to di/dt, becomes smaller. Thus, the current in the circuit approaches $\mathscr{E}/R$ asymptotically.

We can generalize these results as follows: When a switch is opened or closed in a dc circuit, an inductor initially acts to oppose changes in the current through it. A long time later, it acts like ordinary connecting wire that has some resistance R_L.

Now let us analyze the situation quantitatively. With the switch S in Fig. 32-6 thrown to a, the circuit is equivalent to that of Fig. 32-7. Let us apply the loop rule, starting at point x in this figure and moving clockwise around the loop along with current i.

1. *Resistor.* Because we move through the resistor in the direction of current i, the electric potential decreases by iR. Thus, as we move from point x to point y where these points lie *outside* the inductor, we encounter a potential change of $-iR$.

2. *Inductor.* Because current i is changing, there is a self-induced emf $\mathscr{E}_L$ in the inductor. The amount of $\mathscr{E}_L$ is given by Eq. 32-2 as $L\,di/dt$. The direction of $\mathscr{E}_L$ is upward in Fig. 32-7 because current i is downward through the inductor and increasing. Thus, as we move from point y to point z, opposite the direction of $\mathscr{E}_L$, we encounter a potential change of $-L\,di/dt$.

3. *Battery.* As we move from point z back to starting point x, we encounter a potential change of $+\mathscr{E}$ due to the battery's emf.

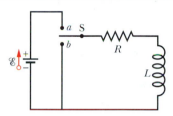

FIGURE 32-6 ■ An *RL* circuit. When switch S is closed on a, the current rises and approaches a limiting value of $\mathscr{E}/R$.

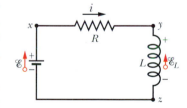

FIGURE 32-7 ■ The circuit of Fig. 32-6 with the switch closed on a. We apply the loop rule for circuits clockwise, starting at x.

Thus, the loop rule gives us

$$-iR - L\frac{di}{dt} + \mathcal{E} = 0$$

or
$$L\frac{di}{dt} + Ri = \mathcal{E} \qquad (RL \text{ circuit}).$$
(32-20)

Equation 32-20 is a differential equation involving the variable i and its first derivative di/dt. To solve it, we seek the function $i(t)$ such that when $i(t)$ and its first derivative are substituted in Eq. 32-20, the equation is satisfied and the initial condition $i(0) = 0$ A is satisfied.

Equation 32-20 and its initial condition are of exactly the form of Eq. 28-38 for an RC circuit, with i replacing q, L replacing R, and R replacing $1/C$. The solution of Eq. 32-20 must then be of exactly the form of Eq. 28-39 with the same replacements. That solution is

$$i = \frac{\mathcal{E}}{R}(1 - e^{-(R/L)t}),$$
(32-21)

which we can rewrite as

$$i = \frac{\mathcal{E}}{R}(1 - e^{-t/\tau_L}) \qquad (\text{rise of current}).$$
(32-22)

Here τ_L, the inductive time constant, is given by

$$\tau_L = \frac{L}{R} \qquad (\text{time constant}).$$
(32-23)

What happens to the current described in Eq. 32-22 between the time the switch is closed (at time $t = 0$ s) and a later time ($t \to \infty$)? If we substitute $t = 0$ s into Eq. 32-22, the exponential becomes $e^{-0} = 1$. Thus, Eq. 32-22 tells us that the current is initially $i = 0$ A, as expected. Next, if we let t go to infinity, then the exponential goes to $e^{-\infty} = 0$. Thus, Eq. 32-22 tells us that the current goes to its equilibrium value of $\mathcal{E}/R$.

We can also examine the potential differences in the circuit. The graphs of Fig. 32-8 show experimental data describing how the potential differences $|\Delta V_R| = iR$ across a resistor and $|\Delta V_L| = L\,di/dt$ across an inductor vary with time for particular

FIGURE 32-8 ■ A computer data acquisition system is used to record the time variation of potential differences (a) ΔV_R across the resistor in Fig. 32-7 and (b) ΔV_L across the inductor in that circuit. The data were obtained at 10 000 samples per second for $R = 9830\ \Omega$, $L \approx 20$ H, and $\mathcal{E} = 5.88$ V. The inductor has a direct current resistance of 167 Ω, so it is not ideal. The data show some different characteristics than those predicted by Eqs. 32-22 and 32-23.

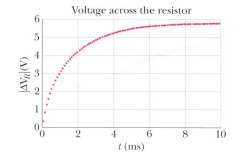

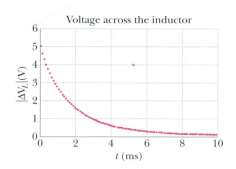

values of $\mathscr{E}$, L, and R. Compare this figure carefully with the corresponding figure for an RC circuit (Fig. 28-23).

To show that the quantity $\tau_L (= L/R)$ has the dimension of time, we convert from henries per ohm as follows:

$$1\frac{H}{\Omega} = 1\frac{H}{\Omega}\left(\frac{1\,V\cdot s}{1\,H\cdot A}\right)\left(\frac{1\,\Omega\cdot A}{1\,V}\right) = 1\,s.$$

The first quantity in parentheses is a conversion factor based on Eq. 32-20, and the second one is a conversion factor based on the relation $\Delta V = iR$.

The physical significance of the time constant follows from Eq. 32-21. If we put $t = \tau_L = L/R$ in this equation, it reduces to

$$i = \frac{\mathscr{E}}{R}(1 - e^{-1}) = 0.63\frac{\mathscr{E}}{R}. \tag{32-24}$$

Thus, the time constant τ_L is the time it takes the current in the circuit to reach about 63% of its final equilibrium value $\mathscr{E}/R$. Since the potential difference ΔV_R across the resistor is proportional to the current i, a graph of the increasing current versus time has the same shape as that of ΔV_R in Fig. 32-8a.

If the switch S in Fig. 32-6 is closed on a long enough for the equilibrium current $\mathscr{E}/R$ to be established and then is thrown to b, the effect will be to remove the battery from the circuit. (The connection to b must actually be made an instant before the connection to a is broken. A switch that does this is called a *make-before-break* switch.)

With the battery gone, the current through the resistor will decrease. However, because of the inductor it cannot drop immediately to zero but must decay to zero over time. The differential equation that governs the decay can be found by putting $\mathscr{E} = 0$ in the RL circuit voltage loop equation (Eq. 32-20):

$$L\frac{di}{dt} + iR = 0. \tag{32-25}$$

By analogy with Eqs. 28-44 and 28-45, the solution of this differential equation that satisfies the initial condition $i(0) = i_0 = \mathscr{E}/R$ is

$$i = \frac{\mathscr{E}}{R}e^{-t/\tau_L} = i_0 e^{-t/\tau_L} \quad \text{(decay of current)}. \tag{32-26}$$

We see that both current rise (Eq. 32-21) and current decay (Eq. 32-26) in an RL circuit are governed by the same inductive time constant, τ_L.

We have used i_0 in Eq. 32-26 to represent the current at time $t = 0$. In our case that happened to be $\mathscr{E}/R$, but it could be any other initial value.

TOUCHSTONE EXAMPLE 32-2: Two Inductors and Three Resistors

Figure 32-9a shows a circuit that contains three identical resistors with resistance $R = 9.0\,\Omega$, two identical ideal inductors with inductance $L = 2.0\,mH$, and an ideal battery with emf $\mathscr{E} = 18\,V$.

(a) What is the current i through the battery just after the switch is closed?

SOLUTION ■ The **Key Idea** here is that just after the switch is closed, the inductor acts to oppose a change in the current through it. Because the current through each inductor is zero before the switch is closed, it will also be zero just afterward. Thus, immediately after the switch is closed, the inductors act as broken wires, as indicated in Fig. 32-9b. We then have a single-loop circuit

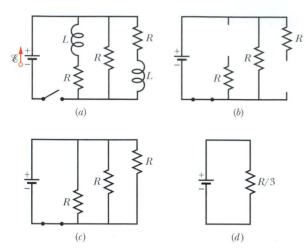

FIGURE 32-9 ■ (*a*) A multiloop *RL* circuit with an open switch. (*b*) The equivalent circuit just after the switch has been closed. (*c*) The equivalent circuit a long time later. (*d*) The single-loop circuit that is equivalent to circuit (*c*).

for which the loop rule gives us

$$\mathscr{E} - iR = 0.$$

Substituting given data, we find that

$$i = \frac{\mathscr{E}}{R} = \frac{18\ \text{V}}{9.0\ \Omega} = 2.0\ \text{A}. \qquad \text{(Answer)}$$

(b) What is the current *i* through the battery long after the switch has been closed?

SOLUTION ■ The **Key Idea** here is that long after the switch has been closed, the currents in the circuit have reached their equilibrium values, and the inductors act as simple connecting wires, as indicated in Fig. 32-9*c*. We then have a circuit with three identical resistors in parallel; from Eq. 27-12, their equivalent resistance is $R^{\text{eq}} = R/3 = (9.0\ \Omega)/3 = 3.0\ \Omega$. The equivalent circuit shown in Fig. 32-9*d* then yields the loop equation $\mathscr{E} - iR^{\text{eq}} = 0.0\ \text{V}$, or

$$i = \frac{\mathscr{E}}{R^{\text{eq}}} = \frac{18\ \text{V}}{3.0\ \Omega} = 6.0\ \text{A}. \qquad \text{(Answer)}$$

32-5 Inductors, Transformers, and Electric Power

In most countries the electrical power used in homes and industries involves voltages and currents that change over time periodically, often sinusoidally. Such power is usually referred to as alternating current or ac electricity. Alternating electrical power is usually generated using induction. An **ac generator** simply consists of a magnet or electromagnet rotating inside an inductor coil or, alternatively, an inductor coil rotating in a magnetic field like that shown in Fig. 32-10.

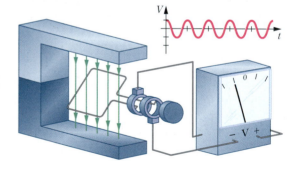

FIGURE 32-10 ■ A simplified diagram of an electric generator showing how a crank can be used to rotate a pickup coil in a magnetic field such that the flux through the coil is changing periodically. Most large generators have a geometry in which the coil rotates outside of an electromagnet.

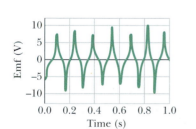

FIGURE 32-11 ■ A periodic emf is induced in a coil that is being turned by a hand crank in the presence of a magnet. The generator is similar to that shown in Fig. 32-10. If the coil were less bulky and the $\vec{B}$-field were more uniform, the emf would vary sinusoidally when the crank is turned steadily.

Generators don't care what form of energy is used to cause the rotation (Fig. 32-11). The shaft can turn when steam produced by a coal-fired or nuclear power plant pushes on propeller-like blades. In hydroelectric plants, falling water can provide the rotational energy. Since the potential difference and current in a generator vary sinusoidally, the voltages and currents are reported as root mean square (or rms) values. The use of rms values is explained in the next chapter, where we deal with alternating-current circuits in more detail.

The Role of Transformers

Generators typically produce power at low voltage, but it is important to transmit this power from generation stations to consumers with minimum energy loss. It turns out that the losses are minimized when ac power is transmitted at high voltages. The reason has to do with how power loss is related to current and voltage. The total power generated is given by Eq. 26-10 as $P^{gen} = i^{gen}\Delta V^{gen}$. If this power is transmitted to consumers with an rms current i^{gen} flowing over long distances, then the power lost in heating transmission lines is given by

$$P^{lost} = (i^{gen})^2 R \qquad \text{(power lost in transmission),} \qquad (32\text{-}27)$$

where R is the total resistance of the wires that make up the transmission lines. The power available to consumers is then $P^{gen} - P^{lost}$. Although we can't get something for nothing, it is obvious from Eq. 32-27 that reorganizing the generated power so that it is transmitted at high voltage and low current would greatly reduce the transmission losses. In other words, we would like to achieve

$$P^{gen} = i^{gen}\Delta V^{gen} = i^{trans}\Delta V^{trans}, \qquad (32\text{-}28)$$

where $\Delta V^{trans} \gg \Delta V^{gen}$ so that $i^{trans} \ll i^{gen}$.

As an example, consider the 735 kV line used to transmit electric energy from the La Grande 2 hydroelectric plant in Quebec to Montreal, 1000 km away. Suppose that the current is 500 A. Then from Eq. 32-28, energy is supplied at the average rate

$$P^{gen} = i^{gen}\Delta V^{gen} = (7.35 \times 10^5 \text{V})(500 \text{ A}) = 368 \text{ MW}.$$

The resistance of the transmission line is about 0.220 Ω/km. Thus, there is a total resistance of about 220 Ω for the 1000 km stretch. Energy is dissipated due to that resistance at a rate of about

$$P^{lost} = (i^{gen})^2 R = (500 \text{ A})^2 (220 \text{ Ω}) = 55.0 \text{ MW},$$

which is nearly 15% of the supply rate.

Imagine what would happen if we could halve the current and double the voltage. Energy would be supplied by the plant at the same average rate of 368 MW as before, but now energy would be dissipated at the much lower rate of about

$$P^{lost} = (i^{gen})^2 R = (250 \text{ A})^2 (220 \text{ Ω}) = 13.8 \text{ MW},$$

This rate of energy loss is *only 4% of the supply rate.* Hence the general energy transmission rule: Transmit at the highest possible voltage and the lowest possible current. There is an upper limit to the voltage that can be used. If the voltage gets too high, the power line insulation and the surrounding air will not be able to prevent the current from passing through them and leaking to the ground.

The Ideal Transformer

The *ideal transformer* in Fig. 32-12 consists of two coils, a *primary* and a *secondary.* These coils have different numbers of turns and are wound around the same iron core. The coils experience mutual induction. The iron core concentrates the flux so that it is the same in both coils. (We will discuss the role iron plays in Section 32-7 on ferromagnetism.) In use, the **primary coil,** of N_p turns, is connected to an alternating-current generator whose emf $\mathscr{E}$ at any time t is given by

$$\mathscr{E} = \mathscr{E}^{max}\sin \omega t. \qquad (32\text{-}29)$$

The **secondary coil,** of N_s turns, is connected to load resistance R, but its circuit is an open circuit as long as switch S is open (which we assume for the present). Thus, there

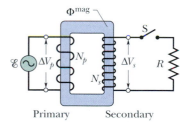

FIGURE 32-12 ■ An ideal transformer (two coils wound on an iron core) in a basic transformer circuit. An ac generator produces current in the coil at the left (the *primary*). The coil at the right (the *secondary*) is connected to the resistive load R when switch S is closed.

can be no current through the secondary coil. We assume further for this ideal transformer that the resistances of the primary and secondary coils (or **windings**) are negligible as are energy losses in the iron core. Well-designed, high-capacity transformers can have energy losses as low as 1%, so our assumptions are reasonable.

For the assumed conditions, the primary winding (or *primary*) is a pure inductance that carries a small alternating primary current i_p. This current induces an alternating magnetic flux Φ^{mag} in the iron core. Because the core extends through the secondary winding (or *secondary*), this induced flux also extends through the turns of the secondary. At any given time the flux in the primary and secondary coils are the same. Therefore, Faraday's law of induction (Eq. 31-7) tells us that the amount of the induced emf per turn, denoted as $\mathscr{E}_{\text{turn}}$ is the same for both the primary and the secondary coils. Also, the voltage ΔV_p across the primary is equal to the emf induced in the primary, and the voltage ΔV_s across the secondary is equal to the emf induced in the secondary. Thus, we can write

$$\mathscr{E}_{\text{turn}} = \frac{d\Phi^{\text{mag}}}{dt} = \frac{\Delta V_p}{N_p} = \frac{\Delta V_s}{N_s},$$

and thus,

$$\Delta V_s = \Delta V_p \frac{N_s}{N_p} \qquad \text{(transformation of voltage).} \qquad (32\text{-}30)$$

If $N_s > N_p$, the transformer is called a **step-up transformer** because it steps the primary's voltage ΔV_p up to a higher voltage ΔV_s. Alternatively, if $N_s < N_p$, the device is a **step-down transformer.**

So far, with switch S open, no energy is transferred from the generator to the rest of the circuit. Now let us close S to connect the secondary to the resistive load R. (In general, the load would also contain inductive and capacitive elements, but here we neglect the capacitance.) We find that now energy is transferred from the generator. To see why, let's explore what happens when we close switch S.

1. An alternating current i_s appears in the secondary circuit, with corresponding energy dissipation rate $i_s^2 R = (\Delta V_s^2)/R$ in the resistive load. Since the emf produced in the secondary coil is a back emf that opposes the direction of the change in current in the primary, the secondary current is out of phase with the primary current.

2. This current produces its own alternating magnetic flux in the iron core, and this flux induces (from Faraday's law and Lenz's law) an opposing emf in the primary windings.

3. The voltage ΔV_p of the primary, however, cannot change in response to this opposing emf because it must always be equal to the emf $\mathscr{E}$ that is provided by the generator; closing switch S cannot change this fact.

In order to relate i_s to i_p, we can apply the principle of conservation of energy. For the ideal transformer without losses in the magnetic core, the power drawn from the primary is equal to the power transferred to the secondary (via the alternating magnetic field linking the two coils). Conservation of energy requires that

$$i_p \Delta V_p = i_s \Delta V_s. \qquad (32\text{-}31)$$

Substituting for ΔV_s from Eq. 32-30, we find that

$$i_s = i_p \frac{N_p}{N_s} \qquad \text{(transformation of currents).} \qquad (32\text{-}32)$$

This equation tells us that the amount of the current i_s in the secondary can be greater than, less than, or the same as the amount of current i_p in the primary, depending on the *ratio of turns (or loops) in the coils given by* N_p/N_s.

Current i_p appears in the primary circuit because of the resistive load R in the secondary circuit. To find i_p, we substitute $i_s = \Delta V_s/R$ into Eq. 32-32 and then we substitute for ΔV_s from Eq. 32-30. We find

$$i_p = \frac{1}{R}\left(\frac{N_s}{N_p}\right)^2 \Delta V_p. \tag{32-33}$$

This equation has the form $i_p = \Delta V_p/R_{eq}$, where equivalent resistance R_{eq} is

$$R_{eq} = \left(\frac{N_p}{N_s}\right)^2 R. \tag{32-34}$$

Here R is the actual resistance in the secondary circuit and R_{eq} is the value of the load resistance as "seen" by the generator. The generator produces the current i_p and voltage ΔV_p as if it were connected to a resistance R_{eq}.

Impedance Matching

Equation 32-34 suggests still another function for the transformer. For maximum transfer of energy from an emf device to a resistive load, the resistance of the emf device and the resistance of the load must be equal. The same relation holds for ac circuits (discussed in Chapter 33) except that the *impedance* (rather than just the resistance) of the generator must be matched to that of the load. Often this condition is not met. For example, in a music-playing system, the amplifier can have high impedance and the speaker set have low impedance. We can match the impedances of the two devices by coupling them through a transformer with a suitable turns ratio N_p/N_s.

READING EXERCISE 32-3: An alternating-current emf device has a smaller resistance than that of the resistive load; to increase the transfer of energy from the device to the load, a transformer will be connected between the two. (a) Should N_s be greater than or less than N_p? (b) Will that make it a step-up or step-down transformer?   ■

32-6 Magnetic Materials—An Introduction

Today, magnets and magnetic materials are ubiquitous. In addition to naturally magnetic lodestones, magnets are also in VCRs, audiocassettes, credit cards, electronic speakers, audio headsets, and even the inks in paper money. In fact, some breakfast cereals that are "iron fortified" contain small bits of magnetic materials (you can collect them from a slurry of cereal and water with a magnet). In this section we are interested in understanding more about why so-called bulk matter, made of billions upon billions of individual atoms, has magnetic properties.

Characteristics of Magnetic Materials

When we speak of magnetism in everyday conversation, we usually have a mental picture of a bar magnet, a disk magnet (probably clinging to a refrigerator door), or even a tiny compass needle. That is, we picture a *ferromagnetic* material made of iron having strong, permanent magnetism. Although most bulk matter does not behave like the familiar iron bar magnets, it turns out that almost all bulk materials have some magnetic behaviors. There are three general types of magnetism: ferromagnetism, paramagnetism, and diamagnetism.

1. **Ferromagnetism** is present if a material produces a strong magnetic field of its own in the presence of an external field, and if its magnetic field partially persists after the external field is removed. We usually use the term *ferromagnetic material,* and also the common term *magnetic material*, to refer to materials that exhibit primarily ferromagnetism. Iron, nickel, and cobalt (and compounds and alloys of these elements) are ferromagnetic.

2. **Paramagnetism** is present if a material that is placed in an external magnetic field is attracted to the region of greater magnetic field and produces a magnetic field of its own—but only while it is in the presence of the external field. The term *paramagnetic material* usually refers to materials that exhibit primarily paramagnetism. This type of magnetism is exhibited by materials such as liquid oxygen and aluminum as well as transition elements, rare earth elements, and actinide elements (see Appendix G).

3. **Diamagnetism** is present if a material that is placed near a magnet is repelled from the region of greater magnetic field. This is opposite to the behavior of the other two types of magnetism. Diamagnetism is exhibited by all common materials, but it is so weak that it is masked if the material exhibits magnetism of either of the other two types. Thus, the term *diamagnetic material* refers to materials that only exhibit diamagnetism. Metals such as bismuth, copper, gold, silver, and lead, as well as many nonmetals such as water and most organic compounds, are diamagnetic. Because people and other animals are made largely of water and organic compounds, they are diamagnetic too.

What causes magnetism? Why are there three types of magnetism? We now believe that magnetism is caused by tiny magnetic dipoles that are intrinsic to the atoms contained in all materials. For this reason, understanding the characteristics of magnetic dipoles is essential to understanding the behavior of magnetic materials. We will conclude this section with a discussion of magnetic dipoles, and then in the next two sections we will explore how the characteristics of atomic magnetic dipoles help us understand the three types of magnetism.

FIGURE 32-13 ■ A bar magnet is a magnetic dipole. The orientations of the iron filings suggest the direction of magnetic field lines.

FIGURE 32-14 ■ If you break a bar magnet, each fragment becomes a smaller magnet, with its own north and south poles. It is impossible to break a fragment into separate north and south poles.

Characteristics of Magnetic Dipoles

Both the bar magnets with which we are familiar and small coils of wire carrying current are magnetic dipoles. Let's review some of the characteristics of magnetic dipoles that we have already discussed. A magnetic dipole:

• Has a magnetic field pattern associated with it similar to that of an electric dipole (like that shown in Fig. 32-13 or that described by the equations derived in Sections 23-6 and 30-7).

• *Always has two poles,* which we have chosen to call north (seeking) and south (seeking) because of the way they behave when placed in the Earth's magnetic field, as shown in Fig. 32-14 (see Section 29-3 for a review).

• Can be described by a magnetic dipole moment $\vec{\mu}$, which is a vector quantity whose *magnitude* tells us how *strong* the magnetic field associated with the dipole is and whose direction tells how the field pattern is oriented. The orientation is along the axis of a bar magnet pointing from its south pole and to its north pole or perpendicular to the plane of a current-carrying coil with a direction determined by the right-hand rule (Section 29-10). *Note:* Our use of conventional notation is unfortunate here. The magnetic moment $\vec{\mu}$ should not be confused with the permeability constant μ_0 or μ that sometimes appears in the same equation.

- Will attempt to align its magnetic dipole moment with an external magnetic field, $\vec{B}$, because the dipole experiences a torque, $\vec{\tau}$, given by $\vec{\tau} = \vec{\mu} \times \vec{B}$ (Section 30-7).

- Has a potential energy U in an external magnetic field given by $U = -\vec{\mu} \cdot \vec{B}$, so that the dipole's potential energy is a minimum when its dipole moment is aligned with an external magnetic field.

Magnetism in Atoms

We believe that the combined effect of tiny magnetic dipole moments in atoms are responsible for all magnetic interactions in bulk matter. Before we attempt to explain why different materials exhibit certain types of magnetism, we need to discuss what is known about atomic magnetism.

The focus of this book is on classical physics. However, understanding atomic phenomena requires some familiarity with quantum physics, which is in general beyond the scope of this book. So, we will present some basic ideas of quantum physics that apply to atomic magnetism without discussing the existing body of experimental evidence.

So far in our classical treatment of magnetism we have already identified two sources of magnetic dipole fields: (1) electric charges that create a current if they move in a loop and (2) magnetic dipoles consisting of a bar or rod of magnetized iron. Also, some effects of atomic magnetism can be explained using a classical model that identifies two types of atomic magnetic dipoles—orbital and spin. First, if we think of electrons as "orbiting" around a nucleus, then an orbit is a current loop with an orbital magnetic moment. Second, we think of the electron as having an intrinsic magnetic dipole moment that we call spin. This model is quite comfortable because it is rather like the familiar picture of the Earth spinning about its own axis as it orbits the Sun. But when we try to predict the magnetic behavior of various types of materials using this classical model, its usefulness is limited and it is completely wrong in many ways.

The bad classical predictions are not surprising since quantum physics, devised to explain atomic behavior, tells us that: (1) We cannot think of electrons as having distinct orbits. Instead we visualize them as swarming about in the vicinity of a nucleus without having distinct paths. So all we can know is something about the probability of finding the electron at various locations in the vicinity of the nucleus and that these probabilities are different for each type of atom or molecule. (2) The spin magnetic moments are a fundamental property of electrons and should not be thought of as being produced by an electron spinning about an internal axis. (3) The spin and orbital magnetic moments associated with atomic electrons are quantized. This means they can only have certain values.

Next let's examine the characteristics of these two types of atomic magnetic moments in more detail.

Spin Magnetic Dipole Moment

An electron has an intrinsic **spin magnetic dipole moment** $\vec{\mu}^{\text{spin}}$. (By *intrinsic*, we mean that $\vec{\mu}^{\text{spin}}$ is a basic characteristic of an electron, like its mass and electric charge.) According to quantum theory,

1. $\vec{\mu}^{\text{spin}}$ itself cannot be measured directly. Only its component along a single axis can be well-defined (and therefore measured) at any one time.

2. A measured component of $\vec{\mu}^{\text{spin}}$ is *quantized*, which is a general term that means it is restricted to certain values.

Let us assume that the component of the spin magnetic moment $\vec{\mu}^{\,\text{spin}}$ is measured along the z axis of a coordinate system you have chosen. Then the measured component μ_z^{spin} can have only the two values given by

$$\mu_z^{\text{spin}} = +\frac{eh}{4\pi m} \quad \text{or} \quad \mu_z^{\text{spin}} = -\frac{eh}{4\pi m}, \tag{32-35}$$

where $h = 6.63 \times 10^{-34}$ J·s and is the well-known Planck constant used often in quantum physics. The constants e and m represent the charge and mass of the electron, respectively. The plus and minus signs given in Eq. 32-35 describe the direction of μ_z^{spin} along the chosen z axis. The plus sign indicates that μ_z^{spin} is parallel to the z axis, and the electron is said to be "spin up." When μ_z^{spin} is antiparallel to the z axis, the minus sign is used and the electron is said to be "spin down."

The combination of constants in Eq. 32-35 is called the *Bohr magneton* μ_B, which can be calculated from the known values of Planck's constant and the electron charge and mass:

$$\mu_B = \frac{eh}{4\pi m} = 9.27 \times 10^{-24} \text{ J/T} \quad \text{(Bohr magneton value for an electron).} \tag{32-36}$$

Spin magnetic dipole moments of electrons and other elementary particles can be expressed in terms of μ_B. In terms of the Bohr magneton, we can substitute in to Eq. 32-35 to rewrite the expression for the two possible values of μ_z^{spin} as

$$\mu_z^{\text{spin}} = +\mu_B \quad \text{or} \quad \mu_z^{\text{spin}} = -\mu_B. \tag{32-37}$$

When an electron is placed in an external magnetic field $\vec{B}^{\text{ext}}$, a potential energy U can be associated with the orientation of the electron's spin magnetic dipole moment $\vec{\mu}^{\,\text{spin}}$ just as a potential energy can be associated with the orientation of the magnetic dipole moment $\vec{\mu}$ of a current loop placed in an external magnetic field $\vec{B}^{\text{ext}}$. From Eq. 29-35, the potential energy for the electron due to its spin orientation has only two possible values

$$U^{\text{spin}} = -\vec{\mu}^{\,\text{spin}} \cdot \vec{B}^{\text{ext}} = -\mu_z^{\text{spin}} B^{\text{ext}} = \pm\, \mu_B B^{\text{ext}}, \tag{32-38}$$

where the z axis is taken to be in the direction of $\vec{B}^{\text{ext}}$.

Again, although we use the word "spin" here, according to quantum theory the fact that electrons have intrinsic magnetic moments does not mean that they spin like tops.

Protons and neutrons also have intrinsic magnetic dipole moments. In fact, these nuclear magnetic moments are a critical element in the development of magnetic resonance imaging—a valuable diagnostic tool in medicine. The masses of protons and neutrons are almost 2000 times that of the electron, so the magnetic moment for these particles is much smaller than that of the electron. For this reason, the contributions of nuclear dipole moments to the magnetic fields of atoms are negligible.

Orbital Magnetic Dipole Moment

An electron that is part of an atom has an additional dipole magnetic moment. This is called its **orbital magnetic moment** $\vec{\mu}^{\,\text{orb}}$. Again, although we use the "orbital" here, electrons do not orbit the nucleus of an atom like planets orbiting the Sun. According to quantum physics, an "orbit" roughly defines a region in space where the electron is most likely to be found. The orientation of this region specifies the direction of the electron's orbital angular momentum. The so-called "outer electrons" in an atom with many electrons will tend to be found further from its nucleus. An outer electron has a larger orbital angular momentum and, hence, a larger magnetic moment. It turns out that in any given atom there are typically more than two possible quantized values for the z-components of orbital magnetic moments. We can express these possible components along a chosen z axis in terms of the Bohr magneton as

$$\mu_z^{orb} = -m_l^{max}\mu_B, \ldots, -3\mu_B, -2\mu_B, -1\mu_B, 0\mu_B, 1\mu_B, 2\mu_B, 3\mu_B, \ldots, m_l^{max}\mu_B, \quad (32\text{-}39)$$

where m_l^{max} is an integer that designates the magnitude of the orbital magnetic moment component an electron can have.

When an atom is placed in an external magnetic field $\vec{B}^{ext}$, an orbital potential energy U^{orb} can be associated with the orientation of the orbital magnetic dipole moment of each electron in the atom. Its value is

$$U^{orb} = -\vec{\mu}_{orb} \cdot \vec{B}^{ext} = -\mu_z^{orb} B^{ext}, \quad (32\text{-}40)$$

where the z axis is taken in the direction of $\vec{B}^{ext}$ so that $\vec{B}^{ext} = \vec{B}^{ext}\hat{k}$.

The Magnetic Dipole Moment of an Atom

Each electron in an atom has an orbital magnetic dipole moment and a spin magnetic dipole moment that combine vectorially. The resultant of these two vector quantities combines vectorially with similar resultants for all other electrons in the atom, and the resultant for each atom combines with those for all the other atoms in a bulk sample of matter. If the combination of all these magnetic dipole moments produces a magnetic field, then the material is magnetic.

We have one more step in preparing to explain the magnetic behavior of bulk material on the basis of the magnetic behavior of its atoms. We need to consider how the spin and orbital magnetic moments associated with all the electrons in a single atom of a certain element (such as iron, arsenic, and so on) could combine to determine its total magnetic moment.

If the spin and orbital magnetic moments of all the electrons in a given atom lined up with each other and then if all the individually aligned atoms lined up with each other in a solid or liquid, we would have an incredibly strong magnet. Most materials are not strongly magnetic because whenever possible it is natural for the electron magnetic moments in an atom to cancel out. Atomic electrons are located in regions around the nucleus called shells. The number of electrons in a shell is governed by a quantum mechanical rule known as the *Pauli exclusion principle*. This exclusion principle requires that no two electrons in the same shell can have both the same components of orbital magnetic moment and the same components of spin magnetic moment. When all different combinations of orbital and spin states have occurred in a shell it is full. Once a shell is full the pairing of electrons in their locations with respect to the nucleus cancel each other. Each successive shell has more possible orbital magnetic moment components than its electrons can have. In addition, the electrons in a given shell have more energy than the ones in the previous shell. If the atom is in its lowest possible energy state, the shells usually fill in order.

In an atom with many electrons, the number of electrons in a full shell is $4n + 2$, where n is 0, 1, 2, 3, and so on. So the first shell can have 2 electrons, the second 6 electrons, the third 10, the fourth 14, and so on. Typically it is unpaired outermost electrons in an atom that determine the magnetic behavior of bulk matter. Magnetism depends critically on how atoms combine with each other as a result of the sharing and interaction of the outermost electrons.

READING EXERCISE 32-4: In this section, we discuss three types of magnetism in materials. Which one is associated with a refrigerator magnet? A standard paper clip? A piece of silver wire? Explain your reasoning. ■

READING EXERCISE 32-5: The figure shows the spin orientations of two particles in an external magnetic field $\vec{B}^{\text{ext}}$. (a) If the particles are electrons, which spin orientation is at lower potential energy? (b) If, instead, the particles are protons, which spin orientation is at lower potential energy? Explain your reasoning.

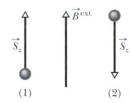

(1)   (2)

■

32-7 Ferromagnetism

Iron, cobalt, nickel, gadolinium, dysprosium, and alloys of these become strongly magnetized in the presence of an external magnetic field. Because they retain this magnetism when the external field is removed, we call them ferromagnetic.

Atomic Magnetic Moments in Ferromagnetism

Although most heavy elements are not ferromagnetic, all ferromagnetic materials are relatively heavy elements with complex electronic structures. The lightest of these is iron, which has 26 electrons. The best explanation to date for iron's ferromagnetism involves the complex behavior of its electrons. The 20 innermost electrons are paired in such a way that their spin and orbital magnetic moments cancel each other. The other 6 electrons behave in a manner that is unusual for most materials. Instead of piling into the third shell that has plenty of room for them, 2 of the 6 electrons move out into the fourth shell. These 2 electrons form outer conduction electrons. The key to the ferromagnetism of iron is that the third shell is unfilled, which allows 4 of 14 electrons in that shell to have spin magnetic moments that end up being aligned. However, these aligned electrons do not participate in chemical bonding with other atoms. The detailed quantum mechanical calculations reveal that this unusual arrangement of electrons gives an individual iron *atom* both a net magnetic moment and a lower energy—a situation that is similar for cobalt and nickel.

Even though individual iron atoms have permanent magnetic dipole moments, we might assume that their orientations relative to each other are random, leaving a bulk sample of iron with no net magnetic moment. We know this is not the case. Although various explanations have been put forth to explain why individual atoms line up in ferromagnetic materials, the situation is not well understood. It is currently believed that the spin-aligned electrons in the third shell, which causes the magnetism, influence the outermost conduction electrons, which are wandering through the material. Because of the Pauli exclusion principle, the spin magnetic moment of a conduction electron will have a tendency to be aligned in a direction opposite to that of the third-shell electrons. This anti-aligned conduction electron could, in turn, influence the alignment of the third shell electrons in a neighboring atom. This interaction could align the third-shell spin magnetic moments in the two neighboring atoms, and so on. The jargon for this quantum physical effect, in which spins of the electrons in one atom interact with those of neighboring atoms via conduction electrons, is called **exchange coupling**. The result is an alignment of the magnetic dipole moments of the atoms, in spite of the randomizing effects of thermal energy that causes atomic collisions. We currently believe that this type of coupling is what gives ferromagnetic materials their permanent magnetism.

Magnetic Domains

Exchange coupling in which spins of the electrons in one atom interact with those of neighboring atoms via conduction electrons, produces strong alignment of adjacent atomic dipoles in a ferromagnetic *material.* So we might expect that all the atoms in a sample of iron would align themselves into a permanent magnet even in the absence of an external magnetic field. This doesn't happen. Instead, a piece of iron, nickel, or cobalt is always made up of a number of *magnetic domains.* Each domain is a region in which the alignment of the atomic dipoles is essentially perfect. The domains, however, are not all aligned. For the sample as a whole, the domains are so oriented that they largely cancel each other as far as their external magnetic effects are concerned.

Two reasons are often given for the existence of domains in ferromagnetic materials. First, calculations reveal that a pure sample with perfectly aligned atoms (known as a *single crystal*) has a lower energy state when there are distinct domains with boundaries between them. Second, most real samples have impurities that can cause even more boundaries between domains to form.

A photograph of a single crystal of nickel is shown in Fig. 32-15. A suspension of powdered iron oxide was sprinkled on the crystal surface. The domain boundaries, which are thin regions in which the alignment of the elementary dipoles changes from a certain orientation in one domain to a different orientation in the other, are the sites of intense, but highly localized and nonuniform, magnetic fields. The suspended iron oxide particles are attracted to some of the more prominent boundaries and show up as the white lines. Although the atomic dipoles in each domain are completely aligned as shown by the arrows, the crystal as a whole has a very small resultant magnetic moment.

Actually, a piece of iron as we ordinarily find it is not a single crystal but an assembly of many tiny crystals, randomly arranged; we call it a polycrystalline solid. Each tiny crystal, however, has its array of variously oriented domains, just as in Fig. 32-15. We can magnetize such a specimen by placing it in an external magnetic field B_z^{ext} of gradually increasing strength, and measuring the magnetization B_z^M of the iron. (The measurement process is explained in the next subsection on Bulk Properties.) A common way to display the results is to plot a magnetization curve. If the piece of iron had all of its magnetic dipoles aligned perfectly with the external field, its magnetization would be a maximum represented by B_{max}^M. The magnetization curve consists of a plot of the ratio B_z^M/B_{max}^M as a function of the external field (shown in Fig. 32-16). Note that B^M/B_{max}^M is always less than one, so the iron does not become perfectly magnetized.

By photographing domain patterns as in Fig. 32-15, we see two microscopic effects that serve to explain the shape of the magnetization curve: One effect is a growth in size of the domains that are oriented along the external field at the expense of those that are not. The second effect is a shift of the orientation of the dipoles within a domain, as a unit, to become closer to the field direction.

Exchange coupling and domain shifting give us the following result:

> A ferromagnetic material placed in an external magnetic field $\vec{B}^{ext}$ develops a strong magnetic dipole moment in the direction of $\vec{B}^{ext}$. If the field is nonuniform, the ferromagnetic material is attracted toward a region of greater magnetic field from a region of lesser field.

Bulk Properties of Ferromagnetic Materials

If the temperature of a ferromagnetic material is raised above a certain critical value, called the **Curie temperature,** the exchange coupling ceases to be effective. Most such materials then become simply paramagnetic. That is, the dipoles still tend to align with an external field but much more weakly, and thermal agitation can now more easily disrupt the alignment. The Curie temperature for iron is 1043 K (= 770°C).

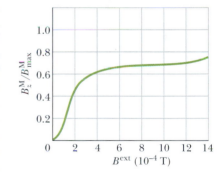

FIGURE 32-15 ■ A photograph of domain patterns within a single crystal of nickel; white lines reveal the boundaries of the domains. The white arrows superimposed on the photograph show the orientations of the magnetic dipoles within the domains and thus the orientations of the net magnetic dipoles of the domains. The crystal as a whole is unmagnetized if the net magnetic field (the vector sum over all the domains) is zero.

FIGURE 32-16 ■ A magnetization curve for a ferromagnetic core material in the Rowland ring of Fig. 32-17. On the vertical axis, 1.0 corresponds to complete alignment (saturation) of the atomic dipoles within the material.

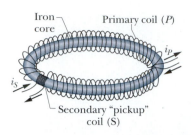

FIGURE 32-17 ■ A toroidal Rowland ring coil in which a current i_P is sent through a primary coil P. This current is used to study the behavior of the ferromagnetic material of the iron core inside the windings. The extent of magnetization of the core determines the total magnetic field $\vec{B}$ within coil P. Field $\vec{B}$ can be measured by means of a secondary or "pickup" coil.

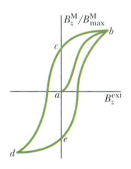

FIGURE 32-18 ■ A magnetization curve (*ab*) for a ferromagnetic specimen and an associated hysteresis loop (*bcdeb*).

We can express the extent to which a given paramagnetic sample is magnetized by finding the ratio of its magnetic dipole moment to its volume V. This vector quantity, the magnetic dipole moment per unit volume, is called the **magnetization** $\vec{M}$ of the sample, and its magnitude is

$$\vec{M} = \frac{\text{measured magnetic moment}}{V}. \tag{32-41}$$

The unit of $\vec{M}$ is the ampere-square meter per cubic meter, or ampere per meter (A/m). Complete alignment of the atomic dipole moments, called **saturation** of the sample, corresponds to the maximum magnetization of magnitude $M^{\text{max}} = N\mu/V$ where N is the number of atoms in the volume V.

The magnetization of a ferromagnetic material such as iron can be studied using a toroidal coil called a *Rowland ring* (Fig. 32-17). A Rowland ring is basically a long solenoid with an iron cylinder at its core, except the whole thing is bent into the shape of a donut. Assume that the ring's primary coil P has n turns per unit length and carries current i_P. If the iron core were not present, the magnitude of the magnetic field inside the coil caused by the "external" solenoid windings (as distinct from the magnetization of a core material inside the windings) would be given by Eq. 30-25,

$$B^{\text{ext}} = n\mu_0|i_P| \qquad \text{(no iron core)}. \tag{32-42}$$

Here μ_0 represents the magnetic constant (or permeability) of air (and is not a magnetic moment).

If an iron core is present, the magnitude of the magnetic field B inside the coil is proportional to B^{ext} but is on the order of 1000 to 10 000 times greater due to the magnetization of the iron core. This magnetization results from the alignment of the atomic dipole moments within the iron. The field $\vec{B}$ inside the coil should be the vector sum of the field $\vec{B}^{\text{ext}}$ contributed by the coil without the core and the field $\vec{B}^{\text{M}}$ contributed by the magnetization of the core. Since the magnitude of $\vec{B}^{\text{ext}}$ field is much smaller than that produced by the core magnetization

$$\vec{B} = \vec{B}^{\text{ext}} + \vec{B}^{\text{M}} \approx \vec{B}^{\text{M}}. \tag{32-43}$$

To determine $\vec{B}^{\text{M}}$ we use a secondary coil S to measure $\vec{B}$ and hence $\vec{B}^{\text{M}}$. If needed, we compute $\vec{B}^{\text{ext}}$ using Eq. 32-42.

Figure 32-18 shows a magnetization curve for a ferromagnetic material in a Rowland ring: the ratio of magnitudes $B^{\text{M}}/B^{\text{M}}_{\text{max}}$ is plotted as a function of B^{ext} (where $B^{\text{M}}_{\text{max}}$ is the maximum possible value of B^{M}, corresponding to saturation). The curve is similar to that for the magnetization curve for a paramagnetic substance shown in Fig. 32-19. Both curves are measures of the extent to which an applied magnetic field can align the atomic dipole moments of a material.

For the ferromagnetic core described by the graph in Fig. 32-16, the alignment of the dipole moments is about 70% complete for $B^{\text{ext}} \approx 1 \times 10^{-3}$ T. If B^{ext} were increased to 1 T, the alignment would be almost complete (but $B^{\text{ext}} = 1$ T, and thus almost complete saturation, is quite difficult to achieve).

Hysteresis

Magnetization curves for ferromagnetic materials are not retraced as we increase and then decrease and then reverse the external magnetic field $\vec{B}^{\text{ext}}$. Let's assume that we choose the z axis to be along the direction of the external magnetic field. Figure 32-18 is a plot of the z-component of the magnetization field B^{M}_z versus the z-component of the external field B^{ext}_z during the following operations with a Rowland ring: (1) Starting with the iron unmagnetized (point *a*), increase the current in the toroid until $B^{\text{ext}}_z = n\mu_0|i|$ has the value corresponding to point *b*; (2) reduce the current in the toroid winding (and thus B^{ext}) back to zero (point c); (3) reverse the toroid current and increase it in amount until B^{ext} has the value corresponding to point d; (4) reduce the current to zero again (point e); (5) reverse the current once more until point b is reached again.

The lack of retraceability shown in Fig. 32-18 is called **hysteresis,** and the curve $bcdeb$ is called a *hysteresis loop*. Note that at points c and e the iron core is magnetized, even though there is no current in the toroid windings; this is the familiar phenomenon of permanent magnetism. In fact when engineers are designing permanent magnets, they look for materials that have a high degree of hysteresis.

Hysteresis can be understood through the concept of magnetic domains. When the magnetic field in the coil due to the current in the solenoid windings, $\vec{B}^{\text{ext}}$, is increased and then decreased back to its initial value, the domains do not return completely to their original configuration but retain some "memory" of their alignment after the initial increase. This memory of magnetic materials is essential for the magnetic storage of information, as on cassette tapes and computer disks.

This memory of the alignment of domains can also occur naturally. When lightning sends currents along multiple tortuous paths through the ground, the currents produce intense magnetic fields that can suddenly magnetize any ferromagnetic material in nearby rock. Because of hysteresis, such rock material retains some of that magnetization after the lightning strike (after the currents disappear) then becomes lodestones.

Inductors and Transformers with Iron Cores

Based on our discussion above of the Rowland ring, it is clear that the use of iron and iron alloys in inductors and transformers can literally increase the performance of these devices by a thousandfold or more.

A great deal of engineering has gone into the design of cores for large inductors and high-performance transformers. For example, these cores should not behave like permanent magnets with large hysteresis. Instead, they should have small hysteresis so that the magnetization of the core can change rapidly in the presence of alternating currents. In addition, transformer cores are not single hunks of iron. Rather, they are built up in layers to prevent eddy currents from being induced in the cores that could reduce the efficiency of the power transfer from the primary to secondary coils in a transformer.

READING EXERCISE 32-6: Iron is a ferromagnetic material. Why then isn't every piece of iron—for example, an iron nail—a naturally strong magnet?   ∎

READING EXERCISE 32-7: What is hysteresis and why does it occur?   ∎

32-8 Other Magnetic Materials

Paramagnetism

In paramagnetic materials, the spin and orbital magnetic dipole moments of the electrons in individual atoms do not cancel but add vectorially to give each *atom* a net (and permanent) magnetic dipole moment $\vec{\mu}$. In the absence of an external magnetic field, these atomic dipole moments are randomly oriented, and the net magnetic dipole moment of the *material* is zero. However, if a sample of the material is placed in an external magnetic field $\vec{B}^{\text{ext}}$, the magnetic dipole moments tend to line up with the field, which gives the sample a net magnetic dipole moment not unlike that found in a ferromagnetic sample. However, paramagnetic materials lack the exchange coupling needed to set up permanent magnetic domains. Paramagnetism is fairly weak compared to ferromagnetism because the forces of alignment from external magnetic

Liquid oxygen is suspended between the two pole faces of a magnet because the liquid is paramagnetic and is magnetically attracted to the magnet.

fields are smaller than the randomizing forces due to thermal motions. Also, paramagnetic materials do not retain their magnetism once an external magnetic field is turned off.

> A paramagnetic material placed in an external magnetic field $\vec{B}^{\text{ext}}$ develops a magnetic dipole moment in the direction of $\vec{B}^{\text{ext}}$. If the field is not uniform, the paramagnetic material is attracted toward a region of greater magnetic field from a region of lesser field.

As is the case for ferromagnetism, we can express the extent to which a given paramagnetic sample is magnetized by measuring the magnetization $\vec{M}$ (defined in Eq. 32-41). In 1895, Pierre Curie discovered that the magnitude of the magnetization of a paramagnetic sample is directly proportional to the external magnetic field magnitude B^{ext} and inversely proportional to the temperature T in kelvins; that is,

$$M = C\frac{B^{\text{ext}}}{T}. \tag{32-44}$$

Equation 32-44 is known as **Curie's law,** and C is called the **Curie constant.** Curie's law is reasonable in that increasing $\vec{B}^{\text{ext}}$ tends to align the atomic dipole moments in a sample and thus to increase $\vec{M}$, whereas increasing T tends to disrupt the alignment via thermal agitation and thus to decrease $\vec{M}$. However, the law is actually an approximation that is valid only when the ratio B^{ext}/T is not too large.

Figure 32-19 shows the ratio M/M^{max} as a function of B^{ext}/T for a sample of the salt potassium chromium sulfate, in which chromium ions are the paramagnetic substance. The plot is called a **magnetization curve.** The straight line for Curie's law fits the experimental data at the left, for B^{ext}/T below about 0.5 T/K. The curve that fits all the data points is based on quantum physics. The data on the right side, near saturation, are very difficult to obtain because they require very strong magnetic fields (about 100 000 times Earth's field), even at the very low temperatures noted in Fig. 32-19.

Diamagnetism

The *atoms* in diamagnetic materials have no net magnetic dipole moments. However, diamagnetic *materials* do undergo a very weak nonpermanent alignment in the presence of an external magnetic field. The strength of the alignments is still proportional to the strength of the external magnetic field (as is the case for both ferro- and paramagnetism). However, the behavior of diamagnetic materials is not very temperature dependent.

The most interesting characteristic of diamagnetism is that in the presence of an external magnetic field that is nonuniform, each atom experiences a net force that is directed *away* from the region of greater magnetic field. Thus, in diamagnetism the

FIGURE 32-19 ■ A *magnetization curve* for potassium chromium sulfate, a paramagnetic salt. The ratio of the magnitudes of the salt magnetization $\vec{M}$ to the maximum possible magnetization $\vec{M}^{\text{max}}$ is plotted versus the ratio of the magnitude of the applied magnetic field B^{ext} to the temperature T. Curie's law fits the data at the left; quantum theory fits all the data. (Based on research by Warren E. Henry, 1909–2001.)

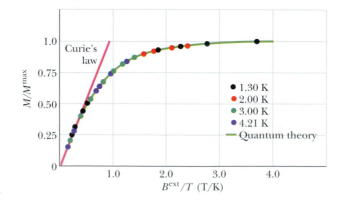

alignment of atomic magnetic moments with an external magnetic field is opposite to that associated with ferromagnetic and paramagnetic materials. In general,

> A diamagnetic material placed in an external magnetic field $\vec{B}^{\text{ext}}$ develops a magnetic dipole moment directed opposite $\vec{B}^{\text{ext}}$. If the field is nonuniform, the diamagnetic material is repelled from a region of greater magnetic field toward a region of lesser field.

Animals like the frog shown in Fig. 32-20 are diamagnetic. This frog has been placed in the diverging magnetic field near the top end of a vertical current-carrying solenoid; every atom in the frog was repelled upward, away from the region of stronger magnetic field at that end of the solenoid. The frog moved upward into weaker and weaker magnetic field until the upward magnetic force balanced the gravitational force on it, and there it hung in midair. People are also diamagnetic, so if we built a large enough solenoid, we could also suspend a person in midair.

FIGURE 32-20 ■ An overhead view of a diamagnetic frog that is being levitated in a magnetic field. The $\vec{B}$-field is produced by current in a vertical solenoid below the frog. The solenoid's upward magnetic force on the frog balances the downward gravitational force on the frog. (The frog is not in discomfort; the sensation is like floating in water, which frogs don't seem to mind.)

READING EXERCISE 32-8: The figure here shows two paramagnetic spheres located near the south pole of a bar magnet. Are (a) the magnetic forces on the spheres and (b) the magnetic dipole moments of the spheres directed toward or away from the bar magnet? (c) Is the magnetic force on sphere 1 greater than, less than, or equal to that on sphere 2?  ■

READING EXERCISE 32-9: The figure shows two diamagnetic spheres located near the south pole of a bar magnet. Are (a) the magnetic forces on the spheres and (b) the magnetic dipole moments of the spheres directed toward or away from the bar magnet? (c) Is the magnetic force on sphere 1 greater than, less than, or equal to that on sphere 2? ■

32-9 The Earth's Magnetism

The Earth has a magnetic field associated with it that behaves approximately like that of a magnetic dipole. In other words, the Earth's magnetic field can be thought of as being produced by a bar magnet that straddles the center of the planet with its axis more or less aligned with the Earth's rotation axis. Figure 32-21 is an idealized depiction of the Earth's dipole field that ignores the distortion of field lines caused by charged particles streaming out of the Sun and other factors.

Characteristics of the Earth's Magnetic Field

For the idealized magnetic field shown in Fig. 32-21, the Earth's magnetic dipole moment $\vec{\mu}$ has a magnitude of 8.0×10^{22} J/T. The point where the Earth's rotation axis intersects the surface is known at the *geographic north pole*. In 2001, the geological survey of Canada placed the direction of the Earth's dipole moment at an angle of $\theta = 8.7°$ from the rotation axis (RR) of the Earth.* The *dipole axis* (MM in Fig. 32-21) lies along $\vec{\mu}$ and intersects the Earth's surface at the *geomagnetic north and south poles*, These days the magnetic north pole is estimated to be somewhere in the Arctic Ocean north of Canada and the south pole is in the Antarctic Ocean. Since the poles are currently moving at about 40 km/yr, the possibility exists that the magnetic north pole could pass north of Alaska and in about fifty years end up in Siberia, although this outcome is not certain.

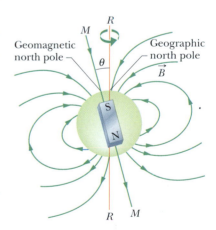

FIGURE 32-21 ■ An idealized view of the Earth's magnetic field as a dipole field. At present, dipole axis MM makes an angle of 8.7° with Earth's rotational axis RR. The "south pole" of the dipole is in Earth's northern hemisphere.

*http://www.geolab.nrcan.gc.ca/geomag/northpole_e.shtml

The lines of the magnetic field $\vec{B}$ generally emerge in the southern hemisphere and reenter Earth in the northern hemisphere. Thus, the magnetic pole that is in the Earth's northern hemisphere and known as a "north magnetic pole" *is really the south pole of the Earth's magnetic dipole*. This means that the north pole of a compass is attracted to the Earth's geographic north pole.

The direction of the magnetic field varies from location to location on the Earth. The field direction at any location on the Earth's surface is commonly specified in terms of two angles. The **field declination** is the angle (left or right) between geographic north (which is toward 90° latitude) and the horizontal component of the field. The **field inclination** is the angle (up or down) between a horizontal plane and the field's direction.

The field's inclination and declination at a given location can be measured with a *compass* and a *dip meter*. A **compass** is simply a needle-shaped magnet that is mounted so it can rotate freely about a vertical axis. When it is held in a horizontal plane, the north-pole end of the needle points, generally, toward the geomagnetic north pole (really a south magnetic pole). The angle between the compass needle and geographic north is the field declination.* A dip meter, used to measure inclination, is simply another needle-shaped magnet mounted so it can rotate freely about a *horizontal* axis. If the plane of the dip meter is aligned with the direction of the compass needle used to measure the declination, then the angle the dip meter needle makes with the horizontal is defined as the inclination angle. The magnetic north pole is defined as the location in the northern hemisphere for which the dip angle is 90°.

Causes of the Earth's Magnetism

The mechanisms that produce the Earth's magnetic field are not completely understood. However, it is helpful to begin our discussion of the latest models with a consideration of what is known about the Earth's formation and structure.

The Earth's Structure: Measurements of the spread of seismic waves tell us that the structure of the Earth is rather like that of a chocolate-covered cherry with gooey liquid between the cherry and the chocolate. This structure makes sense when we consider the currently accepted theory that the Earth was formed five billion years ago as a conglomeration of colliding meteorites and comets. Iron and other dense elements from meteorites were pulled by gravitational forces toward the center of the Earth. Compounds made of lighter elements, as well as the water contained in comets, migrated toward the surface. In between the solid core at the center of the Earth and the solid crust at the Earth's surface there is the gooey liquid consisting of molten lava (Fig. 32-22).

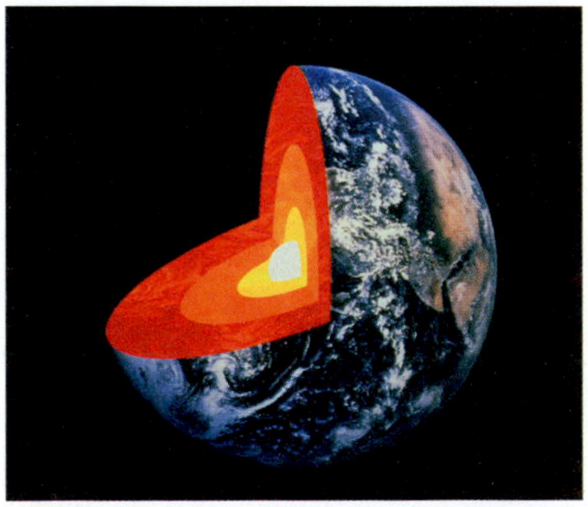

Continuous Molten Lava Currents: Many scientists believe that most of the Earth's magnetic field is produced by electromagnetic interactions that depend on the molten lava acting like a moving electrical conductor. We know from our study of Faraday's law that if even a small magnetic field is present in a region of the core, the electrical currents can be induced in the conducting fluid that travels through it. These induced currents can, in turn, produce magnetic fields that can act on other parts of the liquid core that are also moving. Thus a continuous cycle of induction and magnetic field production can take place as long as the material in the liquid core keeps flowing. In principle, this process is rather like that described for the generator shown in Fig. 32-10.

Two mechanisms have been proposed that explain the flow of molten lava in the liquid core. One possible mechanism is thermal convection produced by the temperature

FIGURE 32-22 ■ Seismic data reveal that the Earth has an **inner core** (white) of solid iron with a radius of about 1200 km, an **outer core** (yellow) of iron rich molten lava about 2200 km thick, a more or less solid **mantle** (orange and red—not to scale) of less dense matter about 2600 km thick, and a very thin **crust** of rocks and soils at the surface with an average thickness of 20 km.

*Inclination is the angle that a magnetic needle makes with the plane of the horizon. It is also called the angle of dip. Declination is the angle between magnetic north and geographic north.

difference between the hot solid inner core and the much cooler mantle. A second proposed mechanism for the flow of lava involves condensation of the heavier elements onto a growing inner core. This causes lighter, less dense, elements to flow toward the Earth's surface. In either case liquid convection currents are produced that are not unlike those in a pot of boiling water.

The Earth's magnetic field depends critically on the existence of *continuous* convection currents that requires the solid core to remain very hot for billions of years. Some scientists believe that nuclear energy in the core is being transformed to thermal energy through the decay of heavy radioactive elements. Other scientists have suggested that thermal energy can be released if the inner core expands by condensing material from the liquid core.

Changes in the Earth's Magnetic Field Over Time: Some mysterious characteristics of the Earth's magnetic field have been gleaned from fossil records and other geomagnetic measurements. The strength of the field and the location of the magnetic poles are constantly changing. For example, in recent years the geographic location of the magnetic poles has changed by an average of about 100 meters a day. These relatively small day-to-day changes are not obvious to someone a long distance from a magnetic pole who uses a compass and dip meter to measure a local field direction. It's another story when longer time scales are involved. We can use simple instruments to detect changes over a time period of a year or more. When even longer time periods are considered, the changes have been dramatic. In fact, the orientations of magnetized minerals imbedded in ancient rocks indicate that the Earth's magnetic field has completely reversed itself many times in the Earth's five billion year history, though reversals seem to take 1000 years or more.

FIGURE 32-23 ■ This image shows one of many configurations of the Earth's magnetic field lines created by the model developed by Glatzmaier and Roberts.

The Glatzmaier/Roberts Model: A few years ago, two scientists, Gary Glatzmaier and Paul Roberts, developed a comprehensive numerical model of the electromagnetic and fluid dynamic processes in the Earth's interior. When this model was run on a CRAY supercomputer for thousands of hours these investigators were able to simulate over 300,000 years of magnetic field conditions. Their results showed many of the key features revealed by geological data, including the existence of a dipole field outside the Earth, a preference for approximate alignment between the Earth's dipole moment and its rotation axis, field strength variations, migration of the magnetic poles over the Earth's surface, and several field reversals. One such configuration of magnetic field lines is seen in Fig. 32-23.

There is still a great deal to be learned about the actual mechanisms responsible for the continual changes in the Earth's magnetic field, but scientists expect to resolve many of their uncertainties within the next few decades.

Magnetic Bacteria

The survival of many organisms depends on their ability to sense the Earth's magnetic field. For example, it is believed that the Earth's dipole field is critical to the navigation of migrating birds and fish as well as certain types of bacteria.

Magnetotactic bacteria are one-celled organisms that can be found almost anywhere in the world where there are ponds, marshes, or muddy lake bottoms. Many species of these bacteria are anaerobic or microanaerobic and must burrow in mud both to get away from oxygen and to feed on nutrients. Notice that on the lower left side of the bacterium shown in the photo at the beginning of this chapter there is a string of tiny 100-nanometer-long particles. These particles, known as magnetosomes, are oriented along the bacterium's long axis. An enlarged view of a set of magnetosomes is shown in Fig. 32-24.

Magnetotactic bacteria synthesize these magnetic particles out of iron-oxygen or iron sulfur compounds. Each magnetosome is just big enough to have a permanent

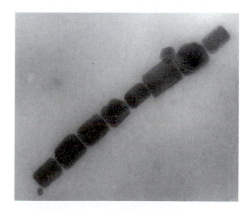

FIGURE 32-24 ■ The type of bacterium shown in the puzzler at the beginning of the chapter is magnetotactic because it contains a chain of dense iron-rich magnetosomes each having a length of about 100 nm. The chain shown in this transmission electron micrograph has a net magnetic moment and tends to align itself with the Earth's magnetic field.

magnetic dipole moment and just small enough to be a single ferromagnetic domain. When strung together like a set of microscopic refrigerator magnets, the array has a net dipole moment. So instead of bumbling around randomly, these bacteria align with the Earth's magnetic field. This allows them to swim naturally along field lines.

Through natural selection, the bacteria that have their magnetosome strings oriented so they swim down along magnetic field lines to the mud at the bottom of a pond or lake will survive and multiply. Those that don't will swim up and die. An examination of the pattern of the Earth's magnetic field lines shown in Fig. 32-21 reveals that "down" is opposite to the direction of the field lines in the southern hemisphere and in the same direction as the field lines in the northern hemisphere. Thus, an Australian bacterium evolved to swim down in its normal habitat would swim up if transported to the United States. Alternatively, a healthy bacterium that evolves in the United States would be preset by evolutionary processes to have its magnetosomes oriented in the opposite direction, so it will swim down.

It is interesting to note that the orientations of bacterial magnetosome strings in fossils have helped scientists piece together evidence for past changes in the Earth's magnetic field.

READING EXERCISE 32-10: Describe the ways in which the Earth's magnetic field varies over its surface. Does the Earth's magnetic field vary in time as well? ∎

Problems

SEC. 32-2 ∎ SELF-INDUCTANCE

1. Close-Packed Coil The inductance of a close-packed coil of 400 turns is 8.0 mH. Calculate the magnetic flux through the coil when the current is 5.0 mA.

2. Circular Coils and Flux A circular coil has a 10.0 cm radius and consists of 30.0 closely wound turns of wire. An externally produced magnetic field of magnitude 2.60 mT is perpendicular to the coil. (a) If no current is in the coil, what is the magnitude of the magnetic flux that links its turns? (b) When the current in the coil is 3.80 A in a certain direction, the net flux through the coil is found to vanish. What is the inductance of the coil?

3. Equal Currents, Opposite Directions Two long parallel wires, both of radius a and whose centers are a distance d apart, carry equal currents in opposite directions. Show that, neglecting the flux within the wires, the inductance of a length l of such a pair of wires is given by

$$L = \frac{\mu_0 l}{\pi} \ln \frac{d-a}{a}$$

(*Hint:* Calculate the flux through a rectangle of which the wires form two opposite sides.)

4. Wide Copper Strip A wide copper strip of width W is bent to form a tube of radius R with two parallel planar extensions, as shown in Fig. 32-25. There is a current i through the strip, distributed uniformly over its width. In this way a "one-turn solenoid" is formed. (a) Derive an expression for the magnitude of the magnetic field $\vec{B}$ in the

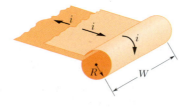

FIGURE 32-25 ∎ Problem 4.

tubular part (far away from the edges). (*Hint:* Assume that the magnetic field outside this one-turn solenoid is negligibly small.) (b) Find the inductance of this one-turn solenoid, neglecting the two planar extensions.

5. Inductor Carries Steady Current A 12 H inductor carries a steady current of 2.0 A. How can a 60 V self-induced emf be made to appear in the inductor?

6. At a Given Instant At a given instant the current and self-induced emf in an inductor are directed as indicated in Fig. 32-26. (a) Is the current increasing or decreasing? (b) The induced emf is 17 V and the rate of change of the current is 25 kA/s; find the inductance.

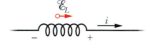

FIGURE 32-26 ∎ Problem 6.

7. Inductors in Series Two inductors L_1 and L_2 are connected in series and are separated by a large distance. (a) Show that the equivalent inductance is given by

$$L_{eq} = L_1 + L_2.$$

(*Hint:* Review the derivations for resistors in series and capacitors in series. Which is similar here?) (b) Why must their separation be large for this relationship to hold? (c) What is the generalization of (a) for N inductors in series?

8. Current Varies with Time The current i through a 4.6 H inductor varies with time t as shown by the graph of Fig. 32-27. The inductor has a resistance of 12 Ω. Find the magnitude of the induced emf $\mathcal{E}$ during the time intervals (a) $t_1 = 0$ to $t_2 = 2$ ms, (b) $t_2 = 2$ ms to $t_3 = 5$ ms, (c) $t_3 = 5$ ms to $t_4 = 6$ ms. (Ignore the behavior at the ends of the intervals.)

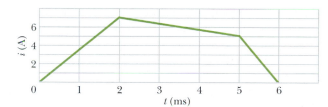

FIGURE 32-27 ▪ Problem 8.

9. At What Rate At time $t = 0$ ms, a 45 V potential difference is suddenly applied to the leads of a coil with inductance $L = 50$ mH and resistance $R = 180$ Ω. At what rate is the current through the coil increasing at $t = 1.2$ ms?

10. Inductors in Parallel Two inductors L_1 and L_2 are connected in parallel and separated by a large distance. (a) Show that the equivalent inductance is given by

$$\frac{1}{L_{eq}} = \frac{1}{L_1} + \frac{1}{L_2}.$$

(*Hint:* Review the derivations for resistors in parallel and capacitors in parallel. Which is similar here?) (b) Why must their separation be large for this relationship to hold? (c) What is the generalization of (a) for N inductors in parallel?

11. What Is L The inductance of a closely wound coil is such that an emf of 3.0 mV is induced when the current changes at the rate of 5.0 A/s. A steady current of 8.0 A produces a magnetic flux of 40 μ Wb through each turn. (a) Calculate the inductance of the coil. (b) How many turns does the coil have?

SEC. 32-3 ▪ MUTUAL INDUCTION

12. Coil 1, Coil 2 Coil 1 in Fig. 32-4 has $L_1 = 25$ mH and $N_1 = 100$ turns. Coil 2 has $L_2 = 40$ mH and $N_2 = 200$ turns. The coils are rigidly positioned with respect to each other; their mutual inductance M is 3.0 mH. A 6.0 mA current in coil 1 is changing at the rate of 4.0 A/s. (a) What magnetic flux $\Phi_{1\to2}$ links coil 2, and what self-induced emf appears there? (b) What magnetic flux $\Phi_{2\to1}$ links coil 1, and what mutually induced emf appears there?

13. Two Coils at Fixed Locations Two coils are at fixed locations. When coil 1 has no current and the current in coil 2 increases at the rate 15.0 A/s, the emf in coil 1 is 25.0 mV. (a) What is their mutual inductance? (b) When coil 2 has no current and coil 1 has a current of 3.60 A, what is the flux linkage in coil 2?

14. Two Solenoids Two solenoids are part of the spark coil of an automobile. When the current in one solenoid falls from 6.0 A to zero in 2.5 ms, an emf of 30 kV is induced in the other solenoid. What is the mutual inductance M of the solenoids?

15. Two Connected Coils Two coils, connected as shown in Fig. 32-28, separately have inductances L_1 and L_2. Their mutual inductance is M. (a) Show that this combination can

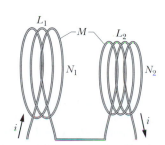

FIGURE 32-28 ▪ Problem 15.

be replaced by a single coil of equivalent inductance given by

$$L_{eq} = L_1 + L_2 + 2M.$$

(b) How could the coils in Fig. 32-28 be reconnected to yield an equivalent inductance of

$$L_{eq} = L_1 + L_2 - 2M?$$

(This problem is an extension of Problem 7, but the requirement that the coils be far apart has been removed.)

16. Coil Around Solenoid A coil C of N turns is placed around a long solenoid S of radius R and n turns per unit length as in Fig. 32-29. Show that the mutual inductance for the coil–solenoid combination is given by $M = \mu_0 \pi R^2 nN$. Explain why M does not depend on the shape, size, or possible lack of close-packing of the coil.

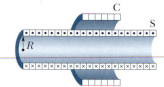

FIGURE 32-29 ▪ Problem 16.

17. Coaxial Solenoid Figure 32-30 shows, in cross section, two coaxial solenoids. Show that the mutual inductance M for a length l of this solenoid–solenoid combination is given by $M = \pi R_1^2 l \mu_0 n_1 n_2$, in which n_1 and n_2 are the respective numbers of turns per unit length and R_1 is the radius of the inner solenoid. Why does M depend on R_1 and not on R_2?

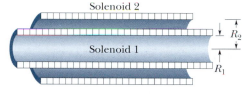

FIGURE 32-30 ▪ Problem 17.

18. Coils Over a Toroid Figure 32-31 shows a coil of N_2 turns wound as shown around part of a toroid of N_1 turns. The toroid's inner radius is a, its outer radius is b, and its height is h. Show that the mutual inductance M for the toroid–coil combination is

$$M = \frac{\mu_0 N_1 N_2 h}{2\pi} \ln \frac{b}{a}.$$

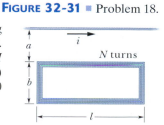

FIGURE 32-31 ▪ Problem 18.

19. Rectangular Loop A rectangular loop of N close-packed turns is positioned near a long straight wire as shown in Fig. 32-32. (a) What is the mutual inductance M for the loop–wire combination? (b) Evaluate M for $N = 100$, $a = 1.0$ cm, $b = 8.0$ cm, and $l = 30$ cm.

SEC. 32-4 ▪ RL CIRCUITS (WITH IDEAL INDUCTORS)

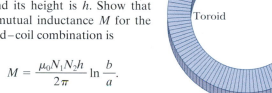

FIGURE 32-32 ▪ Problem 19.

20. Inductive Time Constant The current in an RL circuit builds up to one third of its steady-state value in 5.00 s. Find the inductive time constant.

21. How Long Must We Wait In terms of τ_L, how long must we wait for the current in an RL circuit to build up to within 0.100% of its equilibrium value?

22. In Terms of the emf Consider the RL circuit of Fig. 32-6. In terms of the battery emf $\mathscr{E}$, (a) what is the self-induced emf ΔV_2 when the switch has just been closed on a, and (b) what is ΔV_2 when $t = 2.0\tau_L$? (c) In terms of τ_L, when will ΔV_2 be just one-half the battery emf $\mathscr{E}$?

23. First Second The current in an RL circuit drops from 1.0 A to 10 mA in the first second following removal of the battery from the circuit. If L is 10 H, find the resistance R in the circuit.

24. emf Varies with Time Suppose the emf of the battery in the circuit of Fig. 32-7 varies with time t so that the current is given by $i(t) = 3.0 \text{ A} + (5.0 \text{ A/s})t$, where i is in amperes and t is in seconds. Take $R = 4.0\ \Omega$ and $L = 6.0$ H, and find an expression for the battery emf as function of time. (*Hint:* Apply the loop rule.)

25. Solenoid A solenoid having an inductance of 6.30 μH is connected in series with a 1.20 kΩ resistor. (a) If a 14.0 V battery is inserted into the circuit, how long will it take for the current through the resistor to reach 80.0% of its final value? (b) What is the current through the resistor at time $t = 1.0\tau_L$?

26. Wooden Toroidal Core A wooden torodial core with a square cross section has an inner radius of 10 cm and an outer radius of 12 cm. It is wound with one layer of wire (of diameter 1.0 mm and resistance per meter 0.020 Ω/m). What are (a) the inductance and (b) the inductive time constant of the resulting toroid? Ignore the thickness of the insulation on the wire.

27. Suddenly Applied At time $t = 0$ ms, a 45.0 V potential difference is suddenly applied to a coil with $L = 50.0$ mH and $R = 180\ \Omega$. At what rate is the current increasing at $t = 1.20$ ms?

28. In the Circuit In the circuit of Fig. 32-33, $\mathscr{E} = 10$ V, $R_1 = 5.0\ \Omega$, $R_2 = 10\ \Omega$, and $L = 5.0$ H. For the two separate conditions (I) switch S just closed and (II) switch S closed for a long time, calculate (a) the current i_1 through R_1, (b) the current i_2 through R_2, (c) the current i through the switch, (d) the potential difference across R_2, (e) the potential difference across L, and (f) the rate of change di_2/dt.

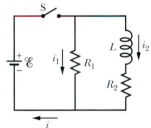

FIGURE 32-33 ■
Problem 28.

29. In the Figure In Fig. 32-34, $\mathscr{E} = 100$ V, $R_1 = 10.0\ \Omega$, $R_2 = 20.0\ \Omega$, $R_3 = 30.0\ \Omega$, and $L = 2.00$ H. Find the values of i_1 and i_2 (a) immediately after closing of switch S, (b) a long time later, (c) immediately after the reopening of switch S, and (d) a long time after the reopening.

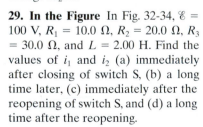

FIGURE 32-34 ■
Problem 29.

30. What Is the Constant Figure 32-35a shows a circuit consisting of an ideal battery with emf $\mathscr{E} = 6.00\ \mu$V, a resistance R, and a small wire loop of area 5.0 cm^2. For the time interval $t_1 = 10$ to $t_2 = 20$ s, an external magnetic field is set up throughout the loop. The field is uniform, its direction is into the page in Fig. 32-35a, and the field magnitude is given by $B = at$, where B is in teslas, a is a

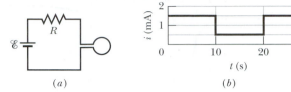

(a)   (b)

FIGURE 32-35 ■ Problem 30.

constant with units of teslas per second, and t is in seconds. Figure 32-35b gives the current i in the circuit before, during, and after the external field is set up. Find a.

31. Once the Switch Is Closed Once the switch S is closed in Fig. 32-36 the time required for the current to reach any obtainable value depends, in part, on the value of resistance R. Suppose the emf $\mathscr{E}$ of the ideal battery is 12 V and the inductance of the ideal (resistanceless) inductor is 18 mH. How much time is needed for the current to reach 2.00 A if R is (a) 1.00 Ω, (b) 5.00 Ω, and (c) 6.00 Ω? (d) Why is there a huge jump between the answers to (b) and (c)? (e) For what value of R is the time required for the current to reach 2.00 A least? (f) What is that least time? (*Hint:* Rethink Eq. 32-21.)

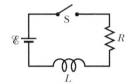

FIGURE 32-36 ■
Problems 31, 34, and 61.

32. Circuit Shown In the circuit shown in Fig. 32-37, switch S is closed at time $t = 0$. Thereafter, the constant current source, by varying its emf, maintains a constant current i out of its upper terminal. (a) Derive an expression for the current through the inductor as a function of time. (b) Show that the current through the resistor equals the current through the inductor at time $t = (L/R)\ln 2$.

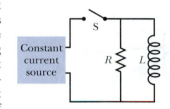

FIGURE 32-37 ■
Problem 32.

33. When Is the Flux Equal In Fig. 32-38a, switch S has been closed on A long enough to establish a steady current in the inductor of inductance $L_1 = 5.00$ mH and the resistor of resistance $R_1 = 25\ \Omega$. Similarly, in Fig. 32-38b, switch S has been closed on A long enough to establish a steady current in the inductor of inductance $L_2 = 3.00$ mH and the resistor of resistance $R_2 = 30\ \Omega$. The ratio Φ_{02}/Φ_{01} of the magnetic flux through a turn in inductor 2 to that in inductor 1 is 1.5. At time $t = 0$, the two switches are closed on B. At what time t is the flux through a turn in the two inductors equal?

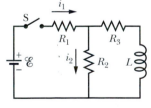 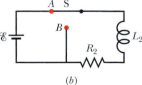

(a)   (b)

FIGURE 32-38 ■ Problem 33.

34. When Is emf Equal Switch S in Fig. 32-36 is closed at time $t = 0$, initiating the buildup of current in the 15.0 mH inductor and the 20.0 Ω resistor. At what time is the emf across the inductor equal to the potential difference across the resistor?

SEC. 32-5 ■ INDUCTORS, TRANSFORMERS, AND ELECTRIC POWER

35. A Transformer A transformer has 500 primary turns and 10 secondary turns. (a) If ΔV_p is 120 V (rms), what is ΔV_s with an open circuit? (b) If the secondary now has a resistive load of 15 Ω, what are the currents in the primary and secondary?

36. A Generator A generator supplies 100 V to the primary coil of a transformer of 50 turns. If the secondary coil has 500 turns, what is the secondary voltage?

37. Audio Amplifier In Fig. 32-39 let the rectangular box on the left represent the (high-impedance) output of an audio amplifier, with $r = 1000\ \Omega$. Let $R = 10\ \Omega$ represent the (low-impedance) coil of a loudspeaker. For maximum transfer of energy to the load R we must have $R = r$, and that is not true in this case. However, a transformer can be used to "transform" resistances, making them behave electrically as if they were larger or smaller than they actually are. Sketch the primary and secondary coils of a transformer that can be introduced between the amplifier and the speaker in Fig. 32-39 to match the impedances. What must be the turns ratio?

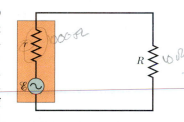

FIGURE 32-39 ■
Problem 37.

38. Autotransformer Figure 32-40 shows an "autotransformer." It consists of a single coil (with an iron core). Three taps T_N are provided. Between taps T_1 and T_2 there are 200 turns, and between taps T_2 and T_3 there are 800 turns. Any two taps can be considered the "primary terminals" and any two taps can be considered the "secondary terminals." List all the ratios by which the primary voltage may be changed to a secondary voltage.

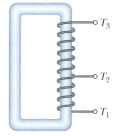

FIGURE 32-40 ■
Problem 38.

SEC. 32-6 ■ MAGNETIC MATERIALS — AN INTRODUCTION

39. Orbital Magnetic Dipole What is the measured component of the orbital magnetic dipole moment of an electron with (a) $m_l = 1$ and (b) $m_l = -2$?

40. Energy Difference What is the energy difference between parallel and antiparallel alignment of the z-component of an electron's spin magnetic dipole moment with an external magnetic field of magnitude 0.25 T, directed parallel to the z axis?

41. Electron in an Atom If an electron in an atom has an orbital angular momentum with $m_l = 0$, (a) what is the component μ_z^{orb}? If the atom is in an external magnetic field $\vec{B}$ of magnitude 35 mT and directed along z axis, what are the potential energies associated with the orientations of (b) the electron's orbital magnetic dipole moment and (c) the electron's spin magnetic dipole moment? (d) Repeat (a) through (c) for $m_l = -3$.

42. Spin Magnetic Moment An electron is placed in a magnetic field $\vec{B}$ that is directed along a z axis. The energy difference between parallel and antiparallel alignments of the z-component of the electron's spin magnetic moment with $\vec{B}$ is 6.00×10^{-25} J. What is the magnitude of $\vec{B}$?

43. How Many Suppose that ± 4 are the limits to the values of m_l for an electron in an atom. (a) How many different values of the z-component μ_z^{orb} of the electron's orbital magnetic dipole moment are possible? (b) What is the greatest magnitude of those possible values? Next, suppose that the atom is in a magnetic field of magnitude 0.250 T, in the positive direction of the z axis. What are (c) the maximum potential energy and (d) the minimum potential energy associated with those possible values of μ_z^{orb}?

44. NMR and MRI Nuclear Magnetic Resonance (NMR) and Magnetic Resonance Imaging (MRI) exploit the interactions between charged particles and very strong magnetic fields in order to produce images (including images of soft tissue). The magnetic field in a certain MRI machine is 0.5 Tesla. What is the maximum difference in energy that one might measure for a single electron placed in this field?

SEC. 32-7 ■ FERROMAGNETISM

45. Saturation Magnetization The saturation magnetization M^{max} of the ferromagnetic metal nickel is 4.70×10^5 A/m. Calculate the magnetic moment of a single nickel atom. (The density of nickel is 8.90 g/cm^3 and its molar mass is 58.71 g/mol.)

46. Iron The dipole moment associated with an atom of iron in an iron bar has magnitude 2.1×10^{-23} J/T. Assume that all the atoms in the bar, which is 5.0 cm long and has a cross-sectional area of 1.0 cm^2, have their dipole moments aligned. (a) What is the magnitude of the dipole moment of the bar? (b) What is the magnitude of the torque that must be exerted to hold this magnet perpendicular to an external field of 1.5 T? (The density of iron is 7.9 g/cm^3.)

47. Earth's Magnetic Moment The magnetic dipole moment of Earth has magnitude 8.0×10^{22} J/T. (a) If the origin of this magnetism were a magnetized iron sphere at the center of the Earth, what would be its radius? (b) What fraction of the volume of the Earth would such a sphere occupy? Assume complete alignment of the dipoles. The density of the Earth's inner core is 14 g/cm^3. The magnetic dipole moment of an iron atom is 2.1×10^{-23} J/T. (*Note:* The Earth's inner core is in fact thought to be in both liquid and solid forms and partly iron, but a permanent magnet as the source of the Earth's magnetism has been ruled out by several considerations. For one, the temperature is certainly above the Curie point.)

48. Mines and Boreholes Measurements in mines and boreholes indicate that the Earth's interior temperature increases with depth at the average rate of 30 C°/km. Assuming a surface temperature of 10°C, at what depth does iron cease to be ferromagnetic? (The Curie temperature of iron varies very little with pressure.)

SEC. 32-8 ■ OTHER MAGNETIC MATERIALS

49. Electron Assume that an electron of mass m and charge magnitude e moves in a circular orbit of radius r about a nucleus. A uniform magnetic field $\vec{B}$ is then established perpendicular to the plane of the orbit. Assuming also that the radius of the orbit does not change and that the change in the speed of the electron due to field $\vec{B}$ is small, find an expression for the change in the orbital magnetic dipole moment of the electron due to the field.

50. Loop Model Figure 32-41 shows a loop model (loop L) for a diamagnetic material. (a) Sketch the magnetic field lines through and about the material due to the bar magnet. (b) What are the directions of the loop's net magnetic dipole moment $\vec{\mu}$ and the conventional current i in the loop? (c) What is the direction of the magnetic force on the loop?

FIGURE 32-41 ■ Problems 50 and 54.

51. Cylindrical Magnet A magnet in the form of a cylindrical rod has a length of 5.00 cm and a diameter of 1.00 cm. It has a uniform magnetization of 5.30×10^3 A/m. What is the magnitude of its magnetic dipole moment?

52. Paramagnetic Gas A magnetic field of magnitude 0.50 T is applied to a paramagnetic gas whose atoms have an intrinsic magnetic dipole moment of magnitude 1.0×10^{-23} J/T. At what temperature will the mean kinetic energy of translation of the gas atoms be equal to the energy required to reverse such a dipole end for end in this magnetic field?

53. Paramagnetic Salt A sample of the paramagnetic salt to which the magnetization curve of Fig. 32-19 applies is to be tested to see whether it obeys Curie's law. The sample is placed in a uniform 0.50 T magnetic field that remains constant throughout the experiment. The magnetization M is then measured at temperatures ranging from 10 to 300 K. Will Curie's law be valid under these conditions?

54. Paramagnetic Material Repeat Problem 50 for the case in which loop L is the model for a paramagnetic material.

55. Electron's Kinetic Energy An electron with kinetic energy K travels in a circular path that is perpendicular to a uniform magnetic field, the electron's motion is subject only to the force due to the field. (a) Show that the magnetic dipole moment of the electron due to its orbital motion has magnitude $\mu = K/|\vec{B}|$ and that it is in the direction opposite that of $\vec{B}$. (b) What are the magnitude and direction of the magnetic dipole moment of a positive ion with kinetic energy K_{ion} under the same circumstances? (c) An ionized gas consists of 5.3×10^{21} electrons/m³ and the same number density of ions. Take the average electron kinetic energy to be 6.2×10^{-20} J and the average ion kinetic energy to be 7.6×10^{-21} J. Calculate the magnetization of the gas when it is in a magnetic field of 1.2 T.

56. Magnetization Curve A sample of the paramagnetic salt to which the magnetization curve of Fig. 32-17 applies is held at room temperature (300 K). At what applied magnetic field will the degree

of magnetic saturation of the sample be (a) 50% and (b) 90%? (c) Are these fields attainable in the laboratory?

SEC. 32-9 ■ THE EARTH'S MAGNETISM

57. New Hampshire In New Hampshire the average horizontal component of the Earth's magnetic field in 1912 was 16 μT and the average inclination or "dip" was 73°. What was the corresponding magnitude of the Earth's magnetic field?

58. Earth's Field Assume the average value of the vertical component of the Earth's magnetic field is 43 μT (downward) for all of Arizona, which has an area of 2.95×10^5 km², and calculate the net magnetic flux through the rest of the Earth's surface (the entire surface excluding Arizona). Is that net magnetic flux outward or inward?

59. Earth's Field Two Use the results of Problem 60 to predict the Earth's magnetic field (both magnitude and inclination) at (a) the geomagnetic equator, (b) a point at geomagnetic latitude 60°, and (c) the north geomagnetic pole.

60. Magnetic Field of Earth The magnetic field of the Earth can be approximated as the magnetic field of a dipole, with horizontal and vertical components, at a point a distance r from the Earth's center, given by

$$B_h = \frac{\mu_0 \mu}{4\pi r^3} \cos \lambda_m, \qquad B_v = \frac{\mu_0 \mu}{2\pi r^3} \sin \lambda_m,$$

where λ_m is the *magnetic latitude* (this type of latitude is measured from the geomagnetic equator toward the north or south geomagnetic pole). Assume that the Earth's magnetic dipole moment is $\mu = 8.00 \times 10^{22}$ A · m². (a) Show that the magnitude of the Earth's field at latitude λ_m is given by

$$B = \frac{\mu_0 \mu}{4\pi r^3} \sqrt{1 + 3 \sin^2 \lambda_m}.$$

(b) Show that the inclination ϕ_i of the magnetic field is related to the magnetic latitude λ_m by

$$\tan \phi_i = 2 \tan \lambda_m.$$

Additional Problems

61. Rate of Energy Transfer In Fig. 32-36, a 12.0 V ideal battery, a 20 Ω resistor, and an ideal inductor are connected by a switch at time $t = 0$ s. At what rate is the battery transferring energy to the inductor's field at $t = 1.61\tau_L$?

62. Compass Needle You place a magnetic compass on a horizontal surface, allow the needle to settle into equilibrium position, and then give the compass a gentle wiggle to cause the needle to oscillate about that equilibrium position. The frequency of oscillation is 0.312 Hz. The Earth's magnetic field at the location of the compass has a horizontal component of 18.0 μT. The needle has a magnetic moment of 0.680 mJ/T. What is the needle's rotational inertia about its (vertical) axis of rotation?

63. Induced Current in a Coil A long narrow coil is surrounded by a short wide coil as shown in Fig. 32-42. Both coils have negligible resistance. The short wide coil has a diameter d_S, n_S turns per unit length, and a length S. Its ends are connected through a resistor of resistance R. The long narrow inner coil has a diameter d_L, n_L turns per unit length, and a length L. Its ends are connected across a variable power source.

For each of the partial sentences below, indicate whether they are correctly completed by the phrase greater than ($>$), less than ($<$), or the same as ($=$). If you cannot determine which is the case from the information given, indicate not sufficient information (NSI).

The current through an inner coil is increased from 0.0 amps to 0.1 amps over a period of 10 seconds in a smooth fashion according to the rule

$$i_L(t) = (0.01 \text{ A/s}) \, t.$$

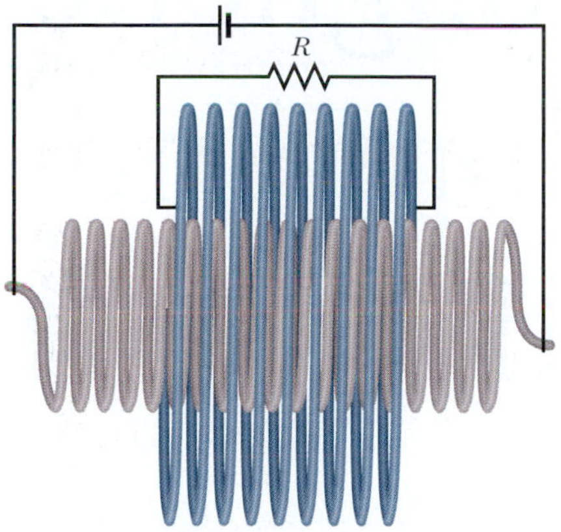

(a) The magnitude of the current in the long narrow coil at time $t = 1$ s is _____ the current in that coil at time $t = 5$ s.

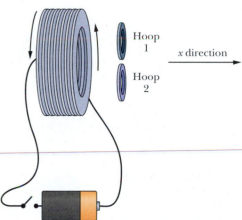

FIGURE 32-42 ■ Problem 63.

(b) The magnitude of the current in the short wide coil at time $t = 1$ s is_____the current in that coil at time $t = 5$ s.
(c) The magnitude of the current in the long narrow coil at time $t = 1$ s is _____ the current in the short wide coil at that same time.
(d) If the long narrow coil was compressed to half its length (without changing its diameter) before the current was turned on, the current in the short wide coil would be _____ it was without the compression.

64. Inducing Current Figure 32-43 shows a solenoid and two hoops. When the switch is closed, the solenoid carries a current in the direction indicated. The planes of the small loops are parallel to the planes of the hoops of the solenoid.

Hoop 1 consists of a single turn of resistive wire that has a resistance per unit length of λ. Hoop 2 consists of N turns of the same wire. Each hoop is a circle of radius r.

(a) The switch is closed and remains closed for a few seconds. Hoop 1 is then moved to the right. Is there a current flow induced? If there is a current, indicate the direction and explain how you figured it out.
(b) The hoops are now returned to their original locations and held fixed. The switch is opened. For a short time, the magnetic field at the hoops decreases like

$$B_x(t) = B_x(0) - \gamma t,$$

FIGURE 32-43 ■ Problem 64.

where γ is a constant with units of gauss per second. Is there a current flow induced in the hoops? If there is a current, indicate the direction.
(c) Calculate the current flow in each hoop for situation (b).
(d) If $B_x(0) = 10$ gauss, $\gamma = 2.5$ gauss/s, the resistivity of the wire is $1 \ \mu\Omega/\text{m}$, and the hoops have a radius of 2 cm, calculate the current induced in hoop 1 as the B-field from the solenoid begins to fall.

33

Electromagnetic Oscillations and Alternating Current

When a high-voltage power transmission line requires repair, a utility company cannot just shut it down, perhaps blacking out an entire city. Repairs must be made while the lines are electrically "hot." The man outside the helicopter in this photograph has just replaced a spacer between 500 kV lines *by hand*, a procedure that requires considerable expertise.

How does he manage this repair without being electrocuted?

The answer is in this chapter.

33-1 Advantages of Alternating Current

So far we have confined our study of electric circuits to **direct current** or **dc** circuits in which the direction of current does not change over time. In reality the vast majority of electric power systems and electrical devices involve **alternating current** or *ac* circuits where the current direction is continuously oscillating back and forth. Why did ac power become so popular? By 1879 the famous American inventor, Thomas Edison, refined the electric lightbulb invented by Humphry Davy in England. Almost overnight, there was high demand in Europe and the United States for the creation of systems for the generation and distribution of electric power.

Edison's quest for a practical lightbulb was apparently motivated by his desire to promote the use of the dc power system that he and his colleagues were developing. Indeed, his dc power station in lower Manhattan quickly become a monopoly, but only temporarily. By 1888 the Serbian immigrant Nikola Tesla (Fig. 33-1) had patented a complete system of alternating current generators, transformers, transmission lines, and induction motors. Shortly thereafter, entrepreneur-inventor George Westinghouse (Fig. 33-2) purchased Tesla's patents. After the Westinghouse Company's ac system was featured at the 1893 Chicago World Fair, more than 80% of all electrical devices were powered by ac circuits.

You already know some of the key factors that render ac power superior to dc. In Section 32-5 we discussed how electricity generated by induction naturally produces alternating current. We discussed how transformers can be used to step up voltages so that power can be transmitted more efficiently over long distances.* There were other factors that favored the Westinghouse system. Alternating current power transmission requires far less copper wire than Edison's dc system. Furthermore, the ac induction motors invented by Tesla were so efficient and easy to manufacture that they quickly became the heart of almost all labor-saving household devices, including water pumps, washing machines, dryers, electric drills, blenders, dishwashers, and garbage disposals.

There are other more recent inventions that we now take for granted that operate on ac circuits. Examples include radio and television transmission (treated in Chapter 34) and reception, computer monitors, and even the graphic equalizers in hi-fi equipment. Thus, without an understanding of ac circuits, it is impossible to understand how modern electrical systems and devices work. So, the major focus of this chapter is to use what you already know about induction and dc circuits to help you understand ac circuits. Resistors, capacitors, and inductors are the basic building blocks of both ac and dc circuits. We have already studied the independent functioning of each. In addition, we learned about dc resistor-capacitor (*RC*) combinations in Chapter 28 and dc resistor-inductor (*RL*) combinations in Chapter 32.

We begin this chapter by deriving equations that quantify the energy and energy density stored in the magnetic field created by current flowing through an inductor. We also review what we learned in Section 28-5 about the energy and energy density stored in a capacitor's electric field due to its charge. This will prepare you to study *electromagnetic oscillations* in several types of ac circuits where energy shuttles back and forth between the magnetic field in an inductor and the electric field in a capacitor.

FIGURE 33-1 ■ Nikola Tesla, an eccentric Serbian-American scientist and electrical engineer, invented the first successful ac power generation system.

FIGURE 33-2 ■ George Westinghouse developed the first ac power distribution system in the United States. It was based on Tesla's design. The system went online at Niagara Falls in 1896. After only a few years it was found to be superior to existing dc systems.

---

*The method of repairing high-voltage lines shown in the opening photograph is patented by Scott H. Yenzer and is licensed exclusively to Haverfield Corporation of Gettysburg, Pennsylvania. As the lineman approaches a hot line, the electric field surrounding the line brings his body to nearly the potential of the line. To match the two potentials, he then extends a conducting "wand" to the line. To avoid being electrocuted, he must be isolated from anything electrically connected to the ground. To ensure that his body is always at a single potential—that of the line he is working on—he wears a conducting suit, hood, and gloves, all of which are electrically connected to the line via the wand.

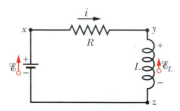

FIGURE 33-3 ▪ The circuit of Fig. 32-6 with the switch closed on *a*. We apply the loop rule for circuits clockwise, starting at *x*.

33-2 Energy Stored in a $\vec{B}$-Field

When we pull two particles with opposite signs of charge away from each other, the resulting electric potential energy is stored in the electric field of the particles. We get this energy back from the field by letting the particles move toward each other again. In the same way we can consider energy to be stored in a magnetic field.

To derive a quantitative expression for that stored energy, consider Fig. 33-3, which shows a source of emf $\mathcal{E}$ connected to a resistor R and an inductor L. After a switch is closed, the growth of current can be described by Eq. 32-20, which is restated here for convenience,

$$\mathcal{E} = L\frac{di}{dt} + iR. \tag{33-1}$$

This differential equation follows immediately from the loop rule for potential differences in single-loop circuits. If we multiply each side of this expression by the current i we obtain

$$\mathcal{E}i = L\,i\frac{di}{dt} + i^2R, \tag{33-2}$$

which has the following physical interpretation in terms of work and energy:

1. If a charge dq passes through the battery of emf $\mathcal{E}$ in Fig. 33-3 in time dt, the battery does work on it in the amount $\mathcal{E}\,dq$. The rate at which the battery does work is $(\mathcal{E}\,dq)/dt$ or $\mathcal{E}i$. Thus, the left side of Eq. 33-2 represents the rate at which the emf device delivers energy to the rest of the circuit.

2. The rightmost term in Eq. 33-2 represents the rate at which energy is transformed to thermal energy in the resistor.

3. Energy that is delivered to the circuit but does not appear as thermal energy must, by the conservation-of-energy hypothesis, be stored in the magnetic field of the inductor. Since Eq. 33-2 represents conservation of energy for RL circuits, the middle term must represent the rate dU^{mag}/dt at which energy is stored in the magnetic field.

Thus

$$\frac{dU^{\text{mag}}}{dt} = L\,i\frac{di}{dt}. \tag{33-3}$$

We can write this as

$$dU^{\text{mag}} = L\,i\,di.$$

Integrating yields

$$\int_0^{U^{\text{mag}}} dU^{\text{mag}} = \int_0^i L\,i\,di$$

or $\qquad\qquad U^{\text{mag}} = \tfrac{1}{2}Li^2 \qquad$ (magnetic energy), $\qquad\qquad\qquad$ (33-4)

which represents the total energy stored in the magnetic field of an inductor L carrying a current i. Note the similarity in form between this expression and the expression for the energy a capacitor stores in its electric field due to its capacitance

C and charge q. That equation is given by Eq. 28-21 and restated here for convenience as

$$U^{\text{elec}} = \tfrac{1}{2}\left(\frac{1}{C}\right)q^2. \tag{33-5}$$

(The variable i corresponds to q, and the constant L corresponds to $1/C$.)

33-3 Energy Density of a $\vec{B}$-Field

Since a typical inductor has the shape of either a solenoid or a toroid (a solenoid bent into a donut shape) it is often useful to know the magnetic field energy per unit volume stored in the magnetic field of this type of inductor. Consider a length l near the middle of a long solenoid of cross-sectional area A carrying current i. The volume associated with this length is Al. The energy U^{mag} stored by the length l of the solenoid must lie entirely within this volume because the magnetic field outside such a solenoid is approximately zero. Moreover, the stored energy must be uniformly distributed within the solenoid because the magnetic field inside a solenoid is also essentially uniform.

Thus, the energy u^{mag} stored per unit of magnetic field volume is given by

$$u^{\text{mag}} = \frac{U^{\text{mag}}}{Al}.$$

But since

$$U^{\text{mag}} = \tfrac{1}{2}L\,i^2,$$

we have

$$u^{\text{mag}} = \frac{Li^2}{2Al} = \left(\frac{L}{l}\right)\frac{i^2}{2A}.$$

Here L/l is the inductance of length l of the solenoid. Since the self-induced emf $\mathcal{E}_L = -L\,di/dt$ (Eq. 32-2) and $\mathcal{E}_L = -\mu_0 An^2 l\,di/dt$ (Eq. 32-1) for an air-filled solenoid, we can replace L/l in the expression above to find

$$u^{\text{mag}} = \tfrac{1}{2}\mu_0 n^2 i^2, \tag{33-6}$$

where n is the number of turns per unit length. By using Eq. 30-25 ($B = n\mu_0|i|$) we can write this *energy density* as

$$u^{\text{mag}} = \frac{1}{2}\frac{B^2}{\mu_0} \qquad \text{(magnetic energy density)}. \tag{33-7}$$

The **magnetic energy density,** u^{mag}, is the density of stored energy at any point where the magnetic field is $\vec{B}$. Even though we derived it by considering a special case, the solenoid, it turns out that Eq. 33-7 holds for all magnetic fields, no matter how they are generated. Equation 33-7 is comparable to Eq. 28-23; namely,

$$u^{\text{elec}} = \tfrac{1}{2}\varepsilon_0 E^2, \tag{33-8}$$

which gives the energy density (in a vacuum) at any point in an electric field. Note that both u^{mag} and u^{elec} are proportional to the square of the appropriate field, $\vec{B}$ or $\vec{E}$.

The table lists the number of turns per unit length, current, and cross-sectional area for three solenoids. Rank the solenoids according to the magnetic energy density within them, greatest first.

Solenoid	Turns per Unit Length	Current	Area
a	$2n_1$	i_1	$2A_1$
b	n_1	$2i_1$	A_1
c	n_1	i_1	$6A_1$

33-4 *LC* Oscillations, Qualitatively

We now turn our attention to how electromagnetic oscillations can occur in various types of circuits. We begin with the consideration of a dc circuit. Of the three circuit elements, resistance R, capacitance C, and inductance L, we have so far discussed dc circuits with the series combinations RC (Section 28-9) and RL (Section 32-4). In these two kinds of circuits we found that under certain circumstances, the charge, current, and potential differences across circuit elements can grow or decay exponentially. The exponential nature of the growth and decay curves is the result of energy losses in the resistor. The time constant associated with the exponential growth or decay is denoted by τ, which is either called capacitive or inductive depending on which circuit element is present.

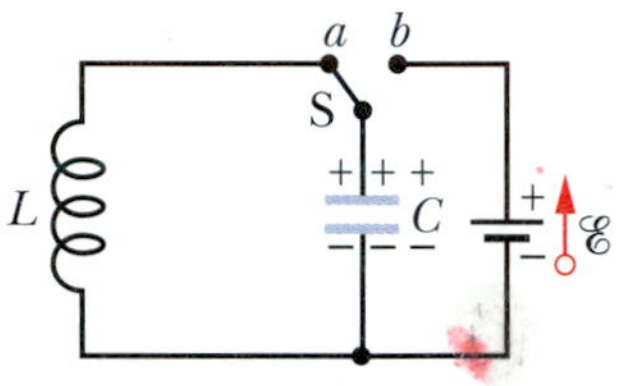

What if there is almost no resistance in a circuit to dissipate energy? We now examine qualitatively the two-element combination LC in a series circuit. Then in Section 33-6 we will derive equations that describe the behavior of the circuit.

We assume that our LC circuit shown in Fig. 33-4 has a negligible resistance. You will see that in this case, the potential difference across the circuit elements is alternately associated with the inductor and capacitor. Why? The sequence of events is as follows:

FIGURE 33-4 ■ A series *LC* circuit. where a switch is thrown from *b* to *a* so that an ideal inductor and a charged capacitor are in series.

- This initial state of the circuit at $t = 0$ as shown in Fig. 33-5a. The bar graphs for energy included there indicate that at this instant, with zero current through the inductor and maximum charge on the capacitor, the energy U^{mag} of the magnetic field is zero and the energy U^{elec} of the electric field is a maximum.

- As soon as the switch is thrown from a to a, the potential difference across the capacitor will start a flow of charge from one capacitor plate through the inductor to the other capacitor plate. The back emf generated by the inductor will slow the rate of capacitor discharge. The energy stored in the capacitor's electric field will decrease while the current through the inductor begins to increase its magnetic field energy as shown in Fig. 33-5b. Eventually as the capacitor is fully discharged, all the energy in its electric field will be transformed into energy stored in the magnetic field. At this point there will be no potential difference across the capacitor as shown in Fig. 33-5c.

- Without the potential difference, the current flowing through the inductor will start to decrease. However, in response to this changing current, a self-induced current will be set up through the inductor: this current will be in the same direction as the original current. Thus excess positive charge will begin to build up on the lower capacitor plate while the upper plate will begin to accumulate negative charge as shown in Fig. 33-5d. This will continue until no current flows through the inductor and the charges on the capacitor plates will be opposite in sign to

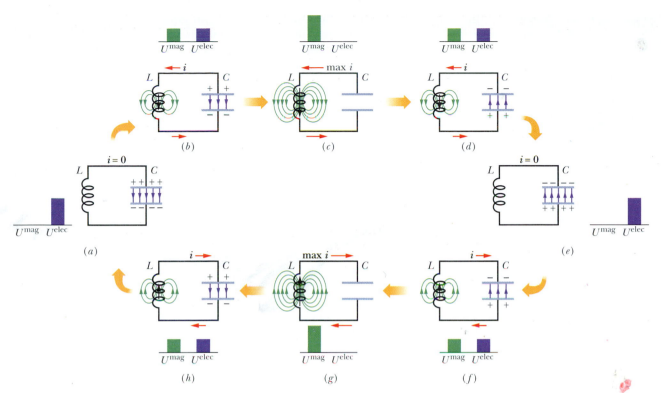

FIGURE 33-5 ■ Stages in a cycle of oscillation of an ideal *LC* circuit. The small bar graphs show levels of stored magnetic and electric energies. The inductor magnetic field lines and the capacitor electric field lines are shown. (*a*) Capacitor with maximum charge, no current. (*b*) Capacitor discharging, current increasing. (*c*) Capacitor fully discharged, current maximum. (*d*) Capacitor charging with opposite polarity to that in (*a*), current decreasing. (*e*) Capacitor with maximum charge with opposite polarity to that in (*a*), no current. (*f*) Capacitor discharging, current increasing with direction opposite that in (*b*). (*g*) Capacitor fully discharged, current maximum. (*h*) Capacitor charging, current decreasing.

that just after the switch was thrown. At this time all the circuit energy is in the capacitor's electric field once again as shown in Fig. 33-5*e*.

- Events 1 and 2 happen again but with the current flowing in the opposite direction until the capacitor is back to its original state (at $t = 0$) as shown in Fig. 33-5*f*, *g*, *h*, and *a*. Without a resistor in the circuit to dissipate energy, the maximum current through the inductor and the maximum amount of charge on the capacitor plates do not decay with time. In theory, for a perfectly ideal inductor, these oscillations can continue forever.

In the next section we do a mathematical analysis that fortunately agrees with observations that the current in the inductor *i* and the charge on the upper capacitor plate *q* vary *sinusoidally* with time as shown in Fig. 33-6. The resulting oscillations of the capacitor's electric field and the inductor's magnetic field are said to be **electromagnetic oscillations.**

Parts *a* through *h* of Fig. 33-5 show succeeding stages of the oscillations in a simple *LC* circuit. We know that the energy stored in the electric field of the capacitor at any time is given by $U^{\text{elec}} = \frac{1}{2}(q^2/C)$ (Eq. 33-5) where *q* is the charge on the capacitor at that time. The energy stored in the magnetic field of the inductor at any time is given by $U^{\text{mag}} = \frac{1}{2}L\,i^2$ (Eq. 33-4) where *i* is the current through the inductor at that time.

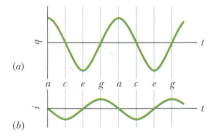

FIGURE 33-6 ■ (*a*) The charge *q* on the upper plate of the capacitor in an ideal *LC* circuit with almost no resistance (Fig. 33-5) as a function of time. (*b*) The current *i* in the circuit of Fig. 33-5 (*q* and *i* are determined from measurements of ΔV_C and ΔV_R across a very small resistor added to the circuit). The letters refer to the correspondingly labeled oscillation stages in Fig. 33-5.

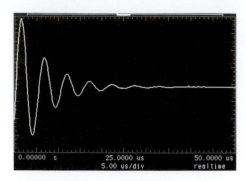

FIGURE 33-7 ■ An oscilloscope trace showing how the oscillations in an RLC circuit actually die away because energy is dissipated in the resistor as thermal energy.

In an actual LC circuit, the oscillations will not continue indefinitely because there is always some resistance present that will drain energy from the electric and magnetic fields and dissipate it as thermal energy (the circuit becomes warmer). The oscillations, once started, will die away as suggested in Fig. 33-7 (which displays a potential difference vs. time for a similar LC circuit with a resistor added to it). Compare this figure with Fig. 16-27, which shows the decay of mechanical oscillations caused by damping forces acting on a physical pendulum.

READING EXERCISE 33-2: A charged capacitor and an inductor are connected in series at time $t = 0$. In terms of the period T of the resulting sinusoidal oscillations shown in Fig. 33-6, determine how much later the following reach their maximums: (a) the charge on the capacitor; (b) the voltage across the capacitor, with its original polarity; (c) the energy stored in the electric field; and (d) the current. ■

TOUCHSTONE EXAMPLE 33-1: LC Oscillation

A 1.5 μF capacitor is charged to 57 V. The charging battery is then disconnected, and a 12 mH coil is connected in series with the capacitor so that LC oscillations occur. What is the maximum current in the coil? Assume that the circuit contains no resistance.

SOLUTION ■ The **Key Ideas** here are these:

1. Because the circuit contains no resistance, the electromagnetic energy of the circuit is conserved as the energy is transferred back and forth between the electric field of the capacitor and the magnetic field of the coil (inductor).

2. At any time t, the energy $U^{\text{mag}}(t)$ of the magnetic field is related to the current $i(t)$ through the coil by Eq. 33-4 ($U^{\text{mag}} = Li^2/2$). When all the energy is stored as magnetic energy, the current is at its maximum value I and that energy is $U^{\text{max}}_{\text{mag}} = LI^2/2$.

3. At any time t, the energy $U^{\text{elec}}(t)$ of the electric field is related to the charge $q(t)$ on the capacitor by Eq. 33-5 ($U^{\text{elec}} = q^2/2C$). When all the energy is stored as electric energy, the charge is at its maximum value Q and that energy is $U^{\text{max}}_{\text{elec}} = Q^2/2C$.

With these ideas, we can now write the conservation of energy as

$$U^{\text{max}}_{\text{mag}} = U^{\text{max}}_{\text{elec}}$$

or

$$\frac{LI^2}{2} = \frac{Q^2}{2C}.$$

Solving for I gives us

$$I = \sqrt{\frac{Q^2}{LC}}.$$

We know L and C, but not Q. However, with Eq. 28-1 ($|q| = C|\Delta V|$) we can relate Q to the maximum potential difference ΔV across the capacitor, which is the initial potential difference of 57 V. Thus, substituting $|Q| = C|\Delta V|$ leads to a maximum current magnitude of

$$I = \Delta V \sqrt{\frac{C}{L}} = (57 \text{ V}) \sqrt{\frac{1.5 \times 10^{-6} \text{ F}}{12 \times 10^{-3} \text{ H}}}$$

$$= 0.637 \text{ A} \approx 640 \text{ mA}. \qquad \text{(Answer)}$$

33-5 The Electrical–Mechanical Analogy

Let us look a little at an analogy between the ideal oscillating LC system like that shown in Fig. 33-5 and an oscillating block–spring system that experiences no friction forces. Two kinds of energy are involved in the block-spring system. One is potential energy of the compressed or extended spring; the other is kinetic energy of the moving block. These two energies are given by the familiar equations in the left energy column in Table 33-1.

The table also shows, in the right energy column, the two kinds of energy involved in LC oscillations. By looking across the table, we can see an analogy between the forms of the two pairs of energies—the mechanical energies of the block–spring system and the electromagnetic energies of the LC oscillator. The equations for velocity and current at the bottom of the table help us see the details of the analogy. They tell us that

TABLE 33-1
Comparison of the Energy in Two Oscillating Systems

Block–Spring System		*LC* Oscillator	
Element	**Energy**	**Element**	**Energy**
Spring	Potential, $\frac{1}{2}kx^2$	Capacitor	Electric, $\frac{1}{2}(1/C)q^2$
Block	Kinetic, $\frac{1}{2}mv^2$	Inductor	Magnetic, $\frac{1}{2}Li^2$
	Block velocity, $v = dx/dt$		Circuit current, $i = dq/dt$

the charge q corresponds to the displacement x and the current i corresponds to the block velocity v (in both equations, the former is differentiated to obtain the latter).

These correspondences suggest that in the energy expressions for an *LC* oscillator, the inverse of the capacitance is mathematically like the spring constant in a block–spring system. It is easy to place charge on the capacitor when $1/C$ is small, just as it's easy to displace a spring when k is small. The inductor is like the block mass. The inductor resists a change of the current in the circuit, and the mass on a spring resists a change in velocity.

In summary,

q corresponds to x, $1/C$ corresponds to k, i corresponds to v, L corresponds to m.

In Section 16-3 we saw that the angular frequency of oscillation of a (frictionless) block–spring system is

$$\omega = \sqrt{\frac{k}{m}} \qquad \text{(block–spring system).} \qquad (33\text{-}9)$$

The correspondences listed above suggest that to find the angular frequency of oscillation for a (resistanceless) *LC* circuit, k should be replaced by $1/C$ and m by L, yielding

$$\omega = \frac{1}{\sqrt{LC}} \qquad (LC \text{ circuit).} \qquad (33\text{-}10)$$

Experimentally, we find this expression is correct. We derive it more formally in the next section.

READING EXERCISE 33-3: What is the standard unit for angular frequency in Eq. 33-10 above? Show that this expression does in fact yield the correct unit. (*Hint:* The unit of ampere can be written as a coulomb per second. How would one write the units of henry and farad in the fundamental units of kilograms, meters, seconds, and coulombs?) ■

33-6 *LC* Oscillations, Quantitatively

Here we want to show explicitly that Eq. 33-10 for the angular frequency of *LC* oscillations is theoretically valid and that the oscillations should be sinusoidal. At the same time, we want to examine even more closely the analogy between *LC* oscillations and block–spring oscillations. We start by extending somewhat our earlier treatment of the mechanical block–spring oscillator.

The Block–Spring Oscillator—A Review

We analyzed block-spring oscillations in Chapter 16 in terms of energy transfers and did not—at that early stage—derive the fundamental differential equation that governs those oscillations. We do so now.

We can write, for the total energy U of a block–spring oscillator with a massless spring at any instant,

$$U = U^{\mathrm{blk}} + U^{\mathrm{spr}} = \tfrac{1}{2}mv^2 + \tfrac{1}{2}kx^2, \tag{33-11}$$

where U^{blk} and U^{spr} are, respectively, the kinetic energy of the moving block and the potential energy of the stretched or compressed spring. If there is no friction—which we assume—the total energy U remains constant with time, even though the values of velocity v and displacement x vary. In more formal language, $dU/dt = 0$. This leads to

$$\frac{dU}{dt} = \frac{d}{dt}(\tfrac{1}{2}mv^2 + \tfrac{1}{2}kx^2) = mv\frac{dv}{dt} + kx\frac{dx}{dt} = 0. \tag{33-12}$$

However, by definition, $v \equiv dx/dt$ and $dv/dt \equiv d^2x/dt^2$. With these substitutions, Eq. 33-12 becomes

$$m\frac{d^2x}{dt^2} + kx = 0 \qquad \text{(block–spring oscillations)}. \tag{33-13}$$

Equation 33-13 is the fundamental differential equation that governs the frictionless block–spring oscillations. It involves the displacement x and its second derivative with respect to time.

The general solution to Eq. 33-13—that is, the function $x(t)$ that describes the block–spring oscillations—is (as we saw in Eq. 16-5)

$$x(t) = X\cos(\omega t + \phi) \qquad \text{(displacement)}, \tag{33-14}$$

where X is the amplitude (or maximum displacement) of the mechanical oscillations undergoing simple harmonic motion, ω is the angular frequency of the oscillations, and ϕ is a phase constant.

The *LC* Oscillator

Now let us analyze the oscillations of an ideal LC circuit with no resistance. We proceed exactly as we just did for the block–spring oscillator. The total energy U present at any instant in an oscillating LC circuit is given by

$$U = U^{\mathrm{mag}} + U^{\mathrm{elec}} = \tfrac{1}{2}L\,i^2 + \frac{1}{2}\left(\frac{1}{C}\right)q^2 \tag{33-15}$$

where U^{mag} is the energy stored in the magnetic field of the inductor and U^{elec} is the energy stored in the electric field of the capacitor. Since we have assumed the circuit resistance to be zero, no energy is transferred to thermal energy and U remains constant with time. In more formal language, dU/dt must be zero. This leads to

$$\frac{dU}{dt} = \frac{d}{dt}\left(\frac{Li^2}{2} + \frac{q^2}{2C}\right) = Li\frac{di}{dt} + \frac{q}{C}\frac{dq}{dt} = 0. \tag{33-16}$$

However, $i = dq/dt$ and $di/dt = d^2q/dt^2$. With these substitutions, Eq. 33-16 becomes

$$L\frac{d^2q}{dt^2} + \frac{1}{C}q = 0 \qquad \text{(\textit{LC} oscillations)}. \tag{33-17}$$

This is the differential equation that describes the oscillations of a resistanceless *LC* circuit. Careful comparison shows that Eqs. 33-17 and 33-13 have exactly the same mathematical form, differing only in the symbols used.

Charge and Current Oscillations

Since the differential equations are mathematically identical, their solutions must also be mathematically identical. Because q corresponds to x, we can write the general solution of Eq. 33-17, giving $q(t)$ as a function of time, by analogy to Eq. 33-14 as

$$q(t) = Q \cos(\omega t + \phi) \quad \text{(charge)}, \qquad (33\text{-}18)$$

where Q is the amplitude or maximum amount of charge on the capacitor during the charge variations, while ω represents the angular frequency of the electromagnetic oscillations, and ϕ is the phase constant.

Taking the first derivative of Eq. 33-18 with respect to time gives us the time-varying current $i(t)$ of the *LC* oscillator:

$$i(t) = \frac{dq}{dt} = -\omega Q \sin(\omega t + \phi) \quad \text{(current)}. \qquad (33\text{-}19)$$

The amplitude I of this sinusoidally varying current is

$$I = \omega Q, \qquad (33\text{-}20)$$

so we can rewrite Eq. 33-19 as

$$i(t) = -I \sin(\omega t + \phi). \qquad (33\text{-}21)$$

Angular Frequencies

We can test whether Eq. 33-18 is a solution of Eq. 33-17 by substituting it and its second derivative with respect to time into Eq. 33-17. The first derivative of Eq. 33-18 is Eq. 33-19. The second derivative is then

$$\frac{d^2q}{dt^2} = -\omega^2 Q \cos(\omega t + \phi).$$

Substituting for q and d^2q/dt^2 in Eq. 33-17, we obtain

$$-L\omega^2 Q \cos(\omega t + \phi) + \left(\frac{1}{C}\right) Q \cos(\omega t + \phi) = 0.$$

Canceling $Q \cos(\omega t + \phi)$ and rearranging lead to

$$\omega = \frac{1}{\sqrt{LC}}.$$

Thus, Eq. 33-18 is indeed a solution of Eq. 33-17 if ω has the constant value $1/\sqrt{LC}$. Note that this expression for ω is exactly that given by Eq. 33-10, which we arrived at by examining correspondences.

The phase constant ϕ in Eq. 33-18 is determined by the conditions that prevail at any certain time—say, $t = 0$. If the conditions yield $\phi = 0$ at $t = 0$, Eq. 33-18 requires that $q = Q$ and Eq. 33-19 requires that $i = 0$; these are the initial conditions represented in Fig. 33-5.

Electric and Magnetic Energy Oscillations

The electric energy stored in the LC circuit at any time t is, from Eqs. 33-5 and 33-18,

$$U^{\text{elec}} = \frac{q^2}{2C} = \frac{Q^2}{2C} \cos^2(\omega t + \phi). \tag{33-22}$$

The magnetic energy is, from Eqs. 33-4 and 33-19,

$$U^{\text{mag}} = \tfrac{1}{2}Li^2 = \tfrac{1}{2}L\omega^2 Q^2 \sin^2(\omega t + \phi).$$

Substituting for ω from Eq. 33-10 then gives us

$$U^{\text{mag}} = \frac{Q^2}{2C} \sin^2(\omega t + \phi). \tag{33-23}$$

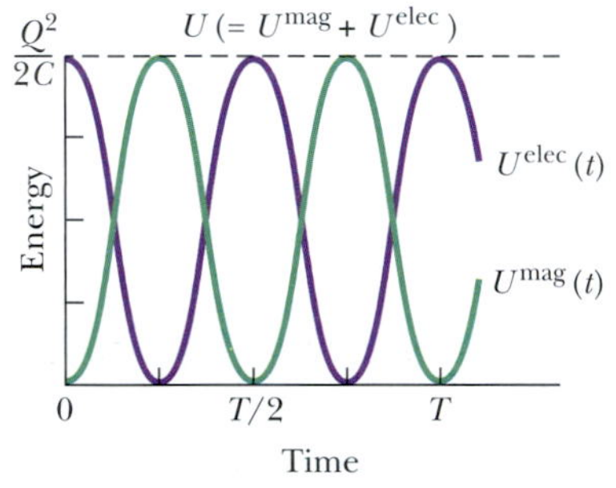

Figure 33-8 shows plots of $U^{\text{elec}}(t)$ and $U^{\text{mag}}(t)$ for the case of $\phi = 0$. Note that

1. The maximum values of U^{elec} and U^{mag} are both $Q^2/2C$.
2. At any instant the sum of U^{elec} and U^{mag} is equal to $Q^2/2C$, a constant.
3. When U^{elec} is maximum, U^{mag} is zero, and conversely.

FIGURE 33-8 ■ The stored magnetic energy and electric energy in the circuit of Fig. 33-5 as a function of time. Note that their sum remains constant. T is the period of oscillation.

READING EXERCISE 33-4: A capacitor in an LC oscillator has a maximum potential difference of 20 V and a maximum energy of 160 μJ. When the capacitor has a potential difference of 5 V and an energy of 10 μJ, what are (a) the emf across the inductor and (b) the energy stored in the magnetic field? ■

TOUCHSTONE EXAMPLE 33-2: *LC* Oscillation Continued

For the situation described in Touchstone Example 33-1, let the coil (inductor) be connected to the charged capacitor at time $t = 0$. The result is an LC circuit like that in Fig. 33-4.

(a) What is the potential difference $\Delta v_L(t)$ across the inductor as a function of time?

SOLUTION ■ One **Key Idea** here is that the current and potential differences of the circuit undergo sinusoidal oscillations. Another **Key Idea** is that we can still apply the loop rule to this oscillating circuit — just as we did for the nonoscillating circuits of Chapter 27. At any time t during the oscillations, the loop rule and Fig. 33-4 give us

$$\Delta v_L(t) = \Delta v_C(t); \tag{33-24}$$

that is, the potential difference Δv_L across the inductor must always be equal to the potential difference Δv_C across the capacitor, so that the net potential difference around the circuit is zero. Thus, we will

find $\Delta v_L(t)$ if we can find $\Delta v_C(t)$, and we can find $\Delta v_C(t)$ from $q(t)$ with Eq. 28-1 $|q| = C|\Delta V|$.

Because the potential difference $\Delta v_C(t)$ is maximum when the oscillations begin at time $t = 0$, the charge q on the capacitor must also be maximum then. Thus, phase constant ϕ must be zero, so that Eq. 33-18 gives us

$$q = Q \cos \omega t. \tag{33-25}$$

(Note that this cosine function does indeed yield maximum q (= Q) when $t = 0$.) To get the potential difference $\Delta v_C(t)$, we divide both sides of Eq. 33-25 by C to write

$$\frac{q}{C} = \frac{Q}{C} \cos \omega t,$$

and then use Eq. 28-1 to write

$$\Delta v_C = \Delta V_C \cos \omega t. \tag{33-26}$$

Here, ΔV_C is the amplitude of the oscillations in the potential difference Δv_C across the capacitor.

Next, substituting $\Delta v_C = \Delta v_L$ from Eq. 33-24, we find

$$\Delta v_L = \Delta V_C \cos \omega t. \qquad (33\text{-}27)$$

We can evaluate the right side of this equation by first noting that the amplitude ΔV_C is equal to the initial (maximum) potential difference of 57 V across the capacitor. Then, using the values of L and C from Touchstone Example 33-1, we find ω with Eq. 33-10:

$$\omega = \frac{1}{\sqrt{LC}} = \frac{1}{[(0.012\ \text{H})(1.5 \times 10^{-6}\ \text{F})]^{0.5}}$$

$$= 7454\ \text{rad/s} \approx 7500\ \text{rad/s}.$$

Thus, Eq. 33-27 becomes

$$\Delta v_L = (57\ \text{V}) \cos(7500\ \text{rad/s})t. \qquad \text{(Answer)}$$

(b) What is the maximum rate $(di/dt)^{\text{max}}$ at which the current i changes in the circuit?

SOLUTION ■ The **Key Idea** here is that, with the charge on the capacitor oscillating as in Eq. 33-18, the current is in the form of Eq. 33-19. Because $\phi = 0$, that equation gives us

$$i = -\omega Q \sin \omega t.$$

Then

$$\frac{di}{dt} = \frac{d}{dt}(-\omega Q \sin \omega t) = -\omega^2 Q \cos \omega t.$$

We can simplify this equation by substituting $C\Delta V_C$ for Q (because we know C and ΔV_C but not Q) and $1/\sqrt{LC}$ for ω according to Eq. 33-10. We get

$$\frac{di}{dt} = -\frac{1}{LC}C\Delta V_C \cos \omega t = -\frac{\Delta V_C}{L}\cos \omega t.$$

This tells us that the current changes at a varying (sinusoidal) rate, with its maximum rate of change being

$$\frac{\Delta V_C}{L} = \frac{57\ \text{V}}{0.012\ \text{H}} = 4750\ \text{A/s} \approx 4800\ \text{A/s}. \qquad \text{(Answer)}$$

33-7 Damped Oscillations in an *RLC* Circuit

A circuit containing resistance, inductance, and capacitance is called an *RLC circuit*. We shall here discuss only *series RLC circuits* like that shown in Fig. 33-9. With a resistance R present, the total *electromagnetic energy* U of the circuit (the sum of the electric energy and magnetic energy) is no longer constant; instead, it decreases with time as energy is transferred to thermal energy in the resistance. Because of this loss of energy, the oscillations of charge, current, and potential difference continuously decrease in amplitude, and the oscillations are referred to as *damped*. As you will see, they are damped in exactly the same way as those of the damped block–spring oscillator of Section 16-8.

To analyze the oscillations of our *RLC* circuit, we write an equation for the total electromagnetic energy U in the circuit at any instant. Because the resistance does not store electromagnetic energy, we can use Eq. 33-15:

$$U = U^{\text{mag}} + U^{\text{elec}} = \frac{Li^2}{2} + \frac{q^2}{2C}. \qquad (33\text{-}28)$$

Now, however, this total energy decreases as energy is transferred to thermal energy. The rate of that transfer is, from Eq. 26-11,

$$\frac{dU}{dt} = -i^2 R, \qquad (33\text{-}29)$$

where the minus sign indicates that U decreases. By differentiating Eq. 33-28 with respect to time and then substituting the result in Eq. 33-29, we obtain

$$\frac{dU}{dt} = Li\frac{di}{dt} + \frac{q}{C}\frac{dq}{dt} = -i^2 R.$$

Substituting dq/dt for i and d^2q/dt^2 for di/dt, we obtain

$$L\frac{d^2q}{dt^2} + R\frac{dq}{dt} + \frac{1}{C}q = 0 \qquad (RLC\ \text{circuit}), \qquad (33\text{-}30)$$

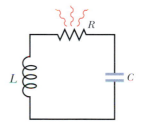

FIGURE 33-9 ■ A series *RLC* circuit. As the charge contained in the circuit oscillates back and forth through the resistance, electromagnetic energy is dissipated as thermal energy, damping (decreasing the amplitude of) the oscillations.

which is the differential equation that describes damped oscillations in an *RLC* circuit. The solution to Eq. 33-30 is

$$q = Qe^{-Rt/2L}\cos(\omega't + \phi),\tag{33-31}$$

where

$$\omega' = \sqrt{\omega^2 - (R/2L)^2},\tag{33-32}$$

with $\omega = 1/\sqrt{LC}$, as with an undamped oscillator. Equation 33-31 tells us how the charge on the capacitor oscillates in a damped *RLC* circuit. That equation is the electromagnetic counterpart of Eq. 16-37, which gives the displacement of a damped block–spring oscillator.

Equation 33-31 describes a sinusoidal oscillation (the cosine function) with an *exponentially decaying amplitude* $Qe^{-Rt/2L}$ (the factor that multiplies the cosine) as shown in Fig. 33-7. The angular frequency ω' of the damped oscillations is always less than the angular frequency ω of the undamped oscillations; however, we shall here consider only situations for which R is small enough for us to replace ω' with ω.

Let us next find an expression for the energy of the electric field in the capacitor, which is given by Eq. 33-5 ($U^{elec} = q^2/2C$). By substituting Eq. 33-31 into Eq. 33-5, we obtain

$$U^{elec} = \frac{q^2}{2C} = \frac{[Qe^{-Rt/2L}\cos(\omega't + \phi)]^2}{2C} = \frac{Q^2}{2C}e^{-Rt/L}\cos^2(\omega't + \phi).\tag{33-33}$$

Thus, the energy of the electric field oscillates according to a cosine-squared term and the amplitude of that oscillation decreases exponentially with time.

If we do a similar derivation for the energy of the magnetic field in the inductor we find that it too oscillates in such a way that its amplitude decreases exponentially in time. Since energy is being traded back and forth between the inductor and the capacitor, the total electromagnetic energy (which is the sum of the electric and magnetic energies) does not oscillate. Instead it just decays exponentially as the total energy is transformed to thermal energy by the total resistance in the circuit.

TOUCHSTONE EXAMPLE 33-3: Decaying Oscillation

A series *RLC* circuit has inductance $L = 12$ mH, capacitance $C = 1.6\ \mu$F, and resistance $R = 1.5$ V.

(a) At what time t will the amplitude of the charge oscillations in the circuit be 50% of its initial value?

SOLUTION ■ The **Key Idea** here is that the amplitude of the charge oscillations decreases exponentially with time t: According to Eq. 33-31, the charge amplitude at any time t is $Qe^{-Rt/2L}$, in which Q is the amplitude at time $t = 0$. We want the time when the charge amplitude has decreased to $0.50Q$—that is, when

$$Qe^{-Rt/2L} = 0.50Q.$$

Canceling Q and taking the natural logarithms of both sides, we have

$$-\frac{Rt}{2L} = \ln 0.50.$$

Solving for t and then substituting given data yield

$$t = -\frac{2L}{R}\ln 0.50 = -\frac{(2)(12 \times 10^{-3}\ \text{H})(\ln 0.50)}{1.5\ \Omega}$$

$$= 0.0111\ \text{s} \approx 11\ \text{ms}.\qquad\text{(Answer)}$$

(b) How many oscillations are completed within this time?

SOLUTION ■ The **Key Idea** here is that the time for one complete oscillation is the period $T' = 2\pi/\omega'$, where the angular frequency for decaying LC oscillations is given by Eq. 33-32 ($\omega' = \sqrt{\omega^2 - (R/2L)^2}$) where

$$\omega^2 = 1/(LC) = 1/[(0.012\ \text{H})(1.6 \times 10^{-6}\ \text{F})]$$

$$= 52.1 \times 10^6\ (\text{rad/s})^2,$$

while

$$(R/2L)^2 = [(1.5 \ \Omega)/(2)(0.012 \ \text{H})]^2$$
$$= 3.91 \times 10^3 (\text{rad/s})^2.$$

Since $(R/2L)^2 \ll \omega^2$, we can neglect $(R/2L)^2$ compared with ω^2, so here $\omega' \cong \omega = \sqrt{52.1 \times 10^6 \ (\text{rad/s}^2)} = 7.22 \times 10^3 \ \text{rad/s}$.

The time for one period of the decaying oscillation is then

$$T' = \frac{2\pi}{\omega'} \cong \frac{2\pi}{\omega} = 2\pi/(7.22 \times 10^3 \ \text{rad/s})$$

$$= 0.871 \times 10^{-3} \ \text{s}.$$

Thus, in the time interval $\Delta t = 0.0111$ s, the number of complete oscillations is

$$\frac{\Delta t}{T'} = \frac{0.0111 \ \text{s}}{0.871 \times 10^{-3} \ \text{s}} \cong 13. \qquad \text{(Answer)}$$

Thus, the amplitude decays by 50% in about 13 complete oscillations. This damping is less severe than that shown in Fig. 33-7, where the amplitude decays by a little more than 50% in one oscillation.

33-8 More About Alternating Current

The oscillations in an *RLC* circuit will not damp out if an external emf device supplies enough energy to make up for the energy dissipated as thermal energy in the resistance *R*. As we discussed at the beginning of the chapter, the United States and most other countries deliver alternating current or ac electricity. These oscillating emfs and currents vary sinusoidally with time, reversing direction (in North America) 120 times per second and thus having frequency $f = 60$ Hz.

At first sight this may seem to be a strange arrangement. We have seen that the drift speed of the conduction electrons in household wiring may typically be 4×10^{-5} m/s. If we now reverse their direction every 1/120th of a second, such electrons can move only about 3×10^{-7} m in a half-cycle. At this rate, a typical electron can drift past no more than about 10 atoms in the wiring before it is required to reverse its direction. How, you may wonder, can the electron ever get anywhere?

Although this question may be worrisome, it is a needless concern. The conduction electrons do not have to "get anywhere." This is similar to the idea that the molecules in a spring do not have to move far longitudinally to transmit energy a long distance. Here, the electrons don't have to move far to have a long-range effect. When we say that the current in a wire is one ampere, we mean that charge passes through any plane cutting across that wire at the rate of one coulomb per second. The speed at which the charge carriers cross that plane does not matter directly; one ampere may correspond to many charge carriers moving very slowly or to a few moving very rapidly.

Furthermore, the signal to the electrons to reverse directions—which originates in the alternating emf provided by the power company's generator—is propagated along the conductor at a speed close to that of light. All electrons, no matter where they are located, get their reversal instructions at about the same instant. Finally, we note that for many devices, such as lightbulbs and toasters, the direction of motion is unimportant as long as the electrons do move so as to transfer energy to the device via collisions with atoms in the device.

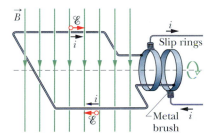

FIGURE 33-10 ■ The basic mechanism of an alternating-current generator is a conducting loop rotated in an external magnetic field. In practice, the alternating emf induced in a coil of many turns of wire is made accessible by means of slip rings attached to the rotating loop. Each ring is connected to one end of the loop wire and is electrically connected to the rest of the generator circuit by a conducting brush against which it slips as the loop (and it) rotates.

Generator Equations

Figure 33-10 shows a simplified model of an ac generator like that shown in Fig. 32-12. As the conducting loop is forced to rotate through the external magnetic field $\vec{B}$, a sinusoidally oscillating emf $\mathcal{E}$ is induced in the loop:

$$\mathcal{E} = \mathcal{E}^{\text{max}} \sin \omega^{\text{dr}} t. \qquad (33\text{-}34)$$

The *angular frequency* ω^{dr} of the emf is equal to the angular speed with which the loop rotates in the magnetic field, the *phase* of the emf is $\omega^{\text{dr}} t$, and the *amplitude* of the emf is $\mathscr{E}^{\,max}$ (where the superscript stands for maximum). When the rotating loop is part of a closed conducting path, this emf produces (*drives*) a sinusoidal (alternating) current along the path with the same angular frequency ω^{dr}, which then is called the **driving angular frequency.** Following Eq. 33-21, we can write the current as

$$i = -I\sin(\omega^{dr}t + \phi) = I\sin(\omega^{dr}t - \phi'), \qquad (33\text{-}35)$$

where I is the amplitude or maximum value of the driven current. (The phase $\omega^{dr}t - \phi'$ of the current is traditionally written with a minus sign instead of as $\omega^{dr}t + \phi$.) We include the phase constant ϕ' in Eq. 33-35 to emphasize that the current i may not be in phase with the emf $\mathscr{E}$. (As you will see, the phase constant depends on the circuit to which the generator is connected.) We can also write the current i in terms of the **driving frequency** f^{dr} of the emf, by substituting $2\pi f^{dr}$ for ω^{dr} in Eq. 33-35.

33-9 Forced Oscillations

We have seen that once started, the charge, potential difference, and current in both undamped LC circuits and damped RLC circuits (with small enough R) oscillate at angular frequency $\omega = 1/\sqrt{LC}$. Such oscillations are said to be *free oscillations* (free of any external emf), and the angular frequency ω is said to be the circuit's **natural angular frequency.**

When the external alternating emf of Eq. 33-34 is connected to an RLC circuit, the oscillations of charge, potential difference, and current are said to be *driven oscillations* or *forced oscillations*. These oscillations always occur at the driving angular frequency ω^{dr}:

> No matter what the natural angular frequency ω of a circuit is, forced oscillations of charge, current, and potential difference in the circuit always occur at the driving angular frequency ω^{dr}.

However, as you will see in Section 33-11, the amplitudes of the oscillations very much depend on how close ω^{dr} is to ω. When the two angular frequencies match—a condition known as resonance—the amplitude I of the current in the circuit is maximum.

33-10 Representing Oscillations with Phasors: Three Simple Circuits

Later in this chapter, we shall connect an external alternating emf device to a series RLC circuit as in Fig. 33-11. We shall then find expressions for the amplitude I and phase constant ϕ of the sinusoidally oscillating current in terms of the amplitude $\mathscr{E}^{\,max}$ and angular frequency ω^{dr} of the external emf. First, however, let us consider three simpler circuits, each having an external emf and only one other circuit element: R, C, or L. We start with a resistive element (a purely *resistive load*). We continue to use the convention here that uppercase letters such as Q, I, and V represent constants, while lowercase letters represent time-varying quantities such as q, i, and v.

A Resistive Load

Figure 33-12*a* shows a circuit containing a resistance element of value R and an ac generator with the alternating emf of Eq. 33-34. By the loop rule, we have

$$\mathscr{E} - \Delta v_R = 0.$$

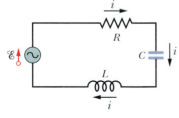

FIGURE 33-11 ■ A single-loop circuit containing a resistor, a capacitor, and an inductor. A generator, represented by a sine wave in a circle, produces an alternating emf that establishes an alternating current; the directions of the emf and current are indicated here at only one instant.

Note that in this context, Δv_R represents a potential difference across the resistance element and not a velocity change. With Eq. 33-34, this gives us

$$\Delta v_R = \mathscr{E}^{\max} \sin \omega^{dr} t.$$

Because the amplitude ΔV_R of the alternating potential difference (or voltage) across the resistance is equal to the amplitude $\mathscr{E}^{\max}$ of the alternating emf, we can write this as

$$\mathscr{E} = \Delta v_R = \Delta V_R \sin \omega^{dr} t. \tag{33-36}$$

From the definition of resistance ($R = \Delta v_R/i$), we can now write the current i_R in the resistor as

$$i_R = \frac{\Delta v_R}{R} = \frac{\Delta V_R}{R} \sin \omega^{dr} t. \tag{33-37}$$

From Eq. 33-35, we can also write this current as

$$i_R = I_R \sin(\omega^{dr} t - \phi), \tag{33-38}$$

where I_R is the amplitude of the current i_R passing through the resistance. Comparing Eqs. 33-37 and 33-38, we see that for a purely resistive load the phase constant $\phi = 0°$. We also see that the voltage amplitude and current amplitude are related by

$$\Delta V_R = I_R R \quad \text{(resistor).} \tag{33-39}$$

Although we found this relation for the circuit of Fig. 33-12a, it applies to any resistance in any ac circuit.

By comparing Eqs. 33-36 and 33-37, we see that the time-varying quantities Δv_R and i_R are both functions of $\sin \omega^{dr} t$ with $\phi = 0°$. Thus, these two quantities are *in phase*, which means that their corresponding maxima (and minima) occur at the same times. Figure 33-12b, which is a plot of $\Delta v_R(t)$ and $i_R(t)$, illustrates this fact. Note that Δv_R and i_R do not decay here, because the generator supplies energy to the circuit to make up for the energy dissipated in R.

Since we have chosen to drive our circuits with a voltage that varies as $\sin(\omega t)$, and constant speed motion around a circle has x- and y-components that vary sinusoidally, it is convenient to use a component of a rotating vector to represent our oscillations. Such a rotating vector representation is called a *phasor*. Recall from Section 17-12 that phasors are vectors that rotate around an origin. Those that represent the voltage across and current in the resistor of Fig. 33-12a are shown in Fig. 33-12c at an arbitrary time t. Such phasors have the following properties:

Angular speed: Both phasors rotate counterclockwise about the origin with an angular speed equal to the angular frequency ω^{dr} of both Δv_R and i_R.

Length: The length of each phasor represents the amplitude of the alternating quantity: ΔV_R for the voltage and i_R for the current.

Projection: The projection of each phasor on the *vertical* axis represents the value of the alternating quantity at time t: Δv_R for the voltage and I_R for the current.

Rotation angle: The rotation angle of each phasor is equal to the phase of the alternating quantity at time t. In Fig. 33-12c, the voltage and current are in phase, so their phasors always have the same phase $\omega^{dr} t$ and the same rotation angle, and thus they rotate together.

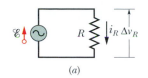
(a)

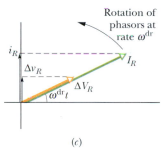
(b)

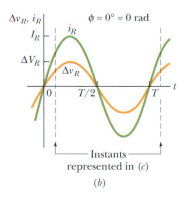
(c)

FIGURE 33-12 ■ (a) A resistor is connected across an alternating-current generator. (b) The current i_R and the potential difference Δv_R across the resistor are plotted on the same graph, both versus time t. They are in phase and complete one cycle in one period T. (c) A phasor diagram shows the same thing as (b).

Mentally follow the rotation. Can you see that when the phasors have rotated so that $\omega^{dr}t = 90°$ (so they point vertically upward), then $\Delta v_R = \Delta V_R$? Equations 33-36 and 33-38 give the same results. For this case, the voltage and the current oscillate together. Hence, it is not especially useful to introduce the phasor representation. However, the value of this representation becomes clearer with more complex situations such as the capacitor and inductor discussed below.

A Capacitive Load

Figure 33-13a shows a circuit containing a capacitance and a generator with the alternating emf of Eq. 33-34. Using the loop rule and proceeding as we did when we obtained Eq. 33-36, we find that the potential difference across the capacitor is

$$\Delta v_C = \Delta V_C \sin\omega^{dr}t, \tag{33-40}$$

where ΔV_C is the amplitude of the alternating voltage across the capacitor. From the definition of capacitance we can also write

$$q_C = C\Delta v_C = C\Delta V_C \sin\omega^{dr}t. \tag{33-41}$$

Our concern, however, is with the current rather than the charge. Thus, we differentiate Eq. 33-41 to find

$$i_C = \frac{dq_C}{dt} = \omega^{dr}C\Delta V_C \cos\omega^{dr}t. \tag{33-42}$$

We now modify Eq. 33-42 in two ways. First, for reasons of symmetry of notation, we introduce the quantity X_C, called the *capacitive reactance* of a capacitor, defined as

$$X_C = \frac{1}{\omega^{dr}C} \qquad \text{(capacitive reactance).} \tag{33-43}$$

Its value depends not only on the capacitance but also on the driving angular frequency ω^{dr}. We know from the definition of the capacitive time constant ($\tau = RC$) that the SI unit for C can be expressed as seconds per ohm. Applying this to Eq. 33-43 shows that the SI unit of X_C is the *ohm*, just as for resistance R.

Second, we replace $\cos\omega^{dr}t$ with a phase-shifted sine:

$$\cos\omega^{dr}t = \sin(\omega^{dr}t - (-90°)).$$

You can verify this identity by shifting a sine curve in the negative direction by 90°.

With these two modifications, Eq. 33-42 becomes

$$i_C = \left(\frac{\Delta V_C}{X_C}\right)\sin(\omega^{dr}t + 90°). \tag{33-44}$$

From Eq. 33-35, we can also write the current i_C in C as

$$i_C = I_C \sin(\omega^{dr}t - \phi), \tag{33-45}$$

where I_C is the amplitude of i_C. Comparing Eqs. 33-44 and 33-45, we see that for a purely capacitive load the phase constant ϕ for the current is $-90°$. We also see that the voltage amplitude and current amplitude are related by

$$\Delta V_C = I_C X_C \qquad \text{(capacitor).} \tag{33-46}$$

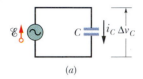

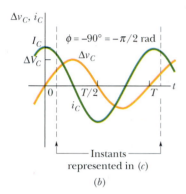

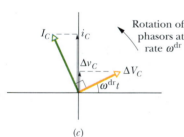

FIGURE 33-13 ■ (a) A capacitor is connected across an alternating-current generator. (b) The current in the capacitor leads the voltage by 90°(= $\pi/2$ rad). (c) A phasor diagram shows the same thing.

Although we found this relation for the circuit of Fig. 33-13*a*, it applies to any capacitance in any ac circuit. Note that using phasors allowed us to write the equation associated with a capacitor in such a way that it looks just like Ohm's law for a resistor. Although this expression is true for the amplitude of the phasor, the full voltage and current don't look like Ohm's law because of the shift in phase. We'll see in the next case that a similar thing happens for the inductor.

Comparison of Eqs. 33-40 and 33-44, or inspection of Fig. 33-13*b*, shows that the quantities Δv_C and i_C are 90°, or one-quarter cycle, out of phase. Furthermore, we see that i_C leads Δv_C, which means that, if you monitored the current i_C and the potential difference Δv_C in the circuit of Fig. 33-13*a*, you would find that i_C reaches its maximum before Δv_C does, by one-quarter cycle.

This relation between i_C and Δv_C is illustrated by the phasor diagram of Fig. 33-13*c*. As the phasors representing these two quantities rotate counterclockwise together, the phasor labeled I_C does indeed lead that labeled ΔV_C, and by an angle of 90°; that is, the phasor I_C coincides with the vertical axis one-quarter cycle before the phasor ΔV_C does. Be sure to convince yourself that the phasor diagram of Fig. 33-13*c* is consistent with Eqs. 33-40 and 33-44.

An Inductive Load

Now let's consider a third situation in which we connect an external alternating emf to a circuit containing just one of our basic circuit elements. Figure 33-14*a* shows a circuit containing an inductance and a generator with the alternating emf of Eq. 33-34. Using the loop rule and proceeding as we did to obtain Eq. 33-36, we find that the potential difference across the inductance is

$$\Delta v_L = \Delta V_L \sin \omega^{\mathrm{dr}} t, \tag{33-47}$$

where ΔV_L is the amplitude of Δv_L. From Eq. 32-2, we can write the potential difference across an inductance L, in which the current is changing at the rate di_L/dt, as

$$\Delta v_L = L \frac{di_L}{dt}. \tag{33-48}$$

If we combine Eqs. 33-47 and 33-48 we have

$$\frac{di_L}{dt} = \frac{\Delta V_L}{L} \sin \omega^{\mathrm{dr}} t. \tag{33-49}$$

Our concern, however, is with the current rather than with its time derivative. We find the former by integrating Eq. 33-49, obtaining

$$i_L = \int di_L = \frac{\Delta V_L}{L} \int \sin \omega^{\mathrm{dr}} t\, dt = -\left(\frac{\Delta V_L}{\omega^{\mathrm{dr}} L}\right) \cos \omega^{\mathrm{dr}} t. \tag{33-50}$$

We now modify this equation in two ways. First, for reasons of symmetry of notation, we introduce the quantity X_L, called the **inductive reactance** of an inductor, which is defined as

$$X_L = \omega^{\mathrm{dr}} L \quad \text{(inductive reactance).} \tag{33-51}$$

The value of X_L depends on the driving angular frequency ω^{dr}. The unit of the inductive time constant τ_L indicates that the SI unit of X_L is the *ohm*, just as it is for X_C and for R.

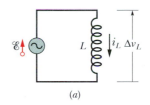

(a)

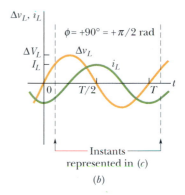

(b)

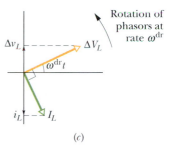

(c)

FIGURE 33-14 ▪ (*a*) An inductor is connected across an alternating-current generator. (*b*) The current in the inductor lags the voltage by 90°(= π/2 rad). (*c*) A phasor diagram shows the same thing.

Second, we replace the function $-\cos\omega^{dr}t$ in Eq. 33-50 with a phase-shifted sine—namely,

$$-\cos\omega^{dr}t = \sin(\omega^{dr}t - (+90°)).$$

You can verify this identity by shifting a sine curve in the positive direction by 90°.

With these two changes, Eq. 33-50 becomes

$$i_L = \left(\frac{\Delta V_L}{X_L}\right)\sin(\omega^{dr}t - 90°). \qquad (33\text{-}52)$$

From Eq. 33-35, we can also write this current in the inductance as

$$i_L = I_L\sin(\omega^{dr}t - \phi), \qquad (33\text{-}53)$$

where I_L is the amplitude of the current i_L. Comparing Eqs. 33-52 and 33-53, we see that for a purely inductive load the phase constant ϕ for the current is $+90°$. We also see that the voltage amplitude and current amplitude are related by

$$\Delta V_L = I_L X_L \qquad \text{(inductor).} \qquad (33\text{-}54)$$

Although we found this relation for the circuit of Fig. 33-14a, it applies to any inductance in any ac circuit.

Comparison of Eqs. 33-47 and 33-52, or inspection of Fig. 33-14b, shows that the quantities i_L and Δv_L are 90° out of phase. In this case, however, i_L *lags* Δv_L; that is, if you monitored the current i_L and the potential difference Δv_L in the circuit of Fig. 33-14a, you would find that i_L reaches its maximum value after Δv_L does, by one-quarter cycle.

The phasor diagram of Fig. 33-14c also contains this information. As the phasors rotate counterclockwise in the figure, the phasor labeled I_L does indeed lag that labeled ΔV_L, and by an angle of 90°. Be sure to convince yourself that Fig. 33-14c represents Eqs. 33-47 and 33-52.

READING EXERCISE 33-5: The figure shows, in (a), a sine curve $S(t) = \sin(\omega^{dr}t)$ and three other sinusoidal curves $A(t)$, $B(t)$, and $C(t)$, each of the form $\sin(\omega^{dr}t - \phi)$. (a) Rank the three other curves according to the value of ϕ, most positive first and most negative last. (b) Which curve corresponds to which phasor in (b) of the figure? (c) Which curve leads the others?

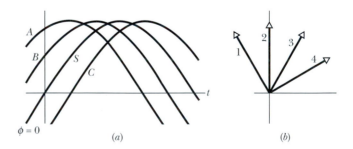

(a)   (b)

33-11 The Series *RLC* Circuit

We are now ready to apply the alternating emf of Eq. 33-34,

$$\mathcal{E} = \mathcal{E}^{max}\sin\omega^{dr}t \qquad \text{(applied emf),} \qquad (33\text{-}55)$$

TABLE 33-2
Phase and Amplitude Relations for Alternating Currents and Voltages

Circuit Element	Symbol	Resistance or Reactance	Phase of the Current	Phase Constant (or Angle ϕ)	Amplitude Relation
Resistor	R	R	In phase with v_R	$0° (= 0$ rad$)$	$\Delta V_R = I_R R$
Capacitor	C	$X_C = 1/\omega^{dr}C$	Leads Δv_C by $90° (= \pi/2$ rad$)$	$-90° (= -\pi/2$ rad$)$	$\Delta V_C = I_C X_C$
Inductor	L	$X_L = \omega^{dr} L$	Lags Δv_L by $90° (= \pi/2$ rad$)$	$+90° (= +\pi/2$ rad$)$	$\Delta V_L = I_L X_L$

to the full *RLC* circuit of Fig. 33-11. Because R, L, and C are in series, the same current

$$i = I \sin(\omega^{dr}t - \phi) \qquad (33\text{-}56)$$

is driven in all three of them. We wish to find the current amplitude I and the phase constant ϕ. The solution is simplified by the use of phasor diagrams.

Table 33-2 summarizes the relations between the current i and the voltage V for each of the three kinds of circuit elements we have considered. When an applied alternating voltage produces an alternating current in them, the current is in phase with the voltage across a resistor, leads the voltage across a capacitor, and lags the voltage across an inductor.

The Current Amplitude

We start with Fig. 33-15a, which shows the phasor representing the current of Eq. 33-56 at an arbitrary time t. The length of the phasor is the current amplitude I, the projection of the phasor on the vertical axis is the current i at time t, and the angle of rotation of the phasor is the phase $\omega^{dr}t - \phi$ of the current at time t.

Figure 33-15b shows the phasors representing the voltages across R, L, and C at the same time t. Each phasor is oriented relative to the angle of rotation of current phasor I in Fig. 33-15a, based on the information in Table 33-2:

Resistor: Here current and voltage are in phase, so the angle of rotation of voltage phasor ΔV_R is the same as that of phasor I.

Capacitor: Here current leads voltage by $90°$, so the angle of rotation of voltage phasor ΔV_C is $90°$ less than that of phasor I.

Inductor: Here current lags voltage by $90°$, so the angle of rotation of voltage phasor ΔV_L is $90°$ greater than that of phasor I.

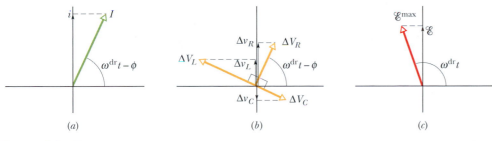
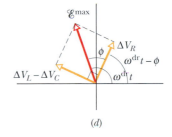

(a) (b) (c) (d)

FIGURE 33-15 ■ (a) A phasor representing the alternating current in the driven *RLC* circuit of Fig. 33-11 at time t. The amplitude I, the instantaneous value i, and the phase $\omega^{dr}t - \phi$ are shown. (b) Phasors representing the voltages across the inductor, resistor, and capacitor, oriented with respect to the current phasor in (a). (c) A phasor representing the alternating emf that drives the current of (a). (d) The emf phasor is equal to the vector sum of the three voltage phasors of (b). Here, voltage phasors ΔV_L and ΔV_C have been added to yield their net phasor ($\Delta V_L - \Delta V_C$).

Figure 33-15b also shows the instantaneous voltages Δv_R, Δv_C, and Δv_L across R, C, and L at time t; those voltages are the projections of the corresponding phasors on the vertical axis of the figure.

Figure 33-15c shows the phasor representing the applied emf of Eq. 33-55. The length of the phasor is the emf amplitude $\mathscr{E}^{max}$, the projection of the phasor on the vertical axis is the emf $\mathscr{E}$ at time t, and the angle of rotation of the phasor is the phase $\omega_d t$ of the emf at time t.

From the loop rule we know that at any instant the sum of the voltages Δv_R, Δv_C, and Δv_L is equal to the applied emf $\mathscr{E}$:

$$\mathscr{E} = \Delta v_R + \Delta v_C + \Delta v_L. \tag{33-57}$$

Thus, at time t the projection $\mathscr{E}$ in Fig. 33-15c is equal to the algebraic sum of the projections Δv_R, Δv_C, and Δv_L in Fig. 33-15b. In fact, as the phasors rotate together, this equality always holds. This means that phasor $\mathscr{E}^{max}$ in Fig. 33-15c must be equal to the vector sum of the three voltage phasors ΔV_R, ΔV_C, and ΔV_L in Fig. 33-15b.

That requirement is indicated in Fig. 33-15d, where phasor $\mathscr{E}^{max}$ is drawn as the sum of phasors ΔV_R, ΔV_L, and ΔV_C. Because phasors ΔV_L and ΔV_C have opposite directions in the figure, we simplify the vector sum by first combining ΔV_L and ΔV_C to form the single phasor $\Delta V_L - \Delta V_C$. Then we combine that single phasor with ΔV_R to find the net phasor. Again, the net phasor must coincide with phasor $\mathscr{E}^{max}$, as shown.

Both triangles in Fig. 33-15d are right triangles. Applying the Pythagorean theorem to either one yields

$$(\mathscr{E}^{max})^2 = \Delta V_R^2 + (\Delta V_L - \Delta V_C)^2. \tag{33-58}$$

From the amplitude information displayed in Table 33-2 we can rewrite this as

$$(\mathscr{E}^{max})^2 = (IR)^2 + (IX_L - IX_C)^2, \tag{33-59}$$

and then rearrange it to the form

$$I = \frac{\mathscr{E}^{max}}{\sqrt{R^2 + (X_L - X_C)^2}}. \tag{33-60}$$

The denominator in Eq. 33-60 is called the **impedance** Z of the circuit for the driving angular frequency ω^{dr}:

$$Z = \sqrt{R^2 + (X_L - X_C)^2} \qquad \text{(impedance defined)}. \tag{33-61}$$

We can then write Eq. 33-60 as

$$I = \frac{\mathscr{E}^{max}}{Z}. \tag{33-62}$$

If we substitute for X_C and X_L from Eqs. 33-43 and 33-51, we can write Eq. 33-60 more explicitly as

$$I = \frac{\mathscr{E}^{max}}{\sqrt{R^2 + (\omega^{dr}L - 1/\omega^{dr}C)^2}} \qquad \text{(current amplitude)} \tag{33-63}$$

We have now accomplished half our goal: We have obtained an expression for the current amplitude I in terms of the sinusoidal driving emf and the circuit elements in a series RLC circuit.

The value of I depends on the difference between $\omega^{dr}L$ and $1/\omega^{dr}C$ in Eq. 33-63 or, equivalently, the difference between X_L and X_C in Eq. 33-60. In either equation, it does not matter which of the two quantities is greater because the difference is always squared.

The current that we have been describing in this section is the *steady-state current* that occurs after the alternating emf has been applied for some time. When the emf is first applied to a circuit, a brief *transient current* occurs. Its duration (before settling down into the steady-state current) is determined by the time constants $\tau_L = L/R$ and $\tau_C = RC$ as the inductive and capacitive elements "turn on." This transient current can be large and can, for example, destroy a motor on startup if it is not properly taken into account in the motor's circuit design.

The Phase Constant

From the right-hand phasor triangle in Fig. 33-15d and from Table 33-2 we can write

$$\tan\phi = \frac{\Delta V_L - \Delta V_C}{\Delta V_R} = \frac{IX_L - IX_C}{IR}, \tag{33-64}$$

which gives us

$$\tan\phi = \frac{X_L - X_C}{R} \quad \text{(phase constant).} \tag{33-65}$$

This is the other half of our goal: an equation for the phase constant ϕ in a sinusoidally driven series *RLC* circuit. In essence, it gives us three different results for the phase constant, depending on the relative values of X_L and X_C:

$X_L > X_C$: The circuit is said to be *more inductive than capacitive*. Equation 33-65 tells us that ϕ is positive for such a circuit, which means that phasor I rotates behind phasor $\mathcal{E}^{max}$ (Fig. 33-16a). A plot of $\mathcal{E}$ and i versus time is like that in Fig. 33-16b. (The phasors in Figs. 33-16c and d were drawn assuming $X_L > X_C$.)

$X_C > X_L$: The circuit is said to be *more capacitive than inductive*. Equation 33-65 tells us that ϕ is negative for such a circuit, which means that phasor I rotates ahead of phasor $\mathcal{E}^{max}$ (Fig. 33-16c). A plot of $\mathcal{E}$ and i versus time is like that in Fig. 33-16d.

$X_C = X_L$: The circuit is said to be in *resonance*, a state that is discussed next. Equation 33-65 tells us that $\phi = 0°$ for such a circuit, which means that phasors $\mathcal{E}^{max}$ and I rotate together (Fig. 33-16e). A plot of $\mathcal{E}$ and i versus time is like that in Fig. 33-16f.

As an illustration, let us reconsider two extreme circuits. In the *purely inductive circuit* of Fig. 33-14a, where X_L is nonzero and $X_C = R = 0$, Eq. 33-65 tells us that $\phi = +90°$ (the greatest value of ϕ), consistent with Fig. 33-14c. In the *purely capacitive circuit* of Fig. 33-13a, where X_C is nonzero and $X_L = R = 0$, Eq. 33-65 tells us that $\phi = -90°$ (the least value of ϕ), consistent with Fig. 33-13c.

Resonance

Equation 33-63 gives the current amplitude I in an *RLC* circuit as a function of the driving angular frequency ω^{dr} of the external alternating emf. For a given resistance R, that amplitude is a maximum when the quantity $\omega^{dr}L - 1/\omega^{dr}C$ in the denominator is zero—that is, when

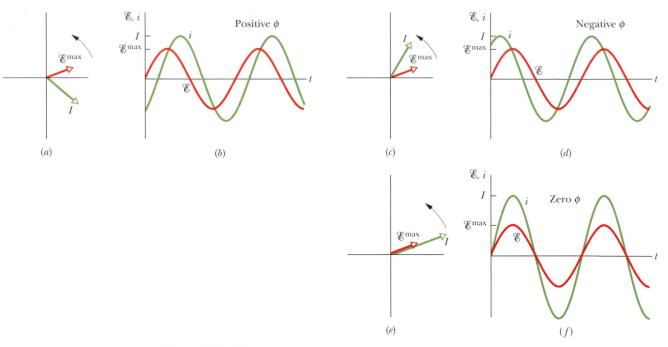

FIGURE 33-16 ■ Phasor diagrams and graphs of the alternating emf and current for the driven *RLC* circuit of Fig. 33-11. In the phasor diagram of (*a*) and the graph of (*b*), the current *i* lags the driving emf $\mathcal{E}$ and the current's phase constant ϕ is positive. In (*c*) and (*d*), the current *i* leads the driving emf $\mathcal{E}$ and its phase constant ϕ is negative. In (*e*) and (*f*), the current *i* is in phase with the driving emf $\mathcal{E}$ and its phase constant ϕ is zero.

$$\omega^{dr} L = \frac{1}{\omega^{dr} C}$$

or $$\omega^{dr} = \frac{1}{\sqrt{LC}} \qquad \text{(for maximum } I\text{).}$$ (33-66)

Because the natural angular frequency ω of the *RLC* circuit is also equal to $1/\sqrt{LC}$, the maximum value of *I* occurs when the driving angular frequency matches the natural angular frequency—that is, at resonance. Thus, in an *RLC* circuit, resonance and maximum current amplitude *I* occur when

$$\omega^{dr} = \omega = \frac{1}{\sqrt{LC}} \qquad \text{(resonance).}$$ (33-67)

Figure 33-17 shows three *resonance curves* for sinusoidally driven oscillations in three series *RLC* circuits differing only in *R*. Each curve peaks at its maximum current amplitude *I* when the ratio ω^{dr}/ω is 1.00, but the maximum value of *I* decreases with increasing *R*. (The maximum *I* is always $\mathcal{E}^{max}/R$; to see why, combine Eqs. 33-61 and 33-62.) In addition, the curves increase in width (measured in Fig. 33-17 at half the maximum value of *I*) with increasing *R*.

To make physical sense of Fig. 33-17, consider how the reactances X_L and X_C change as we increase the driving angular frequency ω^{dr}, starting with a value much less than the natural frequency ω. For small ω^{dr}, reactance $X_L (= \omega^{dr} L)$ is small and reactance $X_C (1/\omega^{dr} C)$ is large. Thus, the circuit is mainly capacitive and the impedance is dominated by the large X_C, which keeps the current low.

As we increase ω^{dr}, reactance X_C remains dominant but decreases while reactance X_L increases. The decrease in X_C decreases the impedance, allowing the current

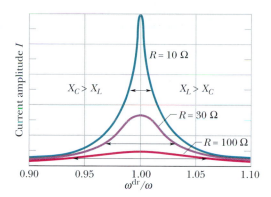

FIGURE 33-17 ■ *Resonance curves* for the driven *RLC* circuit of Fig. 33-11 with $L = 100\ \mu H$, $C = 100$ pF, and three values of R. The current amplitude I of the alternating current depends on how close the driving angular frequency ω^{dr} is to the natural angular frequency ω. The horizontal arrow on each curve measures the curve's width at the half-maximum level, a measure of the sharpness of the resonance. To the left of $\omega^{dr}/\omega = 1.00$, the circuit is mainly capacitive, with $X_C > X_L$ to the right, it is mainly inductive, with $X_L > X_C$.

to increase, as we see on the left side of any resonance curve in Fig. 33-17. When the increasing X_L and the decreasing X_C reach equal values, the current is greatest and the circuit is in resonance, with $\omega^{dr} = \omega$.

As we continue to increase ω^{dr}, the increasing reactance X_L becomes progressively more dominant over the decreasing reactance X_C. The impedance increases because of X_L and the current decreases, as on the right side of any resonance curve in Fig. 33-17. In summary, then: The low-angular-frequency side of a resonance curve is dominated by the capacitor's reactance, the high-angular-frequency side is dominated by the inductor's reactance, and resonance occurs between the two regions.

READING EXERCISE 33-6: Here are the capacitive reactance and inductive reactance, respectively, for three sinusoidally driven series *RLC* circuits: (1) 50 Ω, 100 Ω; (2) 100 Ω, 50 Ω; (3) 50 Ω, 50 Ω. (a) For each, does the current lead or lag the applied emf, or are the two in phase? (b) Which circuit is in resonance? ■

TOUCHSTONE EXAMPLE 33-4: Series *RLC* Circuit

In Fig. 33-11 let $R = 200\ \Omega$, $C = 15.0\ \mu F$, $L = 230$ mH, $f^{dr} = 60.0$ Hz, and $\mathcal{E}^{max} = 36.0$ V.

(a) What is the current amplitude I?

SOLUTION ■ The **Key Idea** here is that current amplitude I depends on the amplitude $\mathcal{E}^{max}$ of the driving emf and on the impedance Z of the circuit, according to Eq. 33-62 ($I = \mathcal{E}^{max}/Z$). Thus, we need to find Z, which depends on the circuit's resistance R, capacitive reactance X_C, and inductive reactance X_L.

The circuit's only resistance is the given resistance R. Its only capacitive reactance is due to the given capacitance. From Table 33-2, ($X_C = 1/\omega^{dr}C$), with $\omega^{dr} = 2\pi f^{dr}$, we can write

$$X_C = \frac{1}{2\pi f^{dr}C} = \frac{1}{(2\pi)(60.0\ \text{Hz})(15.0 \times 10^{-6}\ \text{F})}$$

$$= 177\ \Omega.$$

From Table 33-2 ($X_L = \omega^{dr}L$), with $\omega^{dr} = 2\pi f^{dr}$, we can write

$$X_L = 2\pi f^{dr}L = (2\pi)(60.0\ \text{Hz})(230 \times 10^{-3}\ \text{H})$$

$$= 86.7\ \Omega.$$

Thus, the circuit's impedance is

$$Z = \sqrt{R^2 + (X_L - X_C)^2}$$

$$= \sqrt{(200\ \Omega)^2 + (86.7\ \Omega - 177\ \Omega)^2}$$

$$= 219\ \Omega.$$

We then find

$$|I| = \frac{|\mathcal{E}^{max}|}{Z} = \frac{36.0\ \text{V}}{219\ \Omega} = 0.164\ \text{A}. \qquad \text{(Answer)}$$

(b) What is the phase constant ϕ of the current in the circuit relative to the driving emf?

SOLUTION ■ The **Key Idea** here is that the phase constant depends on the inductive reactance, the capacitive reactance, and the resistance of the circuit, according to Eq. 33-65. Solving that equation for ϕ leads to

$$\phi = \tan^{-1}\frac{X_L - X_C}{R} = \tan^{-1}\frac{86.7\ \Omega - 177\ \Omega}{200\ \Omega}$$

$$= -24.3° = -0.424\ \text{rad}. \qquad \text{(Answer)}$$

The negative phase constant is consistent with the fact that the load is mainly capacitive; that is, $X_C > X_L$.

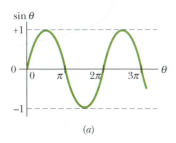

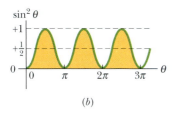

FIGURE 33-18 ■ (a) A plot of sin θ versus θ. The average value over one cycle is zero. (b) A plot of sin² θ versus θ. The average value over one cycle is $\frac{1}{2}$.

33-12 Power in Alternating-Current Circuits

In the *RLC* circuit of Fig. 33-11, the source of energy is the alternating-current generator. Some of the energy that it provides is stored in the electric field in the capacitor, some is stored in the magnetic field in the inductor, and some is dissipated as thermal energy in the resistor. In steady-state operation—which we assume—the average energy stored in the capacitor and inductor together remains constant. The net transfer of energy is thus from the generator to the resistor, where electromagnetic energy is dissipated as thermal energy.

The instantaneous rate at which energy is dissipated in the resistor can be written, with the help of Eqs. 27-26 and 33-35, as

$$P = i^2R = [I\sin(\omega^{dr}t - \phi)]^2 R = I^2R\sin^2(\omega^{dr}t - \phi), \tag{33-68}$$

where *I* is the maximum value of the current.

The *average* rate at which energy is dissipated in the resistor, however, is the average of Eq. 33-68 over time. Although the average value of sin θ, where θ is any variable, is zero (Fig. 33-18a), the average value of sin² θ over one complete cycle is 1/2 (Fig. 33-18b). (Note in Fig. 33-18b how the shaded areas under the curve but above the horizontal line marked +1/2 exactly fill in the unshaded spaces below that line.) Thus, we can write, from Eq. 33-68,

$$\langle P \rangle = \frac{I^2R}{2} = \left(\frac{I}{\sqrt{2}}\right)^2 R. \tag{33-69}$$

The quantity $I/\sqrt{2}$ is defined as the **root-mean-square,** or **rms,** value of the current *i in a cycle.* The square of the rms current is the average of the squares of the instantaneous currents in a cycle. It represents an effective dc current that would produce the same heating effect as an ac current with a maximum value of *I.* So,

$$I^{rms} \equiv \frac{I}{\sqrt{2}} \quad \text{(definition of rms current).} \tag{33-70}$$

We can now rewrite Eq. 33-69 as

$$\langle P \rangle = (I^{rms})^2R \quad \text{(average power).} \tag{33-71}$$

Equation 33-71 looks much like Eq. 26-11 for dc currents ($P = i^2R$); the purpose of defining rms current is that we can use it to compute the average rate of energy dissipation for alternating-current circuits using essentially the same equation we use for dc circuits.

We can also define rms values of voltages and emfs for alternating-current circuits in terms of the maximum values of those quantities in a cycle. So

$$\Delta V^{rms} = \frac{\Delta V}{\sqrt{2}} \quad \text{(rms voltage)} \quad \text{and} \quad \mathcal{E}^{rms} = \frac{\mathcal{E}^{max}}{\sqrt{2}} \quad \text{(rms emf).} \tag{33-72}$$

Alternating-current instruments, such as ammeters and voltmeters, are usually calibrated to read I^{rms}, ΔV^{rms}, and $\mathcal{E}^{rms}$. Thus, if you plug an alternating-current voltmeter into a household electric outlet and it reads 120 V, it represents an rms voltage. The maximum value of the potential difference at the outlet is $\sqrt{2} \times (120 \text{ V})$, or 170 V.

Because the proportionality factor $1/\sqrt{2}$ in Eqs. 33-70 and 33-72 is the same for all three variables, we can write Eqs. 33-62 and 33-60 as

$$I^{\text{rms}} = \frac{\mathscr{E}^{\text{rms}}}{Z} = \frac{\mathscr{E}^{\text{rms}}}{\sqrt{R^2 + (X_L - X_C)^2}}, \tag{33-73}$$

and, indeed, this is the form that we almost always use.

We can use the relationship $I^{\text{rms}} = \mathscr{E}^{\text{rms}}/Z$ to recast Eq. 33-71 in a useful equivalent way. We write

$$\langle P \rangle = \frac{\mathscr{E}^{\text{rms}}}{Z} I^{\text{rms}} R = \mathscr{E}^{\text{rms}} I^{\text{rms}} \frac{R}{Z}. \tag{33-74}$$

From Fig. 33-15d, Table 33-2, and Eq. 33-62, however, we see that R/Z is just the cosine of the phase constant ϕ:

$$\cos\phi = \frac{\Delta V_R}{\mathscr{E}^{\text{max}}} = \frac{IR}{IZ} = \frac{R}{Z}. \tag{33-75}$$

Equation 33-74 then becomes

$$\langle P \rangle = \mathscr{E}^{\text{rms}} I^{\text{rms}} \cos\phi \quad \text{(average power)}, \tag{33-76}$$

in which the term $\cos\phi$ is called the **power factor**. Because $\cos\phi = \cos(-\phi)$, Eq. 33-76 is independent of the sign of the phase constant ϕ.

To maximize the rate at which energy is supplied to a resistive load in an RLC circuit, we should keep the power factor $\cos\phi$ as close to unity as possible. This is equivalent to keeping the phase constant ϕ in Eq. 33-35 as close to zero as possible. If, for example, the circuit is highly inductive, it can be made less so by putting more capacitance in the circuit, connected in series. Adding capacitive reactance counters the excess inductive reactance in the circuit. This makes Z closer in value to R and so reduces the phase constant and increases the power factor in Eq. 33-76. Power companies place series-connected capacitors throughout their transmission systems to get these results.

READING EXERCISE 33-7: (a) If the current in a sinusoidally driven series RLC circuit leads the emf, would we increase or decrease the capacitance to increase the rate at which energy is supplied to the resistance? (b) Will this change bring the resonant angular frequency of the circuit closer to the angular frequency of the emf or move it further away? ■

TOUCHSTONE EXAMPLE 33-5: Power Factor

A series RLC circuit, driven with $\mathscr{E}^{\text{rms}} = 120$ V at frequency $f^{\text{dr}} = 60.0$ Hz, contains a resistance $R = 200\ \Omega$, an inductance with $X_L = 80.0\ \Omega$, and a capacitance with $X_C = 150\ \Omega$.

(a) What are the power factor $\cos\phi$ and phase constant ϕ of the circuit?

SOLUTION ■ The **Key Idea** here is that the power factor $\cos\phi$ can be found from the resistance R and impedance Z via Eq. 33-75 ($\cos\phi = R/Z$). To calculate Z, we use Eq. 33-61:

$$Z = \sqrt{R^2 + (X_L - X_C)^2}$$
$$= \sqrt{(200\ \Omega)^2 + (80.0\ \Omega - 150\ \Omega)^2} = 211.90\ \Omega.$$

Equation 33-75 then gives us

$$\cos\phi = \frac{R}{Z} = \frac{200\ \Omega}{211.90\ \Omega} = 0.9438 \approx 0.944. \quad \text{(Answer)}$$

Taking the inverse cosine then yields

$$\phi = \cos^{-1} 0.944 = \pm 19.3°.$$

Both $+19.3°$ and $-19.3°$ have a cosine of 0.944. To determine which sign is correct, we must consider whether the current leads or lags the driving emf. Because $X_C > X_L$, this circuit is mainly capacitive, with the current leading the emf. Thus, ϕ must be negative:

$$\phi = -19.3°. \quad \text{(Answer)}$$

We could, instead, have found ϕ with Eq. 33-65. A calculator would then have given us the complete answer, with the minus sign.

(b) What is the average rate $\langle P \rangle$ at which energy is dissipated in the resistance?

SOLUTION ■ One way to answer this question is to use this **Key Idea:** Because the circuit is assumed to be in steady-state operation, the rate at which energy is dissipated in the resistance is equal to the rate at which energy is supplied to the circuit, as given by Eq. 33-76 ($\langle P \rangle = \mathcal{E}^{rms} I^{rms} \cos\phi$).

We are given the rms driving emf $\mathcal{E}^{rms}$ and we know $\cos\phi$ from part (a). To find I^{rms} we use the **Key Idea** that the rms current is determined by the rms value of the driving emf and the circuit's impedance Z (which we know), according to Eq. 33-73:

$$I^{rms} = \frac{\mathcal{E}^{rms}}{Z}.$$

Substituting this into Eq. 33-76 then leads to

$$\langle P \rangle = \mathcal{E}^{rms} I^{rms} \cos\phi = \frac{(\mathcal{E}^{rms})^2}{Z} \cos\phi$$

$$= \frac{(120 \text{ V})^2}{211.90 \ \Omega} (0.9438) = 64.1 \text{ W.} \qquad \text{(Answer)}$$

A second way to answer the question is to use the **Key Idea** that the rate at which energy is dissipated in a resistance R depends on the square of the rms current I^{rms} through it, according to Eq. 33-69. We then find

$$\langle P \rangle = (I^{rms})^2 R = \frac{(\mathcal{E}^{rms})^2}{Z^2} R$$

$$= \frac{(120 \text{ V})^2}{(211.90 \ \Omega)^2} (200 \ \Omega) = 64.1 \text{ W.} \qquad \text{(Answer)}$$

(c) What new capacitance C^{new} is needed to maximize $\langle P \rangle$ if the other parameters of the circuit are not changed?

SOLUTION ■ One **Key Idea** here is that the average rate $\langle P \rangle$ at which energy is supplied and dissipated is maximized if the circuit is brought into resonance with the driving emf. A second **Key Idea** is that resonance occurs when $X_C = X_L$. From the given data, we have $X_C > X_L$. Thus, we must decrease X_C to reach resonance. From Eq. 33-43 ($X_C = 1/\omega^{dr} C$), we see that this means we must increase C to the new value C^{new}.

Using Eq. 33-43, we can write the condition $X_C = X_L$ as

$$\frac{1}{\omega^{dr} C^{new}} = X_L.$$

Substituting $2\pi f^{dr}$ for ω^{dr} (because we are given f^{dr} and not ω^{dr}) and then solving for C^{new}, we find

$$C^{new} = \frac{1}{2\pi f^{dr} X_L} = \frac{1}{(2\pi)(60 \text{ Hz})(80.0 \ \Omega)}$$

$$= 3.32 \times 10^{-5} \text{ F} = 33.2 \ \mu\text{F.} \qquad \text{(Answer)}$$

Following the procedure of part (b), you can show that with C^{new}, $\langle P \rangle$ would then be at its maximum value of 72.0 W.

Problems

SEC. 33-2 ■ ENERGY STORED IN A $\vec{B}$-FIELD

1. Current Is Zero Suppose that the inductive time constant for the circuit of Fig. 33-19 is 37.0 ms and the current in the circuit is zero at time $t = 0$ s. At what time does the rate at which energy is dissipated in the resistor equal the rate at which energy is being stored in the inductor?

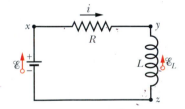

FIGURE 33-19 ■ Problems 1, 2, and 6.

2. Consider the Circuit Consider the circuit of Fig. 33-19. In terms of the inductive time constant, at what instant after the battery is connected will the energy stored in the magnetic field of the inductor be half its steady-state value?

3. Coil Connected in Series A coil is connected in series with a 10.0 k Ω resistor. A 50.0 V battery is applied across the two devices, and the current reaches a value of 2.00 mA after 5.00 ms. (a) Find the inductance of the coil. (b) How much energy is stored in the coil at this same moment?

4. Rates A coil with an inductance of 2.0 H and a resistance of 10 Ω is suddenly connected to a resistanceless battery with $\mathcal{E} = 100$ V. At 0.10 s after the connection is made, what are the rates at which (a) energy is being stored in the magnetic field, (b) thermal energy is appearing in the resistance, and (c) energy is being delivered by the battery?

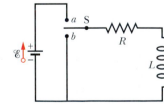

FIGURE 33-20 ■ Problem 5.

5. Prove That Prove that, after switch S in Fig. 33-20 has been thrown from a to b, all the energy stored in the inductor will ultimately appear as thermal energy in the resistor.

6. Energy Delivered For the circuit of Fig. 33-19, assume that $\mathcal{E} = 10.0$ V, $R = 6.70 \ \Omega$, and $L = 5.50$ H. The battery is connected at time $t = 0$ s. (a) How much energy is delivered by the battery during the first 2.00 s? (b) How much of this energy is stored in the magnetic field of the inductor? (c) How much of this energy is dissipated in the resistor?

SEC. 33-3 ■ ENERGY DENSITY OF A $\vec{B}$-FIELD

7. Energy Density A solenoid that is 85.0 cm long has a cross-sectional area of 17.0 cm². There are 950 turns of wire carrying a current of 6.60 A. (a) Calculate the energy density of the magnetic field inside the solenoid. (b) Find the total energy stored in the magnetic field there (neglect end effects).

8. Toroidal Inductor A toroidal inductor with an inductance of 90.0 mH encloses a volume of 0.0200 m³. If the average energy density in the toroid is 70.0 J/m³, what is the current through the inductor?

9. Magnitude of E-Field What must be the magnitude of a uniform electric field if it is to have the same energy density as that possessed by a 0.50 T magnetic field?

10. Interstellar Space The magnetic field in the interstellar space of our galaxy has a magnitude of about 10^{-10} T. How much energy is stored in this field in a cube 10 light-years on edge? (For scale, note that the nearest star is 4.3 light-years distant and the radius of our galaxy is about 8×10^4 light-years.)

11. Length of Copper Wire A length of copper wire carries a current of 10 A, uniformly distributed through its cross section. Calculate the energy density of (a) the magnetic field and (b) the electric field at the surface of the wire. The wire diameter is 2.5 mm, and its resistance per unit length is 3.3 Ω/km.

12. Circular Loop A circular loop of wire 50 mm in radius carries a current of 100 A. (a) Find the magnetic field strength at the center of the loop. (b) Calculate the energy density at the center of the loop.

SEC. 33-4 ■ LC OSCILLATIONS, QUALITATIVELY

13. What Is the Capacitance What is the capacitance of an oscillating LC circuit if the maximum charge on the capacitor is 1.60 μC and the total energy is 140 μJ?

14. Maximum Charge In an oscillating LC circuit, $L = 1.10$ mH and $C = 4.00$ μF. The maximum charge on the capacitor is 3.00 μC. Find the maximum current.

15. Total Energy An oscillating LC circuit consists of a 75.0 mH inductor and a 3.60 μF capacitor. If the maximum charge on the capacitor is 2.90 μC, (a) what is the total energy in the circuit and (b) what is the maximum current?

16. Electric to Magnetic Energy In a certain oscillating LC circuit the total energy is converted from electric energy in the capacitor to magnetic energy in the inductor in 1.50 μs. (a) What is the period of oscillation? (b) What is the frequency of oscillation? (c) How long after the magnetic energy is a maximum will it be a maximum again?

17. Maximum Positive Charge The frequency of oscillation of a certain LC circuit is 200 kHz. At time $t = 0$ s, plate A of the capacitor has maximum positive charge. At what times $t > 0$ s will (a) plate A again have maximum positive charge, (b) the other plate of the capacitor have maximum positive charge, and (c) the inductor have maximum magnetic field?

SEC. 33-5 ■ THE ELECTRICAL–MECHANICAL ANALOGY

18. SHM A 0.50 kg body oscillates in simple harmonic motion on a spring that, when extended 2.0 mm from its equilibrium, has an 8.0 N restoring force. (a) What is the angular frequency of oscillation? (b) What is the period of oscillation? (c) What is the capacitance of an LC circuit with the same period if L is chosen to be 5.0 H?

19. Energy The energy in an oscillating LC circuit containing a 1.25 H inductor is 5.70 μJ. The maximum charge on the capacitor is 175 μC. Find (a) the mass, (b) the spring constant, (c) the maximum displacement, and (d) the maximum speed for a mechanical system with the same period.

SEC. 33-6 ■ LC OSCILLATIONS, QUANTITATIVELY

20. Loudspeakers LC oscillators have been used in circuits connected to loudspeakers to create some of the sounds of electronic music. What inductance must be used with a 6.7 μF capacitor to produce a frequency of 10 kHz, which is near the middle of the audible range of frequencies?

21. Initially a Maximum In an oscillating LC circuit with $L = 50$ mH and $C = 4.0$ μF, the current is initially a maximum. How long will it take before the capacitor is fully charged for the first time?

22. Single Loop A single loop consists of inductors ($L_1, L_2, \ldots$), capacitors ($C_1, C_2, \ldots$), and resistors ($R_1, R_2, \ldots$) connected in series as shown, for example, in Fig. 33-21a. Show that regardless of the sequence of these circuit elements in the loop, the behavior of this circuit is identical to that of the simple LC circuit shown in Fig. 33-21b. (*Hint:* Consider the loop rule and see Problem 7 in Chapter 32.)

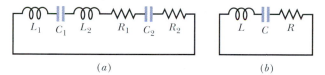

(a)   (b)

FIGURE 33-21 ■ Problem 22.

23. Maximum Voltage An oscillating LC circuit consisting of a 1.0 nF capacitor and a 3.0 mH coil has a maximum voltage of 3.0 V. (a) What is the maximum charge on the capacitor? (b) What is the maximum current through the circuit? (c) What is the maximum energy stored in the magnetic field of the coil?

24. Maximum Potential Difference In an oscillating LC circuit in which $C = 4.00$ μF, the maximum potential difference across the capacitor during the oscillations is 1.50 V and the maximum current through the inductor is 50.0 mA. (a) What is the inductance L? (b) What is the frequency of the oscillations? (c) How much time is required for the charge on the capacitor to rise from zero to its maximum value?

25. Switch Is Thrown In the circuit shown in Fig. 33-22 the switch is kept in position a for a long time. It is then thrown to position b. (a) Calculate the frequency of the resulting oscillating current. (b) What is the amplitude of the current oscillations?

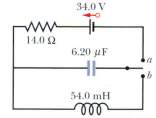

FIGURE 33-22 ■ Problem 25.

26. One Inductor, Two Capacitors You are given a 10 mH inductor and two capacitors, of 5.0 μF and 2.0 μF capacitance. List the oscillation frequencies that can be generated by connecting these elements in various combinations.

27. Variable Capacitor A variable capacitor with a range from 10 to 365 pF is used with a coil to form a variable-frequency LC

circuit to tune the input to a radio. (a) What ratio of maximum to minimum frequencies may be obtained with such a capacitor? (b) If this circuit is to obtain frequencies from 0.54 MHz to 1.60 MHz, the ratio computed in (a) is too large. By adding a capacitor in parallel to the variable capacitor, this range may be adjusted. What should be the capacitance of this added capacitor, and what inductance should be used to obtain the desired range of frequencies?

28. Energy Stored in Magnetic Field In an oscillating LC circuit, 75.0% of the total energy is stored in the magnetic field of the inductor at a certain instant. (a) In terms of the maximum charge on the capacitor, what is the charge there at that instant? (b) In terms of the maximum current in the inductor, what is the current there at that instant?

29. Capacitor Is Charging In an oscillating LC circuit, $L = 25.0$ mH and $C = 7.80$ μF. At time $t = 0$ s the current is 9.20 mA, the charge on the capacitor is 3.80 μC, and the capacitor is charging. (a) What is the total energy in the circuit? (b) What is the maximum charge on the capacitor? (c) What is the maximum current? (d) If the charge on the capacitor is given by $q = |Q| \cos(\omega t + \phi)$, what is the phase angle ϕ? (e) Suppose the data are the same, except that the capacitor is discharging at $t = 0$ s. What then is ϕ?

30. Varied by a Knob An inductor is connected across a capacitor whose capacitance can be varied by turning a knob. We wish to make the frequency of oscillation of this LC circuit vary linearly with the angle of rotation of the knob, going from 2×10^5 to 4×10^5 Hz as the knob turns through 180°. If $L = 1.0$ mH, plot the required capacitance C as a function of the angle of rotation of the knob.

31. Oscillating LC Circuit In an oscillating LC circuit, $L = 3.00$ mH and $C = 2.70$ μF. At $t = 0$ s the charge on the capacitor is zero and the current is 2.00 A. (a) What is the maximum charge that will appear on the capacitor? (b) In terms of the period T of oscillation, how much time will elapse after $t = 0$ until the energy stored in the capacitor will be increasing at its greatest rate? (c) What is this greatest rate at which energy is transferred to the capacitor?

32. Angular Frequency A series circuit containing inductance L_1 and capacitance C_1 oscillates at angular frequency ω. A second series circuit, containing inductance L_2 and capacitance C_2, oscillates at the same angular frequency. In terms of ω, what is the angular frequency of oscillation of a series circuit containing all four of these elements? Neglect resistance. (*Hint:* Use the formulas for equivalent capacitance and equivalent inductance; see Section 28-4 and Problem 7 in Chapter 32.)

33. Current as Function of Time In an oscillating LC circuit with $C = 64.0$ μF, the current as a function of time is given by $i = (1.60$ A$)$ $\sin[(2500$ rad/s$)$ $t + 0.680$ rad$]$, where t is in seconds. (a) How soon after $t = 0$ s will the current reach its maximum value? What are (b) the inductance L and (c) the total energy?

34. Three Identical Inductors Three identical inductors L and two identical capacitors C are connected in a two-loop circuit as shown in Fig. 33-23. (a) Suppose the currents

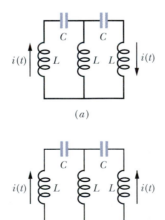

FIGURE 33-23 ■
Problem 34.

are as shown in Fig. 33-23a. What is the current in the middle inductor? Write the loop equations and show that they are satisfied if the current oscillates with angular frequency $\omega = 1/\sqrt{LC}$. (b) Now suppose the currents are as shown in Fig. 33-23b. What is the current in the middle inductor? Write the loop equations and show that they are satisfied if the current oscillates with angular frequency $\omega = 1/\sqrt{3LC}$. Because the circuit can oscillate at two different frequencies, we cannot find an equivalent single-loop LC circuit to replace it.

35. Capacitor One, Capacitor Two In Fig. 33-24, capacitor 1 with $C_1 = 900$ μF is initially charged to 100 V and capacitor 2 with $C_2 = 100$ μF is uncharged. The inductor has an inductance of 10.0 H. Describe in detail how one might charge capacitor 2 to 300 V by manipulating switches S_1 and S_2.

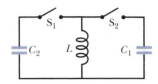

FIGURE 33-24 ■
Problem 35.

SEC. 33-7 ■ DAMPED OSCILLATIONS IN AN *RLC* CIRCUIT

36. Damped LC Consider a damped LC circuit. (a) Show that the damping term $e^{-Rt/2L}$ (which involves L but not C) can be rewritten in a more symmetric manner (involving L and C) as $e^{-\pi R(\sqrt{C/L})t/T}$. Here T is the period of oscillation (neglecting resistance). (b) Using (a), show that the SI unit of $\sqrt{L/C}$ is the ohm. (c) Using (a), show that the condition that the fractional energy loss per cycle be small is $R \ll \sqrt{L/C}$.

37. What Resistance What resistance R should be connected in series with an inductance $L = 220$ mH and capacitance $C = 12.0$ μF for the maximum charge on the capacitor to decay to 99.0% of its initial value in 50.0 cycles? (Assume $\omega' \approx \omega$.)

38. Single-Loop Circuit A single-loop circuit consists of a 7.20 Ω resistor, a 12.0 H inductor, and a 3.20 μF capacitor. Initially the capacitor has a charge of 6.20 μC and the current is zero. Calculate the charge on the capacitor N complete cycles later for $N = 5, 10$, and 100.

39. Oscillating Series *RLC* In an oscillating series RLC circuit, find the time required for the maximum energy present in the capacitor during an oscillation to fall to half its initial value. Assume $q = Q$ at $t = 0$.

40. No Charge on Capacitor At time $t = 0$ s there is no charge on the capacitor of a series RLC circuit but there is current I through the inductor. (a) Find the phase constant ϕ in Eq. 33-31 for the circuit. (b) Write an expression for the charge q on the capacitor as a function of time t and in terms of the current amplitude and angular frequency ω' of the oscillations.

41. Fraction of Energy Lost In an oscillating series RLC circuit, show that the fraction of the energy lost per cycle of oscillation, $\Delta U/U$, is given to a close approximation by $2\pi R/\omega L$. The quantity $\omega L/R$ is often called the Q of the circuit (for *quality*). A high-Q circuit has low resistance and a low fractional energy loss ($= 2\pi/Q$) per cycle.

SEC. 33-10 ■ REPRESENTING OSCILLATIONS WITH PHASORS: THREE SIMPLE CIRCUITS

42. Amplitude A 1.50 μF capacitor is connected as in Fig. 33-13a to an ac generator with $|\mathcal{E}^{max}| = 30.0$ V. What is the amplitude of the resulting alternating current if the frequency of the emf is (a) 1.00 kHz and (b) 8.00 kHz?

43. AC Generator A 50.0 mH inductor is connected as in Fig. 33-14a to an ac generator with $|\mathscr{E}^{max}| = 30.0$ V. What is the amplitude of the resulting alternating current if the frequency of the emf is (a) 1.00 kHz and (b) 8.00 kHz?

44. Frequency of emf Is A 50 Ω resistor is connected as in Fig. 33-12a to an ac generator with $|\mathscr{E}^{max}| = 30.0$ V. What is the amplitude of the resulting alternating current if the frequency of the emf is (a) 1.00 kHz and (b) 8.00 kHz?

45. At What Frequency (a) At what frequency would a 6.0 mH inductor and a 10 μF capacitor have the same reactance? (b) What would the reactance be? (c) Show that this frequency would be the natural frequency of an oscillating circuit with the same L and C.

46. When the Current Is Maximum An ac generator has emf $\mathscr{E} = \mathscr{E}^{max} \sin \omega^{dr} t$, with $\mathscr{E}^{max} = 25.0$ V and $\omega^{dr} = 377$ rad/s. It is connected to a 12.7 H inductor. (a) What is the maximum value of the current? (b) When the current is a maximum, what is the emf of the generator? (c) When the emf of the generator is -12.5 V and increasing in magnitude, what is the current?

47. At What Time An ac generator has emf $\mathscr{E} = \mathscr{E}^{max} \sin(\omega^{dr} t - \pi/4)$, where $\mathscr{E}^{max} = 30.0$ V and $\omega^{dr} = 350$ rad/s. The current produced in a connected circuit is $i(t) = I \sin(\omega^{dr} t - 3\pi/4)$, where $I = 620$ mA. (a) At what time after $t = 0$ does the generator emf first reach a maximum? (b) At what time after $t = 0$ does the current first reach a maximum? (c) The circuit contains a single element other than the generator. Is it a capacitor, an inductor, or a resistor? Justify your answer. (d) What is the value of the capacitance, inductance, or resistance, as the case may be?

48. Generator from Above The ac generator of Problem 46 is connected to a 4.15 μF capacitor. (a) What is the maximum value of the current? (b) When the current is a maximum, what is the emf of the generator? (c) When the emf of the generator is -12.5 V and increasing in magnitude, what is the current?

SEC. 33-11 ■ THE SERIES *RLC* CIRCUIT

49. Find Z, ϕ, and I (a) Find Z, ϕ, and I for the situation of Touchstone Example 33-4 with the capacitor removed from the circuit, all other parameters remaining unchanged. (b) Draw to scale a phasor diagram like that of Fig. 33-15d for this new situation.

50. Find Z, ϕ, and I Two (a) Find Z, ϕ, and I for the situation of Touchstone Example 33-4 with the inductor removed from the circuit, all other parameters remaining unchanged. (b) Draw to scale a phasor diagram like that of Fig. 33-15d for this new situation.

51. Find Z, ϕ, and I Three (a) Find Z, ϕ, and I for the situation of Touchstone Example 33-4 with $C = 70.0$ μF, the other parameters remaining unchanged. (b) Draw a phasor diagram like that of Fig. 33-15d for this new situation and compare the two diagrams closely.

52. Adjustable Frequency In Fig 33-25, a generator with an adjustable frequency of oscillation is connected to a variable resistance R, a capacitor of $C = 5.50$ μF, and an inductor of inductance L. The amplitude of the

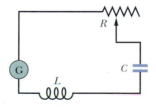

FIGURE 33-25 ■ Problem 52.

current produced in the circuit by the generator is at half-maximum level when the generator's frequency is 1.30 or 1.50 kHz. (a)What is L? (b) If R is increased, what happens to the frequencies at which the current amplitude is at half-maximum level?

53. At Resonance In an *RLC* circuit, can the amplitude of the voltage across an inductor be greater than the amplitude of the generator emf? Consider an *RLC* circuit with $\mathscr{E}^{max} = 10$ V, $R = 10$ Ω, $L = 1.0$ H, and $C = 1.0$ μF. Find the amplitude of the voltage across the inductor at resonance.

54. Emf Is Maximum When the generator emf in Touchstone Example 33-4 is a maximum, what is the voltage across (a) the generator, (b) the resistance, (c) the capacitance, and (d) the inductance? (e) By summing these with appropriate signs, verify that the loop rule is satisfied.

55. Unknown Resistance A coil of inductance 88 mH and unknown resistance and a 0.94 μF capacitor are connected in series with an alternating emf of frequency 930 Hz. If the phase constant between the applied voltage and the current is 75°, what is the resistance of the coil?

56. Capacitive Reactance An ac generator with $\mathscr{E}^{max} = 220$ V and operating at 400 Hz causes oscillations in a series *RLC* circuit having $R = 220$ Ω, $L = 150$ mH, and $C = 24.0$ μF. Find (a) the capacitive reactance X_C, (b) the impedance Z, and (c) the current amplitude I. A second capacitor of the same capacitance is then connected in series with the other components. Determine whether the values of (d) X_C, (e) Z, and (f) I increase, decrease, or remain the same.

57. Half-Width An *RLC* circuit such as that of Fig. 33-11 has $R = 5.00$ Ω, $C = 20.0$ μF, $L = 1.00$ H, and $\mathscr{E}^{max} = 30.0$ V. (a) At what angular frequency ω^{dr} will the current amplitude have its maximum value, as in the resonance curves of Fig. 33-17? (b) What is this maximum value? (c) At what two angular frequencies ω_1^{dr} and ω_2^{dr} will the current amplitude be half this maximum value? (d) What is the fractional half-width $[= (\omega_1^{dr} - \omega_2^{dr})/\omega]$ of the resonance curve for this circuit?

58. Generator in Series An ac generator is to be connected in series with an inductor of $L = 2.00$ mH and a capacitance C. You are to produce C by using capacitors of capacitances $C_1 = 4.00$ μF and $C_2 = 6.00$ μF, either singly or together. What resonant frequencies can the circuit have, depending on how you use C_1 and C_2?

59. Fractional Half-Width Show that the fractional half-width (see Problem 57) of a resonance curve is given by

$$\frac{\Delta\omega^{dr}}{\omega} = \sqrt{\frac{3C}{L}} R,$$

in which ω is the angular frequency at resonance and $\Delta\omega^{dr}$ is the width of the resonance curve at half-amplitude. Note that $\Delta\omega^{dr}/\omega$ increases with R, as Fig. 33-17 shows. Use this formula to check the answer to Problem 57d.

60. Adjustable Frequency Two In Fig. 33-26, a generator with an adjustable frequency of oscillation is connected to resistance $R = 100$ Ω, inductances $L_1 = 1.70$ mH and $L_2 = 2.30$ mH, and capacitances $C_1 = 4.00$ μF,

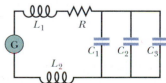

FIGURE 33-26 ■ Problem 60.

$C_2 = 2.50 \; \mu\text{F}$, and $C_3 = 3.50 \; \mu\text{F}$. (a) What is the resonant frequency of the circuit? (*Hint:* See Problem 7 in Chapter 32.) What happens to the resonant frequency if (b) the value of R is increased, (c) the value of L_1 is increased, and (d) capacitance C_3 is removed from the circuit?

SEC. 33-12 ■ POWER IN ALTERNATING-CURRENT CIRCUITS

61. Thermal Energy What direct current will produce the same amount of thermal energy, in a particular resistor, as an alternating current that has a maximum value of 2.60 A?

62. AC Voltmeter An ac voltmeter with large impedance is connected in turn across the inductor, the capacitor, and the resistor in a series circuit having an alternating emf of 100 V(rms); it gives the same reading in volts in each case. What is this reading?

63. AC Voltage What is the maximum value of an ac voltage whose rms value is 100 V?

64. Give or Take (a) For the conditions in Problem 46c, is the generator supplying energy to or taking energy from the rest of the circuit? (b) Repeat for the conditions of Problem 48c.

65. Average Rate of Dissipation Calculate the average rate of energy dissipation in the circuits of Problems 43, 44, 49, and 50.

66. Energy Is Supplied Show that the average rate at which energy is supplied to the circuit of Fig. 33-11 can also be written as $\langle P \rangle = (\mathscr{E}^{\text{rms}})^2 R/Z^2$. Show that this expression for average power gives reasonable results for a purely resistive circuit, for an RLC circuit at resonance, for a purely capacitive circuit, and for a purely inductive circuit.

67. Air Conditioner An air conditioner connected to a 120 V rms ac line is equivalent to a 12.0 Ω resistance and a 1.30 Ω inductive reactance in series. (a) Calculate the impedance of the air conditioner. (b) Find the average rate at which energy is supplied to the appliance.

68. Oscillating RLC In a series oscillating RLC circuit, $R = 16.0 \; \Omega$, $C = 31.2 \; \mu\text{F}$, $L = 9.20$ mH, and $\mathscr{E} = |\mathscr{E}^{\text{max}}| \sin \omega^{\text{dr}} t$ with $|\mathscr{E}^{\text{max}}| = 45.0$ V and $\omega^{\text{dr}} = 3000$ rad/s. For time $t = 0.442$ ms find (a) the rate at which energy is being supplied by the generator, (b) the rate at which the energy in the capacitor is changing, (c) the rate at which the energy in the inductor is changing, and (d) the rate at which energy is being dissipated in the resistor. (e) What is the meaning of a negative result for any of (a), (b), and (c)? (f) Show that the results of (b), (c), and (d) sum to the result of (a).

69. Black Box Figure 33-27 shows an ac generator connected to a "black box" through a pair of terminals. The box contains an RLC circuit, possibly even a multiloop circuit, whose elements and connections we do not know. Measurements outside the

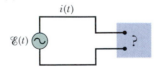

FIGURE 33-27 ■
Problem 69.

box reveal that

$$\mathscr{E}(t) = (75.0 \; \text{V})\sin \omega^{\text{dr}}$$

and

$$i(t) = (1.20 \; \text{A}) \sin (\omega^{\text{dr}} t + 42.0°).$$

(a) What is the power factor? (b) Does the current lead or lag the emf? (c) Is the circuit in the box largely inductive or largely capacitive? (d) Is the circuit in the box in resonance? (e) Must there be a capacitor in the box? An inductor? A resistor ? (f) At what average rate is energy delivered to the box by the generator? (g) Why don't you need to know the angular frequency ω^{dr} to answer all these questions?

70. Average Rate In Fig. 33-28 show that the average rate at which energy is dissipated in resistance R is a maximum when R is equal to the internal resistance r of the ac generator. (In the text discussion we have tacitly assumed that $r = 0$.)

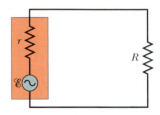

FIGURE 33-28 ■
Problem 70.

71. Energy Is Dissipated In an RLC circuit such as that of Fig. 33.11 assume that $R = 5.00 \; \Omega$, $L = 60.0$ mH $f^{\text{dr}} = 60.0$ Hz, and $|\mathscr{E}^{\text{max}}| = 30.0$ V. For what values of the capacitor would the average rate at which energy is dissipated in the resistance be (a) a maximum and (b) a minimum? (c) What are these maximum and minimum energy dissipation rates? What are (d) the corresponding phase angles and (e) the corresponding power factors?

72. Light Dimmer A typical "light dimmer" used to dim the stage lights in a theater consists of a variable inductor L (whose inductance is adjustable between zero and L^{max}) connected in series with the lightbulb B as shown in Fig. 33-29. The electrical supply is 120 V (rms) at 60.0 Hz; the lightbulb is rated as "120 V, 1000 W." (a) What L^{max} is required if the rate of energy dissipation in the lightbulb is to be varied by a factor of 5 from its upper limit of 1000 W? Assume that the resistance of the lightbulb is independent of its temperature. (b) Could one use a variable resistor (adjustable between zero and R^{max}) instead of an inductor? If so, what R^{max} is required? Why isn't this done?

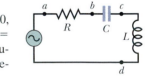

FIGURE 33-29 ■
Problem 72.

73. Sinusoidal Voltage In Fig. 33-30, $R = 15.0 \; \Omega$, $C = 4.70 \; \mu\text{F}$, and $L = 25.0$ mH. The generator provides a sinusoidal voltage of 75.0 V (rms) and frequency $f = 550$ Hz.

(a) Calculate the rms current.
(b) Find the rms voltages ΔV_{ab}, ΔV_{bc}, ΔV_{cd}, ΔV_{bd}, ΔV_{ad}. (c) At what average rate is energy dissipated by each of the three circuit elements?

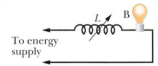

FIGURE 33-30 ■
Problem 73.

The International System of Units (SI)*

1 SI Base Units

1. The SI Base Units

Quantity	Name	Symbol	Definition
length	meter	m	"...the length of the path traveled by light in vacuum in $1/299\ 792\ 458$ of a second." (1983)
mass	kilogram	kg	"...this prototype [a certain platinum–iridium cylinder] shall henceforth be considered to be the unit of mass." (1889)
time	second	s	"...the duration of $9\ 192\ 631\ 770$ periods of the radiation corresponding to the transition between the two hyperfine levels of the ground state of the cesium-133 atom." (1967)
electric current	ampere	A	"...that constant current which, if maintained in two straight parallel conductors of infinite length, of negligible circular cross section, and placed 1 meter apart in vacuum, would produce between these conductors a force equal to 2×10^{-7} newton per meter of length." (1946)
thermodynamic temperature	kelvin	K	"...the fraction $1/273.16$ of the thermodynamic temperature of the triple point of water." (1967)
amount of substance	mole	mol	"...the amount of substance of a system which contains as many elementary entities as there are atoms in 0.012 kilogram of carbon-12." (1971)
luminous intensity	candela	cd	"...the luminous intensity, in a given direction, of a source that emits monochromatic radiation of frequency 540×10^{12} hertz and that has a radiant intensity in that direction of $1/683$ watt per steradian." (1979)

2 The SI Supplementary Units

2. The SI Supplementary Units

Quantity	Name of Unit	Symbol
plane angle	radian	rad
solid angle	steradian	sr

*Adapted from "The International System of Units (SI)," National Bureau of Standards Special Publication 330, 2001 edition. The definitions above were adopted by the General Conference of Weights and Measures, an international body, on the dates shown. In this book we do not use the candela.

3 Some SI Derivations

3. Some SI Derived Units

Quantity	Name of Unit	Symbol	In Terms of other SI Units
area	square meter	m^2	
volume	cubic meter	m^3	
frequency	hertz	Hz	s^{-1}
mass density (density)	kilogram per cubic meter	kg/m^3	
speed, velocity	meter per second	m/s	
rotational velocity	radian per second	rad/s	
acceleration	meter per second per second	m/s^2	
rotational acceleration	radian per second per second	rad/s^2	
force	newton	N	$kg \cdot m/s^2$
pressure	pascal	Pa	N/m^2
work, energy, quantity of heat	joule	J	$N \cdot m$
power	watt	W	J/s
quantity of electric charge	coulomb	C	$A \cdot s$
potential difference, electromotive force	volt	V	W/A
electric field strength	volt per meter (or newton per coulomb)	V/m	N/C
electric resistance	ohm	Ω	V/A
capacitance	farad	F	$A \cdot s/V$
magnetic flux	weber	Wb	$V \cdot s$
inductance	henry	H	$V \cdot s/A$
magnetic flux density	tesla	T	Wb/m^2
magnetic field strength	ampere per meter	A/m	
entropy	joule per kelvin	J/K	
specific heat	joule per kilogram kelvin	$J/(kg \cdot K)$	
thermal conductivity	watt per meter kelvin	$W/(m \cdot K)$	
radiant intensity	watt per steradian	W/sr	

4 Mathematical Notation

Poorly chosen mathematical notation can be a source of considerable confusion to those trying to learn and to do physics. For example, ambiguity in the meaning of a mathematical symbol can prevent a reader from understanding the meaning of a crucial relationship. It is also difficult to solve problems when the symbols used ot represent different quantities are not distinctive. In this text we have taken special care to use mathematical notation in ways that allow important distinctions to be easily visible both on the printed page and in handwritten work.

An excellent starting point for clear mathematical notation is the U.S. National Institute of Standard and Technology's Special Publication 811 (SP 811), *Guide for the Use of the International System of Units (SI)*, available at http://physics.nist.gov/cuu/Units/bibliography.html. In addition to following the National Institute guidelines, we have made a number of systematic choices to facilitate the translation of printed notation into handwritten mathematics. For example:

- Instead of making vectors bold, vector quantities (even in one dimension) are denoted by an arrow above the symbol. So printed equations look like handwritten equations. Example: $\vec{v}$ rather than $\boldsymbol{v}$ is used to denote an instantaneous velocity.

- In general, each vector component has an explicit subscript denoting that it represents the component along a chosen coordinate axis. The one exception is the position vector, $\vec{r}$, whose components are simply written as x, y, and z. For example, $\vec{r} = x\hat{i} + y\hat{j} + z\hat{k}$, whereas, $\vec{v} = v_x\hat{i} + v_y\hat{j} + v_z\hat{k}$.

- To emphasize the distinction between a vector's components and its magnitude, we write the magnitude of a vector, such as $\vec{F}$, as $|\vec{F}|$. However, when it is obvious that a magnitude is being described, we use the plain symbol (such as F with no coordinate subscript) to denote a vector's magnitude.

- We often choose to spell out the names of objects that are associated with mathematical variables—writing, for example, $\vec{v}_{ball}$ and not $\vec{v}_b$ for the velocity of a ball.

- Numerical subscripts most commonly denote sequential times, positions, velocities, and so on. For example, x_1 is the x-component of the position of some object at time t_1, whereas x_2 is the value of that parameter at some later time t_2. We have avoided using the subscript zero to denote initial values, as in x_0 to denote "the initial position along the x axis," to emphasize that *any* time can be chosen as the initial time for consideration of the subsequent time evolution of a system.

- To avoid confusing the numerical time sequence labels with object labels, we prefer to use capital letters as object labels. For example, we would label two particles A and B rather than 1 and 2. Thus, $\vec{p}_{A1}$ and $\vec{p}_{B1}$ would represent the translational momenta of two particles before a collision whereas $\vec{p}_{A2}$ and $\vec{p}_{B2}$ would be their momenta after a collision.

- To avoid excessively long strings of subscripts, we have made the unconventional choice to write all adjectival labels as *super*scripts. Thus, Newton's Second Law is written $\vec{F}^{net} = m\vec{a}$ whereas the sum of the forces acting on a certain object might be written as $\vec{F}^{net} = \vec{F}^{grav} + \vec{F}^{app}$. To avoid confusion with mathematical exponents, an adjectival label is never a single letter.

- Following a usage common in contemporary physics, the time average of a variable $\vec{v}$ is denoted as $\langle \vec{v} \rangle$ and not as $\vec{v}_{avg}$.

- Physical constants such as e, c, g, G, are all **positive** scalar quantities.

5 Significant Figures and the Precision of Numerical Results

Quoting the result of a calculation or a measurement to the correct number of significant figures is merely a way of telling your reader roughly how precise you believe the result to be. Quoting too many significant figures overstates the precision of your result and quoting too few implies less precision than the result may actually possess. So how many significant figures should you quote when reporting your result.

Determining Significant Figures

Before answering the question of how many significant figures to quote, we need to have a clear method for determining how many significant figures a reported number has. The standard method is quite simple:

METHOD FOR COUNTING SIGNIFICANT FIGURES: Read the number from left to right, and count the first nonzero digit and all the digits (zero or not) to the right of it as significant.

Using this rule, 350 mm, 0.000350 km, and 0.350 m each has *three* significant figures. In fact, each of these numbers merely represents the same distance, expressed in different units. As you can see from this example, the number of *decimal places* that a number has is *not* the same as its number of *significant figures*. The first of these distances has zero decimal places, the second has six decimal places, and the third has three, yet all three of these numbers have three significant figures.

One consequence of this method is especially worth noting. Trailing zeros count as significant figures. For example, 2700 m/s has four significant figures. If you really meant it to have only three significant figures, you would have to write it either as 2.70 km/s (changing the unit) or 2.70×10^3 m/s (using scientific notation.)

A Simple Rule for Reporting Significant Figures in a Calculated Result

Now that you know how to count significant figures, how many should the result of a calculation have? A simple rule that will work in most calculations is:

> **SIGNIFICANT FIGURES IN A CALCULATED RESULT:** The common practice is to quote the result of a calculation to the number of significant figures of the *least* precise number used in the calculation.

Although this simple rule will often either understate or (less frequently) overstate the precision of a result, it still serves as a good rule-of-thumb for everyday numerical work. In introductory physics you will only rarely encounter data that are known to better than two, three, or four significant figures. This simple rule then tells you that you can't go very far wrong if you round off all your final results to three significant figures.

There are two situations in which the simple rule should *not* be applied to a calculation. One is when an exact number is involved in the calculation and another is when a calculation is done in parts so that intermediate results are used.

1. *Using Exact Data*   There are some obvious situations in which a number used in a calculation is exact. Numbers based on counting items are exact. For example, if you are told that there are 5 people on an elevator, there are exactly 5 people, not 4.7 or 5.1. Another situation arises when a number is exact by definition. For example, the conversion factor 2.54 cm/inch does *not* have three significant figures because the inch is *defined* to be exactly 2.5400000 . . . cm. *Data that are known exactly should not be included when deciding which of the original data has the fewest significant figures.*

2. *Significant Figures in Intermediate Results* Only the final result at the end of your calculation should be rounded using the simple rule. Intermediate results should never be rounded. Spreadsheet software takes care of this for you, as does your calculator if you store your intermediate results in its memory rather than writing them down and then rekeying them. If you must write down intermediate results, keep a few more significant figures than your final result will have.

Understanding and Refining the Simple Significant Figure Rule

Quoting the result of a calculation or measurement to the correct number of significant figures is a way of indicating its precision. You need to understand what limits the precision of data before you fully understand how to use the simple rule or its exceptions.

Absolute Precision There are two ways of talking about precision. First there is *absolute precision*, which tells you explicitly the smallest scale division of the measurement. It's always quoted in the same units as the measured quantity. For example, saying "I measured the length of the table to the nearest centimeter" states the absolute precision of the measurement. The absolute precision tells you how many *decimal places* the measurement has; it alone does not determine the number of significant figures. Example: if a table is 235 cm long, then 1 cm of absolute precision translates into three significant figures. On the other hand, if a table is for a doll's house and is only 8 cm long, then the same 1 cm of absolute precision has only one significant figure.

Relative Precision Because of this problem with absolute precision, scientists often prefer to describe the precision of data *relative* to the size of the quantity being measured. To use the previous examples, the *relative precision* of the length of the real table in the previous example is 1 cm out of 235 cm. This is usually stated as a ratio (1 part in 235) or as a percentage ($1/235 = 0.004255 \approx 0.4\%$). In the case of the toy table, the same 1 cm of absolute precision yields a relative precision of only 1 part in 8 or $1/8 = 0.125 = 12.5\%$.

Inconsistencies between Significant Figures and Relative Precision There is an inconsistency that goes with using a certain number of significant figures to express relative precision. Quoted to the same number of significant figures, the relative precision of results can be quite different. For example, 13 cm and 94 cm both have two significant figures. Yet the first is specified to only 1 part in 13 or $1/13 \approx 10\%$, whereas the second is known to 1 part in 94 or $1/94 \approx 1\%$. This bias toward greater relative precision for results with larger first significant figures is one weakness of using significant figures to track the precision of calculated results. You can partially address this problem, by including one more significant figure than the simple rule suggests, when the final result of a calculation has a 1 as its first significant figure.

Multiplying and Dividing When multiplying or dividing numbers, the *relative* precision of the result cannot exceed that of the least precise number used. Since the number of significant figures in the result tells us its relative precision, the simple rule is all that you need when you multiply or divide. For example, the area of a strip of paper of measured size is 280 cm by 2.5 cm would be correctly reported, according to the simple rule, as 7.0×10^2 cm^2. This result has only two significant figures since the less precise measurement, 2.5 cm, that went into the calculation had only two significant figures. Reporting this result as 700 cm^2 would not be correct since this result has three significant figures, exceeding the relative precision of the 2.5 cm measurement.

Addition and Subtraction When adding or subtracting, you line up the decimal points before you add or subtract. This means that it's the *absolute* precision of the least precise number that limits the precision of the sum or the difference. This can lead to some exceptions to the simple rule. For example, adding 957 cm and 878 cm yields 1835 cm. Here the result is reliable to an absolute precision of about 1 cm since both of the original distances had this reliability. But the result then has four significant figures whereas each of the original numbers had only three. If, on the other hand, you take the difference between these two distances you get 79 cm. The difference is still reliable to about 1 cm, but that absolute precision now translates into only two significant figures worth of relative precision. So, you should be careful when adding or subtracting, since addition can actually increase the relative precision of your result and, more important, subtraction can reduce it.

Evaluating Functions What about the evaluation of functions? For example, how many significant figures does the $\sin(88.2°)$ have? You can use your calculator to answer this question. First use your calculator to note that $\sin(88.2°) = 0.999506$. Now add 1 to the least significant decimal place of the argument of the function and evaluate it again. Here this gives $\sin(88.3°) = 0.999559$. Take the last significant figure in the result to be *the first one from the left that changed* when you repeated the calculation. In this example the first digit that changed was the 0; it became a 5 (the second 5) in the recalculation. So, using the empirical approach gives you five significant figures.

Some Fundamental Constants of Physics*

Constant	Symbol	Computational Value	Best (1998) Value	
			Value[a]	Uncertainty[b]
Speed of light in a vacuum	c	3.00×10^8 m/s	2.997 924 58	exact
Elementary charge	e	1.60×10^{-19} C	1.602 176 462	0.039
Gravitational constant	G	6.67×10^{-11} m³/s²·kg	6.673	1500
Universal gas constant	R	8.31 J/mol·K	8.314 472	1.7
Avogadro constant	N_A	6.02×10^{23} mol⁻¹	6.022 141 99	0.079
Boltzmann constant	k_B	1.38×10^{-23} J/K	1.380 650 3	1.7
Stefan–Boltzmann constant	σ	5.67×10^{-8} W/m²·K⁴	5.670 400	7.0
Molar volume of ideal gas at STP[d]	V_m	2.27×10^{-2} m³/mol	2.271 098 1	1.7
Electric constant (permittivity)	ϵ_0	8.85×10^{-12} C²/N·m²	8.854 187 817 62	exact
Coulomb constant	$k = 1/4\pi\epsilon_0$	8.99×10^9 N·m²/C²	8.987 551 787	5×10^{-10}
Magnetic constant (permeability)	μ_0	1.26×10^{-6} N/A²	1.256 637 061 43	exact
Planck constant	h	6.63×10^{-34} J·s	6.626 068 76	0.078
Electron mass[c]	m_e	9.11×10^{-31} kg	9.109 381 88	0.079
		5.49×10^{-4} u	5.485 799 110	0.0021
Proton mass[c]	m_p	1.67×10^{-27} kg	1.672 621 58	0.079
		1.0073 u	1.007 276 466 88	$1.3 \times .10^{-4}$
Ratio of proton mass to electron mass	m_p/m_e	1840	1836.152 667 5	0.0021
Electron charge-to-mass ratio	e/m_e	1.76×10^{11} C/kg	1.758 820 174	0.040
Neutron mass[c]	m_n	1.68×10^{-27} kg	1.674 927 16	0.079
		1.0087 u	1.008 664 915 78	5.4×10^{-4}
Hydrogen atom mass[c]	m_{1H}	1.0078 u	1.007 825 031 6	0.0005
Deuterium atom mass[c]	m_{2H}	2.0141 u	2.014 101 777 9	0.0005
Helium atom mass[c]	m_{4He}	4.0026 u	4.002 603 2	0.067
Muon mass	m_μ	1.88×10^{-28} kg	1.883 531 09	0.084
Electron magnetic moment	μ_e	9.28×10^{-24} J/T	9.284 763 62	0.040
Proton magnetic moment	μ_p	1.41×10^{-26} J/T	1.410 606 663	0.041
Bohr magneton	μ_B	9.27×10^{-24} J/T	9.274 008 99	0.040
Nuclear magneton	μ_N	5.05×10^{-27} J/T	5.050 783 17	0.040
Bohr radius	r_B	5.29×10^{-11} m	5.291 772 083	0.0037
Rydberg constant	R	1.10×10^7 m⁻¹	1.097 373 156 854 8	7.6×10^{-6}
Electron Compton wavelength	λ_C	2.43×10^{-12} m	2.426 310 215	0.0073

[a]Values given in this column should be given the same unit and power of 10 as the computational value.
[b]Parts per million.
[c]Masses given in u are in unified atomic mass units, where 1 u = 1.660 538 73 × 10⁻²⁷ kg.
[d]STP means standard temperature and pressure: 0°C and 1.0 atm (0.1 MPa).

*The values in this table were selected from the 1998 CODATA recommended values (www.physics.nist.gov).

APPENDIX C

Some Astronomical Data

Some Distances from Earth

To the Moon*	3.82×10^8 m	To the center of our galaxy	2.2×10^{20} m
To the Sun*	1.50×10^{11} m	To the Andromeda Galaxy	2.1×10^{22} m
To the nearest star (Proxima Centauri)	4.04×10^{16} m	To the edge of the observable universe	$\sim 10^{26}$ m

* Mean distance.

The Sun, Earth, and the Moon

Property	Unit	Sun		Earth	Moon
Mass	kg	1.99×10^{30}		5.98×10^{24}	7.36×10^{22}
Mean radius	m	6.96×10^8		6.37×10^6	1.74×10^6
Mean density	kg/m³	1410		5520	3340
Free-fall acceleration at the surface	m/s²	274		9.81	1.67
Escape velocity	km/s	618		11.2	2.38
Period of rotation[a]	—	37 d at poles[b]	26 d at equator[b]	23 h 56 min	27.3 d
Radiation power[c]	W	3.90×10^{26}			

[a] Measured with respect to the distant stars, [b] The Sun, a ball of gas, does not rotate as a rigid body; [c] Just outside Earth's atmosphere solar energy is received, assuming normal incidence, at the rate of 1340 W/m².

Some Properties of the Planets

	Mercury	Venus	Earth	Mars	Jupiter	Saturn	Uranus	Neptune	Pluto
Mean distance from Sun, 10^6 km	57.9	108	150	228	778	1430	2870	4500	5900
Period of revolution, y	0.241	0.615	1.00	1.88	11.9	29.5	84.0	165	248
Period of rotation,[a] d	58.7	-243^b	0.997	1.03	0.409	0.426	-0.451^b	0.658	6.39
Orbital speed, km/s	47.9	35.0	29.8	24.1	13.1	9.64	6.81	5.43	4.74
Equatorial diameter, km	4880	12 100	12 800	6790	143 000	120 000	51 800	49 500	2300
Mass (Earth = 1)	0.0558	0.815	1.000	0.107	318	95.1	14.5	17.2	0.002
Surface value of g,[c] m/s²	3.78	8.60	9.78	3.72	22.9	9.05	7.77	11.0	0.5
Escape velocity,[c] km/s	4.3	10.3	11.2	5.0	59.5	35.6	21.2	23.6	1.1

[a] Measured with respect to the distant stars.
[b] Venus and Uranus rotate opposite their orbital motion.
[c] Gravitational acceleration measured at the planet's equator.

Conversion Factors

Conversion factors may be read directly from these tables. For example, 1 degree = 2.778×10^{-3} revolutions, so $16.7° = 16.7 \times 2.778 \times 10^{-3}$ rev. The SI units are fully capitalized. Adapted in part from G. Shortley and D. Williams, *Elements of Physics*, 1971, Prentice-Hall, Englewood Cliffs, N.J.

Solid Angle

1 sphere
= 4π steradians
= 12.57 steradians

Plane Angle

	°	′	″	RADIAN	rev
1 degree =	1	60	3600	1.745×10^{-2}	2.778×10^{-3}
1 minute =	1.667×10^{-2}	1	60	2.909×10^{-4}	4.630×10^{-5}
1 second =	2.778×10^{-4}	1.667×10^{-2}	1	4.848×10^{-6}	7.716×10^{-7}
1 RADIAN =	57.30	3438	2.063×10^{5}	1	0.1592
1 revolution =	360	2.16×10^{4}	1.296×10^{6}	6.283	1

Length

cm	METER	km	in.	ft	mi
1 centimeter = 1	10^{-2}	10^{-5}	0.3937	3.281×10^{-2}	6.214×10^{-6}
1 METER = 100	1	10^{-3}	39.37	3.281	6.214×10^{-4}
1 kilometer = 10^{5}	1000	1	3.937×10^{4}	3281	0.6214
1 inch = 2.540	2.540×10^{-2}	2.540×10^{-5}	1	8.333×10^{-2}	1.578×10^{-5}
1 foot = 30.48	0.3048	3.048×10^{-4}	12	1	1.894×10^{-4}
1 mile = 1.609×10^{5}	1609	1.609	6.336×10^{4}	5280	1

1 angström = 10^{-10} m 1 fermi = 10^{-15} m 1 light-year = 9.460×10^{12} km 1 fathom = 6 ft 1 yard = 3 ft 1 mil = 10^{-3} in.
1 nautical mile = 1852 m 1 parsec = 3.084×10^{13} km 1 Bohr radius = 5.292×10^{-11} m 1 rod = 16.5 ft 1 nm = 10^{-9} m
= 1.151 miles = 6076 ft

Area

METER²	cm²	ft²	in.²
1 SQUARE METER = 1	10^{4}	10.76	1550
1 square centimeter = 10^{-4}	1	1.076×10^{-3}	0.1550
1 square foot = 9.290×10^{-2}	929.0	1	144
1 square inch = 6.452×10^{-4}	6.452	6.944×10^{-3}	1

key: 1 square mile = 2.788×10^{7} ft² = 640 acres; 1 barn = 10^{-28} m²; 1 acre = 43 560 ft²; 1 hectare = 10^{4} m² = 2.471 acres.

Volume

METER³	cm³	L	ft³	in.³
1 CUBIC METER = 1	10^{6}	1000	35.31	6.102×10^{4}
1 cubic centimeter = 10^{-6}	1	1.000×10^{-3}	3.531×10^{-5}	6.102×10^{-2}
1 liter = 1.000×10^{-3}	1000	1	3.531×10^{-2}	61.02
1 cubic foot = 2.832×10^{-2}	2.832×10^{4}	28.32	1	1728
1 cubic inch = 1.639×10^{-5}	16.39	1.639×10^{-2}	5.787×10^{-4}	1

key: 1 U.S. fluid gallon = 4 U.S. fluid quarts = 8 U.S. pints = 128 U.S. fluid ounces = 231 in.³ 1 British imperial gallon = 277.4 in.³ = 1.201 U.S. fluid gallons.

Mass

Quantities in the colored areas are not mass units but are often used as such. When we write, for example, 1 kg "=" 2.205 lb, this means that a kilogram is a *mass* that *weighs* 2.205 pounds at a location where g has the standard value of 9.80665 m/s^2.

g	KILOGRAM	slug	u	oz	lb	ton
1 gram = 1	0.001	6.852×10^{-5}	6.022×10^{23}	3.527×10^{-2}	2.205×10^{-3}	1.102×10^{-6}
1 KILOGRAM = 1000	1	6.852×10^{-2}	6.022×10^{26}	35.27	2.205	1.102×10^{-3}
1 slug = 1.459×10^4	14.59	1	8.786×10^{27}	514.8	32.17	1.609×10^{-2}
1 atomic mass unit = 1.661×10^{-24}	1.661×10^{-27}	1.138×10^{-28}	1	5.857×10^{-26}	3.662×10^{-27}	1.830×10^{-30}
1 ounce = 28.35	2.835×10^{-2}	1.943×10^{-3}	1.718×10^{25}	1	6.250×10^{-2}	3.125×10^{-5}
1 pound = 453.6	0.4536	3.108×10^{-2}	2.732×10^{26}	16	1	0.0005
1 ton = 9.072×10^5	907.2	62.16	5.463×10^{29}	3.2×10^4	2000	1

1 metric ton = 1000 kg

Time

y	d	h	min	SECOND
1 year = 1	365.25	8.766×10^3	5.259×10^5	3.156×10^7
1 day = 2.738×10^{-3}	1	24	1440	8.640×10^4
1 hour = 1.141×10^{-4}	4.167×10^{-2}	1	60	3600
1 minute = 1.901×10^{-6}	6.944×10^{-4}	1.667×10^{-2}	1	60
1 SECOND = 3.169×10^{-8}	1.157×10^{-5}	2.778×10^{-4}	1.667×10^{-2}	1

Speed

ft/s	km/h	METER/SECOND	mi/h	cm/s
1 foot per second = 1	1.097	0.3048	0.6818	30.48
1 kilometer per hour = 0.9113	1	0.2778	0.6214	27.78
1 METER per SECOND = 3.281	3.6	1	2.237	100
1 mile per hour = 1.467	1.609	0.4470	1	44.70
1 centimeter per second = 3.281×10^{-2}	3.6×10^{-2}	0.01	2.237×10^{-2}	1

1 knot = 1 nautical mi/h = 1.688 ft/s   1 mi/min = 88.00 ft/s = 60.00 mi/h

Force

dyne	NEWTON	lb	pdl
1 dyne = 1	10^{-5}	2.248×10^{-6}	7.233×10^{-5}
1 NEWTON = 10^5	1	0.2248	7.233
1 pound = 4.448×10^5	4.448	1	32.17
1 poundal = 1.383×10^4	0.1383	3.108×10^{-2}	1

1 ton = 2000 lb

Pressure

atm	dyne/cm²	inch of water	cm Hg	PASCAL	lb/in.²	lb/ft²
1 atmosphere = 1	1.013×10^6	406.8	76	1.013×10^5	14.70	2116
1 dyne per centimeter² = 9.869×10^{-7}	1	4.015×10^{-4}	7.501×10^{-5}	0.1	1.405×10^{-5}	2.089×10^{-3}
1 inch of water[a] at 4°C = 2.458×10^{-3}	2491	1	0.1868	249.1	3.613×10^{-2}	5.202
1 centimeter of mercury[a] at 0°C = 1.316×10^{-2}	1.333×10^4	5.353	1	1333	0.1934	27.85
1 PASCAL = 9.869×10^{-6}	10	4.015×10^{-3}	7.501×10^{-4}	1	1.450×10^{-4}	2.089×10^{-2}
1 pound per inch² = 6.805×10^{-2}	6.895×10^4	27.68	5.171	6.895×10^3	1	144
1 pound per foot² = 4.725×10^{-4}	478.8	0.1922	3.591×10^{-2}	47.88	6.944×10^{-3}	1

[a] Where the acceleration of gravity has the standard value of 9.80665 m/s².

1 bar = 10^6 dyne/cm² = 0.1 MPa   1 millibar = 10^3 dyne/cm² = 10^2 Pa   1 torr = 1 mm Hg

Energy, Work, Heat

Btu	erg	ft · lb	hp · h	JOULE	cal	kW · h	eV	MeV
1 British thermal unit = 1	1.055×10^{10}	777.9	3.929×10^{-4}	1055	252.0	2.930×10^{-4}	6.585×10^{21}	6.585×10^{15}
1 erg = 9.481×10^{-11}	1	7.376×10^{-8}	3.725×10^{-14}	10^{-7}	2.389×10^{-8}	2.778×10^{-14}	6.242×10^{11}	6.242×10^5
1 foot-pound = 1.285×10^{-3}	1.356×10^7	1	5.051×10^{-7}	1.356	0.3238	3.766×10^{-7}	8.464×10^{18}	8.464×10^{12}
1 horsepower-hour = 2545	2.685×10^{13}	1.980×10^6	1	2.685×10^6	6.413×10^5	0.7457	1.676×10^{25}	1.676×10^{19}
1 JOULE = 9.481×10^{-4}	10^7	0.7376	3.725×10^{-7}	1	0.2389	2.778×10^{-7}	6.242×10^{18}	6.242×10^{12}
1 calorie = 3.969×10^{-3}	4.186×10^7	3.088	1.560×10^{-6}	4.186	1	1.163×10^{-6}	2.613×10^{19}	2.613×10^{13}
1 kilowatt hour = 3413	3.600×10^{13}	2.655×10^6	1.341	3.600×10^6	8.600×10^5	1	2.247×10^{25}	2.247×10^{19}
1 electron-volt = 1.519×10^{-22}	1.602×10^{-12}	1.182×10^{-19}	5.967×10^{-26}	1.602×10^{-19}	3.827×10^{-20}	4.450×10^{-26}	1	10^{-6}
1 million electron-volts = 1.519×10^{-16}	1.602×10^{-6}	1.182×10^{-13}	5.967×10^{-20}	1.602×10^{-13}	3.827×10^{-14}	4.450×10^{-20}	10^{-6}	1

Power

Btu/h	ft · lb/s	hp	cal/s	kW	WATT
1 British thermal unit per hour = 1	0.2161	3.929×10^{-4}	6.998×10^{-2}	2.930×10^{-4}	0.2930
1 foot-pound per second = 4.628	1	1.818×10^{-3}	0.3239	1.356×10^{-3}	1.356
1 horsepower = 2545	550	1	178.1	0.7457	745.7
1 calorie per second = 14.29	3.088	5.615×10^{-3}	1	4.186×10^{-3}	4.186
1 kilowatt = 3413	737.6	1.341	238.9	1	1000
1 WATT = 3.413	0.7376	1.341×10^{-3}	0.2389	0.001	1

Magnetic Field

gauss	TESLA	milligauss
1 gauss = 1	10^{-4}	1000
1 TESLA = 10^4	1	10^7
1 milligauss = 0.001	10^{-7}	1

Magnetic Flux

maxwell	WEBER
1 maxwell = 1	10^{-8}
1 WEBER = 10^8	1

1 tesla = 1 weber/meter²

Mathematical Formulas

Geometry

Circle of radius r: circumference $= 2\pi r$; area $= \pi r^2$.

Sphere of radius r: area $= 4\pi r^2$; volume $= \frac{4}{3}\pi r^3$.

Right circular cylinder of radius r and height h:
area $= 2\pi r^2 + 2\pi rh$; volume $= \pi r^2 h$.

Triangle of base a and altitude h: area $= \frac{1}{2}ah$.

Quadratic Formula

If $ax^2 + bx + c = 0$, then $x = \dfrac{-b \pm \sqrt{b^2 - 4ac}}{2a}$.

Trigonometric Functions of Angle θ

$\sin\theta = \dfrac{y}{r}$ $\qquad \cos\theta = \dfrac{x}{r}$

$\tan\theta = \dfrac{y}{x}$ $\qquad \cot\theta = \dfrac{x}{y}$

$\sec\theta = \dfrac{r}{x}$ $\qquad \csc\theta = \dfrac{r}{y}$

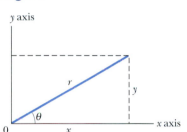

Pythagorean Theorem

In this right triangle,
$$a^2 + b^2 = c^2$$

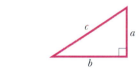

Triangles

Angles are A, B, C

Opposite sides are a, b, c

Angles $A + B + C = 180°$

$\dfrac{\sin A}{a} = \dfrac{\sin B}{b} = \dfrac{\sin C}{c}$

$c^2 = a^2 + b^2 - 2ab\cos C$

Exterior angle $D = A + C$

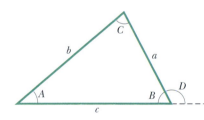

Mathematical Signs and Symbols

$=$ equals

$\approx$ equals approximately

$\sim$ is the order of magnitude of

$\neq$ is not equal to

$\equiv$ is identical to, is defined as

$>$ is greater than ($\gg$ is much greater than)

$<$ is less than ($\ll$ is much less than)

$\geq$ is greater than or equal to (or, is no less than)

$\leq$ is less than or equal to (or, is no more than)

$\pm$ plus or minus

$\propto$ is proportional to

Σ the sum of

$\langle x \rangle$ the average value of x

Trigonometric Identities

$\sin(90° - \theta) = \cos\theta$

$\cos(90° - \theta) = \sin\theta$

$\sin\theta/\cos\theta = \tan\theta$

$\sin^2\theta + \cos^2\theta = 1$

$\sec^2\theta - \tan^2\theta = 1$

$\csc^2\theta - \cot^2\theta = 1$

$\sin 2\theta = 2\sin\theta\cos\theta$

$\cos 2\theta = \cos^2\theta - \sin^2\theta = 2\cos^2\theta - 1 = 1 - 2\sin^2\theta$

$\sin(\alpha \pm \beta) = \sin\alpha\cos\beta \pm \cos\alpha\sin\beta$

$\cos(\alpha \pm \beta) = \cos\alpha\cos\beta \mp \sin\alpha\sin\beta$

$\tan(\alpha \pm \beta) = \dfrac{\tan\alpha \pm \tan\beta}{1 \mp \tan\alpha\tan\beta}$

$\sin\alpha \pm \sin\beta = 2\sin\frac{1}{2}(\alpha \pm \beta)\cos\frac{1}{2}(\alpha \mp \beta)$

$\cos\alpha + \cos\beta = 2\cos\frac{1}{2}(\alpha + \beta)\cos\frac{1}{2}(\alpha - \beta)$

$\cos\alpha - \cos\beta = -2\sin\frac{1}{2}(\alpha + \beta)\sin\frac{1}{2}(\alpha - \beta)$

Binomial Theorem

$$(1 + x)^n = 1 + \frac{nx}{1!} + \frac{n(n-1)x^2}{2!} + \cdots \qquad (x^2 < 1)$$

Exponential Expansion

$$e^x = 1 + x + \frac{x^2}{2!} + \frac{x^3}{3!} + \cdots$$

Logarithmic Expansion

$$\ln(1 + x) = x - \tfrac{1}{2}x^2 + \tfrac{1}{3}x^3 - \cdots \qquad (|x| < 1)$$

Trigonometric Expansions (θ in radians)

$$\sin\theta = \theta - \frac{\theta^3}{3!} + \frac{\theta^5}{5!} - \cdots$$

$$\cos\theta = 1 - \frac{\theta^2}{2!} + \frac{\theta^4}{4!} - \cdots$$

$$\tan\theta = \theta + \frac{\theta^3}{3} + \frac{2\theta^5}{15} + \cdots$$

Cramer's Rule

Two simultaneous equations in unknowns x and y,

$$a_1x + b_1y = c_1 \quad \text{and} \quad a_2x + b_2y = c_2,$$

have the solutions

$$x = \frac{\begin{vmatrix} c_1 & b_1 \\ c_2 & b_2 \end{vmatrix}}{\begin{vmatrix} a_1 & b_1 \\ a_2 & b_2 \end{vmatrix}} = \frac{c_1b_2 - c_2b_1}{a_1b_2 - a_2b_1}$$

and

$$y = \frac{\begin{vmatrix} a_1 & c_1 \\ a_2 & c_2 \end{vmatrix}}{\begin{vmatrix} a_1 & b_1 \\ a_2 & b_2 \end{vmatrix}} = \frac{a_1c_2 - a_2c_1}{a_1b_2 - a_2b_1}.$$

Products of Vectors

Let $\hat{i}, \hat{j},$ and $\hat{k}$ and be unit vectors in the $x, y,$ and z directions. Then

$$\hat{i} \cdot \hat{i} = \hat{j} \cdot \hat{j} = \hat{k} \cdot \hat{k} = 1, \quad \hat{i} \cdot \hat{j} = \hat{j} \cdot \hat{k} = \hat{k} \cdot \hat{i} = 0,$$

$$\hat{i} \times \hat{i} = \hat{j} \times \hat{j} = \hat{k} \times \hat{k} = 0,$$

$$\hat{i} \times \hat{j} = \hat{k}, \quad \hat{j} \times \hat{k} = \hat{i}, \quad \hat{k} \times \hat{i} = \hat{j}.$$

Any vector $\vec{a}$ with components $a_x, a_y,$ and a_z along the $x, y,$ and z axes can be written as

$$\vec{a} = a_x\hat{i} + a_y\hat{j} + a_z\hat{k}.$$

Let $\vec{a}, \vec{b},$ and $\vec{c}$ be arbitrary vectors with magnitudes $a, b,$ and c. Then

$$\vec{a} \times (\vec{b} + \vec{c}) = (\vec{a} \times \vec{b}) + (\vec{a} \times \vec{c})$$

$$(s\vec{a}) \times \vec{b} = \vec{a} \times (s\vec{b}) = s(\vec{a} \times \vec{b}) \quad (s = \text{a scalar}).$$

Let θ be the smaller of the two angles between $\vec{a}$ and $\vec{b}$. Then

$$\vec{a} \cdot \vec{b} = \vec{b} \cdot \vec{a} = a_xb_x + a_yb_y + a_zb_z = ab \cos \theta$$

$$\vec{a} \times \vec{b} = -\vec{b} \times \vec{a} = \begin{vmatrix} \hat{i} & \hat{j} & \hat{k} \\ a_x & a_y & a_z \\ b_x & b_y & b_z \end{vmatrix}$$

$$= \hat{i}\begin{vmatrix} a_y & a_z \\ b_y & b_z \end{vmatrix} - \hat{j}\begin{vmatrix} a_x & a_z \\ b_x & b_z \end{vmatrix} + \hat{k}\begin{vmatrix} a_x & a_y \\ b_x & b_y \end{vmatrix}$$

$$= (a_yb_z - b_ya_z)\hat{i} + (a_zb_x - b_za_x)\hat{j} + (a_xb_y - b_xa_y)\hat{k}$$

$$|\vec{a} \times \vec{b}| = ab \sin \theta$$

$$\vec{a} \cdot (\vec{b} \times \vec{c}) = \vec{b} \cdot (\vec{c} \times \vec{a}) = \vec{c} \cdot (\vec{a} \times \vec{b})$$

$$\vec{a} \times (\vec{b} \times \vec{c}) = (\vec{a} \cdot \vec{c})\vec{b} - (\vec{a} \cdot \vec{b})\vec{c}$$

Derivatives and Integrals

In what follows, the letters u and v stand for any functions of x, and a and m are constants. To each of the indefinite integrals should be added an arbitrary constant of integration. The *Handbook of Chemistry and Physics* (CRC Press Inc.) gives a more extensive tabulation.

Derivatives

1. $\dfrac{dx}{dx} = 1$

2. $\dfrac{d}{dx}(au) = a\dfrac{du}{dx}$

3. $\dfrac{d}{dx}(u + v) = \dfrac{du}{dx} + \dfrac{dv}{dx}$

4. $\dfrac{d}{dx}x^m = mx^{m-1}$

5. $\dfrac{d}{dx}\ln x = \dfrac{1}{x}$

6. $\dfrac{d}{dx}(uv) = u\dfrac{dv}{dx} + v\dfrac{du}{dx}$

7. $\dfrac{d}{dx}e^x = e^x$

8. $\dfrac{d}{dx}\sin x = \cos x$

9. $\dfrac{d}{dx}\cos x = -\sin x$

10. $\dfrac{d}{dx}\tan x = \sec^2 x$

11. $\dfrac{d}{dx}\cot x = -\csc^2 x$

12. $\dfrac{d}{dx}\sec x = \tan x \sec x$

13. $\dfrac{d}{dx}\csc x = -\cot x \csc x$

14. $\dfrac{d}{dx}e^u = e^u\dfrac{du}{dx}$

15. $\dfrac{d}{dx}\sin u = \cos u\dfrac{du}{dx}$

16. $\dfrac{d}{dx}\cos u = -\sin u\dfrac{du}{dx}$

Integrals

1. $\int dx = x$

2. $\int au\, dx = a \int u\, dx$

3. $\int (u + v)\, dx = \int u\, dx + \int v\, dx$

4. $\int x^m dx = \dfrac{x^{m+1}}{m + 1} \quad (m \neq -1)$

5. $\int \dfrac{dx}{x} = \ln |x|$

6. $\int u\, \dfrac{dv}{dx}\, dx = uv - \int v\, \dfrac{du}{dx}\, dx$

7. $\int e^x dx = e^x$

8. $\int \sin x\, dx = -\cos x$

9. $\int \cos x\, dx = \sin x$

10. $\int \tan x\, dx = \ln |\sec x|$

11. $\int \sin^2 x\, dx = \tfrac{1}{2}x - \tfrac{1}{4} \sin 2x$

12. $\int e^{-ax} dx = -\dfrac{1}{a} e^{-ax}$

13. $\int x e^{-ax} dx = -\dfrac{1}{a^2} (ax + 1)e^{-ax}$

14. $\int x^2 e^{-ax} dx = -\dfrac{1}{a^3} (a^2 x^2 + 2ax + 2)e^{-ax}$

15. $\int_0^{\infty} x^n e^{-ax} dx = \dfrac{n!}{a^{n+1}}$

16. $\int_0^{\infty} x^{2n} e^{-ax^2} dx = \dfrac{1 \cdot 3 \cdot 5 \cdots (2n - 1)}{2^{n+1} a^n} \sqrt{\dfrac{\pi}{a}}$

17. $\int \dfrac{dx}{\sqrt{x^2 + a^2}} = \ln(x + \sqrt{x^2 + a^2})$

18. $\int \dfrac{x\, dx}{(x^2 + a^2)^{3/2}} = -\dfrac{1}{(x^2 + a^2)^{1/2}}$

19. $\int \dfrac{dx}{(x^2 + a^2)^{3/2}} = \dfrac{x}{a^2(x^2 + a^2)^{1/2}}$

20. $\int_0^{\infty} x^{2n+1} e^{-ax^2} dx = \dfrac{n!}{2a^{n+1}} \quad (a > 0)$

21. $\int \dfrac{x\, dx}{x + d} = x - d \ln(x + d)$

APPENDIX F | Properties of Common Elements

All physical properties are for a pressure of 1 atm unless otherwise specified.

Element	Symbol	Atomic Number Z	Molar Mass, g/mol	Density, g/cm³ at 20°C	Melting Point, °C	Boiling Point, °C	Specific Heat, J/(g·°C) at 25°C
Aluminum	Al	13	26.9815	2.699	660	2450	0.900
Antimony	Sb	51	121.75	6.691	630.5	1380	0.205
Argon	Ar	18	39.948	1.6626×10^{-3}	−189.4	−185.8	0.523
Arsenic	As	33	74.9216	5.78	817 (28 atm)	613	0.331
Barium	Ba	56	137.34	3.594	729	1640	0.205
Beryllium	Be	4	9.0122	1.848	1287	2770	1.83
Bismuth	Bi	83	208.980	9.747	271.37	1560	0.122
Boron	B	5	10.811	2.34	2030	—	1.11
Bromine	Br	35	79.909	3.12 (liquid)	−7.2	58	0.293
Cadmium	Cd	48	112.40	8.65	321.03	765	0.226
Calcium	Ca	20	40.08	1.55	838	1440	0.624
Carbon	C	6	12.01115	2.26	3727	4830	0.691
Cesium	Cs	55	132.905	1.873	28.40	690	0.243
Chlorine	Cl	17	35.453	3.214×10^{-3} (0°C)	−101	−34.7	0.486
Chromium	Cr	24	51.996	7.19	1857	2665	0.448
Cobalt	Co	27	58.9332	8.85	1495	2900	0.423
Copper	Cu	29	63.54	8.96	1083.40	2595	0.385
Fluorine	F	9	18.9984	1.696×10^{-3} (0°C)	−219.6	−188.2	0.753
Gadolinium	Gd	64	157.25	7.90	1312	2730	0.234
Gallium	Ga	31	69.72	5.907	29.75	2237	0.377
Germanium	Ge	32	72.59	5.323	937.25	2830	0.322
Gold	Au	79	196.967	19.32	1064.43	2970	0.131
Hafnium	Hf	72	178.49	13.31	2227	5400	0.144
Helium	He	2	4.0026	0.1664×10^{-3}	−269.7	−268.9	5.23
Hydrogen	H	1	1.00797	0.08375×10^{-3}	−259.19	−252.7	14.4
Indium	In	49	114.82	7.31	156.634	2000	0.233
Iodine	I	53	126.9044	4.93	113.7	183	0.218
Iridium	Ir	77	192.2	22.5	2447	(5300)	0.130
Iron	Fe	26	55.847	7.874	1536.5	3000	0.447
Krypton	Kr	36	83.80	3.488×10^{-3}	−157.37	−152	0.247
Lanthanum	La	57	138.91	6.189	920	3470	0.195
Lead	Pb	82	207.19	11.35	327.45	1725	0.129
Lithium	Li	3	6.939	0.534	180.55	1300	3.58
Magnesium	Mg	12	24.312	1.738	650	1107	1.03
Manganese	Mn	25	54.9380	7.44	1244	2150	0.481
Mercury	Hg	80	200.59	13.55	−38.87	357	0.138
Molybdenum	Mo	42	95.94	10.22	2617	5560	0.251
Neodymium	Nd	60	144.24	7.007	1016	3180	0.188

Element	Symbol	Atomic Number Z	Molar Mass, g/mol	Density, g/cm³ at 20°C	Melting Point, °C	Boiling Point, °C	Specific Heat, J/(g·°C) at 25°C
Neon	Ne	10	20.183	0.8387×10^{-3}	−248.597	−246.0	1.03
Nickel	Ni	28	58.71	8.902	1453	2730	0.444
Niobium	Nb	41	92.906	8.57	2468	4927	0.264
Nitrogen	N	7	14.0067	1.1649×10^{-3}	−210	−195.8	1.03
Osmium	Os	76	190.2	22.59	3027	5500	0.130
Oxygen	O	8	15.9994	1.3318×10^{-3}	−218.80	−183.0	0.913
Palladium	Pd	46	106.4	12.02	1552	3980	0.243
Phosphorus	P	15	30.9738	1.83	44.25	280	0.741
Platinum	Pt	78	195.09	21.45	1769	4530	0.134
Plutonium	Pu	94	(244)	19.8	640	3235	0.130
Polonium	Po	84	(210)	9.32	254	—	—
Potassium	K	19	39.102	0.862	63.20	760	0.758
Radium	Ra	88	(226)	5.0	700	—	—
Radon	Rn	86	(222)	9.96×10^{-3} (0°C)	(−71)	−61.8	0.092
Rhenium	Re	75	186.2	21.02	3180	5900	0.134
Rubidium	Rb	37	85.47	1.532	39.49	688	0.364
Scandium	Sc	21	44.956	2.99	1539	2730	0.569
Selenium	Se	34	78.96	4.79	221	685	0.318
Silicon	Si	14	28.086	2.33	1412	2680	0.712
Silver	Ag	47	107.870	10.49	960.8	2210	0.234
Sodium	Na	11	22.9898	0.9712	97.85	892	1.23
Strontium	Sr	38	87.62	2.54	768	1380	0.737
Sulfur	S	16	32.064	2.07	119.0	444.6	0.707
Tantalum	Ta	73	180.948	16.6	3014	5425	0.138
Tellurium	Te	52	127.60	6.24	449.5	990	0.201
Thallium	Tl	81	204.37	11.85	304	1457	0.130
Thorium	Th	90	(232)	11.72	1755	(3850)	0.117
Tin	Sn	50	118.69	7.2984	231.868	2270	0.226
Titanium	Ti	22	47.90	4.54	1670	3260	0.523
Tungsten	W	74	183.85	19.3	3380	5930	0.134
Uranium	U	92	(238)	18.95	1132	3818	0.117
Vanadium	V	23	50.942	6.11	1902	3400	0.490
Xenon	Xe	54	131.30	5.495×10^{-3}	−111.79	−108	0.159
Ytterbium	Yb	70	173.04	6.965	824	1530	0.155
Yttrium	Y	39	88.905	4.469	1526	3030	0.297
Zinc	Zn	30	65.37	7.133	419.58	906	0.389
Zirconium	Zr	40	91.22	6.506	1852	3580	0.276

The values in parentheses in the column of molar masses are the mass numbers of the longest-lived isotopes of those elements that are radioactive. Melting points and boiling points in parentheses are uncertain. The data for gases are valid only when these are in their usual molecular state, such as H_2, He, O_2, Ne, etc. The specific heats of the gases are the values at constant pressure. *Primary source*: Adapted fron J. Emsley, *The Elements*, 3rd ed., 1998, Clarendon Press, Oxford (www.webelements.com). Data on newest elements are current.

Periodic Table of the Elements

	Metals
	Metalloids
	Nonmetals

THE HORIZONTAL PERIODS

Alkali metals
IA

Transition metals

VIIIB

Noble gases
0

Period	IA	IIA	IIIB	IVB	VB	VIB	VIIB		VIIIB		IB	IIB	IIIA	IVA	VA	VIA	VIIA	0
1	1 H																	2 He
2	3 Li	4 Be											5 B	6 C	7 N	8 O	9 F	10 Ne
3	11 Na	12 Mg											13 Al	14 Si	15 P	16 S	17 Cl	18 Ar
4	19 K	20 Ca	21 Sc	22 Ti	23 V	24 Cr	25 Mn	26 Fe	27 Co	28 Ni	29 Cu	30 Zn	31 Ga	32 Ge	33 As	34 Se	35 Br	36 Kr
5	37 Rb	38 Sr	39 Y	40 Zr	41 Nb	42 Mo	43 Tc	44 Ru	45 Rh	46 Pd	47 Ag	48 Cd	49 In	50 Sn	51 Sb	52 Te	53 I	54 Xe
6	55 Cs	56 Ba	57-71 *	72 Hf	73 Ta	74 W	75 Re	76 Os	77 Ir	78 Pt	79 Au	80 Hg	81 Tl	82 Pb	83 Bi	84 Po	85 At	86 Rn
7	87 Fr	88 Ra	89-103 †	104 Rf	105 Db	106 Sg	107 Bh	108 Hs	109 Mt	110 Ds	111 Uua	112 Uub	113	114 Uuq	115	116	117	118

Inner transition metals

Lanthanide series *	57 La	58 Ce	59 Pr	60 Nd	61 Pm	62 Sm	63 Eu	64 Gd	65 Tb	66 Dy	67 Ho	68 Er	69 Tm	70 Yb	71 Lu
Actinide series †	89 Ac	90 Th	91 Pa	92 U	93 Np	94 Pu	95 Am	96 Cm	97 Bk	98 Cf	99 Es	100 Fm	101 Md	102 No	103 Lr

The names of elements 104 through 109 (Rutherfordium, Dubnium, Seaborgium, Bohrium, Hassium, and Meitnerium, respectively) were adopted by the International Union of Pure and Applied Chemistry (IUPAC) in 1997. As of May 2003, elements 110, 111, 112, and 114 have been discovered. See www.webelements.com for the latest information and newest elements.

Answers to Reading Exercises and Odd-Numbered Problems

(Answers that involve a proof, graph, or otherwise lengthy solution are not included.)

Chapter 22

RE 22-1: Electric stove, microwave, lights, car starter motor, toothbrush, computer, tape recorder, CD player, FM radio, amplifier, etc.

RE 22-2: (a) Since the two tapes have identical histories, they should have like charges and repel. (b) The observations were consistent with my predictions. The two tapes repelled each other.

RE 22-3: (a) If woodolin was a new type of charge then two wooden rods charged with linen would repel each other. A wooden rod would have to attract *both* the charged amber (or plastic) rod *and* the charged glass rod. (b) According to the text statements, this observation has never been made. It has always been the case that a suspected new type of charge (such as woodolin) always repels either a charged amber rod or a charged glass rod and attracts the other type rod. This makes it the same as one of the existing charges.

RE 22-4: A very simple explanation is that in a solid, all parts are stiff. But since one can melt ice into water and then boil water into a gas (water vapor) the atomic explanation seems quite plausible.

RE 22-5: I would discharge one of the spheres by touching it. Then I would allow the two spheres to touch each other. They should share the charge q equally so each sphere has charge $q/2$. If I *repeat* the process, then each sphere will have charge $(q/2)/2 = q/4$.

RE 22-6: (a) If the paper bits are uncharged, then there is no mutual attraction or repulsion. (b) "Induction" always causes the neutral object to be *attracted* toward the charged object, independent of the sign of the charge on the charged object. So, no, you can't tell the sign of the charge on the charged object in this way.

RE 22-7: (1) *A, B* is attractive (unlike charges), (2) *A, A* is repulsive (like charges), (3) *B, B* is repulsive (like charges), (4) *B, C* attract (by induction), (5) *C, C* nonexistent forces (both neutral), and (6) *C, A* attract (by induction).

RE 22-8: (a) Scotch tape acts like an insulator since charge doesn't draw away as you handle the tape at its ends. (b) A balloon behaves like an insulator, because when you charge it, it can stick by induction to a wall rather than touch and pull away.

RE 22-9: (a) No, since charges on it are not mobile. (b) If we start with a positively charged glass plate as the bottom plate in Fig. 22-10 and perform all the same steps, the aluminum pie plate will be negatively charged.

RE 22-10: All of these assertions are inconsistent with the experimental results in the text.

RE 22-11: (a) The central proton is attracted toward the electron, so this force is to the left. (b) The central proton is repelled by the other proton, so this force is also to the left. (c) Thus the net force on the central proton is to the left. (d) There are no locations along the line connecting the charges where the force on the former central proton can be zero. Since the magnitudes of the charges on the proton and electron are the same, the only location where the force magnitudes on the other proton are zero is halfway between the first two, but we know the forces don't cancel there.

Problems

1. -1.32×10^{13} C **3.** 6.3×10^{11} **5.** 122 mA **7.** (a) positron; (b) electron **9.** 1.38 m **11.** (a) 4.9×10^{-7} kg; (b) 7.1×10^{-11} C **13.** (a) 0.17 N; (b) -0.046 N **15.** either -1.00 μC and $+3.00$ μC or $+1.00$ μC and -3.00 μC **17.** (a) charge $-4q/9$ must be located on the line joining the two positive charges, a distance $L/3$ from charge $+q$. **19.** $q = Q/2$ **21.** (a) 3.2×10^{-19} C; (b) two **23.** (a) 0; (b) 1.9×10^{-9} N **25.** (a) 6.05 cm; (b) 6.05 cm from central bead **27.** $+13e$ **29.** (a) positive; (b) $+9$ **31.** 9.0 kN **33.** $1.72a$, directly rightward **35.** -11.1 μC **37.** $q = 0.71Q$ **39.** (b) $1e, 0.654$ rad; $2e, 0.889$ rad; $3e, 0.988$ rad; $4e, 1.047$ rad; $5e, 1.088$ rad **41.** (a) Let $J = qQ/4\pi\varepsilon_0 d^2$. For $\alpha < 0, F = -J[\alpha^{-2} + (1 + |\alpha|)^{-2}]$; for $0 < \alpha < 1, F = J[\alpha^{-2} - (1 - \alpha)^{-2}]$; for $1 < \alpha, F = J[\alpha^{-2} + (\alpha - 1)^{-2}]$ **43.** (a) 5.7×10^{13} C, no; (b) 6.0×10^5 kg **45.** (b) $\pm 2.4 \times 10^{-8}$ C

47. (a) $\dfrac{L}{2}\left(1 + \dfrac{1}{4\pi\varepsilon_0}\dfrac{qQ}{Wh^2}\right)$; (b) $\sqrt{3qQ/4\pi\varepsilon_0 W}$

Chapter 23

RE 23-1: $F^{\text{elec}} \propto 1/r^2$. Thus at 4 cm, F^{elec} would be $(1/2)^2 = 1/4$ of its value at 2 cm, or 9 mm. At 6 cm, F^{elec} would be $(1/3)^2 = 1/9$ of its value at 2 cm, or 4 mm.

RE 23-2: Since the force on the test object to the sources, $\vec{F}_{s \to t}$, varies from point to point in space, the test object must be small enough spatially to test the "local" value rather than the average value over too large a volume of space.

RE 23-3: The type of test charge makes no difference! For a negative test charge we would still use Eq. 23-9 to determine the electric field vector. But, the new $(\vec{F}^{\text{elec}})' = -\vec{F}^{\text{elec}}$ and the new negative charge $q_t' = -q_t$. So $\vec{E}_s'$ will equal E_s' (no change).

RE 23-4: (a) Rightward, (b) leftward, (c) leftward, (d) rightward (p and e have the same charge magnitude and p is farther).

RE 23-5: (a) To the left, (b) to the left in a parabolic path, (c) its speed decreases at first, then increases. It will move in a straight line first rightward, then leftward.

RE 23-6: All four experience the same magnitude torque.

RE 23-7: Near a positive charge, $\vec{E}$ points always *away* from the charge; near a negative charge, $\vec{E}$ points always *toward* the charge.

RE 23-8: Just as for the two equidistant point charges in Fig. 23-10, we can "pair up" equal patches of charge equidistant from the point at which we are calculating $\vec{E}$ for all such patches of charge on the sheet, canceling the contributions to $\vec{E}$ parallel to the sheet.

Problems

1. 56 pC **3.** 3.07×10^{21} N/C, radially outward **5.** 50 cm from q_A and 100 cm from q_B **7.** 0 **9.** 1.02×10^5 N/C, upward **11.** (a) 47 N/C; (b) 27 N/C **13.** $4kQ/3d^2$ or $Q/3\pi\varepsilon_0 d^2$ **15.** 1.38×10^{-10} N/C, $180°$ from $+x$ **17.** 6.88×10^{-28} C · m **23.** $q/\pi^2\varepsilon_0 r^2$, vertically downward **25.** (a) $-q/L$; (b) $q/4\pi\varepsilon_0 a(L+a)$ **29.** (a) -1.72×10^{-15} C/m; (b) -3.82×10^{-14} C/m²; (c) -9.56×10^{-15} C/m²; (d) -1.43×10^{-12} C/m³ **31.** $E = 2k|Q|(\sin \theta/2)/\theta R^2$ **33.** $217°$ **35.** 3.51×10^{15} m/s² **37.** 6.6×10^{-15} N **39.** (a) 1.5×10^3 N/C; (b) 2.4×10^{-16} N, up; (c) 1.6×10^{-26} N; (d) 1.5×10^{10} **41.** (a) 1.92×10^{12} m/s²; (b) 1.96×10^5 m/s **43.** (a) 2.7×10^6 m/s; (b) 1000 N/C **45.** 27 μm **47.** (a) yes; (b) upper plate, 2.73 cm **49.** (a) 27 km/s; (b) 50 μm **51.** 5.2 cm **53.** (a) 0; (b) 8.5×10^{-22} N · m; (c) 0 **55.** $(1/2\pi)\sqrt{pE/I}$ **57.** 1.92×10^{-21} J **59.** (a) 6.4×10^{-18} N; (b) 20 N/C **63.** (a) to the right in the figure; (b) $(2kqQ \cos 60°)/a^2$

Chapter 24

RE 24-1: (a) $\phi = \vec{v} \cdot \Delta\vec{A} = (3 \text{ m/s})(2 \times 10^{-4} \text{ m}^2) \cos 60° = (3 \times 10^{-4} \text{ m}^3/\text{s})$. Whatever fluid that is represented by this vector velocity field is flowing through this surface area dA. (b) $\phi = \vec{E} \cdot \Delta\vec{A} = (3 \text{ N/C})(2 \times 10^{-4} \text{ m}^2) \cos 60° = 3 \times 10^{-4} \text{ N} \cdot \text{m}^2/\text{C}$. Nothing is flowing through the small area. Instead, the flux represents the product of the E-field component normal to the area.

RE 24-2: To find the answers we simply sum the flux through all six faces. We get $\phi^{\text{net}}_{\text{cube 1}} = 0 \text{ N} \cdot \text{m}^2/\text{C}$, $\phi^{\text{net}}_{\text{cube 2}} = +5 \text{ N} \cdot \text{m}^2/\text{C}$, and $\phi^{\text{net}}_{\text{cube 3}} = -3 \text{ N} \cdot \text{m}^2/\text{C}$. (a) Cube 2, (b) cube 3, and (c) cube 1.

RE 24-3: The central charge always acts along the central line. For each noncentral charge (for example, the one to the left) that acts at a point on this central line, there is a conjugate charge (in this example, the one to the right of center) that is exactly the same distance from the point as the original point. The E-field vectors have the same magnitude. The E-components perpendicular to the plane act in the same direction and add vectorially. The parallel components act in opposite directions and cancel.

RE 24-4: Since Gauss' law states that $\phi^{\text{net}} = q^{\text{enc}}/\varepsilon_0$ as long as the same charge is enclosed by the new Gaussian surfaces, ϕ^{net} is unchanged.

RE 24-5: Negative charges would be induced on the inside surface of the cavity so that $q^{\text{enc}} = q^{\text{induced}} + q^{\text{center}} = 0$. Thus, the net flux at the cavity's Gaussian surface would be zero.

Problems

1. (a) 0; (b) $-3.92 \text{ N} \cdot \text{m}^2/\text{C}$; (c) 0; (d) 0 for each field **3.** $2.0 \times 10^5 \text{ N} \cdot \text{m}^2/\text{C}$ **5.** (a) $8.23 \text{ N} \cdot \text{m}^2/\text{C}$; (b) $8.23 \text{ N} \cdot \text{m}^2/\text{C}$; (c) 72.8 pC in each case **7.** 3.54 μC **9.** 0 through each of the three faces meeting at q, $q/24\varepsilon_0$ through each of the other faces **11.** -7.5 nC **15.** -1.04 nC **19.** (a) $E = (q/4\pi\varepsilon_0 a^3)r$; (b) $E = q/4\pi\varepsilon_0 r^2$; (c) 0; (d) 0; (e) inner, $-q$; outer, 0 **21.** $q/2\pi a^2$ **23.** $6K\varepsilon_0 r^3$ **25.** 5.0 μC/m **27.** (a) $E = q/2\pi\varepsilon_0 LR$, radially inward; (b) $-q$ on both inner and outer surfaces; (c) $E = q/2\mu\varepsilon_0 Lr$, radially outward **29.** (a) 2.3×10^6 N/C, radially out; (b) 4.5×10^5 N/C, radially in **31.** (b) $\rho R^2/2\varepsilon_0 r$ **33.** (a) 5.3×10^7 N/C; (b) 60 N/C **35.** 5.0 nC/m² **37.** 0.44 mm **39.** 2.0 μC/m² **41.** (a) 37 μC; (b) $4.1 \times 10^6 \text{ N} \cdot \text{m}^2/\text{C}$

Chapter 25

RE 25-1: Question 1: Because charges that are infinitely far apart exert no forces on each other. Question 2: Zero separation between particles would involve infinite attractive or repulsive forces.

RE 25-2: (a) If we assume the E-field does not change as a result of the reconfiguration of the charge then the positive charge displacement is opposite to the direction of the E-field, so the E-field does negative work. (b) It takes external work to move the charge against the field so ΔU increases, and (c) because we are interested in the *change* of electric potential between points 1 and 2.

RE 25-3: (a) The external force does positive work. (b) The proton moves to a higher potential so $V_2 > V_1$.

RE 25-4: (a) The E-field acts from left to right. (b) Positive external work is done on the electron in paths 1, 2, 3, and 5. Negative work is done on Path 4. (c) $\Delta V_3 > \Delta V_1 = \Delta V_2 = \Delta V_5 > \Delta V_4$.

RE 25-5: Given the charge distribution, we can simply add the contribution to the potential at a point P due to each of the charges, taken separately, using Eq. 25-25. If all we know is $\vec{E}(\vec{r})$ then we must calculate $V(\vec{r})$ using Eq. 25-17.

RE 25-6: V at P is the same for all three of these configurations. The potential at P due to each proton only depends on how far away that proton is from P and not on the direction.

RE 25-7: Using Eq. 25-29 for case (a) $\theta = 0$ and so $\cos\theta = +1$, for case (b) $\theta = 180°$ and $\cos\theta = -1$, and for case (c) $\theta = 90°$ so $\cos\theta = 0$. All other terms remain constant, so ranked from most to least positive, $V_a > V_c > V_b$.

RE 25-8: (a) $E_2 > E_1 = E_3$, (b) Pair 3. (c) It accelerates leftward.

RE 25-9: Since potential energy is a scalar quantity, its superposition involves only scalar addition while the superposition of electric fields requires adding vectors.

RE 25-10: (a) A is wrong since it originates on $-$ and terminates on $+$. B is wrong since it is not perpendicular to the plate near the plate. C is wrong since it has a kink. D is wrong for the same reason as A. E is wrong since it both originates and terminates on a $+$ charge. F is ok. (b) A correct drawing would have curves like A, D, and F but with arrows pointing toward the negatively-charged sphere.

RE 25-11: Because her skin is a conductor and thus an equipotential surface. Charges will redistribute so they have a higher density near the top of her head, which has more curvature than the sides of her head. The strength of the electric field is higher where the charges bunch so the equipotential surfaces are closer together than they were.

Problems

1. (a) 3.0×10^5 C; (b) 3.6×10^6 J **3.** (a) 3.0×10^{10} J; (b) 7.7 km/s; (c) 9.0×10^4 kg **5.** 8.8 mm **7.** (a) 136 MV/m; (b) 8.82 kV/m **9.** (b) because $V = 0$ point is chosen differently; (c) $q/(8\pi\varepsilon_0 R)$; (d) potential differences are independent of the choice for the $V = 0$ point

11. (a) $Q/4\pi\varepsilon_0 r$; (b) $\dfrac{\rho}{3\varepsilon_0}\left(\dfrac{3}{2}r_2^2 - \dfrac{1}{2}r^2 - \dfrac{r_1^3}{r}\right)$, $\rho = \dfrac{Q}{\frac{4\pi}{3}(r_2^3 - r_1^3)}$;

(c) $\dfrac{\rho}{2\varepsilon_0}(r_2^2 - r_1^2)$, with ρ as in (b); (d) yes **13.** (a) -4.5 kV; (b) -4.5 kV

15. $x = d/4$ and $x = -d/2$ **17.** (a) 0.54 mm; (b) 790 V **19.** 6.4×10^8 V **21.** $2.5q/4\pi\varepsilon_0 d$ **23.** $-0.21q^2/\varepsilon_0 a$ **25.** (a) $+6.0 \times 10^4$ V; (b) -7.8×10^5 V; (c) 2.5 J; (d) increase; (e) same; (f) same

27. $W = \dfrac{qQ}{8\pi\varepsilon_0}\left(\dfrac{1}{r_1} - \dfrac{1}{r_2}\right)$ **29.** 2.5 km/s **31.** (a) 0.225 J; (b) A, 45.0 m/s²; B, 22.5 m/s²; (c) A, 7.75 m/s, B, 3.87 m/s **33.** 0.32 km/s **35.** 1.6×10^{-9} m **39.** $(c/4\pi\varepsilon_0)[L - d \ln(1 + L/d)]$ **41.** 17 V/m at 135° counterclockwise from $+x$ **45.** (a) $\dfrac{Q}{4\pi\varepsilon_0 d(d + L)}$, leftward; (b) 0 **47.** 2.5×10^{-8} C **49.** (a) -180 V; (b) 2700 V, -8900 V **51.** (a) -0.12 V; (b) 1.8×10^{-8} N/C, radially inward

Chapter 26

RE 26-1: Volta probably felt a tingling sensation or perhaps a shock or jolt that would cause him to let go of the terminals.

RE 26-2: "Circuit" means a full round trip around some route. This is just what the electric charge does.

RE 26-3: (a) If the overall circuit had $q^{net} = 0$ before the switch was closed, it will remain charge neutral after the switch is closed since the circuit is a closed system and charge is neither created nor destroyed, but merely flows around the circuit. (b) Individual wires in the circuit can and do acquire a (small) net positive or negative charge, but this charge must come from other parts of the circuit.

RE 26-4: Electrical current *is* the net transport of charge past a given point in a circuit in a given time. If equal amounts of positive charge moving, say, right, and negative charge moving right go past the same point, there is no net transport of charge past that point, so $i = 0$ A.

RE 26-5: Let's assume that the unknown current i flows from right to left. Then the net current flowing into $(+)$ or out of $(-)$ the *middle* node is $(+ 2 + 3 + 4 - 1 + 2 - 2 + i)$ A. But currents must all add to zero at this node. Thus $i = -8$ A, meaning our assumption was wrong and that $i = 8$ A flowing from left to right.

RE 26-6: A voltmeter is attached *across* a circuit element because it is designed to measure the potential difference *between* the ends of the circuit. An ammeter is inserted in a branch of a circuit because it is designed to measure the current *through* that part of the circuit. In a series circuit where there are no branches or alternate paths for current to flow, it doesn't matter whether the ammeter is placed before or after a series circuit element.

RE 26-7: Device 1 is ohmic since $(\Delta V/i) = 2.25\ \Omega = $ constant and $i = 0$ A when $\Delta V = 0$ V. Device 2 is nonohmic since $(\Delta V/i) \neq$

constant. Device 3 is nonohmic. Although a plot of ΔV vs. i is a straight line, i is nonzero at $\Delta V = 0$, so i is not proportional to ΔV.

RE 26-8: If the cross-sectional area of the Nichrome wire is cut in half, its resistance will double, so the slope of the i vs. ΔV graph which is $1/R$ will be cut in half.

RE 26-9: (a) $R \propto 1/r^2$ for most wires, suggesting the current flows through the whole cross-sectional area of the wire, not just on its surface as indicated in Eq. 26-8. (b) If the current flowed only in a thin layer near the surface then I'd expect $R \propto 1/r$ since the circumference is $2\pi r$ for a wire with a circular cross section.

RE 26-10: Since $R = \rho L/A$, $(a) = (c) > (b)$.

RE 26-11: $(a) = (b) > (d) > (c)$.

RE 26-12: Only the cross-sectional area A matters in comparing current densities, so $(a) = (d) > (b) = (c)$.

RE 26-13: Since the current density is $(I/A) = (\Delta V/(RA))$ and $RA = \rho L$, we see here that the current density is just inversely proportional to the length of each wire. So $(b) = (d) > (a) = (c)$.

Problems

1. (a) 1200 C; (b) 7.5×10^{21} **3.** 5.6 ms **5.** 100 V **7.** $2.0 \times 10^{-8}\ \Omega \cdot$ m **9.** 2.4 Ω **11.** 54 Ω **13.** 3.0 **15.** (a) 0.43%, 0.0017%, 0.0034% **17.** 560 W **19.** (a) 1.0 kW; (b) 25 ¢ **21.** 0.135 W **23.** (a) 10.9 A; (b) 10.6 Ω; (c) 4.5 MJ **25.** 660 W **27.** (a) 3.1×10^{11}; (b) 25 μA; (c) 1300 W, 25 MW **29.** (a) 17 mV/m; (b) 243 J **31.** (a) 6.4 A/m², north; (b) no, cross-sectional area **33.** 0.38 mm **35.** (a) 2×10^{12}; (b) 5000; (c) 10 MV **37.** 13 min **39.** $8.2 \times 10^{-4}\ \Omega \cdot$ m **41.** (a) 0.67A; (b) toward the negative terminal **43.** (a) 1.73 cm/s; (b) 3.24 pA/m²

Chapter 27

RE 27-1: $R = \rho L/A$; $A = \pi r^2 = \frac{1}{4}\pi d^2$

$\rho_{Cu} = 1.7 \times 10^{-8}\ \Omega \cdot$m; $d = 2.4 \times 10^{-4}$ m; $L = 0.30$ m

$\therefore R = (1.7 \times 10^{-8}\ \Omega \cdot\text{m})(0.30\ \text{m})/(\frac{1}{4}\pi (2.4 \times 10^{-4}\ \text{m})^2)$

$= 0.113\ \Omega.$

RE 27-2: (a) If all the current were "used up" in the first bulb, the second and third bulbs would be dark. (b) If most of the current were "used up" in the first bulb, the second bulb would glow more dimly than the first and the third bulb would glow more dimly than the second. (c) If only a small amount were "used up" in the first bulb, the third would be dimmer than the second, and the second would be a bit dimmer than the first.

RE 27-3: $i_a = i_b = i_c$ and $V_b > V_c > V_a$.

RE 27-4: (a) $i_1 = i_2 = i_3$. (b) $\Delta V_1 > \Delta V_2 > \Delta V_3$.

RE 27-5: Since the ammeter is wired *in series* with the resistors, its resistance *adds* to theirs. (a) With no ammeter, the current will be largest. (b) with $R_A \ll R_1 + R_2$ the current will be reduced, but only a little. (c) With $R_A = R_1 + R_2$ the current would be cut in half. Thus a good ammeter should have as *small* a resistance as possible.

RE 27-6: (a) $R_1 = R_2$ in series so $i_1 = i_2$ and $\Delta V_1 = \Delta V_2 = \frac{1}{2}\Delta V_B$ and so $i = i_1 = i_2 = \Delta V_B/(R_1 + R_2)$. (b) $R_1 = R_2$ in parallel so $i_1 = i_2$ and $\Delta V_1 = \Delta V_2 = \Delta V_B$ so now $i = i_1 + i_2 = 2\Delta V_B/R_1 = 2\Delta V_B/R_2$.

RE 27-7: Note that R_V is in parallel with R_1. Thus if $R_V \ll R_1$, the effective resistance between d and e in Fig. 27-7 would be dramatically decreased from R_1 to less than R_V. This would "pull down" ΔV_{de} to a smaller value that it had before I installed the voltmeter. However, if $R_V \gg R_1$, then the effective resistance between d and e remains just about R_1 and the value of ΔV_{de} is about what it was without the meter present. Thus $R_V \gg R$ gives more accurate measurements of potential differences.

RE 27-8: Since the bulbs are identical and wired in parallel, $i_1 = i_2 = i_3$. If only one bulb were connected to the battery its brightness would be the same as before, since the potential difference across it is still ΔV_B.

Problems

1. (a) 30 Ω; (b) clockwise; (c) A **3.** (a) 45 Ω; (b) 0.33 A each; (c) 0.33 A **5.** $V_1 = 3.5$ V; $V_2 = 4.3$ V; $V_3 = 7.2$ V **7.** 8.0 Ω **9.** (a) 0; (b) 1.25 A, downward **11.** (a) 120 Ω; (b) $i_1 = 51$ mA, $i_2 = i_3 = 19$ mA, $i_4 = 13$ mA **13.** 20 Ω **15.** (a) bulb 2; (b) bulb 1 **17.** 0.45 A **19.** $i_1 = -50$ mA, $i_2 = 60$ mA, $V_{ab} = 9.0$ V **21.** (a) Cu: 1.11 A, Al: 0.893 A; (b) 126 m **23.** 5.56 A **25.** $3d$ **29.** nine **31.** providing energy, 360 W **33.** (a) 3.0 A, downward; (b) 1.6 A, downward; (c) 6.4 W, supplying; (d) 55.2 W, supplying **35.** (a) 12 eV (1.9×10^{-18} J); (b) 6.5 W **39.** (a) 7.50 A, leftward; (b) 10.0 A, leftward; (c) 87.5 W, supplied **41.** (a) 0.33 A, rightward; (b) 720 J **43.** (a) \$320; (b) 4.8 cents **45.** 14 h 24 min **47.** (a) 0.50 A; (b) $P_1 = 1.0$ W, $P_2 = 2.0$ W; (c) $P_1 = 6.0$ W supplied, $P_2 = 3.0$ W absorbed **49.** (a) $V_T = -ir + \mathcal{E}$; (b) 13.6 V; (c) 0.060 Ω **51.** (a) 14 V; (b) 100 W; (c) 600 W; (d) 10 V, 100 W **53.** (a) 50 V; (b) 48 V; (c) B is connected to the negative terminal **55.** (a) $r_1 - r_2$; (b) battery with r_1 **59.** (a) $R = r/2$; (b) $P^{max} = \varepsilon^2/2r$ **61.** (a) 0.346 W; (b) 0.050 W; (c) 0.709 W; (d) 1.26 W; (e) -0.158 W **63.** (a) battery 1, 0.67 A down; battery 2, 0.33 A up; battery 3, 0.33 A up; (b) 3.3 V

Chapter 28

RE 28-1: The capacitance of a capacitor remains the same, whatever the amount of excess charge on its plates and whatever potential difference is applied across it. Doubling $|q|$ doubles ΔV_c while tripling ΔV_c triples $|q|$.

RE 28-2: Each of these three types of capacitors becomes electrically isolated when removed from a battery so the excess charge on each of the "plates" does not change.

RE 28-3: In these cases ΔV is constant and C and hence $|q|$ must change when spacings change, so $|q|$ (a) decreases, (b) increases, and (c) decreases.

RE 28-4: Each capacitor initially has the same $|q|$ and the same $|\Delta V|$. (a) Wiring them in parallel, positive plate to positive and negative to negative, leaves $|q|$ and $|\Delta V|$ on each unchanged. Wiring them in parallel, positive to negative, makes $|\Delta V| = 0$ and so $|q| = 0$ on each. (b) Wiring them in series leaves these quantities unchanged.

RE 28-5: (a) Since $i_0 = |\Delta V_B|/R$, $(i_0)_1 > (i_0)_2 > (i_0)_4 > (i_0)_3$. (b) Since $t_{(1/2)}$ is proportional to $\tau = RC$, $(t_{(1/2)})_4 > (t_{(1/2)})_1 = (t_{(1/2)})_2 > (t_{(1/2)})_3$.

Problems

1. 7.5 pC **3.** 3.0 mC **5.** (a) 140 pF; (b) 17 nC **7.** $5.04\pi\varepsilon_0 R$ **11.** 9090 **13.** 3.16 μF **17.** 43 pF **19.** (a) 50 V; (b) 5.0×10^{-5} C; (c) 1.5×10^{-4} C

21. $q_1 = \dfrac{C_1C_2 + C_1C_3}{C_1C_2 + C_1C_3 + C_2C_3} C_1\Delta V_0$,

$q_2 = q_3 = \dfrac{C_2C_3}{C_1C_2 + C_1C_3 + C_2C_3} C_1\Delta V_0$

23. 72 F **25.** 0.27 J **27.** (a) 2.0 J **29.** (a) $2\Delta V$; (b) $U_i = \varepsilon_0 A\Delta V^2/2d$, $U_f = 2U_i$; (c) $\varepsilon_0 A\Delta V^2/2d$ **35.** Pyrex **37.** 81 pF/m **39.** 0.63 m^2 **43.** (a) 10 kV/m; (b) 5.0 nC; (c) 4.1 nC

45. (a) $C = 4\pi\varepsilon_0\kappa\left(\dfrac{ab}{b - a}\right)$; (b) $q = 4\pi\varepsilon_0\kappa\Delta V\left(\dfrac{ab}{b - a}\right)$; (c) $q' = q(1 - 1/\kappa)$ **47.** 4.6 **49.** (a) 2.41 μs; (b) 161 pF **51.** (a) 2.17 s; (b) 39.6 mV **53.** (a) 1.0×10^{-3} C; (b) 1.0×10^{-3} A; (c) $\Delta V_C = 1.0 \times 10^3$ e^{-t} V, $\Delta V_R = 1.0 \times 10^3$ e^{-t} V; (d) P $= $ e^{-2t} W

Chapter 29

RE 29-1: (a) z axis, (b) $-x$ axis, (c) no direction since $\vec{F} = 0$ N.

RE 29-2: (a) The electron, because it's less massive and "bends" more easily in the presence of a perpendicular force, (b) the electron travels clockwise.

RE 29-3: $\vec{F}^{net} = \vec{F}^{elec} + \vec{F}^{mag}$. The force exerted on the charge by the E-field is the same in all 4 cases and points out of the page. In cases 1 and 3, $\vec{B}$ and $\vec{v}$ are parallel so there is no magnetic force on the charged particle. In cases 2 and 4, $\vec{B}$ and $\vec{v}$ are perpendicular with magnetic forces out of and into the page respectively. (a) In terms of force magnitude $|\vec{F}_2^{net}| > |\vec{F}_1^{net}| = |\vec{F}_3^{net}|$. $|\vec{F}_4^{net}|$ can take on any value from zero to larger than $|\vec{F}^{net}|$ and so can not be ranked. (b) A zero net force is only possible for case 4.

RE 29-4: The equation $|\vec{F}^{mag}| = |i\vec{L} \times \vec{B}|$ is a maximum for a given $|\vec{B}|$ when $\vec{B}$ is perpendicular to both $\vec{F}^{mag}$ and $\vec{L}$. This is true whenever $\vec{B} = \pm|\vec{B}|\hat{j}$. Trying each direction, the right-hand rule yields $\vec{B}$ pointing along the $-y$ axis.

RE 29-5: (a) $\tau = |\vec{\mu}||\vec{B}| \sin\phi$ where $\phi = \theta$ for cases 2 and 3 and $\phi = \pi - \theta$ for cases 1 and 4. But $\sin\theta = \sin(\pi - \theta)$ so τ is the same for all 4 cases. (b) $U(\phi) = -\vec{\mu}\cdot\vec{B} = -|\vec{\mu}||\vec{B}| \cos\phi$. Now for cases 2 and 3, $\phi = \theta < \pi/2$ so $\cos\theta > 0$, and for cases 1 and 4 $\phi = \pi - \theta > \pi/2$ $\cos\phi = -\cos\theta < 0$, thus $U_1 = U_4 > U_3 = U_2$.

Problems

1. (a) 6.2×10^{-18} N; (b) 9.5×10^8 m/s^2; (c) remains equal to 550 m/s **3.** (a) 400 km/s; (b) 835 eV **5.** (a) east; (b) 6.28×10^{14} m/s^2; (c) 2.98 mm **7.** 21 μT **9.** (a) 2.05×10^7 m/s; (b) 467 μT; (c) 13.1 MHz; (d) 76.3 ns **11.** (a) 0.978 MHz; (b) 96.4 cm **15.** (a) 1.0 MeV; (b) 0.5 MeV **17.** (a) 495 mT; (b) 22.7 mA; (c) 8.17 MJ **19.** (a) 0.36 ns; (b) 0.17 nm; (c) 1.5 mm **21.** (a) 3.4×10^{-4} T, horizontal and to the left as viewed along $\vec{v}_1$; (b) yes, if its velocity is the same as the electron's velocity **23.** 0.27 mT **25.** 680 kV/m **27.** (b) 2.84×10^{-3} **29.** 38.2 cm/s **31.** 28.2N, horizontally west **33.** 467 mA, from left to right **35.** 0.10 T, at 31° from the vertical **37.** 4.3×10^{-3} N·m, negative y **41.** 2 $\pi aiB \sin\theta$, normal to the plane of the loop (up) **43.** 2.45 A **45.** (a) 12.7 A; (b) 0.0805 N·m **47.** (a) 0.30 J/T; (b) 0.024 N·m **49.** (a) 2.86 A·m^2; (b) 1.10 A·m^2 **51.** (a) (8.0×10^{-4} N·m)$(-1.2\hat{i} - 0.90\hat{j} + 1.0\hat{k})$; (b) -6.0×10^{-4} J

Chapter 30

RE 30-1: (a) $\vec{B}$ is to the left at point 1, (b) $\vec{B}$ is up at point 2, (c) $\vec{B}$ is to the right at point 1, (d) $\vec{B}$ is down at point 2.

RE 30-2: (a) If $\vec{B}^{net} = 0$ at point 1 then the current in the wire is coming *out* of the page. (b) Since $\vec{B}^{net} = \vec{B}^{ext} + \vec{B}^{wire}$, and since $\vec{B}^{wire}$ at point 2 points straight down and has the same magnitude as $\vec{B}^{ext}$, $\vec{B}^{net}$ is directed 45 degrees down and toward the right at point 2 and its magnitude is $\sqrt{2}\, B^{ext}$.

RE 30-3: $F_b > F_c > F_a$.

RE 30-4: $\oint \vec{B} \cdot d\vec{s} = \mu_0 i^{enc}$ where i^{enc} is the *net* current flowing *through* the loop. Therefore, $\left| \dfrac{1}{\mu_0} \oint \vec{B} \cdot d\vec{s} \right| = i$ for case (a)

$$= 0 \text{ for case (b)}$$
$$= i \text{ for case (c)}$$
$$= 2i \text{ for case (d)}.$$

(d) > (a) = (c) > (b).

RE 30-5: For $z \gg R$, $|\vec{B}|$ due to any *one* loop is proportional to $|\vec{\mu}| = iA$. Since all the i's are equal, $|\vec{B}| \propto A$ for *each* loop. Taking the directions of the currents into account and calling B_1 the magnetic field magnitude for one *small* loop, and $B_2 = 4B_1$, the magnetic field magnitude for one *large* loop, $B_a = 2B_1$; $B_b = 0$; $B_c = 0$; $B_d = 2B_1 + B_2 = 2B_1 + 4B_1 \therefore |\vec{B}_d| > |\vec{B}_a| > |\vec{B}_b| = |\vec{B}_c| = 0$.

Problems

1. (a) 3.3 μT; (b) yes **3.** (a) 16 A; (b) west to east **5.** $\mu_0 qvi/2\pi d$, antiparallel to i; (b) same magnitude, parallel to i **7.** 2 rad

9. $\dfrac{\mu_0 i \theta}{4\pi} \left(\dfrac{1}{b} - \dfrac{1}{a} \right)$, out of page. **19.** $(\mu_0 i/2\pi w)$ $\ln(1 + w/d)$, up

21. 256 nT **23.** (a) it is impossible to have other than $B = 0$ midway between them; (b) 30 A **25.** 4.3 A, out of page **27.** 80 μT, up the page **29.** $0.791 \mu_0 i^2/\pi a$, 162° counterclockwise from the horizontal **31.** 3.2 mN, toward the wire **33.** (a) $(-2.0 \text{ A})\mu_0$; (b) 0 **37.** $\mu_0 J_0 r^2/3a$ **43.** 0.30 mT **45.** (a) 533 μT; (b) 400 μT **49.** (a) 4.77 cm; (b) 35.5 μT **51.** 0.47 A·m² **53.** (a) 2.4 A·m²; (b) 46 cm **59.** (a) 79 μT; (b) 1.1×10^{-6} N·m

Chapter 31

RE 31-1: They were trying to relate induction to the presence of a magnetic field rather than to a changing field.

RE 31-2: Since the magnetic field is uniform, the left and right segments are polarized symmetrically as shown in the diagram. There is no favored direction in which current can flow.

RE 31-3: This case is similar to the one shown in Fig. 31-7. However, now the polarization will always be stronger on the *right* side of the coil than it is on the left side, so the current will flow continuously in a *counter* clockwise direction.

RE 31-4: Yes, since observations show that the $\vec{v}$ in the magnetic force law ($\vec{F} = q\vec{v} \times \vec{B}$) turns out to be the relative velocity between the object producing the B-field and the charge.

RE 31-5: The magnet is accelerating downward as it falls at $\vec{a} = (-9.8 \text{ m/s}^2)\hat{\jmath}$. By the time its rear end is passing through the area subtended by the loop, it is traveling faster than the front pole was as it passed by, so the rate of change of the B-field is greater at $t = 0.20$ s than it was at $t = 0.10$ s and the amount of induced current is also greater.

RE 31-6: (a) $b > d = e > a = c$. (b) The magnitude $|dB/dt|$ determines that the amount of induced emf is greatest when the slope is greatest.

RE 31-7: (a) into the page to add to the decreasing field, (b) out of the page to subtract from the decreasing field.

RE 31-8: In each semicircular area $|d\Phi^{mag}/dt|$ is identical. The only issue is the "sense" of the induced emf contributed by each semicircle. Using Lenz's law, loop (a) has a nonzero, clockwise (CW) induced current. Loop (b) has a counterclockwise (CCW) current in both the upper and lower halves, so $|i_a| = |i_b|$. In loop (c), the induced emfs in the upper and lower half circles cancel one another out so $|i_c| = 0$ so $|i_a| = |i_b| > |i_c|$.

RE 31-9: As each loop enters or leaves the region where $B \neq 0$, $|\mathcal{E}| = |d\Phi^{mag}/dt| \propto (h)(v)$ where h = height of the loop and v is its speed. Since $v =$ constant for each, $|\mathcal{E}_c| = |\mathcal{E}_d| = 2|\mathcal{E}_a| = 2|\mathcal{E}_b|$.

RE 31-10: (a) Out (given), (b) out since path 3 has $|\mathcal{E}| = 3(\text{mag})$, (c) out since path 3 has $|\mathcal{E}| = 3(\text{mag})$, (e) in, since path 4 has $|\mathcal{E}| = 0$, (d) in since path 2 has $|\mathcal{E}| = 2(\text{mag})$.

RE 31-11: When we pointed a right thumb in the direction of the current our fingers wrapped around the wire in the direction of the magnetic field. This is consistent with the direction of the magnetic field shown in Fig. 31-24.

RE 31-12: The quantity $i^{dis} = \mathcal{E}_0 d\Phi^{elec}/dt$ has the units of current. We can use the right hand rule to find the direction of $\vec{B}$ and we can use it to find the magnitude of $\vec{B}$ induced by a capacitor.

RE 31-13: (a) $|\Phi_d| > |\Phi_b| > |\Phi_c| > |\Phi_a|$. Since $\oint \vec{B} \cdot d\vec{A} = 0$ (Eq. 31-49),

$$\Phi^{net} = \underbrace{\oint \vec{B} \cdot d\vec{A}}_{ends} + \underbrace{\oint \vec{B} \cdot d\vec{A}}_{curve} \text{ so } \Phi_{curve} = -\underbrace{\oint \vec{B} \cdot d\vec{A}}_{ends}.$$

RE 31-14: They both involve the integration of a field vector over a closed Gaussian surface. Each integral determines a net flux at the closed surface that is proportional to the net electric or magnetic charge enclosed by the surface. The major difference between the electric and magnetic situation is that the net magnetic charge enclosed is always zero (that is, north and south poles always appear together), and the net electric charge enclosed can be positive, negative, or zero.

RE 31-15: A statement of Faraday's law is that a changing magnetic field produces an electric field. The Ampère-Maxwell law states that a changing electric field produces a magnetic field. So there is a mathematical symmetry between the two fields.

Problems

1. 1.5 mV **3.** (a) 31 mV; (b) right to left **5.** (a) 1.1×10^{-3} Ω; (b) 1.4 T/s **7.** 30 mA **9.** 2.9 mV **11.** (a) $\mu_0 i R^2 \pi r^2/2x^3$; (b) $3\mu_0 i \pi R^2 r^2 v/2x^4$; (c) in the same direction as the current in the large loop **13.** (b) no **15.** 29.5 mC **17.** (a) 21.7 V; (b) counterclockwise **19.** (b) design it so that $Nab = (5/2\pi)$ m² **21.** 5.50 kV **23.** 80 μV, clockwise **25.** (a) 13 μWb/m; (b) 17%; (c) 0 **27.** 3.66 μW **29.** (a) 48.1 mV; (b) 2.67 mA; (c) 0.128 mW **31.** (a) 600 mV, up the page; (b) 1.5 A, clockwise; (c) 0.90 W; (d) 0.18 N; (e) same as (c) **33.** (a) 240 μV; (b) 0.600 mA; (c) 0.144 μW; (d) 2.88 $\times 10^{-8}$ N; (e) same as (c) **35.** (a) 71.5 μV/m; (b) 143 μV/m **39.** 2.4 $\times 10^{13}$ V/m·s **41.** (a) 1.18×10^{-19} T; (b) 1.06×10^{-19} T **43.** (a) $5.01 \times$

10^{-22} T; (b) 4.51×10^{-22} T **45.** 52 nT · m **51.** (a) 0.63 μT; (b) 2.3 $\times$ 10^{12} V/m·s **53.** (a) 710 mA; (b) 0; (c) 1.1 A **55.** (A) 2.0 A; (b) 2.3 $\times$ 10^{11} V/m·s; (c) 0.50 A; (d) 0.63 μT·m **57.** (a) 75.4 nT; (b) 67.9 nT **59.** (a) 27.9 nT; (b) 15.1 nT **61.** (b) sign is minus; (c) no, there is compensating positive flux through open end near magnet **63.** 47.4 μWb, inward

Chapter 32

RE 32-1: Combine Eqs. 32-1 and 32-2 to get $L = \mu_0 A n^2 l$. (a) If n doubles $L \rightarrow 4L$. (b) If l doubles $A \rightarrow 2A$.

RE 32-2: (d) decreasing rightward or (e) increasing and leftward.

RE 32-3: (a) $R_{eq} = (N_p/N_s)^2 R$ we want R_{eq} seen by the generator to be smaller. So N_s must be greater than N_p. (b) This would be a step up transformer.

RE 32-4: A refrigerator magnet is ferromagnetic; a standard paper clip is also ferromagnetic, since it is made of steel, a ferromagnetic material; a silver wire is diamagnetic (the book says so).

RE 32-5: (a) Spin down or (2). (b) Since the proton has the opposite sign of charge, spin up or (1).

RE 32-6: A ferromagnetic material must have well more than 50% of its domains aligned with each other to act like a strong magnet. If no one alignment of the domains dominates, then it is not a permanent magnet.

RE 32-7: Hysteresis is a lack of retraceability of a magnetization curve. It occurs because the reorientation of domains are not completely reversible.

RE 32-8: (a) $\vec{F}^{\,\text{mag}}$ is directed *toward* the magnet. (b) The dipole moments are also directed *toward*. (c) The force on sphere 1 is *less*.

RE 32-9: (a) $\vec{F}^{\,\text{mag}}$ is directed *away* from the magnet. (b) The dipole moments are also directed *away*. (c) The force on sphere 1 is *less*.

RE 32-10: The Earth's B-field has a different declination and inclination at different locations at any one time. But, it also varies in time. Currently the geographic poles are moving daily. They can also reverse themselves in time periods on the scale of 1000 years.

Problems

1. 0.10 μWb **5.** let the current change at 5.0 A/s **7.** (b) so that the changing magnetic field of one does not induce current in the other; (c) $L_{eq} = \sum_{j=1}^{N} L_j$ **9.** 12 A/s **11.** (a) 0.60 mH; (b) 120 **13.** (a) 1.67 mH; (b) 6.00 mWb **15.** (b) have the turns of the two solenoids wrapped in opposite directions **17.** magnetic field exists only within the cross section of solenoid 1 **19.** (a) $\dfrac{\mu_0 N l}{2\pi} \ln\left(1 + \dfrac{b}{a}\right)$; (b) 13 μH **21.** 6.91τ_L **23.** 46 Ω **25.** (a) 8.45 ns; (b) 7.37 mA **27.** 10.6 A/s **29.** (a) $i_1 = i_2 = $ 3.33 A; (b) i_1 = 4.55 A; i_2 = 2.73 A; (c) i_1 = 0, i_2 = 1.82 A (reversed); (d) $i_1 = i_2 = 0$ **31.** (a) 3.28 ms; (b) 6.45 ms; (c) infinite time; (d) for R = 6.0 Ω, the current of the 2.00 A is the equilibrium current, given by $\mathcal{E}/R$ = (12 V)/(6.0 Ω); it takes an infinite time to reach. For R = 5.00 Ω, the current of 2.00 A is less than the equilibrium current and

requires a finite time to reach. (e) 0; (f) 3 ms **33.** 81.1 μs **35.** (a) 2.4 V; (b) 3.2 mA, 0.16 A **37.** 10 **39.** (a) -9.3×10^{-24} J/T; (b) 1.9×10^{-23} J/T **41.** (a) 0; (b)0; (c) 0; (d) $\pm 3.2 \times 10^{-25}$ J; (e) -3.2×10^{-34} J·s, 2.8×10^{-23} J/T, $+9.7 \times 10^{-25}$ J, $\pm 3.2 \times 10^{-25}$ J **43.** (a) nine; (b) 4 μ_B = 3.71 $\times 10^{-23}$ J/T; (c) $+9.27 \times 10^{-24}$ J; (d) -9.27×10^{-24} J **45.** 5.15×10^{-24} A·m^2 **47.** (a) 180 km; (b) 2.3×10^{-5} **49.** $\Delta\mu = e^2 r^2 B/4m$ **51.** 20.8 mJ/T **53.** yes **55.** (b) K_i/B, opposite to the field; (c) 310 A/m **57.** 55 μT **59.** (a) 31.0 μT, 0°; (b) 55.9μT, 73.9°; (c) 62.0 μT, 90°

Chapter 33

RE 33-1: Using Eq. 33-6, $a = b > c$. (Note that coil area doesn't matter here.)

RE 33-2: At $t = 0$ s, U^{elec} = max and U^{mag} = 0. T = period = $1/f$. (a) $|q(t)|$ is a maximum again at $t = T/2$. (b) Δv_C is next the same at $t = T$. (c) U^{elec} is next a maximum at $t = T/2$. (d) i is next a maximum at $t = T/4$.

RE 33-3: The unit for ω is [rad/s]. Since $L = \mathcal{E}_L/(di/dt)$, we get [H] = [V/(A/s)]. Since $C = q/\Delta V$ we get [F] = [Q/V]. $\omega = 1/\sqrt{LC}$ and the units of $1/\sqrt{LC}$ are $[1/(V\cdot s/A)(Q/V)]^{1/2}$ but [A] = [Q/s] so $[1/s^2]^{1/2}$ or [1/s]. This matches the ω unit of [rad/s].

RE 33-4: (a) According to the loop rule, $\Delta v_C + \Delta v_L = 0$. Since $\mathcal{E}_L = \Delta v_L$, $\mathcal{E}_L = -5$ V. (b) $U^{\text{mag}} = U - U^{\text{elec}} = 160$ μJ $- 10$ μJ $= 150$ μJ.

RE 33-5: (a) C > B > A. (b) 1 & A, 2 & B, 3 & S, 4 & C. (c) A.

RE 33-6: (a) (1) lags, (2) leads, (3) in phase. (b) (3) ($\omega^{\text{dr}} = \omega$ when $X_L = X_C$).

RE 33-7: (a) Increase since the circuit is mainly capacitive; increase C to decrease X_C to be closer to resonance for maximum $\langle P \rangle$. (b) Closer.

Problems

1. 25.6 ms **3.** (a) 97.9 H; (b) 0.196 mJ **7.** (a) 34.2 J/m^3; (b) 49.4 mJ **9.** 1.5×10^8 V/m **11.** (a) 1.0 J/m^3; (b) 4.8×10^{-15} J/m^3 **13.** 9.14 nF **15.** (a) 1.17 μJ; (b) 5.58 mA **17.** with n a positive integer: (a) $t = n(5.00 \ \mu s)$; (b) $t = (2n - 1)(2.50 \ \mu s)$; (c) $t = (2n - 1)(1.25 \ \mu s)$ **19.** (a) 1.25 kg; (b) 372 N/m; (c) 1.75×10^{-4} m; (d) 3.02 mm/s **21.** 7.0×10^{-4} s **23.** (a) 3.0 nC; (b) 1.7 mA; (c) 4.5 nJ **25.** (a) 275 Hz; (b) 364 mA **27.** (a) 6.0:1; (b) 36 pF, 0.22 mH **29.** (a) 1.98 μJ; (b) 5.56 μC; (c) 12.6 mA; (d) $-46.9°$; (e) $+46.9°$ **31.** (a) 0.180 mC; (b) $T/8$; (c) 66.7 W **33.** (a) 356 μs; (b) 2.50 mH; (c) 3.20 mJ **35.** Let T_2 (= 0.596 s) be the period of the inductor plus the 900 μF capacitor and let T_1 (= 0.199 s) be the period of the inductor plus the 100 μF capacitor. Close S_2, wait $T_2/4$; quickly close S_1, then open S_2; wait $T_1/4$ and then open S_1. **37.** 8.66 mΩ **39.** (L/R) ln 2 **43.** (a) 0.0955 A; (b) 0.0119 A **45.** (a) 0.65 kHz; (b) 24 Ω **47.** (a) 6.73 ms; (b) 11.2 ms; (c) inductor; (d) 138 mH **49.** (a) $X_C = 0$, $X_L = 86.7$ Ω, $Z = 218$ Ω, $I = 165$ mA, $\phi = 23.4°$ **51.** (a) $X_C = 37.9$ Ω, $X_L = 86.7$ Ω, $Z = 206$ Ω, $I = 175$ mA, $\phi = 13.7°$ **53.** 1000 V **55.** 89 Ω **57.** (a) 224 rad/s; (b) 6.00 A; (c) 228 rad/s, 219 rad/s; (d) 0.040 **61.** 1.84 A **63.** 141 V **65.** 0, 9.00 W, 2.73 W, 1.82 W **67.** (a) 12.1 Ω; (b) 1.19 kW **69.** (a) 0.743; (b) leads; (c) capacitive; (d) no; (e) yes, no, yes; (f) 33.4 W **71.** (a) 117 μF; (b) 0; (c) 90.0 W, 0; (d) 0°, 90°; (e) 1, 0 **73.** (a) 2.59 A; (b) 38.8 V, 159 V, 224 V, 64.2 V, 75.0 V; (c) 100 W for R, 0 for L and C.

Photo Credits

Chapter 22

Opener: Fundamental Photographs.
Page 634: Vaughan Fleming/Photo
Researchers. Page 644: Johann Gabriel
Doppelmayr, Neuentdeckte
Phaenenomena von Bewünderswurdigen
Würckungen der Natur, Nuremberg 1744.
Page 645: Courtesy Priscilla Laws.

Chapter 23

Opener: Courtesy Paula Brakke.

Chapter 24

Opener: Peter Menzel.

Chapter 25

Opener: Larry Lee/Corbis Images. Page
717: Courtesy PASCO scientific.
Pages 738 (top) and 739: Courtesy NOAA.
Page 738 (bottom): Courtesy
Westinghouse Corporation.

Chapter 26

Opener: Corbis-Bettmann. Page 752:
Courtesy Priscilla Laws. Page 755: The
Image Works. Page 758: Tim Flach/Stone/
Getty Images. Page 767: Courtesy
Shoji Tonaka/International
Superconductivity Technology Center,
Tokyo, Japan.

Chapter 27

Opener: George Grall/National
Geographic Society.

Chapter 28

Opener: Photo by Harold E. Edgerton.
©The Harold and Esther Edgerton
Family Trust, courtesy of Palm Press, Inc.
Page 800 (top): Lester V. Bergman/Corbis
Images. Pages 800 (bottom) and 801:
Courtesy Priscilla Laws. Page 806:
Courtesy Timothy Settlemyer. Page 815:
The Royal Institution, England/Bridgeman
Art Library/NY.

Chapter 29

Opener: EFDA-JET/Photo Researchers.
Page 830: Jeremy Walker/Photo
Researchers. Page 836: Lawrence Berkeley
Laboratory/Photo Researchers.
Page 837: Courtesy Dr. Richard Cannon,
Southeast Missouri State University, Cape
Girardeau. Page 840: Courtesy John Le P.
Webb, Sussex University, England.
Page 841: Courtesy EFDA-JET,
www.jet.efda.org.

Chapter 30

Opener: NASDA/Gamma-Presse, Inc.
Page 863: Courtesy Education
Development Center.

Chapter 31

Opener: Copyright General Motors
Corporation. Page 889: Science Photo
Library/Photo Researchers. Pages 894 and
897 (top): Courtesy PASCO scientific.
Page 897 (bottom): Joseph Sia/Archive
Photos/Hulton Archive/Getty Images.

Chapter 32

Opener: Courtesy Dr. Timothy St. Pierre,
University of Western Australia.
Page 923: The Royal Institution,
England/Bridgeman Art Library/NY.
Pages 936 and 943: Yoav Levy/Phototake.
Page 941: Courtesy Ralph W. DeBlois.
Page 945: Courtesy Andre Geim,
University of Manchester, U.K. Page 946:
Mehau Kulyk/Photo Researchers. Page 947
(top): Courtesy Greg Foss, Pittsburgh
Supercomputing Center; research and
data: Gary Glatzmaier, USC; Earth map
provided by NOAA/NOS. Page 947
(bottom): Courtesy Dr. Timothy St. Pierre,
University of Western Australia.

Chapter 33

Opener: Photo by Rick Diaz, provided
courtesy Haverfield Helicopter Co.
Page 955 (top): Corbis Images. Page 955
(bottom): Bettmann/Corbis Images.
Page 960: Courtesy Agilent Technologies.

Data

Pages 645, 803, and 892: Courtesy Priscilla
Laws.

Index

Page references followed by italic *table* indicate material in tables.
Page references followed by italic *n* indicate material in footnotes.